D1619327

Edited by
Yahachi Saito

Carbon Nanotube and Related Field Emitters

Related Titles

Kim, Dae Mann

Introductory Quantum Mechanics for Semiconductor Nanotechnology

2010

ISBN: 978-3-527-40975-4

Krüger, Anke

Carbon Materials and Nanotechnology

2010

ISBN: 978-3-527-31803-2

Guldi, Dirk M., Martín, Nazario. (eds.)

Carbon Nanotubes and Related Structures
Synthesis, Characterization, Functionalization, and Applications

2010

ISBN: 978-3-527-32406-4

Hierold, C. (ed.)

Carbon Nanotube Devices
Properties, Modeling, Integration and Applications

2008

ISBN: 978-3-527-31720-2

Tsimring, S. E.

Electron Beams and Microwave Vacuum Electronics

2006

ISBN: 978-0-470-04816-0

Oks, E.

Plasma Cathode Electron Sources
Physics, Technology, Applications

2005

ISBN: 978-3-527-40634-0

Edited by Yahachi Saito

Carbon Nanotube and Related Field Emitters

Fundamentals and Applications

WILEY-VCH Verlag GmbH & Co. KGaA

The Editor

Prof. Yahachi Saito
Nagoya University
Dept. of Quantum Engineering
Furo-cho, Chikusa-ku
Nagoya 464-8603
Japan

Library of Congress Card No.: applied for

British Library Cataloguing-in-Publication Data
A catalogue record for this book is available from the British Library.

Bibliographic information published by the Deutsche Nationalbibliothek
The Deutsche Nationalbibliothek lists this publication in the Deutsche Nationalbibliografie; detailed bibliographic data are available on the Internet at
<http://dnb.d-nb.de>.

Cover Design Formgeber, Eppelheim
Typesetting Laserwords Private Limited, Chennai, India
Printing and Binding Strauss GmbH, Mörlenbach

Printed in the Federal Republic of Germany
Printed on acid-free paper

ISBN: 978-3-527-32734-8

Contents

Preface *XVII*
List of Contributors *XIX*

Part I Preparation and Characterization of Carbon Nanotubes *1*

1 **Structures and Synthesis of Carbon Nanotubes** *3*
Yahachi Saito
1.1 Structures of Carbon Nanotubes *3*
1.1.1 Single-Wall CNTs *3*
1.1.2 Multiwall CNTs *6*
1.1.3 Thin-Walled CNTs *7*
1.2 Synthesis of Carbon Nanotubes *7*
1.2.1 Arc Discharge *7*
1.2.2 Chemical Vapor Deposition *8*
1.2.2.1 Thermal CVD *9*
1.2.2.2 Plasma-Enhanced CVD *9*
1.3 Electrical and Mechanical Properties of Carbon Nanotubes *10*
1.3.1 Electronic Structure *10*
1.3.2 Electric Properties *11*
1.3.3 Mechanical Properties *12*
1.3.4 Heat-Transport Properties *13*
References *13*

2 **Preparation of CNT Emitters** *15*
Yahachi Saito
2.1 Introduction *15*
2.2 CNT Point Emitters *15*
2.2.1 Manual Attachment of a CNT Bundle *15*
2.2.2 Mounting Inside an SEM *16*
2.2.3 Electrophoric and Magnetophoretic Methods *16*
2.2.4 Direct Growth on the Apex of a Tip *18*
2.2.5 Other Methods *19*

Carbon Nanotube and Related Field Emitters: Fundamentals and Applications. Edited by Yahachi Saito

ISBN: 978-3-527-32734-8

2.3 CNT Film Emitters 19
2.3.1 Spray Coating 19
2.3.2 Screen Printing 19
2.3.3 Electrophoresis 20
2.3.4 CVD Method 20
References 21

3 Preparation of Patterned CNT Emitters 23
Mark Mann, William Ireland Milne, and Kenneth Boh Khin Teo
3.1 Background 23
3.2 Growth of Carbon Nanotubes from Patterned Catalysts 25
3.2.1 Patterned Growth from Catalyst Film Edges 25
3.2.2 Patterned Growth from Catalyst Thin Films on a Diffusion Barrier 27
3.3 Single Nanotube Growth – Requirements and Uniformity 28
3.4 Nanotube Growth without Surface Carbon 32
3.4.1 Analysis of Substrate Surfaces Exposed to the Plasma 32
3.4.2 Analysis of Substrate Surfaces Shielded from the Plasma 37
3.5 Summary 38
Acknowledgments 40
References 40

Part II Field Emission from Carbon Nanotubes 41

4 Field Emission Theory 43
Seungwu Han
4.1 Fowler–Nordheim Theory 43
4.2 Field Emission from CNTs 44
4.2.1 Computational Methods to Calculate the Emission Currents from Carbon Nanotubes 46
4.2.1.1 The Integration of Time-Dependent Schrödinger Equation 46
4.2.1.2 Transfer Matrix Method 47
4.2.1.3 Other Quantum Mechanical Methods 49
4.2.1.4 Semiclassical Approaches 49
4.2.2 Current–Voltage Characteristics of Field Emission Currents from Carbon Nanotubes 51
4.3 Concluding Remarks 52
References 52

5 Field Emission from Graphitic Nanostructures 55
Kazuyuki Watanabe and Masaaki Araidai
5.1 Introduction 55
5.2 Method and Model 56
5.3 Results 57
5.3.1 Graphitic Ribbons: H Termination and Field Direction 57
5.3.2 Graphene Arrays: Interlayer Interaction 61

5.3.3 Graphene Sheet: Defects *61*
5.3.4 Diamond Surfaces: Impurities *62*
5.4 Conclusion *64*
Acknowledgments *64*
References *65*

6 The Optical Performance of Carbon Nanotube Field Emitters *67*
Niels de Jonge
6.1 Introduction *67*
6.2 Making an Electron Source from an Individual Carbon Nanotube *68*
6.3 The Emission Process *69*
6.3.1 The Fowler–Nordheim Model *69*
6.3.2 Measurement of the Fowler–Nordheim Plot *70*
6.3.3 The Energy Spread *71*
6.3.4 Measurement of Energy Spectra *72*
6.3.5 Comparing the Measured Tunneling Parameter with Theory *74*
6.3.6 Determining the Work Function *74*
6.4 The Brightness *74*
6.4.1 Measuring the Brightness *74*
6.4.2 New Model for the Brightness *75*
6.4.3 Discussion of the New Model *76*
6.4.4 The Total Figure of Merit for Carbon Nanotube Electron Sources *77*
6.5 Conclusions *78*
Acknowledgments *78*
References *78*

7 Heat Generation and Losses in Carbon Nanotubes during Field Emission *81*
Stephen T. Purcell, Pascal Vincent, and Anthony Ayari
7.1 Introduction *81*
7.2 Heat Diffusion Equation for Nanotubes *83*
7.3 Simulations *85*
7.4 Experiments *88*
7.5 Conclusion *92*
References *92*

8 Field Emission Microscopy of Multiwall CNTs *95*
Yahachi Saito
8.1 Introduction *95*
8.2 FEM of Carbon Nanotubes *96*
8.2.1 FEM Measurement *96*
8.2.2 MWNTs with Clean Surfaces *97*
8.2.3 FEM Patterns Depending on Tip Radius *98*
8.3 Field Emission from Adsorbates on an MWNT *99*
8.3.1 Molecules *99*

8.3.1.1 Hydrogen 99
8.3.1.2 Nitrogen 99
8.3.1.3 Oxygen 101
8.3.1.4 Carbon Monoxide 101
8.3.1.5 Carbon Dioxide 101
8.3.1.6 Methane 102
8.3.1.7 Comparison with Related Theoretical Studies 103
8.3.2 Aluminum Clusters 103
8.4 Resolution in FEM and Possible Observation of Atomic Detail 105
8.5 Concluding Remarks 106
References 107

9 *In situ* Transmission Electron Microscopy of CNT Emitters 109
Koji Asaka and Yahachi Saito
9.1 Introduction 109
9.2 Degradation and Failure of Nanotubes at Large Emission Current Conditions 110
9.3 Effect of Tip Structure of Nanotubes on Field Emission 112
9.4 Relationship between Field Emission and Gap Width 113
9.5 Other Studies by *In situ* TEM of CNT Emitters 114
References 116

10 Field Emission from Single-Wall Nanotubes 119
Kenneth A. Dean
10.1 Introduction 119
10.2 Single-Wall Nanotubes and Field Emission 119
10.3 Measuring the Properties of a Single SWNT 120
10.4 Field Emission from a Clean SWNT Surface 121
10.4.1 Clean SWNT Field Emission Microscope Images 122
10.4.2 Clean SWNT $I-V$s 124
10.4.3 Thermal Field Emission 126
10.4.4 High Current and Field Evaporation 128
10.4.5 Anomalous High-Temperature Behavior 130
10.5 SWNT-Adsorbate Field Emission 131
10.5.1 Field Emission Microscopy 131
10.5.2 Electron Energy Distributions 133
10.5.3 Current Saturation and Field-Emission-Induced Surface Cleaning 134
10.6 Field Emission Stability 136
10.6.1 Current Fluctuation 137
10.6.2 Current Degradation 137
10.7 Conclusions 140
References 140

11 Simulated Electric Field in an Array of CNTs *143*
Hidekazu Murata and Hiroshi Shimoyama
11.1 Introduction *143*
11.2 Simulation Method *143*
11.3 Computational Model *145*
11.4 Field Analysis for the VA-CNT System *148*
11.4.1 Dependence of Numerical Accuracy in Electric Field Calculation on the Discretization Number *148*
11.4.2 Appropriateness of 9×9 CNT Computational Model *149*
11.5 Field Analysis for VA-CNT System with Uniform Length *150*
11.5.1 Dependence of the Electric Field Strength at the CNT Apex on Geometrical Parameters of the CNTs *152*
11.5.2 Universal Curve *153*
11.6 Field Analysis for VA-CNT System with Nonuniform Length *154*
11.7 Effect of Shape of CNT Apex *157*
11.8 Effect of CNT Length *158*
11.9 Electric Field Analysis of Network-Structured CNT System *160*
References *162*

12 Surface Coating of CNT Emitters *163*
Yoshikazu Nakayama
12.1 Effects of Surface Coating of CNT Emitters *164*
12.1.1 Parameters Determining Field Emission Properties *164*
12.1.2 Lowering of the Potential Barrier *165*
12.1.2.1 Coating Layer with Low Work Function *165*
12.1.2.2 Coating Layer with Wide Band Gap *165*
12.1.3 Stabilization of Emission Current *167*
12.2 Field Emission from Individual CNT Coated with BN *167*
12.3 Field Emission from Brush-Like CNTs Coated with MgO *169*
12.4 Field Emission from Brush-Like CNTs Coated with TiC *172*
References *174*

Part III Field Emission from Related Nanomaterials *177*

13 Graphite Nanoneedle Field Emitter *179*
Takahiro Matsumoto and Hidenori Mimura
13.1 Introduction *179*
13.2 Fabrication and Structure Characterization *179*
13.3 Field Emission Characteristics *181*
13.4 Applications *182*
13.4.1 Pulse X-ray Generation and Time-Resolved X-ray Radiography *182*
13.4.2 Construction of a Compact FE Scanning Electron Microscope (FE-SEM) System *184*
13.4.3 Stabilization of the FE-SEM System by Thermal Field Operation *186*
13.5 Stochastic Model *188*

13.6 Summary *191*
References *191*

14 Field Emission from Carbon Nanowalls *193*
Masaru Hori and Mineo Hiramatsu
14.1 General Description of Carbon Nanowalls *193*
14.2 Synthesis of Carbon Nanowall Films *194*
14.2.1 Synthesis Techniques *194*
14.2.2 Characterization *195*
14.2.3 Morphology of Carbon Nanowall Film *197*
14.3 Field Emission Properties of Carbon Nanowalls *199*
14.4 Surface Treatment for Improvement of Field Emission Properties *200*
14.4.1 Metal Nanoparticle Deposition *200*
14.4.2 N_2 Plasma Treatment *202*
14.5 Prospects for the Future *203*
References *203*

15 Flexible Field Emitters: Carbon Nanofibers *205*
Masaki Tanemura and Shu-Ping Lau
15.1 Introduction *205*
15.2 Room Temperature Fabrication of Ion-Induced Carbon Nanofibers *205*
15.3 Applications to Field Electron Emission Sources *208*
15.3.1 Current–Voltage ($I-V$) Characteristics *208*
15.3.2 Lifetime *209*
15.3.3 Flexible CNF Cathode *211*
15.4 Summary *215*
References *215*

16 Diamond Emitters *219*
Shozo Kono
16.1 Field Emission from Intrinsic or p-Type Diamonds *219*
16.2 Field Emission from Nitrogen-Doped n-Type Diamonds *220*
16.3 Field Emission from Phosphorus-Doped n-Type Diamonds *221*
16.4 Electron Emission from pn-Junction Diamond Diodes *225*
16.5 Other Application of Diamond Emitter *228*
16.5.1 Diamond Cold-Discharge Cathodes for Cold-Cathode Fluorescent Lamps *228*
16.5.2 Low-Temperature Thermionic Emitters Based on N-Incorporated Diamond Films *229*
References *229*

17 ZnO Nanowires and Si Nanowires *231*
Baoqing Zeng and Zhi Feng Ren
17.1 Introduction *231*

17.2 Synthesis of ZnO and Si Nanowires or Nanobelts *231*
17.2.1 Vapor–Liquid–Solid Nanowire Growth *232*
17.2.2 Controlled Growth of Si Nanowires and ZnO Nanowires *236*
17.2.2.1 Diameter Control *236*
17.2.2.2 Orientation Control *237*
17.2.2.3 Positional Control *238*
17.2.3 Hydrothermal-Based Chemical Approach *240*
17.3 Field Emission of Si and ZnO Nanowires *241*
17.3.1 ZnO Nanowires *244*
17.3.2 Si Nanowires *248*
17.4 Summary *253*
Acknowledgment *253*
References *253*

Part IV Applications of Carbon Nanotubes *259*

18 Lamp Devices and Character Displays *261*
Sashiro Uemura
18.1 Introduction *261*
18.2 Lamp Devices for Light Sources *261*
18.2.1 Structure of the Lighting Element *261*
18.2.2 Carbon Nanotube Emitter for the Lighting Element *263*
18.2.3 Performance of the Lighting Elements *265*
18.3 Super-High-Luminance Light Source Device *266*
18.3.1 Device Structure of the Super-High-Luminance Light Source Device *267*
18.3.2 Performance of the Super-High-Luminance Light Source Device *268*
18.4 Summary of Lamp Devices *271*
18.5 Carbon Nanotube Field Emission Displays for Low-Power Character Displays *272*
18.5.1 Panel Structure and Rib Design *273*
18.5.2 Pixel Design *274*
18.5.3 CNT Electrode for the Display Panel *275*
18.5.4 Uniform Emission from the CNT Electrode *275*
18.5.5 Preparation of CNT Selectively Deposited Lead Frame *277*
18.5.6 Fabrication Process for the Display Panel *279*
18.5.7 Performance of the Display Panel *280*
18.6 Summary of the Display Panel *282*
Acknowledgments *284*
References *284*

19 Screen-Printed Carbon Nanotube Field Emitters for Display Applications *287*
Yong Churl Kim, In Taek Han, and Jong Min Kim
19.1 Introduction *287*

19.2 Formulation of Photoimageable CNT Paste *292*
19.3 Posttreatment *295*
19.4 Field Emission Display Based on Printed CNTs *300*
19.4.1 Cathode *300*
19.4.2 Anode: Phosphors and Phosphor Plate *305*
19.5 Conclusion *306*
References *307*

20 Nanotube Field Emission Displays: Nanotube Integration by Direct Growth Techniques *311*
Kenneth A. Dean
20.1 Introduction *311*
20.2 Field Emission Display Design and Drive Voltage *312*
20.3 Fabricating the Display *316*
20.3.1 Building the Structure *316*
20.3.2 Growth of Carbon Nanotubes on Glass *316*
20.4 Luminance Uniformity and Control and Nanotube Distributions *321*
20.5 Display Performance *323*
20.5.1 Luminance *323*
20.5.2 Color Purity *325*
20.6 Sealing *327*
20.7 Operating Lifetime *328*
20.8 Conclusions *329*
References *330*

21 Transparent-Like CNT-FED *333*
Takeshi Tonegawa, Masateru Taniguchi, and Shigeo Itoh
21.1 Diode-Type CNT-FED *333*
21.2 Structure of Diode-Type CNT-FED *333*
21.3 Characteristics of CNT-FED *335*
21.4 Relation between Gap and Emission *337*
21.5 Property of CNT-FED *338*
21.6 Nonevaporable Getter *338*
21.7 Summary *340*
References *341*

22 CNT-Based FEL for BLU in LCD *343*
Yoon-Ho Song, Jin-Woo Jeong, and Dae-Jun Kim
22.1 Introduction *343*
22.2 CNT-FEL Structure *346*
22.3 CNT Cathode *348*
22.4 Anode *356*
22.5 Vacuum Packaging *358*
22.6 Driving and Characterization *360*
22.7 Future Works *368*

Acknowledgments 368
References 368

23 High-Current-Density Field Emission Electron Source 373
Shigeki Kato and Tsuneyuki Noguchi
23.1 Introduction 373
23.2 Guiding Principles and Practical Methods for High-Performance Emitter 374
23.2.1 Elicitation of Inherent Emission Properties of Individual CNTs 375
23.2.2 Increase in Field Enhancement Factor at the CNT Surface 375
23.2.3 Optimization of Electric Field Distribution on Film Emitter Surface 376
23.2.4 Reduction of Work Function of Emitter 376
23.2.5 Improvement of Thermal Conduction and Mitigation of Joule Heating at CNT Junction 377
23.2.6 Restraint of CNT Disappearance 377
23.2.7 Mitigation of Ion Sputtering and Reactive Etching of Emitter Surface 378
23.3 Impregnation of RuO_2 and OsO_2 379
23.3.1 Properties of RuO_2 and OsO_2 379
23.3.2 The Method of RuO_2 Impregnation 379
23.3.3 Observation of CNTs with Impregnation of RuO_2 379
23.4 CNT Rooting 380
23.5 Effect of Impregnation on Field Emission Properties 381
23.6 Effect of Rooting on Field Emission Properties 384
23.7 Influence of Residual Gas 386
References 388

24 High-Resolution Microfocused X-ray Source with Functions of Scanning Electron Microscope 389
Koichi Hata and Ryosuke Yabushita
24.1 Introduction 389
24.2 Multiwalled CNT Field Emission Cathode 390
24.3 Construction of High-Resolution Transmission X-ray Microscope Equipped with the Function of SEM 392
24.4 Characteristic Evaluation of High-Resolution X-ray Microscope Provided with SEM Function 394
24.4.1 Resolution of SEM 394
24.4.2 Resolution of Transmission X-ray Microscope 395
24.5 Factors Limiting Resolution of X-ray Transmission Image 396
24.5.1 Lateral Distribution d_s of X-ray Generating Region 396
24.5.2 Blurring δ_F Caused by Fresnel Diffraction 397
24.5.3 Evaluation of Theoretical Resolution δ_X 398
24.6 Conclusion 398
References 399

25 **Miniature X-ray Tubes** *401*
Fumio Okuyama
25.1 Introduction *401*
25.2 Our Technical Basis for Miniaturizing X-ray Tubes *402*
25.3 The Pd Emitter *404*
25.4 Devising X-ray Tubes with Miniature Dimensions *405*
25.4.1 The 10-mm-Diameter Tube *405*
25.4.2 The 5-mm-Diameter Tube *409*
25.5 Status Quo of Our MXT Technique *413*
25.5.1 DSB *413*
25.5.2 Apoptosis *415*
25.6 Future Prospect of MXTs in Radiation Therapy *416*
References *416*

26 **Carbon Nanotube-Based Field Emission X-ray Technology** *417*
Otto Zhou and Xiomara Calderon-Colon
26.1 Introduction *417*
26.1.1 Current Thermionic X-ray Technology *417*
26.1.2 Previous Studies of Field Emission X-ray *418*
26.1.3 Carbon Nanotube-Based Field Emission X-ray *418*
26.2 Fabrication of CNT Cathodes for X-ray Generation *420*
26.2.1 Fabrication Process *420*
26.2.2 Field Emission Properties *424*
26.3 Field Emission Microfocus X-ray Tube *425*
26.3.1 Tube Design *425*
26.3.2 Tube Current and Lifetime *426*
26.3.3 Focal Spot Size *427*
26.4 Distributed Multibeam Field Emission X-ray *428*
26.5 Imaging Systems *430*
26.5.1 Dynamic Micro-Computed Tomography *430*
26.5.2 Stationary Digital Breast Tomosynthesis *431*
26.6 Summary and Outlook *434*
Acknowledgments *434*
References *435*

27 **Microwave Amplifiers** *439*
Pierre Legagneux, Pierrick Guiset, Nicolas Le Sech, Jean-Philippe Schnell, Laurent Gangloff, William I. Milne, Costel S. Cojocaru, and Didier Pribat
27.1 Introduction *439*
27.2 State of the Art of Thermionic Cathodes and Methodology to Review CNT Cathodes *441*
27.2.1 State of the Art of Thermionic Cathodes Used in Traveling-Wave Tubes *441*
27.2.2 Interest in Cathodes Delivering a High-Frequency Modulated Electron Beam *442*

27.2.3 Methodology of Reviewing CNT Cathodes *443*
27.3 CNT-Based Electron Guns as High Current Electron Sources *444*
27.3.1 Current Density at Cathode Level *444*
27.3.1.1 Currents Emitted by Individual CNTs *444*
27.3.1.2 Current Density Emitted by CNT Cathodes *444*
27.3.2 Convergence Factors Obtained with CNT-Based Electron Gun *446*
27.3.2.1 Simulation of a CNT-Based Electron Gun *446*
27.3.2.2 Design of Cathodes Delivering Low Transverse Electron Velocities *448*
27.3.3 Potential of CNT Electron Guns as High Current Electron Sources *450*
27.4 CNT Cathodes Delivering a Modulated Electron Beam *450*
27.4.1 Modulation of the Applied Electric Field *450*
27.4.1.1 Modulation of the Applied Electric Field with an Integrated Grid *450*
27.4.1.2 Modulation of the Applied Electric Field with an External Grid Electrode *451*
27.4.1.3 Modulation of the Applied Electric Field with a Resonant Cavity *452*
27.4.1.4 Conclusion about the Approach Consisting in Modulating the Applied Field *456*
27.4.2 Optical Modulation of the Current Supplied to the CNTs *457*
27.4.2.1 Design of a CNT Photocathode *457*
27.4.2.2 Demonstration of a 300 MHz CNT Photocathode *462*
27.4.2.3 Development of High-Frequency CNT Photocathodes for Microwave Amplifiers *463*
27.4.3 Optical Modulation of the Electric Field at Nanotube Apex: THz Cathodes *465*
27.5 Conclusion *466*
References *468*

Index *471*

Preface

The discovery of carbon nanotubes (CNTs) in 1991 has not only opened a rich field in fundamental science but also given a scope of potential technological applications. CNT is a new class of materials that possess extraordinary properties and propagate nanotechnology and nanoscience. Among the numerous numbers of application proposals of CNTs, the most promising one is a tiny, nanometer-scale field electron emitter that works by applying a low electric voltage in a moderate vacuum. A field emission display (FED) is considered to be the most influential industrial product in which the nanotechnology, CNT, is utilized as a key material, since its commercial market is huge and consumers in general directly experience the technology. At the beginning of the CNT-based FED development, there is an atmosphere that it does not take so much time to realize the products. Through a few R & D projects on CNT-FED, however, it was cognized that it was not so easy to make the high-definition FED with CNT emitters. There are still a number of technical problems to be overcome to put CNT-based FEDs to practical use. Current reduction in retail prices of LCD (liquid crystal display) and PDP (plasma display panel) also makes it difficult to forward the development of FED for TV monitors. But, applications of CNT electron emitters are not limited to TV monitors; character display, digital sinage, back light unit, electron sources for various vacuum electronic devices such as miniature X-ray source, and microwave amplifiers. For example, CNT-based character displays are actively developed and are practically used as public signs. Fundamental studies of CNT and related emitters and also continuing development will make CNT electron sources indispensable and core elements in various fields from consumer devices, medical to industry, and space aviation.

This book is the first, comprehensive monograph dealing with CNT and related field emitters covering from the fundamental to the applications. The fundamental part includes structures and preparations of CNTs, electron emission mechanism, characteristics of CNT electron sources, dynamic behaviour of CNTs during operation and so on. Applications of CNT emitters to vacuum electronic devices include displays, electron sources in electron microscopes, X-ray sources, and microwave amplifiers. The book has sought to bring leading researchers in the respective fields to summarize, using tutorial style, the important advances and to suggest promising future research directions. Authors of the chapters are from

Carbon Nanotube and Related Field Emitters: Fundamentals and Applications. Edited by Yahachi Saito

ISBN: 978-3-527-32734-8

different groups worldwide, including academic and industrial circles, guaranteeing a broad view of the topic. I am thankful to the authors who produced excellent chapters that will greatly benefit many readers interested in CNT and related field emitters, and also to John-Wiley for cooperating with us in implementing the book project.

Nagoya, March 2010 *Yahachi Saito*

List of Contributors

Masaaki Araidai
Tohoku University
WPI Advanced Institute for
Materials Research
2-1-1 Katahira
Aoba-ku
Sendai
Miyagi 980-8577
Japan

Koji Asaka
Nagoya University
Department of Quantum
Engineering
Furo-cho, Chikusa-ku
Nagoya 464-8603
Japan

Anthony Ayari
Université Claude Bernard Lyon 1
Lab. de Physique de la Matière
Condensée et Nanostructures
UMR CNRS 5586
43 Blvd 11 Novembre
F-69622
Villeurbanne Cedex
France

Xiomara Calderon-Colon
The University of North Carolina
Department of Physics and
Astronomy and
Curriculum in Applied and
Materials Science
Chapel Hill
NC 27599
USA

Costel S. Cojocaru
University of Cambridge
Electrical Engineering Division
Centre for Advanced Photonics
and Electronics
9 JJ Thomson Avenue
Cambridge
CB3 0FA
UK

Kenneth A. Dean
Motorola Inc.
2100 East Elliot Road
Tempe
AZ 85284
USA

Carbon Nanotube and Related Field Emitters: Fundamentals and Applications. Edited by Yahachi Saito

ISBN: 978-3-527-32734-8

Laurent Gangloff
THALES-Ecole Polytechnique
1, Av. Augustin Fresnel
NANOCARB
91767
Palaiseau Cedex
France

Pierrick Guiset
THALES-Ecole Polytechnique
1, Av. Augustin Fresnel
NANOCARB
91767
Palaiseau Cedex
France

In Taek Han
Samsung Advanced
Institute of Technology
Samsung Electronics
Gheung-gu
Yongin-si
Gyeonggi-do 446-712
Korea

Seungwu Han
Seoul National University
Department of Materials
Science and Engineering
Seoul 151-744
Korea

Koichi Hata
Mie University
Department of Electrical and
Electronic Engineering
1577 Kurima-machiya-cho
Tsu 514-8507
Japan

Mineo Hiramatsu
Meijo University
Department of Electrical and
Electronic Engineering
1-501 Shiogamaguchi
Tempaku
Nagoya 468-8502
Japan

Masaru Hori
Nagoya University
Department of Electrical
Engineering and Computer
Science
Furo-cho
Chikusa
Nagoya 464-8603
Japan

Shigeo Itoh
Futaba Corporation
R & D Center
Chosei-mura
Chosei-gun
Chiba 299-4395
Japan

Jin-Woo Jeong
Convergence Components &
Materials Research Laboratory
Electronics and
Telecommunications Research
Institute
138 Gajeongno
Yuseong-Gu
Daejeon 305-700
Korea

Niels de Jonge
Oak Ridge National Laboratory
Materials Science and Technology
Division
1 Bethel Valley Road
Oak Ridge
TN 37831-6064
USA

and

Vanderbilt University
Medical Center
Department of Molecular
Physiology and Biophysics
2215 Garland Ave.
Nashville
37232-0615
USA

Shigeki Kato
High Energy Accelerator
Research Organization
Accelerator Laboratory
Tsukuba
Ibaraki 305-0801
Japan

Dae-Jun Kim
Nano Convergence
Device Team
R&D Center
VSI, 461-34
Jeonmin-dong
Yuseong-gu
Daejeon 305-811
Korea

Jong Min Kim
Samsung Advanced
Institute of Technology
Samsung Electronics
Gheung-gu
Yongin-si
Gyeonggi-do 446-712
Korea

Yong Churl Kim
Samsung Advanced
Institute of Technology
Samsung Electronics
Gheung-gu
Yongin-si
Gyeonggi-do 446-712
Korea

Shozo Kono
Tohoku University
Institute of Multidisciplinary
Research for Advanced Materials
Katahira 2-1-1
Aoba-ku
Sendai 980-8577
Japan

Shu-Ping Lau
Department of Applied Physics
The Hong Kong Polytechnic
University
Hung Hom, Kowloon
Hong Kong

Nicolas Le Sech
THALES-Ecole Polytechnique
1, Av. Augustin Fresnel
NANOCARB
91767
Palaiseau Cedex
France

Pierre Legagneux
THALES-Ecole Polytechnique
1, Av. Augustin Fresnel
NANOCARB
91767
Palaiseau Cedex
France

Mark Mann
University of Cambridge
Electrical Engineering Division
Centre for Advanced Photonics
and Electronics
9 JJ Thomson Avenue
Cambridge
CB3 0FA
UK

Takahiro Matsumoto
Research and Development
Center
Stanley Electric Corporation
5-9-5 Tokodai
Tsukuba
300-2635
Japan

and

Research Institute of Electronics
Shizuoka University
3-5-1 Johoku
Hamamatsu
432-8011
Japan

Hidenori Mimura
Research Institute of Electronics
Shizuoka University
3-5-1 Johoku
Hamamatsu
432-8011
Japan

William I. Milne
University of Cambridge
Electrical Engineering Division
Centre for Advanced Photonics
and Electronics
9 JJ Thomson Avenue
Cambridge
CB3 0FA
UK

Hidekazu Murata
Meijo University
Department of Electrical and
Electronic Engineering
Faculty of Science and
Technology
1-501 Shiogamaguchi
Tempaku-ku
Nagoya 468-8502
Japan

Yoshikazu Nakayama
Osaka University
Department of Mechanical
Engineering
2-1 Yamadaoka
Suita
Osaka 565-0871
Japan

Tsuneyuki Noguchi
KAKEN Inc.
1044, Holimachi
Mito
Ibaraki 310-0903
Japan

Fumio Okuyama
Kawauchi-Sanjyunin-Machi
49-53, Aoba-ku
Sendai 980-0866
Japan

Didier Pribat
THALES-Ecole Polytechnique
1, Av. Augustin Fresnel
NANOCARB
91767
Palaiseau Cedex
France

Stephen T. Purcell
Université Claude Bernard Lyon l
Lab. de Physique de la Matière
Condensée et Nanostructures
UMR CNRS 5586
43 Blvd 11 Novembre
F-69622
Villeurbanne Cedex
France

Zhi Feng Ren
Department of Physics
Boston College
140 Commonwealth Ave.
Chestnut Hill MA 02467
USA

Yahachi Saito
Nagoya University
Department of Quantum
Engineering
Furo-cho, Chikusa-ku
Nagoya 464-8603
Japan

Jean-Philippe Schnell
THALES-Ecole Polytechnique
1, Av. Augustin Fresnel
NANOCARB
91767
Palaiseau Cedex
France

Hiroshi Shimoyama
Meijo University
Department of Electrical and
Electronic Engineering
Faculty of Science and
Technology
1-501 Shiogamaguchi
Tempaku-ku
Nagoya 468-8502
Japan

Yoon-Ho Song
Convergence Components &
Materials Research Laboratory
Electronics and
Telecommunications Research
Institute
138 Gajeongno
Yuseong-Gu
Daejeon 305-700
Korea

Masaki Tanemura
Department of Frontier Materials
Graduate School of Engineering
Nagoya Institute of Technology
Gokiso-cho, Showa-ku
Nagoya, 466-8555
Japan

Masateru Taniguchi
Futaba Corporation
R & D Center
Chosei-mura
Chosei-gun
Chiba 299-4395
Japan

Kenneth Boh Khin Teo
AIXTRON Nanoinstruments
Buckingway Business Park
Anderson Road
Cambridge
CB24 4FQ
UK

Takeshi Tonegawa
Futaba Corporation
R & D Center
Chosei-mura
Chosei-gun
Chiba 299-4395
Japan

Sashiro Uemura
Noritake Company Ltd.
728-23 Tsumura-cho
Ise 516-1103
Japan

Pascal Vincent
Université Claude Bernard Lyon l
Lab. de Physique de la Matière
Condensée et Nanostructures
UMR CNRS 5586
43 Blvd 11 Novembre
F-69622
Villeurbanne Cedex
France

Kazuyuki Watanabe
Tokyo University of Science
Department of Physics and
Research Institute for
Science and Technology
1-3 Kagurazaka
Shinjuku-ku
Tokyo 162-8601
Japan

Ryosuke Yabushita
Mie University
Department of Electrical and
Electronic Engineering
1577 Kurima-machiya-cho
Tsu 514-8507
Japan

Baoqing Zeng
School of Physical Electronics
University of Electronic Science
and Technology of China
Chengdu 610054
China

Otto Zhou
The University of North Carolina
Department of Physics and
Astronomy and
Curriculum in Applied and
Materials Science
Chapel Hill
NC 27599
USA

Part I
Preparation and Characterization of Carbon Nanotubes

Carbon Nanotube and Related Field Emitters: Fundamentals and Applications. Edited by Yahachi Saito

ISBN: 978-3-527-32734-8

1
Structures and Synthesis of Carbon Nanotubes

Yahachi Saito

1.1
Structures of Carbon Nanotubes

Carbon nanotubes (CNTs) are hollow cylinders made of seamlessly rolled graphene (honeycomb lattice of carbon atoms, Figure 1.1) with diameters ranging from about 1 over 50 nm, depending on the number of walls comprising the nanotubes. Structurally well-ordered CNTs were discovered and their structures were well characterized by Iijima in 1991 [1]. CNTs composed of one sheet of graphene are called *single-wall carbon nanotubes* (abbreviated SWCNTs or SWNTs) [2–4], and those made of more than two sheets, multiwall carbon nanotubes (abbreviated MWCNTs or MWNTs). The length of nanotubes exceeds 10 μm, and the longest ones are reportedly on the order of a centimeter [5, 6].

1.1.1
Single-Wall CNTs

There are numerous ways of rolling a sheet of the honeycomb pattern into a seamless cylinder, which give rise to a vast range of diameters and various helical structures. The structure of a CNT is specified (except for handedness) by a vector connecting two lattice points in an unrolled honeycomb lattice (e.g., points O and A in Figure 1.1). When the honeycomb sheet is rolled so as to make the two points coincide, a cylinder whose circumference corresponding to the line AO is formed. Such a vector connecting the crystallographically equivalent lattice points in an unrolled lattice and being perpendicular to the tube axis is called the *chiral vector* (or wrapping vector), which is expressed using two fundamental translational vectors $\boldsymbol{a}_1, \boldsymbol{a}_2$:

$$\boldsymbol{C}_\mathrm{h} = n\boldsymbol{a}_1 + m\boldsymbol{a}_2 \equiv (n, m) \tag{1.1}$$

where n and m are integers with $0 \leq |m| \leq n$. The sets of two integers (n, m) are called *chiral indices*, which are used to specify the structure of SWNTs. The diameter

Carbon Nanotube and Related Field Emitters: Fundamentals and Applications. Edited by Yahachi Saito

ISBN: 978-3-527-32734-8

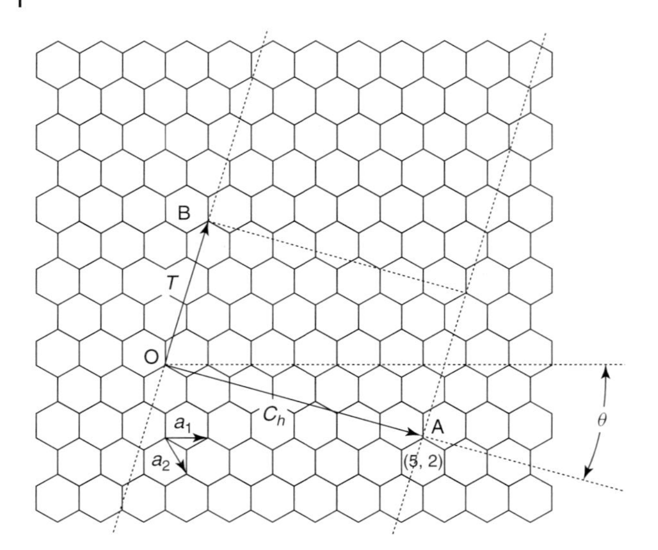

Figure 1.1 Graphene (a sheet of honeycomb lattice of carbon atoms). A carbon nanotube is a seamless cylinder made of a rolled graphene by superimposing the two lattice points, for example, O and A. $\boldsymbol{C}_h$ and $\boldsymbol{T}$ for chiral indices (5, 2) are illustrated.

d_t and chiral angle θ (Figure 1.1) can be expressed in terms of n and m

$$d_t = a\sqrt{n^2 + nm + m^2}\Big/\pi \tag{1.2}$$

$$\theta = \cos^{-1}\left(\frac{2n + m}{2\sqrt{n^2 + nm + m^2}}\right) \quad (|\theta| \leq \pi/6) \tag{1.3}$$

where $a = |\boldsymbol{a}_1| = |\boldsymbol{a}_2| = 0.246\,\text{nm}$.

A SWNT has translational symmetry along the tube axis. The basic translational vector T is represented by

$$T = \{(2m + n)\boldsymbol{a}_1 - (2n + m)\,\boldsymbol{a}_2)\}/d_R \tag{1.4}$$

where d_R is defined by using the greatest common divisor D_G of n and m as follows:

$$d_R = \begin{cases} D_G & \text{if } n - m \text{ is not a multiplie of } 3D_G \\ 3D_G & \text{if } n - m \text{ is a multiplie of } 3D_G \end{cases} \tag{1.5}$$

CNTs with $n = m(\theta = \pi/6)$ and $m = 0\,(\theta = 0)$, called *armchair* type and *zigzag* type, respectively, do not have helicity. CNTs with other chiral indices ($n \neq m \neq 0$) have helical structures and are called *chiral type*. In Figure 1.2a–c, structure models of armchair, zigzag, and chiral type CNTs, respectively, are shown.

The diameter of SWNTs actually synthesized ranges from 0.7 to about 3 nm, depending on synthesis conditions especially on the diameter of catalyst particles employed for the synthesis. Figure 1.3 shows a transmission electron microscope (TEM) image of SWNTs, with diameter ranging between 1.0 and 1.3 nm, produced

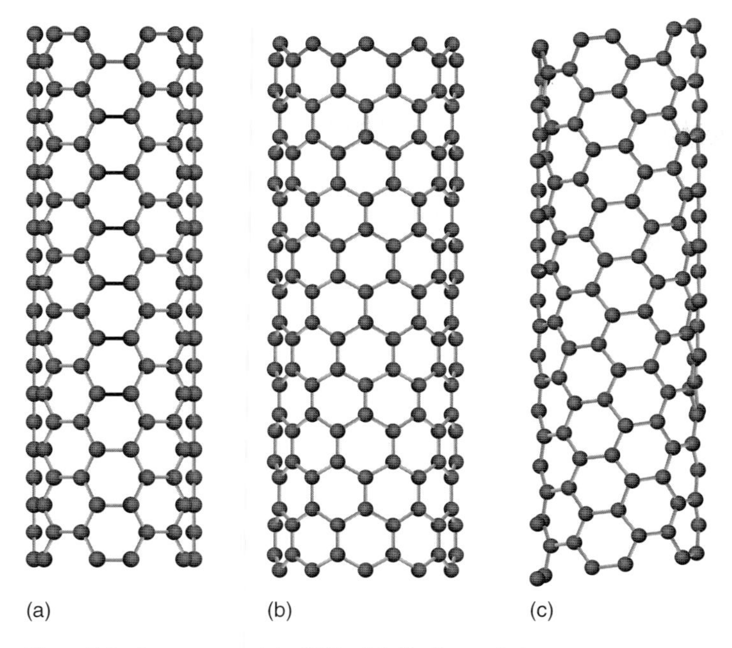

Figure 1.2 Structure models CNTs: (a) (5, 5) armchair, (b) (10, 0) zigzag, and (c) (4, 6) chiral type.

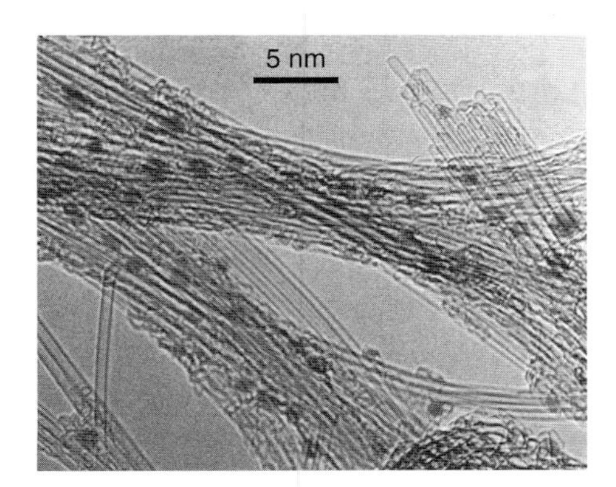

Figure 1.3 TEM picture of SWNTs produced by arc discharge with Ni catalyst.

by arc discharge with a Ni catalyst. The ends of a tube made from a rolled graphene remain open, but CNTs actually synthesized are capped by a curved graphene (Figure 1.3) or by a metal particle used as catalyst in the synthesis process. Pentagons (five-membered rings) have to be inserted into a graphene sheet in order to render a positive curvature to the sheet. Since one pentagon introduces a $+\pi/6$ wedge disclination in the honeycomb lattice, six pentagons are required to from a hemispherical dome of a curved graphene, like the half of the C_{60} fullerene, and to close one end of a CNT. The caps of CNTs with small diameters (less than about 4 nm in diameter) such as SWNTs look spherical (see the ends of SWNTs

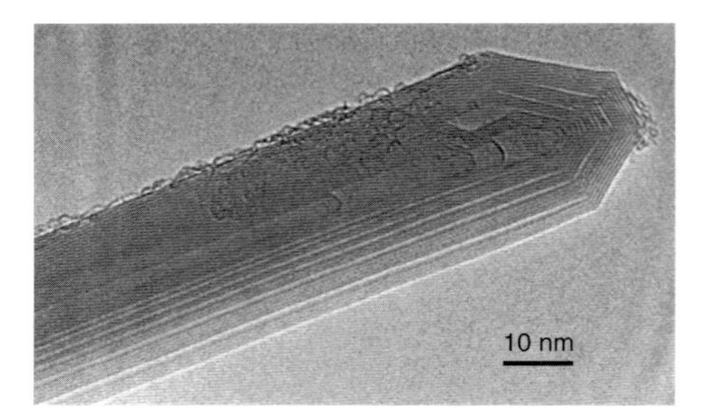

Figure 1.4 TEM picture of MWNTs produced by arc discharge.

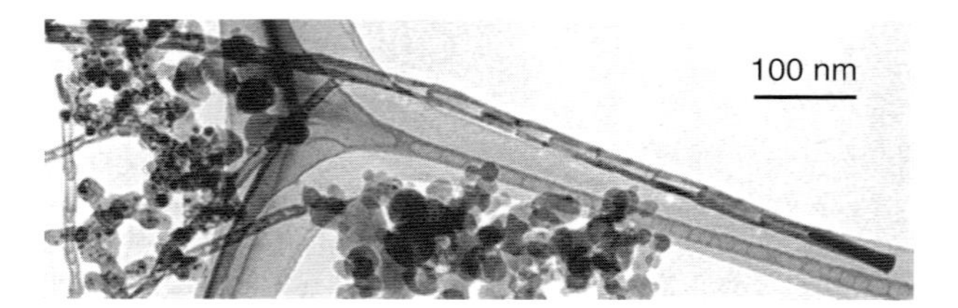

Figure 1.5 TEM picture of a bamboo-shaped MWNT.

in Figure 1.3), while those of thick CNTs (e.g., MWNTs) exhibit polyhedral shapes because the pentagon sites extrude like corners of the polyhedra (see Figure 1.4 in the next section).

1.1.2 Multiwall CNTs

An ideal MWNT consists of graphene sheets stacked in a concentric way. MWNTs produced by the arc-discharge technique (see below) have a high structural perfection and are straight, as shown in Figure 1.4. The number of sheets ranges from 2 to about 40. For thick MWNTs composed of more than three or four layers, the interlayer spacing between rolled graphene sheets is 0.344 nm on average [7]. The spacing is wider by a small percentage than the ideal graphite value (0.3354 nm), being characteristic of the turbostratic carbon (graphitic layers that are stacked in parallel but without translational and rotational relations between the adjacent layers) [8]. The diameters of MWNTs are in the range 4–50 nm, and the lengths are over 10 μm. MWNTs contain narrow cavities (approximately 2–10 nm in diameter) in the center.

On the other hand, MWNTs synthesized by thermal decomposition of hydrocarbon gases (so-called chemical vapor deposition (CVD), see below) are in general comprised of defective graphene sheets and are sometimes curved and curled. Bamboo-like structures composed of a series of compartments with a metal catalyst at their tips are formed as shown in Figure 1.5 [9]. The graphite layers in a bamboo-structured MWNT are not parallel to the tube axis but are inclined to form stacked cones.

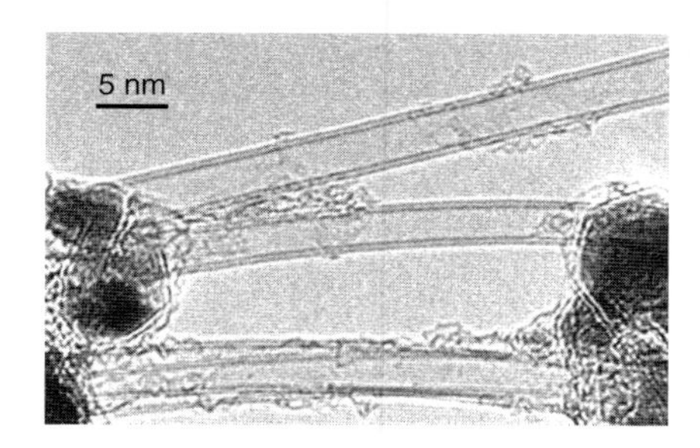

Figure 1.6 TEM image of DWCNTs produced by arc discharge.

1.1.3 Thin-Walled CNTs

Double-wall carbon nanotubes (DWNTs or DWCNTs) consisting of two layers of graphene can be selectively prepared by arc discharge and CVD. Figure 1.6 shows a typical TEM image of DWCNTs produced by arc discharge in a He/H_2 gas mixture with Fe–Co–Ni as catalyst and sulfur as promoter [10]. The diameter of DWCNTs is in the range 3–4 nm, that is, intermediate between those of SWNTs and thick MWNTs. The interlayer spacing between the outer and inner walls ranges from 0.37 to 0.39 nm, being about 10% wider than that of thick MWNTs. Due to the small diameter of DWCNTs, the electric voltage required for field emission from DWCNTs is as low as that for SWNTs, though DWCNTs are more robust against degradation during field emission than SWNTs because of the smaller curvature (i.e., more stable with the smaller strains) and the larger cross section (two layers) for electron flow.

1.2 Synthesis of Carbon Nanotubes

Production methods of CNTs are roughly divided into three categories: (i) electric arc discharge between carbon electrodes, (ii) laser vaporization of a carbon target, and (iii) thermal decomposition of hydrocarbon gases or CVD. The arc discharge and CVD techniques are briefly described here. Concerning the laser ablation method, which is mainly used for the preparation of high-purity SWNTs, the details can be found in [11].

1.2.1 Arc Discharge

A direct current (DC) arc is almost exclusively used because it is amenable and provides a high yield of nanotubes, even though an alternating current (AC) arc can also evaporate carbon electrodes. Both the anode and the cathode are made of graphite rods for producing MWNTs, while a metal (or metal oxide) powder is impregnated into the anode for the production of SWNTs. The metal catalyzes the formation of SWNTs. Typical catalysts are Fe–Ni, Co–Ni, Y–Ni, and Rh–Pt [12].

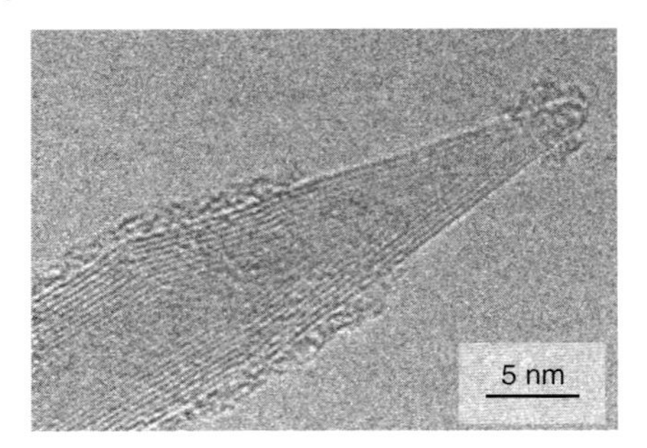

Figure 1.7 TEM picture of a MWNT with a cone-shaped tip. The cone angle is 19.2°, indicating the presence of five pentagons at the apex.

Carbon electrodes are evaporated in a buffer gas (usually helium) at a desired pressure (200–600 Torr for helium gas). Since the anode surface is heated to a higher temperature (~4000 K) than the cathode surface (~3500 K), the anode is selectively consumed in the arc. The position of the anode tip must be adjusted in order to maintain the proper spacing (about 1 mm) between the electrodes. Approximately half of the evaporated carbon condenses on the tip of the cathode, forming a cylindrical hard deposit. MWNTs are obtained inside the cylindrical cathode deposit even without metal catalysts. The remaining carbon vapor condenses in a gas phase, forming soot. Fullerenes such as C_{60} and C_{70} are grown in the soot. When the catalyst metal is co-evaporated with carbon, SWNTs are formed and found in the soot deposited on the walls of the reaction chamber and on the surface of the cathode.

For the production of MWNTs, pure carbon is evaporated mainly in helium gas. Hydrogen gas and even air can also be employed as the working gas. In the latter case, relatively "clean" MWNTs are produced (i.e., the amount of byproducts such as carbon nanoparticles is very small [13]).

Radio frequency (RF) plasma heating of graphite in argon gas can be used to synthesize MWNTs, which are characterized by their cone-shaped tip [14]. The cone angle is 19.2°, indicating the presence of five pentagons at the apex (Figure 1.7). The outer diameter of the cylinder part is approximately 10 nm.

1.2.2
Chemical Vapor Deposition

Thermal or plasma-assisted decomposition of gaseous carbon molecules can be used to produce a variety of CNTs ranging from SWNTs to MWNTs. The commonly used carbon sources are methane, acetylene, alcohol, and carbon monoxide. The decomposition of molecules is assisted on the surface of small catalytic particles, and it is believed that CNTs nucleate and grow using the catalyst as a scaffold. The most effective catalysts are Fe, Ni, and Co. This technique enables the formation of CNTs directly on solid substrates on which metal catalyst are deposited, and is therefore called *catalytic chemical vapor deposition*. CVD is the term used to describe heterogeneous reactions in which both solid and gaseous products are formed from a gaseous precursor through chemical reaction. CVD methods have several advantages over the other methods such as arc discharge: CNTs can be grown on various solid substrates, and a wide range of process parameters makes the control over the morphology and structure of the products easy. Moreover, since the CVD method is compatible with the present-day microfabrication technique for Si-based integrated circuits, catalytic CVD is now attracting considerable attention for the fabrication of CNT-based nanoelectronic devices as well as emitter arrays in field-emission displays.

In this subsection we focus on a growth method on a substrate by which CNTs are formed in a controlled position (patterned area) on a solid surface and also aligned vertically to the substrate. Concerning vapor-phase growth or the floating catalyst methods in which both carbon-bearing gases and catalysts are injected into a reaction chamber without a substrate, the reader can consult reviews on the subject [15].

1.2.2.1 Thermal CVD

The apparatus for CNT growth by thermal CVD ranges from a simple homemade reactor to automated large ones for industrial production. The simple apparatus consists of a quartz tube (20–50 mm in diameter and about 700 mm in length) inserted into a tubular electric furnace capable of maintaining a high temperature (up to about 1000 °C) over a length about 20 cm. This type of system is a hot-wall reactor, inside which solid substrates, typically about 10 square millimeters or larger, are placed and the carbon feedstock gas is supplied with or without dilution through mass flow controllers. Catalyst particles have to be deposited on the substrates. Physical vapor deposition techniques (sputtering and vacuum evaporation) as well as wet chemistry can be used to prepare catalysts including transition metals onto the substrates. The temperature for the synthesis of CNTs by CVD is generally in the range 650–900 °C.

Cold-wall reactors, where the substrate is heated by resistive, inductive, or infrared (IR) radiation heaters, can also be employed for low-pressure operations. It is reported that the employment of a hot filament enables MWNTs to grow at the substrate temperature of 400 °C [16].

CNTs synthesized by CVD generally contain metal particles as residue of the catalyst on their tips or roots. Formation of CNTs aligned vertically to the substrate surface is another characteristic of the CVD technique. This alignment occurs when CNTs grow densely on the surface because the adjacent CNTs support each other from falling down.

1.2.2.2 Plasma-Enhanced CVD

The plasma-enhanced chemical vapor deposition (PECVD) method involves a glow discharge in a reaction chamber through a high-frequency voltage applied to the electrodes. PECVD was first introduced in the fabrication of microelectronic devices in order to enable the CVD process to proceed at reduced temperatures of a substrate because the substrate cannot tolerate the elevated temperature of some thermal CVD processes. The common PECVD processes proceed at substantially lower substrate temperatures (room temperature to 100 °C). Low-temperature operation is possible because the dissociation of the precursor for the deposition of semiconductor, metal, and insulating films is enabled by the high-energy electrons in a cold plasma.

In CNT growth by CVD, on the other hand, dissociation of the precursor (carbon-containing molecules) on the surface of catalytic particles is the critical process, and therefore precursor dissociation in the gas phase is not necessary though some activation or excitation of molecules may contribute to the reduction of the substrate temperature. When precursor dissociation occurs in the gas phase, an excessive amount of amorphous carbon is produced. There may be a minimum temperature to which the substrate has to be heated because precursor dissociation on the catalyst surface is the key to CNT growth. MWNTs are reportedly grown at 430 and 500 °C by PECVD [17, 18]. CNTs grown by PECVD contain more defects in their structures than those grown by arc discharge and thermal CVD because of damages by ion bombardment.

1.3 Electrical and Mechanical Properties of Carbon Nanotubes

1.3.1 Electronic Structure

CNTs are predicted to be metallic or semiconductive depending on their chiral indices (n, m). When $n - m = 3j$, where j is an integer, the tube is metallic. Strictly, the effect of tube curvature is to open a very tiny gap (on the order of millielectron volts) at the Fermi level for metallic tubes with $n - m = 3j$, while armchair-type tubes $(n = m)$ are always metallic because they are independent of curvature. All other tubes (i.e., $n - m \neq 3j$) are true semiconductors. A slight change in the wrapping angle of a hexagonal lattice drastically changes their conductive properties. The energy gap of semiconducting CNTs is approximately inversely proportional to the tube diameter, and to be independent of the chiral angle (wrapping angle), according to the tight binding calculation, the energy gap E_g is given by

$$E_g = 2\gamma a_{C-C}/d_t \tag{1.6}$$

where γ is the nearest neighbor interaction energy, $a_{C-C} = 0.142\,\text{nm}$ being the carbon–carbon bond distance, and d_t is the nanotube diameter [19]. The $1/d_t$ dependence is really observed experimentally for SWNTs with diameters between 1 and 2 nm. The reported values of the tight binding parameter γ are in the range 2.4–2.9 eV [20].

The electronic structures of SWNTs have been extensively studied by resonance Raman scattering spectroscopy [21], photo-absorption and luminescence spectroscopy [22], scanning tunneling microscopy (STM) [23, 24], and so on. The electronic density of states (DOS) of SWNTs are characterized by the presence of

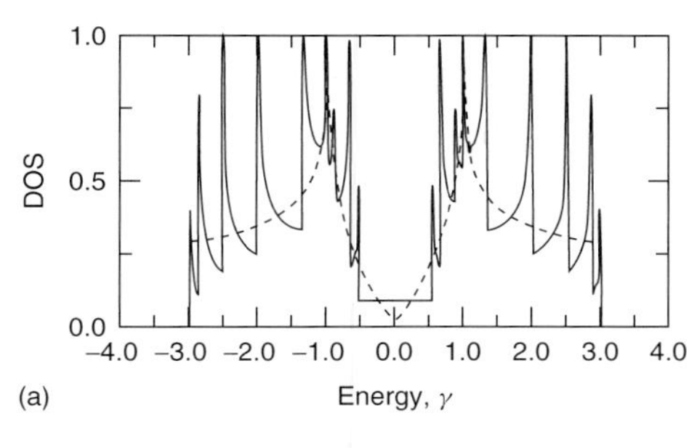

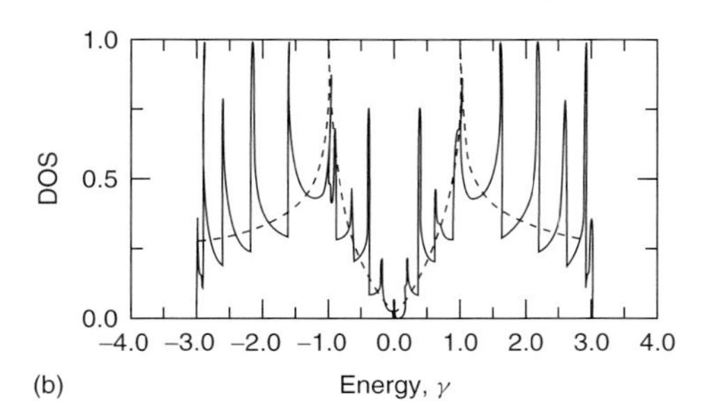

Figure 1.8 The electronic density of states (DOS) of (a) a metallic (9, 0) and (b) a semiconducting (10, 0) SWNT. Spikelike peaks called *van Hove singularities* are characteristic of one-dimensional material. Broken lines represent the DOS of a graphene sheet. (Reprinted with permission from Saito, R. *et al.* (1992) *Appl. Phys. Lett.*, **60**, 2204. Copyright 1992, American Institute of Physics.)

a series of spikelike peaks called *van Hove singularities* as shown in Figure 1.8, reflecting one-dimensional (1D) materials [25].

1.3.2 Electric Properties

CNTs exhibit quantum mechanical electric transport phenomena owing to their small size and structural perfection, and the ballistic transport of electrons over 1 μm distance is also expected because the backscattering of conduction electrons by lattice defects (such as impurity ions) with long-range potentials is annihilated in CNTs [26]. Metallic CNTs can act as tiny wires, and semiconducting ones can be envisioned to act as transistors. The two-terminal conductance of a metallic CNT is given by Landauer formula for 1D conductors [27]:

$$G = \left(2e^2/h\right) \cdot \sum_{i}^{N} T_i \tag{1.7}$$

where $2e^2/h = G_0 \approx 1/\left(12.9\,\text{k}\Omega\right)$ is the quantum unit of conductance, and T_i is the transmission of the ith conducting channel. When $T_i = 1$, corresponding to the case of no scattering inside the CNT and at the contacts to electrodes, a metallic SWNT is expected to have a resistance $R = 1/2G_0 \approx 6.5\,\text{k}\Omega$, because there are two channels of conductance near the Fermi level [26].

In the case of scattering within a CNT, an effective mean free path is used to describe the scattering probability of conduction electrons. Elastic scattering caused by a potential irregularity such as impurity ions and inelastic scattering by phonons contribute to the effective mean free path λ_{eff}:

$$\frac{1}{\lambda_{\text{eff}}} = \frac{1}{\lambda_{\text{el}}} + \frac{1}{\lambda_{\text{ac}}} + \frac{1}{\lambda_{\text{op}}} \tag{1.8}$$

where λ_{el} is the mean free path for elastic scattering and λ_{ac} and λ_{op} are the mean free paths for scattering by acoustic and optical phonons, respectively. Due to the backscattering constraint, the elastic scattering is drastically reduced to $\lambda_{el} \geq 1\,\mu m$. Acoustic phones contribute weakly to inelastic scattering as well, with $\lambda_{ac} \approx 1\,\mu m$ [28]. Thus, conduction electrons in metallic CNTs at low energies (under low electric field) transport without scattering (i.e., ballistic) for a long distance of $1\,\mu m$ or so. On the other hand, optical phone can scatter efficiently once the energy of conduction electrons exceed optical phonon energies (~180 meV), with $\lambda_{op} \approx 20-30\,nm$ [29, 30]. This results in current saturation at elevated biases and can lead to Joule heating and breakdown of CNTs. For semiconducting CNTs, there are indications that at low energies λ_{eff} is on the order of a few hundred nanometers [29, 30].

The low scattering probability together with strong chemical bonding and high thermal conductivity of CNTs allows them to withstand extremely high current densities up to $\sim 10^9\ A\,cm^{-2}$ [31].

1.3.3
Mechanical Properties

Since the basal-plane elastic modulus of graphite is larger than that of any other known materials, CNTs that are formed by seamlessly rolling graphitic sheets are predicted to be extraordinary stiff and strong. Theoretical estimates predict that CNTs have a Young's modulus of the order of 1 TPa [32–34]. Their high stiffness, coupled with their low density, implies that the nanotube might be useful as nanoscale mechanical parts. Young's moduli of isolated MWNTs were measured to be in a range from ~0.1 to ~2 TPa by various methods [35–37], which include observations of thermal vibration and resonance vibration in TEM and stress–strain measurements by an atomic force microscope (AFM). The variation of the measured moduli is considered to be due the difference in the CNT structure (e.g., layer numbers and presence/absence of the bamboo structure), length of a CNT cantilever, and error of temperature measurement. The moduli at lower bound are possibly caused by the occurrence of kinks on the compressed side (inner arc) of a bent CNT. Since the kinks are liable to occur at a smaller bent for a larger diameter CNT, the moduli show a decreasing trend with the increase of diameter [37].

Tensile strength of arc-grown MWNTs has been measured to be in a range 11–63 GPa by a tension test in TEM [36] and 150 ± 45 GPa by AFM [39]. For SWNTs, a high value of tensile strength, about 200 GPa, has been reported [38]. However, a bundle of CVD-grown MWNTs is reported to give much lower strength, 1.72 GPa [39], which is due to structural defects and sliding between adjacent MWNTs.

The most distinguished mechanical property of nanotubes is their unusual strength in the large-strain region. CNTs can be bent repeatedly through large angles (up to ~110°) without undergoing catastrophic fracture, despite the occurrence of kinks [40], being remarkably flexible and resilient.

1.3.4
Heat-Transport Properties

Carbon materials such as diamond and graphite have the highest thermal conductivity near room temperature among materials because of the strong carbon–carbon chemical bonds and light mass of carbon atoms. Phonons (lattice vibrations) are the major and exclusive carriers of heat for graphite and diamond, respectively. Thermal conductivity of a test piece strongly depends on the grain sizes in the specimen. Therefore, the high crystallinity and directionality of CNTs promise their thermal conductivity to be equal to or higher than those of graphite parallel to the basal planes (maximum of ~3000 W$(m\cdot K)^{-1}$ at 100 K, ~2000 W$(m\cdot K)^{-1}$ at 300 K) [41] and natural diamond (maximum ~10^4 W $(m\cdot K)^{-1}$ at 80 K, ~2000 W $(m\cdot K)^{-1}$ at 300 K) [42].

Thermal conductivities of an isolated MWNT with diameter of 14 nm and bundles of MWNTs (bundle diameters of 80 and 200 nm) have been measured using a microfabricated suspended device in the temperature range 8–370 K [43]. The isolated, suspended MWNTs exhibit thermal conductivity of more than 3000 W $(m\cdot K)^{-1}$ at room temperature, which is 1–2 orders of magnitude higher than those for aligned bundles or macroscopic mat samples. Anisotropic thermal conductivities in aligned SWNT films have been observed [44]: 40–60 W $(m\cdot K)^{-1}$ along the fiber direction and about 11 W $(m\cdot K)^{-1}$ perpendicular to the fiber direction.

References

1. Iijima, S. (1991) *Nature*, **354**, 56.
2. Iijima, S. and Ichihashi, T. (1993) *Nature*, **363**, 603.
3. Bethune, D.S., Kiang, C.H., de Vries, M.S., Gorman, G., Savoy, R., Vazquez, J., and Beyers, R. (1993) *Nature*, **363**, 605.
4. Saito, Y., Yoshikawa, T., Okuda, M., Fujimoto, N., Sumiyama, K., Suzuki, K., Kasuya, A., and Nishina, Y. (1993) *J. Phys. Chem. Solids*, **54**, 1849.
5. Zheng, L.X., O'Connel, M.J., Doorn, S.K., Liao, X.Z., Zhao, Y.H., Akhadov, E.A., Hoffbauer, M.A., Roop, B.J., Jia, Q.X., Dye, R.C., Peterson, D.E., Huang, S.W., and Liu, J. (2004) *Nat. Mater.*, **3**, 673.
6. Yun, Y., Shanov, V., Tu, Y., Subramaniam, S., and Schulz, M.J. (2006) *J. Phys. Chem. B*, **110**, 23920.
7. Saito, Y., Yoshikawa, T., Bandow, S., Tomita, M., and Hayashi, T. (1993) *Phys. Rev. B*, **48**, 1907.
8. See for example Kelly, B.T. (1981) *Physics of Graphite*, Chapter 1, Applied Science Publishers, London.
9. Saito, Y. and Yoshikawa, T. (1993) *J. Cryst. Growth*, **134**, 154.
10. Saito, Y., Nakahira, T., and Uemura, S. (2003) *J. Phys. Chem. B*, **107**, 931.
11. Thess, A., Lee, R., Nikolaev, P., Dai, H., Petit, P., Robert, J., Xu, C., Lee, Y.H., Kim, S.G., Rinzler, A.G., Colbert, D.T., Scuseria, G.E., Tomanek, D., Fischer, J.E., and Smalley, R.E. (1996) *Science*, **273**, 483.
12. Saito, Y. (1999) *New Diamond Front. Carbon Technol.*, **9**, 1.
13. JFE Engineering Corporation (2004) JEF Technical Report No. 3, March 2004, CNT Tape of High Purity, p. 78 (in Japanese).
14. Koshio, A., Yudasaka, M., and Iijima, S. (2002) *Chem. Phys. Lett.*, **356**, 595.
15. See for example Govindaraj, A. and Rao, C.N.R. (2006) *Carbon*

Nanotechnol, Chapter 2 (ed. L.Dai), Elsevier, Amsterdam.
16. (a) Nihei, M., Kawabata, A., Hyakushima, T., Sato, S., Nozue, T., Kondo, D., Shioya, H., Iwai, T., Ohfuchi, M., and Awano, Y. (2006) Proceedings of the International Conference on Solid State Devices and Materials, p. 140; (b) Ishikawa, Y. and Ishizuka, K. (2009) *Appl. Phys. Express*, **2**, 045001.
17. Sakuma, N., Katagiri, M., Sakai, T., Suzuki, M., Sato, S., Nihei, M., Nihei, Awano, Y., and Kawarada, H. (2007) New Diamond and Nano Carbon, Abstract Book, p. 195.
18. Honda, S., Katayama, M., Lee, K.-Y., Ikuno, T., Ohkura, S., Oura, K., Furuta, H., and Hirao, T. (2003) *Jpn. J. Appl. Phys.*, **42**, L441.
19. White, C.T. and Mintmire, J.W. (1998) *Nature*, **394**, 29.
20. Odom, T.W., Huang, J.-L., Kim, P., and Lieber, C.M. (1998) *Nature*, **391**, 62.
21. Saito, R. and Kataura, H. (2001) *Carbon Nanotubes; Synthesis, Structure, Properties, and Applications* (ed. M. S. Dresselhaus *et al.*), Springer, Berlin, pp. 213–246.
22. O'Connell, M., Bachilo, S.M., Huffman, C.B., Moore, V., Strano, M.S., Haroz, E., Rialon, K., Boul, P.J., Noon, W.H., Kittrell, C., Ma, J., hauge, R.H., Weisman, R.B., and Smalley, R.E. (2002) *Science*, **297**, 593.
23. Tan, S.J., Devoret, M.H., Dai, H., Thess, A., Smalley, R.E., Greerligs, L.J., and Dekker, C. (1997) *Nature*, **386**, 474.
24. Wildoer, J.W.G., Venema, L.C., Rinzler, A.G., Smalley, R.E., and Dekker, C. (1998) *Nature*, **391**, 59.
25. Saito, R., Fujita, M., Dresselhaus, G., and Dresselhaus, M.S. (1992) *Appl. Phys. Lett.*, **60**, 2204.
26. Ando, T. (2005) *J. Phys. Soc. Jpn.*, **74**, 777.
27. Landauer, R. (1970) *Philos. Mag.*, **21**, 863.
28. Avouris, P. (2004) *MRS Bull.*, **29**, 403.
29. Wind, S.J., Appenzeller, J., and Avouris, P. (2003) *Phys. Rev. Lett.*, **91**, 058301.
30. Yaish, Y., Park, J.Y., Rosenblatt, S., Sazonova, V., Brink, M., and McEuen, P.L. (2004) *Phys. Rev. Lett.*, **92**, 046401.
31. Wang, Z.L., Gao, R.P., de Heer, W.A., and Poncharal, P. (2002) *Appl. Phys. Lett.*, **80**, 856.
32. Lu, J.P. (1997) *Phys. Rev. Lett.*, **79**, 1297.
33. Zhang, P., Huang, Y., Geubelle, P.H., Klein, P.A., and Hwang, K.C. (2002) *Int. J. Solids Struct.*, **39**, 3893.
34. Li, C. and Chou, T.W. (2003) *Compos. Sci. Technol.*, **63**, 1517.
35. Treacy, M.M.J., Ebbesen, T.W., and Gibson, J.M. (1996) *Nature*, **381**, 678.
36. Yu, M.F., Lourie, O., Dyer, M.J., Moloni, K., Kelly, T.F., and Ruoff, R.S. (2000) *Science*, **287**, 637.
37. Poncharal, P., Wang, Z.L., Ugarte, D., and de Heer, W.A. (1999) *Science*, **283**, 1513.
38. Kong, J., Soh, H.T., Cassell, A.M., Quante, C.F., and Dai, H. (1998) *Nature*, **395**, 878.
39. Pan, Z.W., Xie, S.S., Lu, L., Chang, B.H., Sun, L.F., Zhou, W.Y., Wang, G., and Zhang, D.L. (1999) *Appl. Phys. Lett.*, **74**, 3152.
40. Iijima, S., Brabec, C., Maiti, A., and Bernholc, J. (1996) *J. Chem. Phys.*, **104**, 2089.
41. Kelly, B.T. (1981) *Physics of Graphite*, Chapter 4, Applied Science Publisher, London.
42. Barman, S. and Srivastava, G.P. (2007) *J. Appl. Phys.*, **101**, 123507.
43. Kim, P., Shi, L., Majumdar, A., and McEuen, P.L. (2001) *Phys. Rev. Lett.*, **87**, 215502.
44. Fischer, J.E., Zhou, W., Vavro, J., Llaguno, M.C., Guthy, C., Haggenmueller, R., Casavant, M.J., Walters, D.E., and Smalley, R.E. (2003) *J. Appl. Phys.*, **93**, 2157.

2
Preparation of CNT Emitters

Yahachi Saito

2.1
Introduction

Carbon nanotubes (CNTs) as electron emitters may be used in two forms: one is a pointed electron source consisting of a single CNT or a tiny, thin bundle of CNTs; and the other is a planar emitter consisting of a CNT film or a patterned film. The former is suitable for the formation of a finely focused electron beam applicable to electron microscopes, electron beam lithography, micro X-ray sources, and so on, and the latter is applicable to display devices, light sources (e.g., lamp and back light), high-power vacuum microwave amplifiers, industrial and clinical e-beam apparatus for surface treatment, disinfection, and so forth.

Currently, efficient and reliable methods to fabricate CNT pointed emitters are lacking and their fabrication rely on trial-and-error methods, though many such methods have been reported. On the other hand, film emitters can be produced with fairly good reproducibility in pilot pants, but further studies for realizing high emission site density and low driving voltage are required. In this chapter, fabrication methods of CNT point emitters and surface emitters are reviewed.

2.2
CNT Point Emitters

Substrates employed so far for supporting an isolated CNT or a thin bundle of CNTs are electrochemically etched needles of a metal, hairpin-shaped filaments of a refractory metal, carbon fibers, and so on. CNTs are either mounted or directly grown on the apex of a supporting substrate.

2.2.1
Manual Attachment of a CNT Bundle

The simplest and easiest method is to glue a thread of CNTs on the tip of a needle-shaped support or on the top of a hairpin-shaped filament on which a tiny

Carbon Nanotube and Related Field Emitters: Fundamentals and Applications. Edited by Yahachi Saito

ISBN: 978-3-527-32734-8

amount of conductive adhesive is applied [1]. A fibrous agglomerate (a bundle with a diameter of approximately 50 μm) of as-grown multiwall carbon nanotubes (MWCNTs) produced by arc discharge, a piece of CNT film (sometimes called *mat* made of purified single-wall nanotubes (SWNTs or MWNTs), and so on, which can be observed by the naked eye and handled manually, are used for this simple method. Since this type of emitters is a bundle consisting of numerous CNTs, many individual and bundled CNTs protrude out from a lump of CNTs. When an emission experiment is carried out, however, only a limited number of CNTs emit electrons since the electric field strong enough for electron tunneling appears on only a small number of CNT tips because of the shielding effect.

In order to mount a single MWNT or a very thin bundle consisting of a limited number of MWNTs, the gluing procedure is carried out under an optical microscope using dark field illumination and operated at typical magnifications of 600–1250 [2, 3]. The microscope is equipped with one or two 3-axes micromanipulators which are used to move a tip of the support material (typically, a tungsten tip) and/or an ensemble of MWNTs. Individual CNTs are first attached to the tungsten tip, the side of which is coated with an acrylic adhesive by sticking it to an adhesive carbon tape, and then the tip (or the CNT assembly) is pulled away, leaving a single MWNT (or a very thin bundle with a single CNT at the end) extending from the tip.

2.2.2 Mounting Inside an SEM

A technique for mounting a single CNT on a support material using a scanning electron microscope (SEM) equipped with two independent piezo-driven nanomanipulators has been developed. An ensemble of MWNTs (bundled or aligned CNTs) and a support material (e.g., a tungsten tip) are mounted on the stages (or tips) of the manipulators [4, 5]. When a single MWNT projecting out of the nanotube ensemble is attached to the sidewall of the supporting tip, the MWNT gets stuck to the substrate surface by van der Waals force. In order to fix the CNT firmly, a metal (e.g., tungsten) or amorphous carbon (so-called contamination) deposit is spotted at a few points along the CNT contacting with the substrate by using a focused electron beam (electron beam–induced deposition). The source material for the tungsten deposition is sublimed $W(CO)_6$ vapor, and that for carbon deposition is hydrocarbon vapor inside the SEM chamber. A typical example of MWNT emitters so prepared is shown in Figure 2.1. The CNT emitters thus prepared are usually covered with contaminants (hydrocarbon deposit) during the SEM observation.

2.2.3 Electrophoric and Magnetophoretic Methods

Electrophoresis is a method to attach an aligned, single CNT or a very thin bundle of CNTs to the tip of a metal needle (typically, a W needle), even though the control of the number of CNTs in an attached bundle is difficult. The procedure

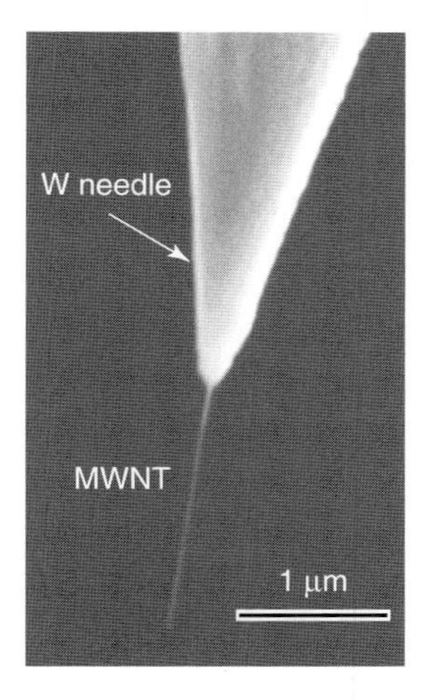

Figure 2.1 SEM image of a single MWNT mounted by nanomanipulation inside an SEM.

of electrophoresis attachment is as follows. First, CNTs are suspended in distilled (or deionized) water by sonication for 10–15 min, and a droplet containing CNTs is placed on a metal (e.g., copper) plate as an electrode. The W needle, which was prepared by electrolytic polishing, is dipped into the droplet on the electrode. As shown in Figure 2.2a, an alternating current (AC) field of 10 V at 2–5 MHz is applied between the W needle and the electrode to induce the electrophoresis. Polarized CNTs are aligned along the electric field because of their anisotropic properties and are attracted toward the W tip where the electric field is strongest. Various types of CNTs (MWNT, DWNT, SWNT) can be fixed onto the W tip by the van der Waals force. CNTs attached to the tip form typically a thin fibril (0.5–10 μm in length and 10–50 nm in diameter near a tip) comprising multiple CNTs, as shown in Figure 2.2b. Individual CNTs are bound strongly with each other and adhere to the needle surface.

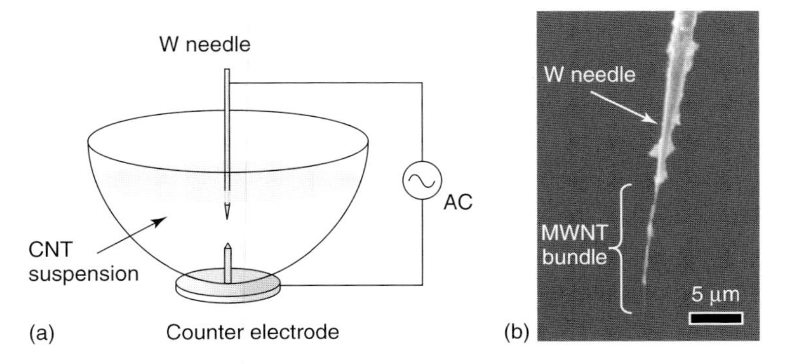

Figure 2.2 (a) Schematic illustrating electrophoresis to stick oriented CNTs onto a W tip. (b) SEM image of a thin bundle of MWNTs extending from the W tip.

Orientation of the attached CNTs is approximately parallel to the direction of the tip axis (confined within a 12° cone angle) [6]. The length of the attached CNTs can be fairly controlled by the withdrawal of the tip from the suspension surface under observation through an optical microscope.

CNTs can also be aligned along magnetic field lines as a result of their anisotropic nature. A magnetophoretic method, where an alternating magnetic field at 60 Hz is applied to the MWCNTs suspended in dichloromethane to attach aligned CNTs onto the tip of an atomic force microscope (AFM), has been reported [7].

2.2.4
Direct Growth on the Apex of a Tip

A technique for growing individual CNT probe tips directly by chemical vapor deposition (CVD) at the end of a silicon tip, which is used as an AFM probe, has been developed [8], and the improvement in the spatial resolution, wear resistance, and the probing depth by the employment of CNT scanning probes has been demonstrated. For the direct growth of individual CNTs on an AFM probe, Hafner *et al.* [9] developed the following process called *pore growth*. They first anodized a commercial silicon (Si) tip, which was previously flattened at its apex by high-load AFM scanning on a hard substrate, in hydrogen fluoride to create nanopores of 50–100 nm diameter along the tip axis. Then, iron catalyst is electrodeposited into the pores, and thin MWNTs are grown by CVD with ethylene and hydrogen at 750 °C. The orientated pore structure was chosen for the catalyst support in order to control the direction of growth. Since the MWNTs grown are usually too long to be used as probes, they are shortened by an *in situ* AFM technique (e.g., application of pulses of electric current).

A simple alternative to fabricate CVD-grown SWNT tips, called *surface growth*, has been reported [8], in which the anodization process is omitted and CNTs grow on the Si pyramidal surface, guided along the edges toward the tip apex. The reason why SWNTs grow along the surface of pyramids toward the apex and extend out of the apex is as follows: When a growing CNT reaches an edge of the pyramid, it can either bend to align with the edge or protrude from the surface. If the energy required to bend the CNT and follow the edge is less than the attractive CNT–substrate interaction energy, then the CNT will follow the pyramid edge to the apex; that is, CNTs are steered toward the tip apex by the pyramid edges. At the apex, the CNT must protrude along the tip axis since the energetic cost of bending is too high.

The direct-growth technique developed for AFM probes is considered to be applicable to the fabrication of a CNT pointed emitter, but any obvious success in growing a single CNT on the apex of a metal needle has not been reported yet. Growth of CNTs (both SWNT and MWNT) at the apexes of multiple Si microcones on a wafer, on the other hand, was successfully carried out and the reduction of threshold voltages was reported [10].

2.2.5 Other Methods

A process consisting of electric field alignment and attachment of CNTs followed by spot welding, which was used for the fabrication of CNT-based AFM probes [11], may be applicable to CNT pointed emitters. Picking up a CNT during AFM imaging of vertically aligned CNTs grown from planar substrate surfaces is an alternative method to create CNT probes [12], which may also be applicable to pointed emitters. A unique method based on wire drawing, with which it is difficult to create a single emitter though, has been reported by Kuzumaki *et al.* [13].

2.3 CNT Film Emitters

Several methods for preparing CNT films have been reported. MWNT films left on a filter after a purification process using filtration were the first ones reported for the preparation of CNT films [14]. In this section, (i) spray coating, (ii) screen printing, (iii) electrophoresis, and (iv) CVD method are described. The first three methods prepare films using CNTs produced beforehand by arc discharge, laser ablation, or CVD.

2.3.1 Spray Coating

CNT films with various densities can be formed by spraying a CNT suspension onto solid substrates with a conventional air brush [15]. CNTs (MWNTs, DWNTs, or SWNTs) are first dispersed in a solvent (e.g., ethanol), and then films are prepared by spraying the suspended solution onto substrate surfaces.

2.3.2 Screen Printing

Crude lumps of CNTs recovered from synthesis vessels are crushed and mixed with binders and surfactants to prepare pastes or slurries. The concentration of CNTs can be adjusted by adding conductive fillers and/or binders to the paste in this method. After printing, MWNT films are heated at 450–500 °C in air to vaporize the organic binders and to form electrical and mechanical contact between CNTs and the metal electrode on the substrate. Since the CNTs are buried under the surface of films, the film surfaces have to be activated to expose CNT tips from the surface by an appropriate surface treatment [16] such as laser irradiation, taping, and blast etching, after the calcination. Surface materials such as graphite and fillers are removed by an appropriate surface treatment [16].

2.3.3
Electrophoresis

Alignment and purification of MWNTs by using AC electrophoresis in isopropyl alcohol (IPA) were first reported by Nakayama *et al.* [17]. MWNTs produced by arc discharge were ultrasonically dispersed in IPA, and large debris carbon particles were removed by centrifugation. The suspension thus obtained was dropped onto coplanar aluminum electrodes with a gap of 0.4 mm on a glass substrate, and an AC of 80 V at 10 Hz to 10 MHz was applied until the IPA evaporated completely at room temperature. They also attempted to deposit MWNTs on patterned conductive layers by electrophoresis [18].

Choi *et al.* deposited SWNTs by electrophoresis onto a patterned metal cathode for a triode-type field emission display (FED) [19]. SWNTs purified by acid treatment and filtration were uniformly mixed with distilled water during sonication. $Mg(NO_3)_2 \cdot 6H_2O$ of 10^{-6}–10^{-2} mol was added to the suspension to give electric charges to the SWNT surfaces. A negative bias of direct current (DC) in the range 10–50 V was applied on the patterned metal plate in this process.

2.3.4
CVD Method

CVD is an alternative to the arc discharge and laser ablation methods to produce CNTs. This method is based on the decomposition of a hydrocarbon gas over a catalytic metal to grow CNTs. In general, MWNTs grown by this method have a larger diameter compared to those made by arc discharge, and are not perfectly graphitized, that is, highly defective, or composed of grains [20]. Catalytic CVD method allows one to fabricate patterns of CNTs on substrates such as glass plates by prepatterning the substrate with a catalyst and subsequently growing CNTs on the catalyst pattern [21–25]. Fe, Co, Ni, and their alloys are commonly used as catalysts, and hydrocarbon gases such as CH_4, C_2H_2, and C_2H_4 are used as carbon feedstock, sometimes with NH_3 or H_2 as carrier (or dilution) gas [26].

Placing CNTs in the required regions on a substrate is a prerequisite to use them as electron emitters in some microelectronic devices. In the case of the direct CVD growth of CNTs, this can be realized by patterning the catalyst on the support substrate and subsequently growing the CNTs. Standard lithographic techniques, such as photolithography or electron beam lithography, can be employed for catalyst patterning. Soft lithography, a technique of stamping liquid-phase catalyst precursors, is another method to pattern surfaces with carbon nanotubes. Microcontact printing of catalyst precursors for the patterned growth of CNTs has been demonstrated by Bonard's group [24].

A comparable method for growing only one nanotube on each catalyst island was developed by Ren *et al.* [27]. Well-separated single carbon nanotubes, tapered with a diameter of 150 nm at the base, were grown vertically on catalyst dots 150 nm in diameter. The structure of the nanotubes resembles more that of carbon fibers since the graphitic planes are inclined to the needle axis and no hollow space in the core is found. Fabrication and structural properties of well-separated arrays of carbon nanotubes, together with their application to field emission devices, are described in chapters 3 and 27 of this book.

References

1. Saito, Y. and Uemura, S. (2000) *Carbon*, **38**, 169.
2. Dai, H., Hafner, J.H., Rinzler, A.G., Colbert, D.T., and Smalley, R.E. (1996) *Nature*, **384**, 147.
3. Fransen, M.J., van Rooy, Th.L., and Kruit, P. (1999) *Appl. Surf. Sci.*, **146**, 312.
4. de Jonge, N., Lamy, Y., Schoots, K., and Oosterkamp, T.H. (2002) *Nature*, **420**, 393.
5. Akita, S., Nishijima, H., Nakayama, Y., Tokumasu, F., and Takeyasu, K. (1999) *J. Phys. D: Appl. Phys.*, **32**, 1044.
6. Tang, J., Yang, G., Zhang, Q., Parhat, A., Maynor, B., Liu, J., Qin, L.C., and Zhou, O. (2005) *Nano Lett.*, **5**, 11.
7. Hall, A., Matthews, W.G., Superfine, R., Falvo, M.R., and Washburn, S. (2003) *Appl. Phys. Lett.*, **82**, 2506.
8. Hafner, J.H., Cheung, C.L., Woolley, A.T., and Lieber, C.M. (2001) *Prog. Biophys. Mol. Biol.*, **77**, 73.
9. Hafner, J.H., Cheung, C.L., and Lieber, C.M. (1999) *Nature*, **398**, 761.
10. Matsumoto, K., Kinosita, S., Gotoh, Y., Uchiyama, T., Manalis, S., and Quate, C. (2001) *Appl. Phys. Lett.*, **78**, 539.
11. Stevens, R., Nguyen, C., Cassell, A., Delzeit, L., Meyyappan, M., and Han, J. (2000) *Appl. Phys. Lett.*, **77**, 3453.
12. Hafner, J.H., Cheung, C.L., Oosterkamp, T.H., and Lieber, C.M. (2001) *J. Phys. Chem. B*, **105**, 743.
13. Kuzumaki, T., Takamura, Y., Ichinose, Hi., and Horiike, Y. (2001) *Appl. Phys. Lett.*, **78**, 3699.
14. de Heer, W.A., Bacsa, W.S., Chatelain, A., Gerfin, T., Humphrey-Baker, R., Forro, L., and Ugarte, D. (1995) *Science*, **268**, 845.
15. Saito, Y. (2003) *J. Nanosci. Nanotech.*, **3**, 39.
16. Uemura, S., Yotani, J., Nagasako, T., Saito, Y., and Yumura, M. (1999) Proceedings of Euro Display 19th IDRC, pp. 93–96.
17. Yamamoto, K., Akita, S., and Nakayama, Y. (1998) *J. Phys. D*, **31**, L34.
18. Nakayama, Y., Akita, S., Pan, L., Yokoyama, S., and Chen, C. (1999) Proceedings of the Japan-Korea Joint Symposium on Imaging Materials and Technology, Osaka Prefecture University, December 4, 1999, Osaka, Japan.
19. Choi, W.B., Jin, Y.W., Kim, H.Y., Lee, S.J., Yun, M.J., Kang, J.H., Choi, Y.S., Park, N.S., Lee, N.S., and Kim, J.M. (2001) *Appl. Phys. Lett.*, **78**, 1547.
20. Dai, H. (2001) in *Carbon Nanotubes Synthesis, Structure, Properties, and Applications* (eds M.S. Dresselhaus *et al.*), Springer, Berlin, pp. 29–53.
21. Ren, Z.F., Huang, Z.P., Xu, J.W., Wang, J.H., Bush, P., Siegal, M.P., and Provencio, P.N. (1998) *Science*, **282**, 1105.
22. Fan, S., Chapline, M.G., Franklin, N.R., Tombler, T.W., Cassell, A.M., and Dai, H. (1999) *Science*, **283**, 512.
23. Murakami, H., Hirakawa, M., Tanaka, C., and Yamakawa, H. (2000) *Appl. Phys. Lett.*, **76**, 1776.
24. Kind, H., Bonard, J.M., Emmenegger, C., Nilsson, L.O., Hernadi, K., Maillard-Schaller, E., Schlapbach, L., Forró, L., and Kern, K. (1999) *Adv. Mater.*, **11**, 1285.
25. Franklin, N.R. and Dai, H. (2000) *Adv. Mater.*, **12**, 890.
26. Choi, G.S., Cho, Y.S., Hong, S.Y., Park, J.B., Son, K.H., and Kim, D.J. (2002) *J. Appl. Phys.*, **91**, 3847.
27. Ren, Z.F., Huang, Z.P., Wang, D.Z., Wen, J.G., Xu, J.W., Wang, J.H., Calvet, L.E., Chen, J., Klemic, J.F., and Reed, M.A. (1999) *Appl. Phys. Lett.*, **75**, 1086.

3
Preparation of Patterned CNT Emitters

Mark Mann, William Ireland Milne, and Kenneth Boh Khin Teo

3.1
Background

The selective placement of carbon nanotubes (CNTs) is extremely important for technological applications such as individually addressable electron emission sources in field emission displays or parallel electron beam lithography systems. Using chemical vapor deposition (CVD) or plasma-enhanced chemical vapor deposition (PECVD) as the growth method, selective growth of vertically aligned CNTs can be achieved by patterning the catalyst prior to growth. For patterned field emission devices to give a uniform, reproducible current, they must have the following:

- The same dimensions: CNTs of differing dimensions will have differing fields induced around them upon the application of a voltage. This would result in a different current coming from each CNT.
- A constant pitch: This is particularly important to maintain the same current density. Also, optimizing the spacing for maximum current density for a given voltage is important for the power consumption of the device.
- One CNT at each location: A secondary CNT perturbs the field around the emitting CNT and may shield the emitting CNT enough to prevent it from emitting.

Some of the earliest reported work to achieve these is shown in Figure 3.1 [1–3]. The work by Fan *et al.* [1] showed that extremely uniform "towers" of nanotubes could be deposited using thermal CVD where the van der Waals interactions between adjacent nanotubes aligned their growth vertically. However, this arrangement of nanotubes is not optimal for field emission because there is extensive electric field shielding between the closely packed nanotubes. Ren *et al.* [2] improved upon this structure by showing that individual CNTs could be deposited from 100-nm-sized patterned dots of Ni catalyst on a Si substrate. However, for CNTs to work effectively (as shown in Figure 3.1b), there must be greater uniformity so that the fields around the apex of the CNTs are broadly similar. In this instance, the poor uniformity could be attributed to there not

Carbon Nanotube and Related Field Emitters: Fundamentals and Applications. Edited by Yahachi Saito

ISBN: 978-3-527-32734-8

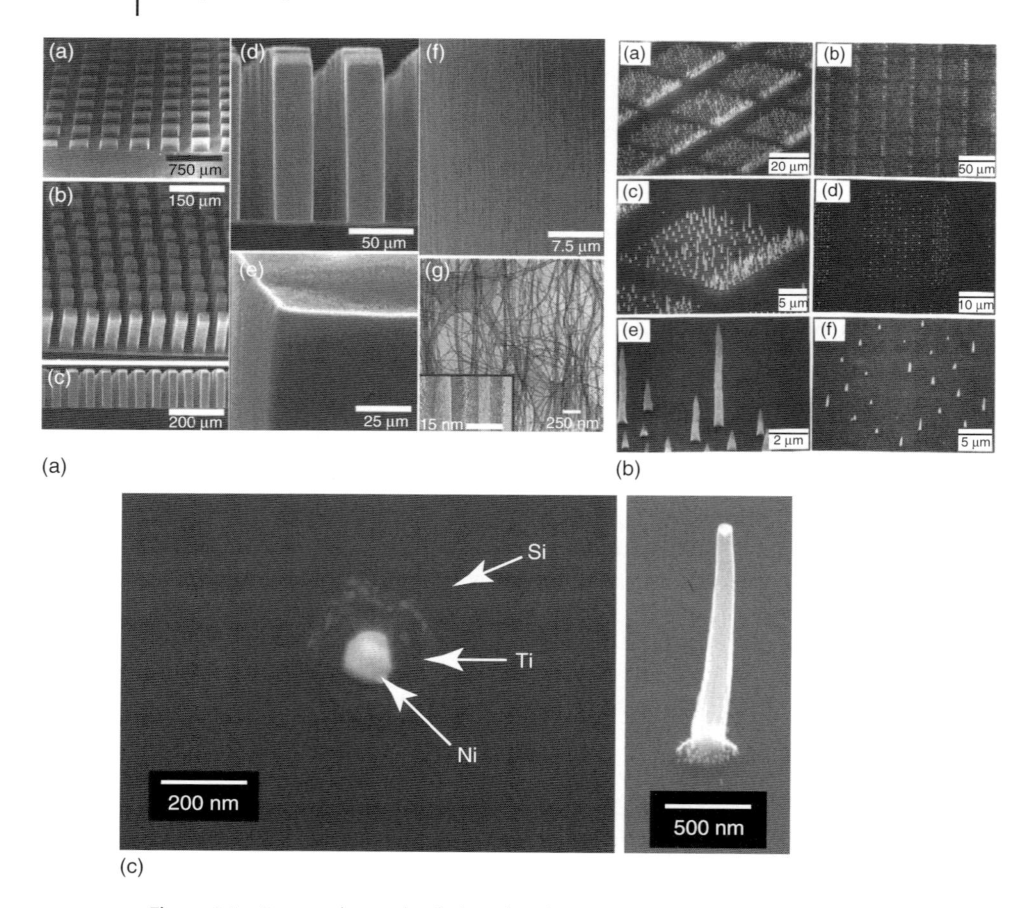

Figure 3.1 Patterned growth of aligned carbon nanotubes: (a) thermal CVD [1], (b) PECVD [2], and (c) PECVD [3].

being a diffusion barrier between the Ni catalyst and the Si substrate, which can cause some of the catalyst to react with the substrate. Merkulov *et al.* [3] utilized a Ti diffusion barrier, achieved better height uniformity, and determined that the critical catalyst dot size to obtain the growth of a single nanotube was 350 nm. However, as seen in Figure 3.1c, although single nanotubes were nucleated from some catalysts, there were also instances of multiple nanotube growth.

It is interesting to note from Figures 3.1b and c that the growth of individual nanotubes by PECVD led to conical structures. These were probably the result of amorphous carbon buildup on the sidewalls of the nanotube (which could be deliberate, as in [4]). Amorphous carbon removal from the flat substrate, which was exposed directly to the plasma, was more efficient than from the sidewalls of the vertical nanotube. Another possibility was that, as the nanotube grew taller, the field enhancement of the structure increased, which caused a focusing of the plasma species onto the tip, thereby producing the "sharpening effect."

This leaves a few unanswered questions: What are the lithographic requirements to achieve *truly* single nanotube growth? What is the actual height and diameter uniformity of the nanotubes (this is important to have a narrow range of field enhancement factors for electron emission)? What is the actual yield of single/multiple nanotubes as a function of lithographic dot size? Furthermore, unlike nanotube growth which only occurs at patterned catalyst sites, unwanted amorphous carbon byproducts are condensed over the rest of the substrate as a result of the plasma decomposition of the PECVD gases – hence, are there growth conditions that result in nanotube growth without surface carbon? This chapter details the process developments at Cambridge University Engineering Department and quantitatively answers these questions. Though the methods described are not the only possible solutions, they have resulted in control over CNT morphology to a degree not reported thus far elsewhere [5].

3.2 Growth of Carbon Nanotubes from Patterned Catalysts

3.2.1 Patterned Growth from Catalyst Film Edges

In this work, the first attempt to grow patterned arrays of CNTs was based on preparing thin film Ni metal edges. To do this, a photoresist was spin-coated onto a Si substrate, baked, positively exposed, and developed to produce an array of 1-µm-diameter holes. The resist profile had a positive slope as seen in Figure 3.2a (micrograph Figure 3.3a). The structure was then coated with 10 nm of Ni using thermal evaporation. When the resist was dissolved in acetone for liftoff, 1-µm-diameter rings of Ni edges or "wings" [6] were formed as shown schematically in Figure 3.2 (micrograph in Figure 3.3b).

CNT growth was then performed under typical PECVD conditions of 700 °C and 40 : 200 sccm of $C_2H_2 : NH_3$ with a substrate bias of −600 V.

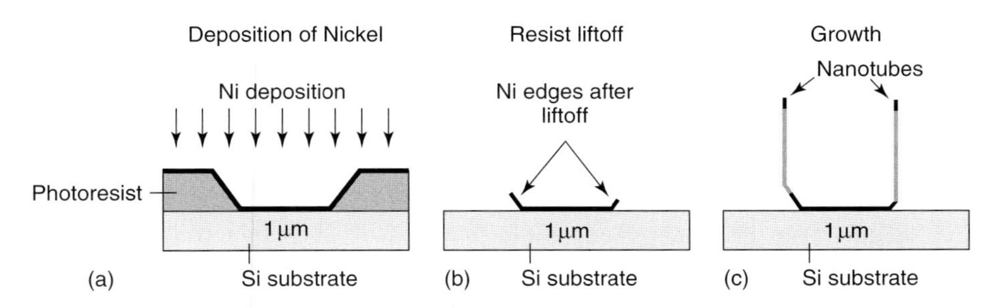

Figure 3.2 Process used to produce Ni metal edges for nanotube growth.

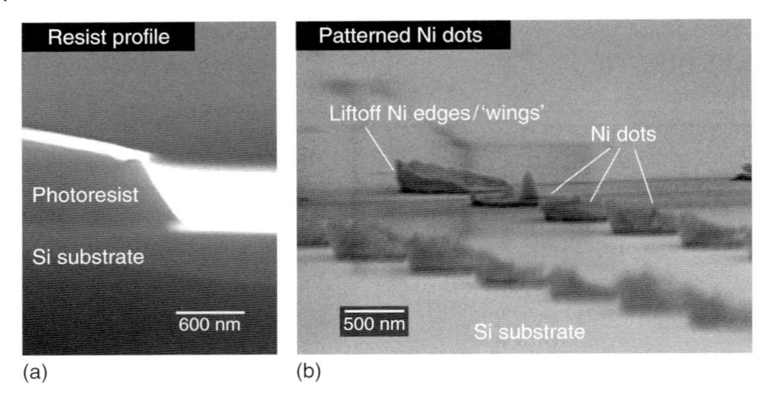

Figure 3.3 (a) Positive resist profile used to produce Ni metal edges during liftoff and (b) the actual Ni dots after liftoff with uneven edges that are 200 nm high.

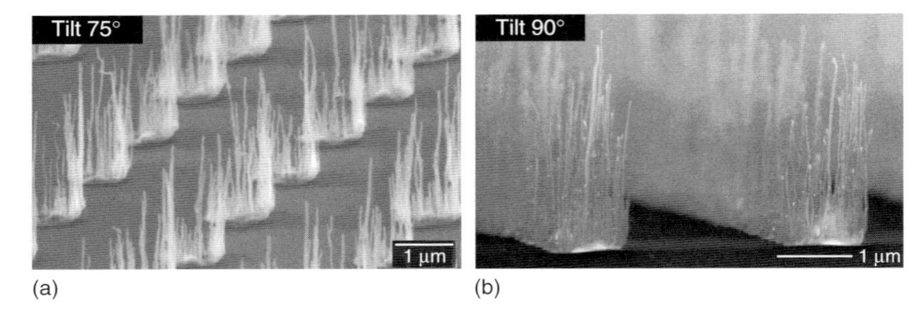

Figure 3.4 Arrays of carbon nanotubes nucleated from the Ni edges of Figure 3.3b.

The high-contrast dot at the tip of the CNTs indicates a tip-based growth mechanism (Figure 3.4b). The presence of Ni at the head of the nanotube was also confirmed by Auger electron spectroscopy (AES). The CNTs were preferentially deposited around the edges of the Ni dot. No CNTs nucleated in the center of the Ni dot because, as confirmed with AES, the Ni was in contact with the Si substrate and hence diffused into the substrate during growth at 700 °C. Indeed, this is a phenomenon widely reported in the literature, where Ni-catalyzed CNT growth cannot occur directly on silicon because of the formation of NiS_x upon the application of heat. Note, finally, that the nanotubes in Figure 3.4 are rather non-uniform in height because the original Ni edges (Figure 3.3b) were jagged and non-uniform.

"Edge" growth could be utilized to fabricate single-file rows of nanotubes as shown in Figure 3.5. First, Si wedges were fabricated using anisotropic etching of masked micrometer-sized horizontal lines on a Si <100> substrate with KOH solution. The tips of these wedges were then coated with Ni by angled evaporation, and then the wedges drawn back slightly by wet etching to reveal the Ni edge. CNT growth then selectively occurred at the exposed Ni edges. Note that in this procedure, although much larger micrometer-sized lithography was initially used

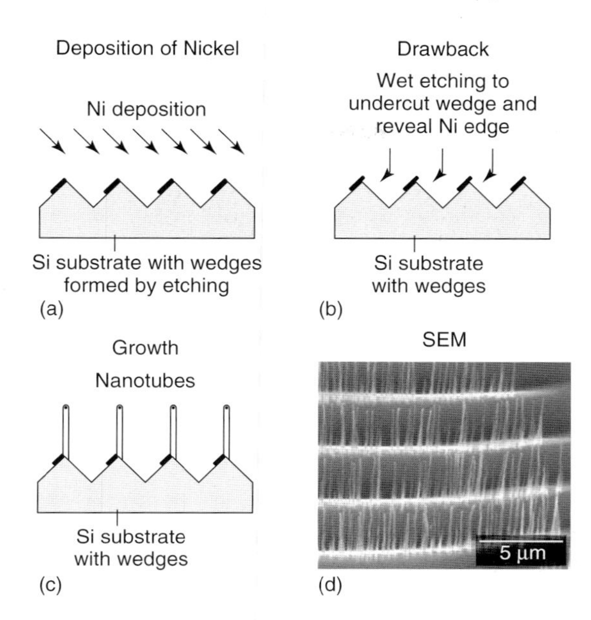

Figure 3.5 Method for the growth of nanotubes on wedges using catalyst metal edges.

to produce the wedges, spaced lines of single-file nanotubes which would be useful for electron emission purposes could be deposited with relative ease. This method has been extended to produce nanotubes on Si tips [7].

3.2.2 Patterned Growth from Catalyst Thin Films on a Diffusion Barrier

Diffusion barriers prevent the diffusion of the catalyst upon the application of heat into the silicon. A suitable barrier remains solid at typical CNT growth temperatures, with the dewetted catalyst sitting upon it before the carbon feedstock is let in. Both photolithography and electron beam lithography can be used for patterning the CNT catalyst prior to nanotube growth. A typical patterning technique for electron beam lithography is described below.

Diffusion barriers typically used are indium tin oxide [8], silica [9], and titanium nitride [10]. This is deposited first, either by sputtering or by PECVD. Typically, a resist such as poly(methyl methacrylate) (PMMA) (4% PMMA in anisol) is spin-coated onto substrates and baked. High-resolution electron beam lithography is then used to define the patterns. The resist can be deliberately overexposed[1] in order to produce a slightly undercut profile (Figure 3.6a) to ease the liftoff later in the fabrication process. The resist is then developed in a 3 : 1 solution of isopropyl alcohol : methylisobutylketone [11], followed by ashing in oxygen plasma. The

1) The electron beam shape and the backscattered electrons from the substrate exposed the underside of the resist more, which led to the undercut resist profile. See also [5].

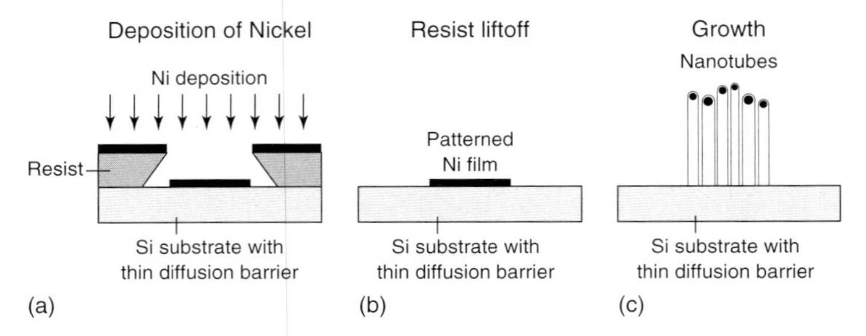

Figure 3.6 Patterned growth from catalyst thin films.

Figure 3.7 Example of patterned nanotube growth, from initial artwork to after growth.

catalyst film, in this case Ni, is deposited on top, which can be done by either sputtering or evaporation. To complete the patterning, liftoff of the unwanted Ni is performed by dissolving the resist in acetone. Note that both the diffusion barrier and the catalyst can be patterned together using this process. An example of patterned nanotube growth using this method is shown in Figure 3.7.

3.3
Single Nanotube Growth – Requirements and Uniformity

When micrometer-sized patterned areas of thin film Ni catalyst are used, multiple nanotubes are nucleated because the catalyst thin film dewets and coalesces into multiple clusters when the film is annealed at 700 °C as shown in Figure 3.8a. If the width of the catalyst, w, is reduced below a critical value, a single nanocluster will form as a result of surface tension effects. A single nanocluster leads to the nucleation of a single nanotube as shown in Figure 3.8b.

In the following experiment, the influence of the patterned width of the catalyst (w) was studied by varying w from 100 to 800 nm. Arrays of square Ni dots, 5 μm in

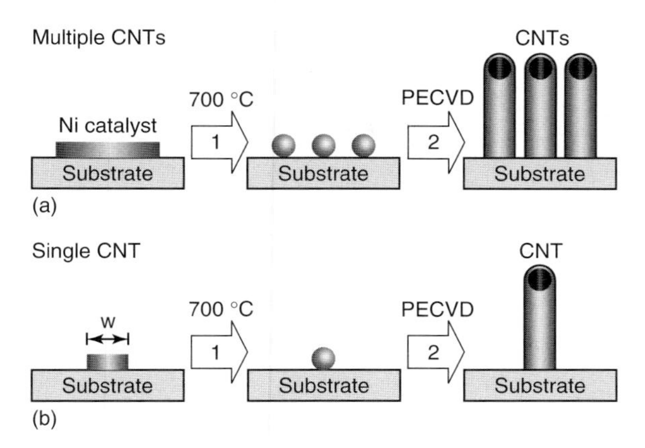

Figure 3.8 Nucleation of (a) multiple nanotubes from "large" patterned catalysts and (b) single nanotubes from "small" patterned catalysts. The dimension *w* was varied here.

pitch, were patterned using electron beam lithography. The Ni film thickness was fixed at 7 nm.

Figure 3.9 shows CNTs nucleated on Ni dot arrays ranging from 100 to 800 nm in size. Clearly, single nanotubes form when the Ni dot size is less than 300 nm, which is in good agreement with the value of 350 nm found by Merkulov *et al.* [3]. However, for both the 300- and 200-nm-sized catalysts, there were still some instances of multiple (double) nanotubes from the catalyst dots (12 and 8%, respectively). A statistical study was undertaken to determine the effect of the catalyst dot size on the number of CNTs nucleated.

The average number of nanotubes and standard deviation were plotted as a function of catalyst dot size, as shown in Figure 3.10. The figure shows that, to a certain extent, it is possible to control the number of nanotubes per catalyst dot by controlling the dot dimensions. It was found that only at 100 nm single nanotubes were deterministically (100%) obtained. However, the distribution in the number of nanotubes nucleated was also found to increase with the catalyst dot size, as shown in the standard deviation plot of Figure 3.10b. This makes it difficult to obtain the desired number of nanotubes per dot for large dot sizes. Note also from the high-resolution SEM images of Figure 3.9 that, when multiple nanotubes are nucleated, the nanotubes have a significant distribution in both diameter and height.

Next, the tip diameter and height distribution of the arrays of individual nanotubes were investigated. The diameters of single nanotubes, which were produced using 100–300-nm-sized catalyst dots, were measured, and histograms showing their distributions are shown in Figure 3.11a–c. For a particular dot size, a tight distribution in the nanotube diameters was observed, and the standard deviation was ~4% of the average nanotube diameter. For comparison, Figure 3.11d shows the distribution in diameters from a 100-nm patterned catalyst line which nucleated multiple nanotubes. In this case, due to the "random" coalescence of the

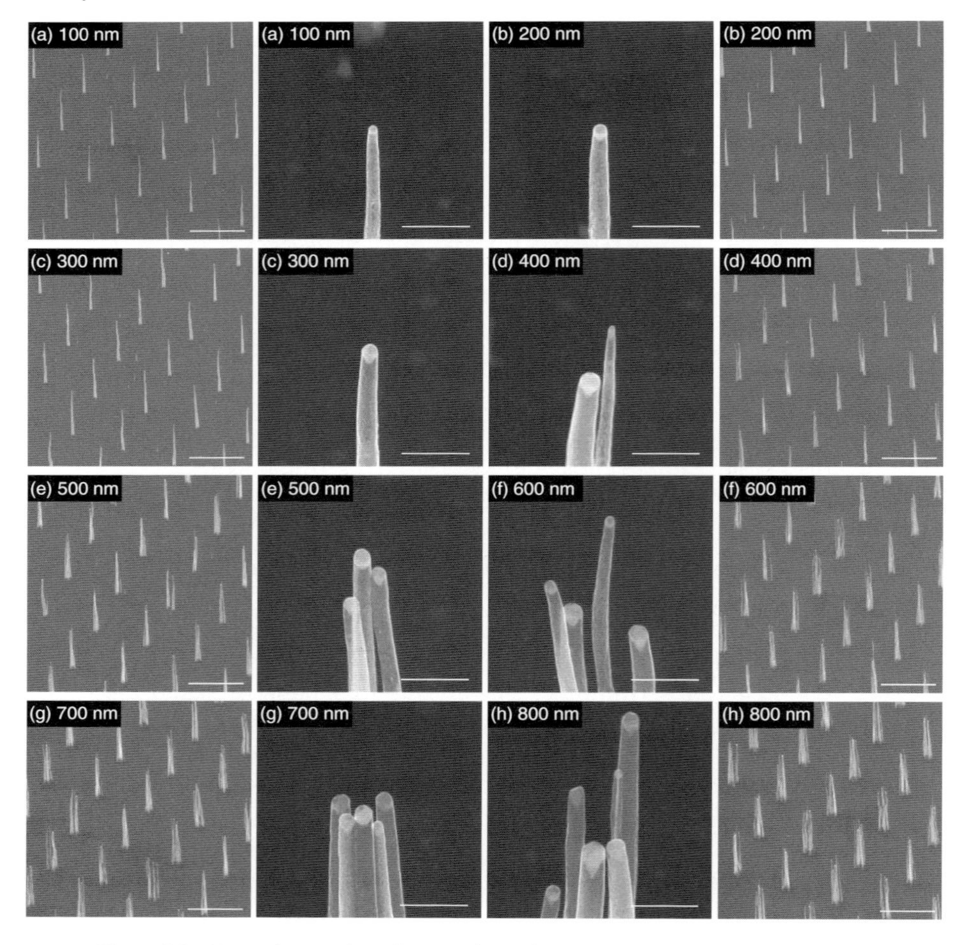

Figure 3.9 Nanotubes nucleated on Ni dots whose width was varied from (a) 100 nm to (h) 800 nm. The Ni thickness was fixed at 7 nm. For low-resolution images, the sample was tilted at 40° with 5 μm scale bar. For high-resolution images, the scale bar is 400 nm long and the sample tilt was 40°.

catalyst, the standard deviation was much larger at 18% of the average nanotube diameter.

Note that in all cases the observed diameter of the nanotube was actually smaller than the catalyst dot size. This is because the catalyst forms a nanocluster of equal volume during growth. If it is assumed that the catalyst becomes a spherical nanocluster, the expected diameters of each nanotube after growth could be calculated by equating the volume of the patterned catalyst with the volume of a sphere because of the conservation of the catalyst. The observed diameters, especially when small, correspond well to the calculated diameters, but

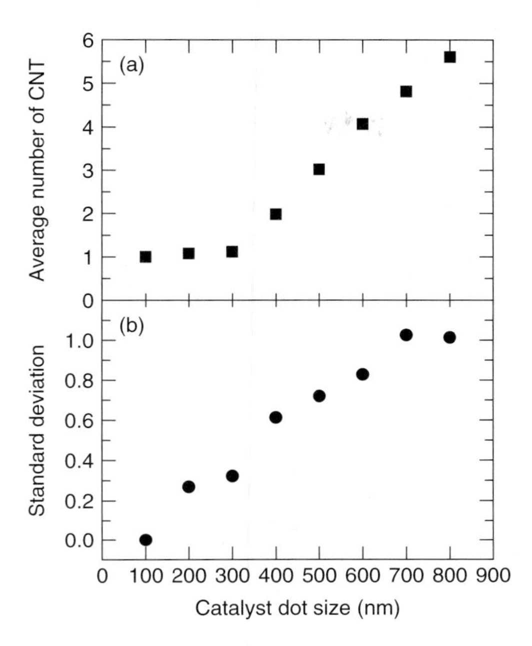

Figure 3.10 Plot of the (a) average number of nanotubes and its (b) standard deviation versus the catalyst dot size. The inset is a high-resolution micrograph of the top section of a nanotube obtained with a 100-nm catalyst dot.

the observed diameters are actually slightly smaller than the calculated diameters because the nanocluster tends to elongate to form a droplet rather than maintain a spherical shape. For larger diameters, the deviation is larger.

The distributions in nanotube heights for single nanotubes nucleated from 100- to 300-nm catalyst dots are shown in Figure 3.12a–c. A relatively tight distribution was observed and the standard deviation was ~6% with an average height of ~5.8 μm. For comparison, the height distribution of 60 nanotubes deposited from a 100-nm catalyst line is shown in Figure 3.12d and e. Note that the distribution is significantly wider (with a standard deviation 18% of the average) in this case because of the "random" nature in which the patterned catalyst line broke up to form multiple nanoclusters.

To summarize, the nucleation and growth of an individual nanotube from a single catalyst dot is a deterministic process which can produce highly uniform nanotubes in terms of height and diameter (Figure 3.13). By contrast, the formation of multiple nanotubes from "large" catalyst patterns produces a relatively large distribution in the number of nanotubes per catalyst dot, nanotube diameter, and nanotube height.

Finally, the standard deviation of the field enhancement factor (β = height/radius [12]) of single nanotubes was estimated to be 7% of the average from the observed standard deviations of diameter (4%) and height (6%).

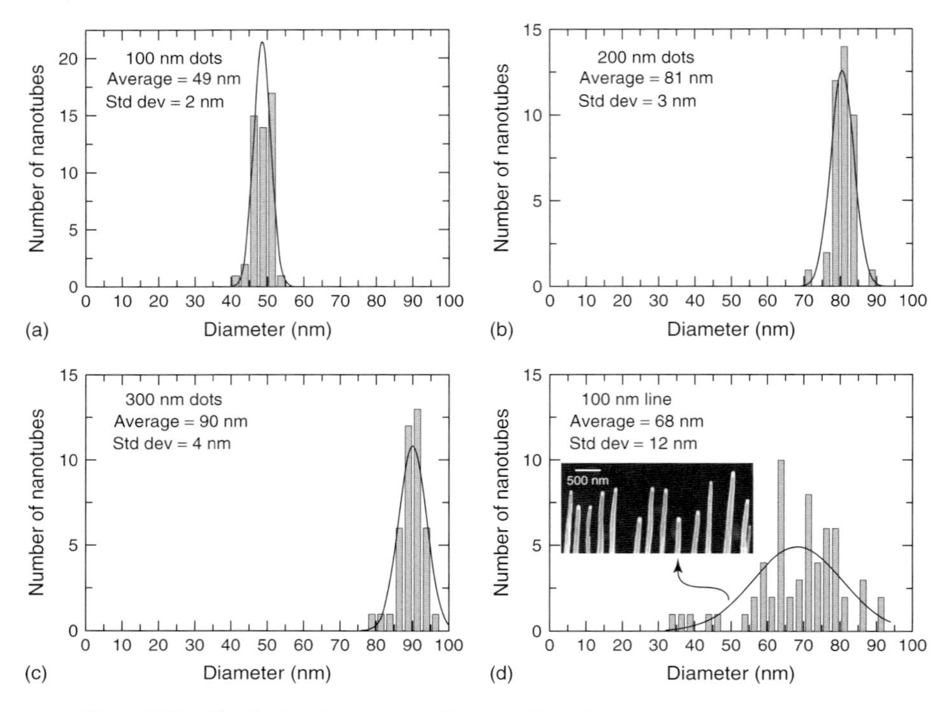

Figure 3.11 Distribution in nanotube diameters from (a) 100 nm, (b) 200 nm, and (c) 300 nm catalyst dot size. The distribution from multiple nanotubes nucleated from a 100-nm-wide catalyst line is provided in (d).

3.4 Nanotube Growth without Surface Carbon

3.4.1 Analysis of Substrate Surfaces Exposed to the Plasma

During CNT growth, byproducts condense on the substrate surface by the plasma decomposition of C_2H_2 and NH_3. If a high concentration C_2H_2 is used, amorphous carbon will condense on the substrate surface in areas without Ni catalyst. Conversely, if a high NH_3 concentration is used, the substrate surface would be free from amorphous carbon as it is etched away by N^+ and H^+ species in the plasma; however, too much NH_3 would also attack the Si substrate.

The substrate surface was examined under various deposition conditions ranging from 15 to 75% $C_2H_2 : NH_3$ (Table 3.1) in order to find conditions in which a balance would be established between the deposition and etching of amorphous carbon, thereby leaving the surface free from amorphous carbon. The temperature, substrate bias, and deposition time were fixed. Si substrates patterned with Ni catalyst and diffusion barrier thin films were used. The non-nanotube areas of the

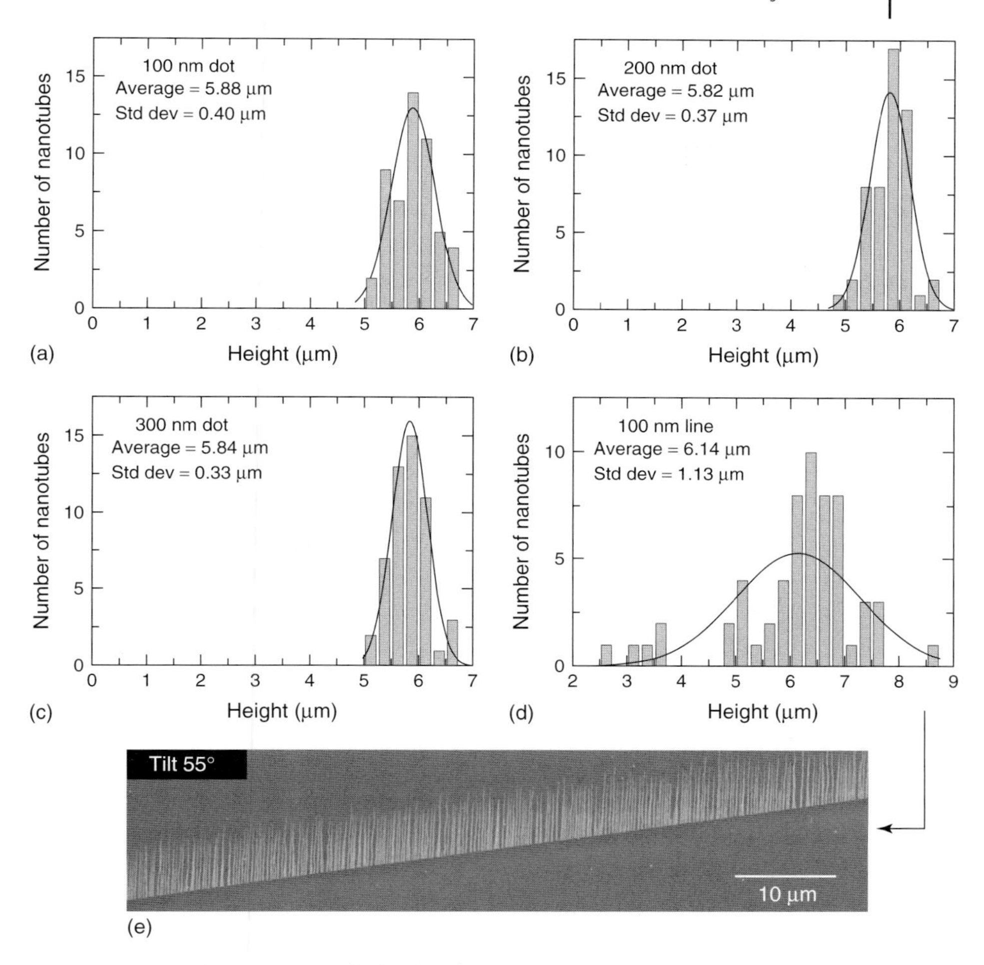

Figure 3.12 Distribution in nanotube heights for single nanotubes grown from (a) 100 nm, (b) 200 nm, (c) 300 nm catalyst dots, and (d and e) multiple nanotubes nucleated from a 100-nm-wide catalyst line.

substrate were characterized after growth by AES. Figure 3.14 shows the evolution of the nonnanotube areas of the substrate under increasing $C_2H_2:NH_3$ ratio. For 15% (Figure 3.14a), anisotropic etching of the silicon substrate by NH_3 was observed, whereas for 75% a thick film was found delaminating from the silicon substrate (Figure 3.14e).

The AES results for 50% $C_2H_2:NH_3$ and 20% $C_2H_2:NH_3$ illustrate the difference in amorphous carbon deposition. The scanning electron micrographs of Figure 3.14b and d show that the 20% condition yielded an apparently clean silicon surface on the nonnanotube areas, whereas some byproducts were condensed on the substrate for the 50% condition. A full Auger elemental survey of these

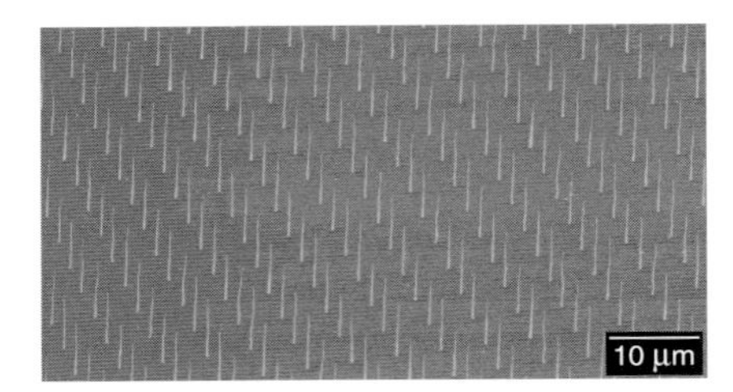

Figure 3.13 Highly uniform growth of individual nanotubes. Sample tilt 55°.

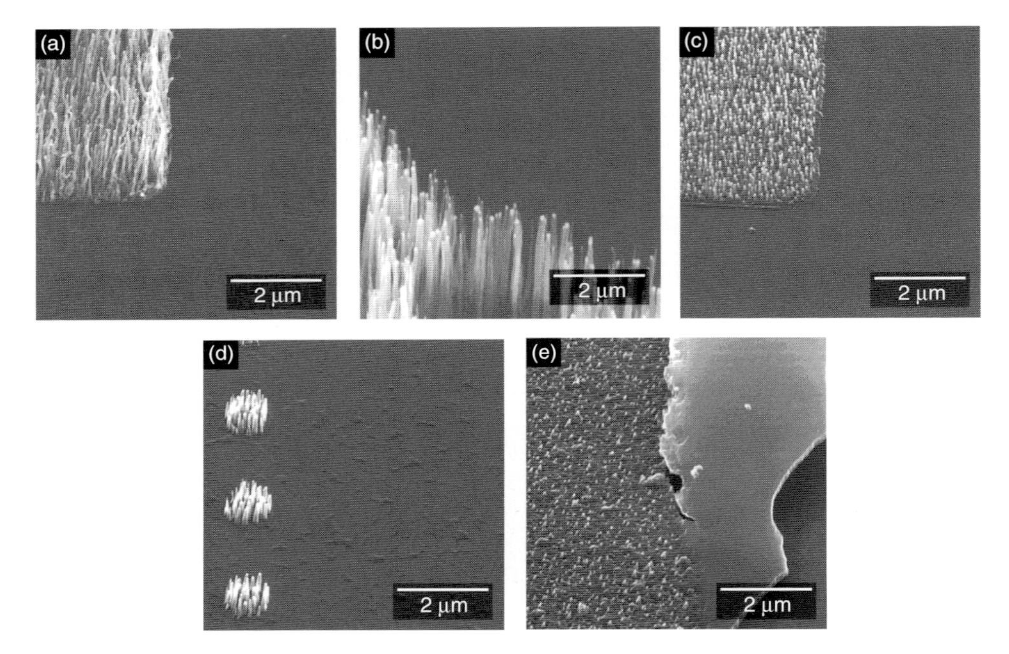

Figure 3.14 Evolution of the Si substrate areas as a function of increasing C_2H_2 : NH_3 gas ratio. (a) At 15%, the Si substrate was anisotropically etched. (b) The substrate appears "smooth" at 20%. (c) At 30%, byproducts began to be deposited on the substrate. (d) At 50%, there were significant byproducts on the substrate surface. (e) At 75%, a thick film of the byproduct was observed to delaminate from the substrate surface.

noncatalyzed/nonnanotube areas of the substrate surface was acquired as shown in Figure 3.15. The "clean" Si substrate surface (gas composition 20% $C_2H_2 : NH_3$) contained C, N, O, Si, and a trace amount of W, whereas the condensed byproduct layer (gas composition 50% $C_2H_2 : NH_3$) consisted of mainly carbon with a trace of N. The Auger (KVV) spectrum in the inset of Figure 3.15 indicates that the carbon is amorphous and highly disordered sp^2.

Depth-profiled elemental analysis was then performed by sputtering the substrate surfaces layer by layer using a 2-keV Ar gun. The composition depth profile for

Table 3.1 Summary of gas flow ratios used in the experiment.

C_2H_2 flow rate (sccm)	NH_3 flow rate (sccm)
30	200
40	200
50	200
60	200
100	200
150	200

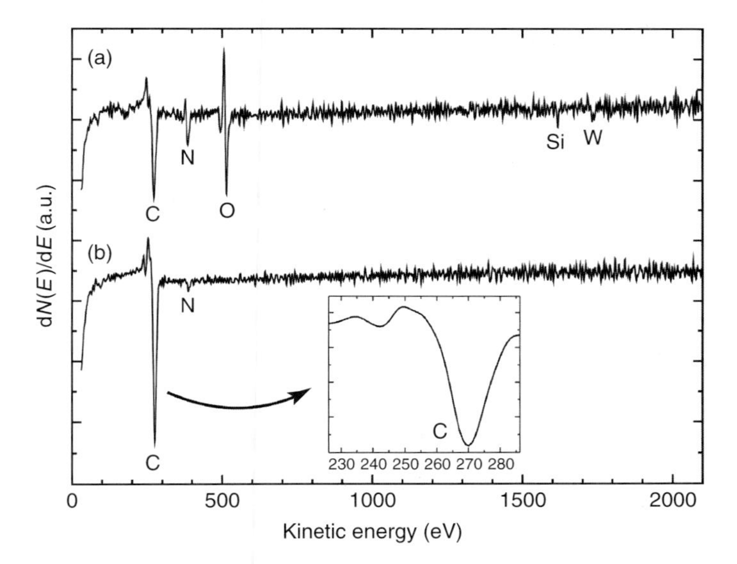

Figure 3.15 Auger chemical composition analysis of substrate surface for C_2H_2 : NH_3 deposition gas ratios of (a) 20% and (b) 50%. The inset plot of the Auger C (KVV) peak confirms that the carbon formed at 50% C_2H_2 : NH_3 deposition gas ratio was amorphous in nature.

20% $C_2H_2 : NH_3$ (Figure 3.16a) shows the presence of a Si/C/N/O/W surface layer which extended a short distance into the silicon substrate. The Si (KLL) peak is resolvable into two contributions: namely Si–x (x for C, N, O, W) bonding for the surface layer and Si–Si bonding for the silicon substrate. The C and N atoms are products from the plasma decomposition of $C_2H_2 : NH_3$ and their subsequent chemical reaction with the Si surface. The O peak is believed to be from the native oxide on the Si substrate surface. The heater in the system was made from W and so its presence on the substrate surface was due to some heater erosion by the plasma.

The composition depth profile for 50% $C_2H_2 : NH_3$ (Figure 3.16b) clearly shows a thick amorphous carbon : nitrogen (a-C : N) layer, followed by an interfacial layer, and finally the silicon substrate. The Auger N (KLL) peak was resolvable into two

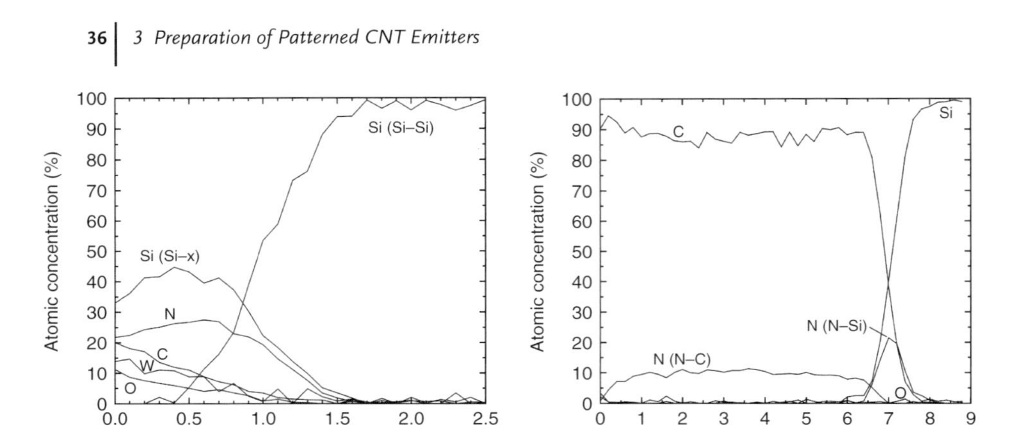

Figure 3.16 Auger depth profile of the substrate surface for C_2H_2 : NH_3 deposition gas ratios of (a) 20% and (b) 50%.

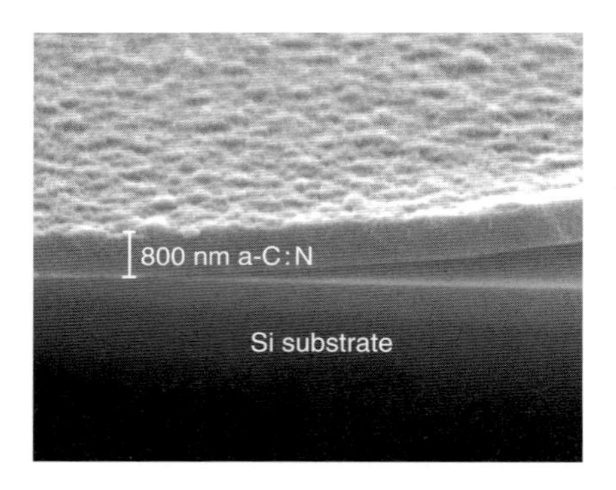

Figure 3.17 Cross section SEM to measure the thickness of the a-C:N layer for deposition under 75% C_2H_2:NH_3 gas ratio.

contributions: N–C bonding in the carbon layer and N–Si bonding at the interface. The concentration of N in the interfacial layer was higher than in the carbon layer, indicating that the N species preferably reacted with the silicon surface. Nonquantitative secondary ion mass spectroscopy (SIMS) experiments have shown that this amorphous carbon layer also contained H. A small amount of O was detected at the interface, which can be attributed to the native oxide on the silicon substrate.

Under conditions of 75% C_2H_2:NH_3, the a-C:N layer was so thick that it delaminated from the substrate surface in some places (Figure 3.14e). This sample was cut in half and imaged, and the a-C:N layer was determined to be 800 nm thick as shown in Figure 3.17.

The depth profile compositions of the substrate surfaces for C_2H_2:NH_3 gas ratios ranging from 15 to 75% are summarized in Figure 3.18. The layer thicknesses presented here were determined by calibrating the sputter rate to the depth of the 800-nm a-C:N film deposited at 75% C_2H_2:NH_3 gas ratio. The sputter rate

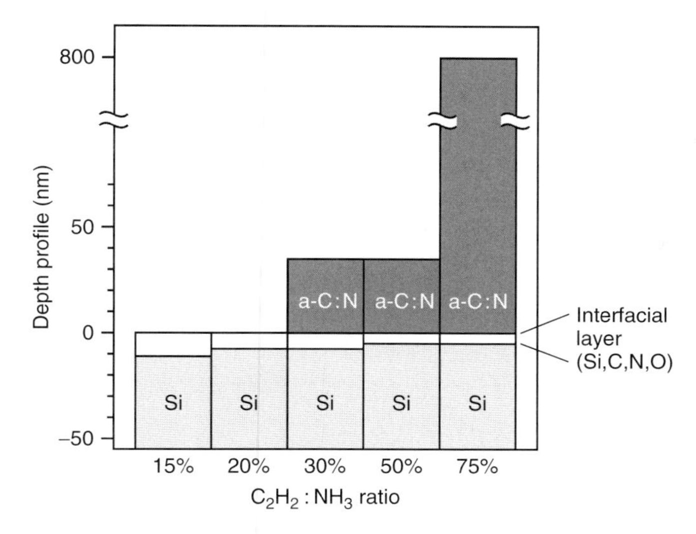

Figure 3.18 Summary of depth profiles performed on the substrate surfaces of Figure 3.14. For $C_2H_2 : NH_3$ deposition gas ratios of 30% or greater, an amorphous carbon : nitrogen (a-C : N) film was detected on the surface of the substrate. In all cases, a thin (5–10 nm) interfacial layer was detected.

for all the layers was assumed to be the same. For $C_2H_2 : NH_3$ gas ratios of 30% and above, an a-C : N layer was clearly found on top of the interface layer. This a-C : N layer rapidly increases in thickness with increasing gas ratio. For amorphous carbon-free growth of nanotubes, a $C_2H_2 : NH_3$ gas ratio of 20% or below must be used. An ~5-nm "interfacial" layer was always present on the surface because of the exposure of the Si substrate to the C/N species in the plasma. At a gas ratio of 15%, the interfacial layer appeared to have extended to 10 nm. This was probably an artifact of the Auger measurement due to the increase of the surface roughness induced by the anisotropic etching of the silicon surface by the plasma (Figure 3.14a).

3.4.2
Analysis of Substrate Surfaces Shielded from the Plasma

The thin interfacial layer has the potential to cause problems in device operation, such as degrading the performance of insulating layers on the substrate. To prevent its formation, it is necessary to occlude/shield the substrate from the plasma. A CNT growth was performed under 20% $C_2H_2 : NH_3$ gas composition (optimal for amorphous carbon-free deposition) but with part of the substrate protected from the plasma by means of a silicon mask, as shown in the inset of Figure 3.19. After deposition, the mask was removed and the shielded area was examined using AES. As shown in the depth profile of Figure 3.19, the substrate surface contained mainly Si and O (from native oxide on the surface) with a low amount of C and N.

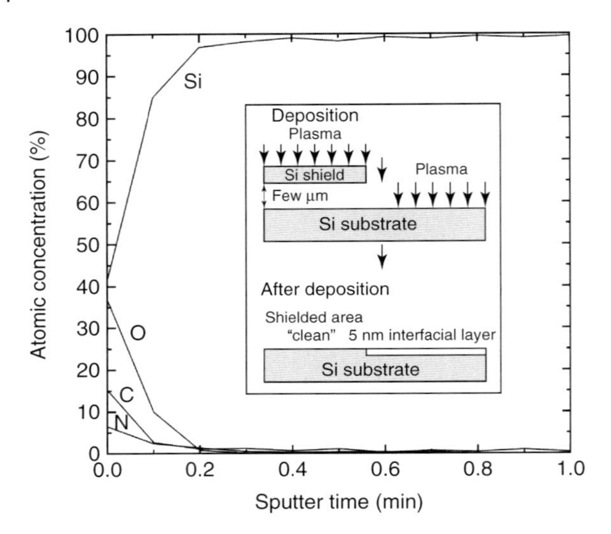

Figure 3.19 The depth profile of a surface shielded from the plasma revealed that the surface was essentially unmodified (surface C/N attributed to adsorbed contaminants, O from native oxide on Si). The inset shows how the silicon shield was used to physically protect the part of the substrate surface during deposition. The shield was subsequently removed for the depth profile measurement.

This C and N residue was rapidly sputtered away during the depth profile, which corresponded to a surface layer depth of only ~1 nm. Thus, the shielded substrate surface was actually "clean" and the C and N were surface contaminants. The shielded surface also contained no W contamination. Clearly, this shows that the silicon mask was effective in shielding the surface from the plasma.

The device structure presented in Figure 3.20 is a practical demonstration of using occlusion to protect certain areas of a device from the deposition plasma. The silicon dioxide insulator was deliberately undercut beneath the polysilicon gate. The gate thus protected the sidewalls of the silicon dioxide from being coated during the plasma deposition. A vertically aligned CNT was then selectively grown *in situ* inside the device.

3.5 Summary

Patterned growth of CNTs has been demonstrated by a number of methods. CNT growth from silicon can occur only if a diffusion barrier is deposited or grown between the silicon and the catalyst particle. This is because catalyst diffuses into the Si substrate at the growth temperature, which inhibits CNT growth.

When a patterned (Ni) catalyst thin film on a diffusion barrier is used, CNTs grow from wherever the catalyst is deposited and with good uniformity and 100% nucleation. By reducing the patterned Ni width to less than 300 nm, a high

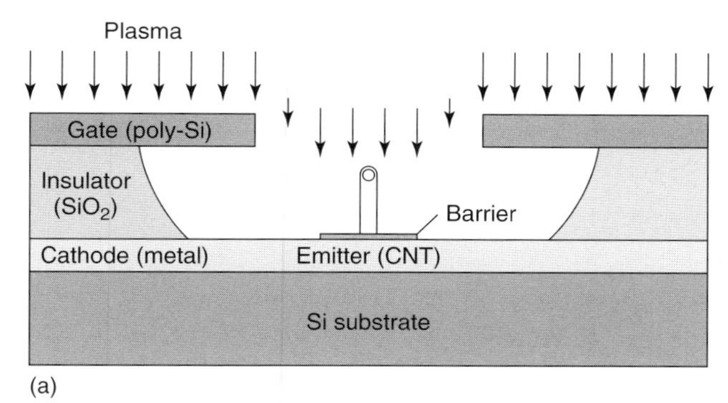

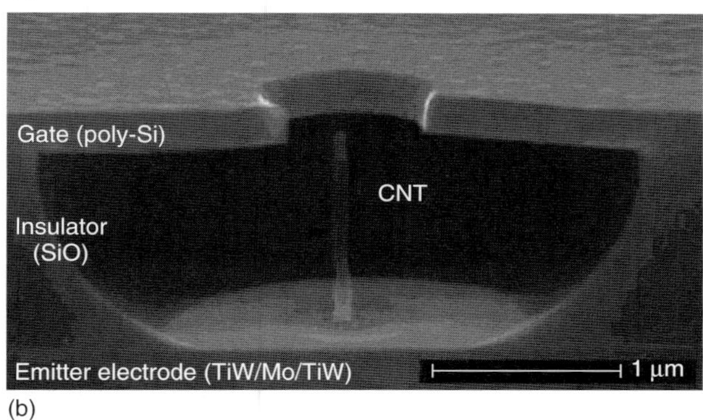

Figure 3.20 (a) The schematic of a field emission device in which the insulator is shielded from plasma during CNT deposition by a polysilicon gate above it. (b) A cross section image of the actual field emission device showing a CNT selectively grown inside the gated structure.

yield (88% or more) of single nanotubes per catalyst can be obtained. To obtain truly deterministic/100% growth of individual nanotubes, the patterned catalyst dimension should be reduced to ~100 nm. Any value less than this results in the liftoff becoming more difficult.

Arrays of individual nanotubes exhibited excellent uniformity in terms of tip diameter and nanotube height, and the standard deviation of these were 4 and 6% of the average, respectively. On the contrary, the growth of multiple nanotubes from large-area patterned catalyst produces a relatively large distribution in the number of CNTs per catalyst dot, nanotube diameter, and nanotube height. For example, multiple nanotube nucleation from a patterned catalyst line produces a standard deviation of 18% of the average nanotube diameter and height.

The optimal gas ratio for surface amorphous carbon-free growth of nanotubes was determined to be 20% $C_2H_2 : NH_3$. This is by no means the only way to prevent the amorphous carbon, since various concentrations of reducing gases such as hydrogen [13] have been reported. However, in this case, the optimal recipe produced a thin, 5-nm interfacial layer containing the plasma species mixed with the substrate material because of the exposure of the substrate to the plasma. In areas where this interfacial layer is not desirable, it was shown that occluding these areas from the plasma was effective in preventing the formation of the interfacial layer, leaving these areas essentially untouched/pristine.

Acknowledgments

The authors would like to acknowledge Didier Pribat at Ecole Polytechnique and Pierre Legagneux, Gilles Pirio, Frederic Wyczisk, and J. Oliver of Thales, France, for their input into this work.

References

1. Fan, S., Chapline, M., Franklin, N., Tombler, T., Cassell, A., and Dai, H. (1999) *Science*, **283**, 512.
2. Ren, Z.F., Huang, Z.P., Wang, D.Z., Wen, J.G., Xu, J.W., Wang, J.H., Calvet, L.E., Chen, J., Klemic, J.F., and Reed, M.A. (1999) *Appl. Phys. Lett.*, **75**, 1086.
3. Merkulov, V.I., Lowndes, D.H., Wei, Y.Y., Eres, G., and Voelkl, E. (2000) *Appl. Phys. Lett.*, **76**, 3555. Picture from *http://www.ornl.gov/ment/docs/transition-modes.pdf.*
4. Merkulov, V.I., Guillorn, M.A., Lowndes, D.H., Simpson, M.L., and Voekl, E. (2001) *App. Phys. Lett.*, **79**, 1178.
5. Teo, K.B.K., Lee, S.B., Chhowalla, M., Semet, V., Binh, V.T., Groening, O., Castignolles, M., Loiseau, A., Pirio, G., Legagneux, P., Pribat, D., Hasko, D.G., Ahmed, H., Amaratunga, G.A.J., and Milne, W.I. (2003) *Nanotechnology*, **14**, 204.
6. Hill, G. (2002) *Experimental Techniques of Semiconductor Research Course Notes,* Institute of Physics, 6 March 2002, p. 15.
7. Bell, M.S., Teo, K.B.K., and Milne, W.I. (2007) *J. Phys. D: Appl. Phys.*, **40**, 2285.
8. Mann, M., Teo, K.B.K., Milne, W.I., and Tessner, T. (2006) *Nano: Brief Rep. Rev.*, **1**, 35.
9. Chhowalla, M., Teo, K.B.K., Ducati, C., Rupesinghe, N.L., Amaratunga, G.A.J., Ferrari, A.C., Roy, D., Robertson, J., and Milne, W.I. (2001) *J. Appl. Phys.*, **90**, 5308.
10. Minoux, E., Groening, O., Teo, K.B.K., Dalal, S.H., Gangloff, L., Schnell, J.-P., Hudanski, L., Bu, I.Y.Y., Vincent, P., Legagneux, P., Amaratunga, G.A.J., and Milne, W.I. (2005) *Nanoletters*, **5**, 2135.
11. Chen, W. and Ahmed, H. (1993) *Appl. Phys. Lett.*, **62**, 1499.
12. Utsumi, T. (1991) *IEEE Trans. Electron Dev.*, **38**, 2276.
13. Meyyappan, M., Delzeit, L., Cassell, A., and Hash, D. (2003) *Plasma Sources Sci. Technol.*, **12**, 205–216.

Part II
Field Emission from Carbon Nanotubes

Carbon Nanotube and Related Field Emitters: Fundamentals and Applications. Edited by Yahachi Saito

ISBN: 978-3-527-32734-8

4
Field Emission Theory

Seungwu Han

4.1
Fowler–Nordheim Theory

Fowler and Nordheim (FN) first derived a semiclassical theory of field emission currents from cold metals in 1928 [1]. In this theory, the system is simplified as a one-dimensional structure along the direction of the external field. The emission tip is modeled as a semi-infinite quantum well with the work function of ϕ, and the local electric field (F) is approximated as a linear potential (Figure 4.1).

By employing the Wentzel–Kramers–Brillouin (WKB) approximation, the following FN equation was obtained:

$$J = 6.2 \times 10^{-6} \frac{\mu^{1/2}}{(\phi + \mu)^{1/2} \phi^{1/2}} F^2 \exp\left(-2.1 \times 10^8 \phi^{3/2}/F\right) \tag{4.1}$$

where J is in amperes per square centimeter of emitting surface, μ and ϕ are in volts, and F in volts per centimeter. Reordering the equation gives the FN plot as follows:

$$\log\left(\frac{J}{F^2}\right) = a - \frac{2.1 \times 10^8 \phi^{3/2}}{F} \quad \text{or} \quad \log\left(\frac{J}{V^2}\right) = a' - \frac{b'}{V} \tag{4.2}$$

where V is the applied bias voltage which is proportional to the local electric field F. The emission tip is geometrically sharp and the electric field is intensified at the tip end, producing a much higher local electric field than the macroscopic applied field. The ratio between the local and applied electric fields (F and E_a, respectively) is called the *field enhancement ratio* (β). For the nanotubes, β typically ranges between hundreds and thousands [2]. Since $\beta = F/E_a$, Eq. (4.2) can be rewritten as follows:

$$\log\left(\frac{J}{E_a^2}\right) = a'' - \frac{2.1 \times 10^8 \phi^{3/2}}{\beta E_a} \tag{4.3}$$

The above FN equation assumes a mathematically sharp surface–vacuum interface. In reality, the electron clouds from the metal do not terminate so sharply. In addition, the escaping electrons feel the image potential V_{im} exerted by the free electrons in the metal. In the classical expression, $V_{im} = -e^2/4x$, where x is

Carbon Nanotube and Related Field Emitters: Fundamentals and Applications. Edited by Yahachi Saito

ISBN: 978-3-527-32734-8

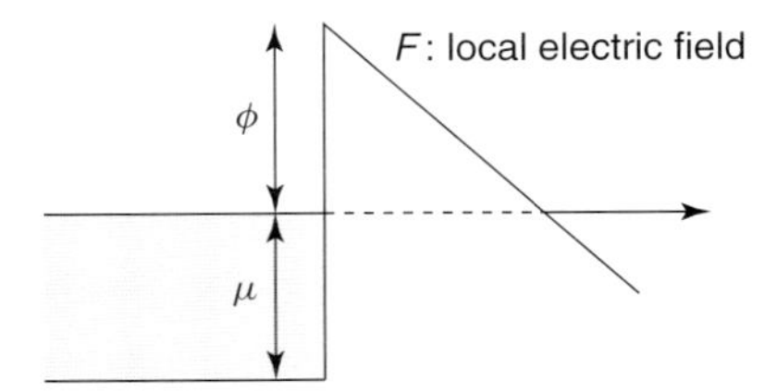

Figure 4.1 Schematic diagram to show the simplified model in the Fowler–Nordheim theory. The dashed line indicates the tunneling region.

the distance from the surface. (Note that, within a few angstroms from the tip, this formula should be modified to reflect the exchange-correlation effects.) The image potential effectively adds a multiplication factor in the exponent of Eq. (4.1) as a slowly varying function of F and can be regarded as a constant in most applications [3].

The FN theory has been successfully applied to numerous metallic systems even when the nonfree electrons such as d-band states are contributing to the carrier density. This is because the nonfree-electron-like states decay much faster than s-like free electrons because of the additional potential barrier associated with the spatial symmetry [4].

When the radii of the emitter tips are of the order of nanometers, the geometrical effects from the nonplanar shape should be taken into account. In [5], the image potentials depending on the tip geometry such as hyperboloid and cone were considered in calculating the one-dimensional tunneling currents within FN theory and the deviation from the linear FN behavior was noted (Figure 4.2). The resulting curve was well fitted to a formula including $1/V^2$ in the exponent as follows:

$$J = AV^2 \exp\left(-\frac{B}{V} - \frac{C}{V^2}\right) \tag{4.4}$$

4.2
Field Emission from CNTs

From the beginning of its discovery, the carbon nanotube has been regarded as an ideal material to make field emitters because of its unusually high aspect ratio as well as the mechanical and chemical stability. The semiclassical FN theory presented in the previous section has been very successful in many systems, not only for the planar geometry but also for small tips with the size of micrometers. As long as the radius of curvature is much higher than the wavelength of the electrons, the one-dimensional picture is a good approximation. However, in the nanostructures where the size of the tip is at most several times of the electron wavelength, such a simple scheme is not validated easily for the following reasons: First, the boundary of the tip is not a well-defined physical quantity in nanosize systems and the potential lines obtained through solving the Poisson equation will not be

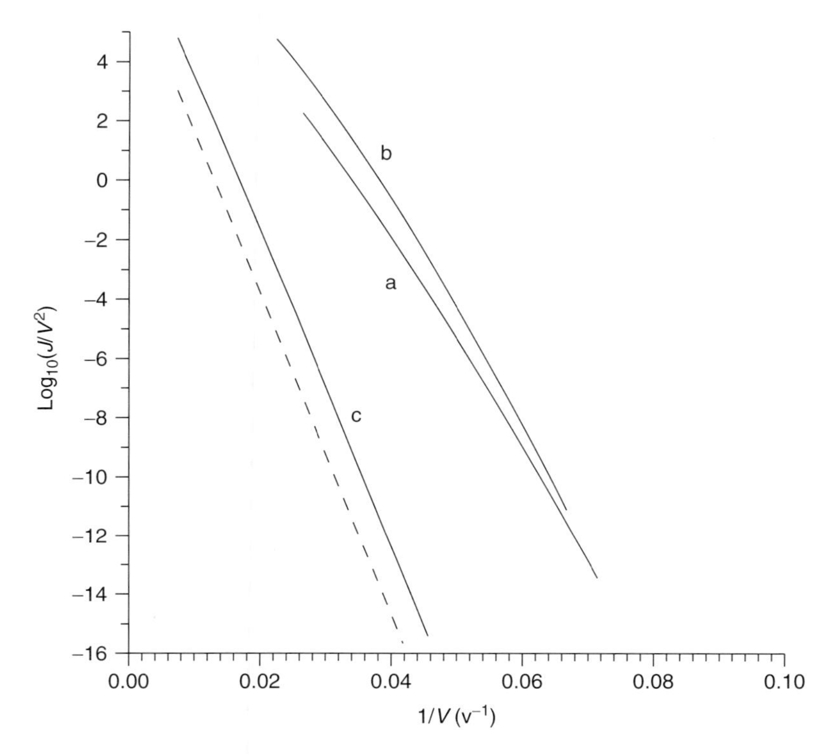

Figure 4.2 Fowler–Nordheim plots for the (a) hyperboloidal, (b) conical, and (c) planar tips. Nonlinearity is noticed for nonplanar geometries. The dashed line is the result from the FN formula. (Adapted from [5].)

valid at the atomic scale. In considering the emission process, the one-dimensional WKB approximation neglects the *xy* variations of the wave function as well as the potential. For example, the additional barriers felt by d-orbitals have been addressed in many experiments and theories [4]. The situation becomes more complicated in the nanotip where the dimension in the *xy*-plane is on the nanometer scale. It is also well known that the localized state induced by the adsorbates at the tip changes the currents dramatically [4]. In the nanotip, such localized states become relatively important as the number of channels for the metallic states is reduced because of the atomistic scale of the cross-sectional area. For carbon nanotubes, the localized states exist at the emission tip because of the topological constraint, and it was suggested that the defects can play an important role during the field emission [6]. The consideration of the localized states is rather clumsy in the semiclassical approach because they are not the current-carrying state.

Therefore, to quantitatively describe the field emission currents from the carbon nanotubes, a quantum mechanical approach considering the realistic three-dimensional atomic structure is required. So far, several methods have been proposed to calculate the emission currents of the nanotubes. Below, we introduce them one by one with representative results.

4.2.1 Computational Methods to Calculate the Emission Currents from Carbon Nanotubes

4.2.1.1 The Integration of Time-Dependent Schrödinger Equation

In this method [7], the tunneling process is directly simulated by monitoring the time evolution of the wave functions. First, the first-principles calculations are performed on the nanotip under the electric field and the self-consistent potential and the wave functions are obtained. To confine the electronic charge within the emission tip during the self-consistency cycle, either a large barrier is applied temporarily outside the tip or, a localized basis set can be employed with the vacuum region is free of any basis. With the computed eigenstates as initial electronic configurations, the temporal wave functions are determined by solving the time-dependent Schrödinger equation. During the time evolution, the change of the electronic density is ignored to a first approximation and, therefore, the exchange and correlation potential is fixed with respect to the time. The transition rate of each state in the model nanotube tip is then evaluated from the amount of charge flowing out of the tip per unit time. The total current is obtained by summing up the product of the transition rate and the occupation number of individual states (Figure 4.3).

One advantage of this approach is that it can consider the tunneling process of both extended and localized states. (However, the extended states could be affected by the finite size effect of the model system.) It is noted that the localized states

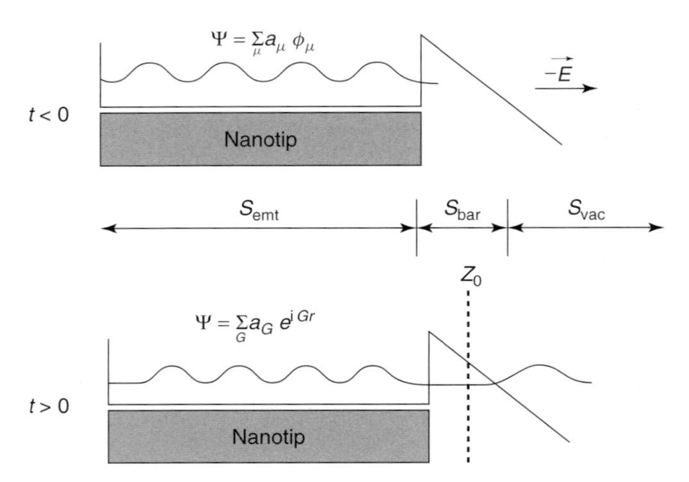

Figure 4.3 Schematic diagram to show the simulation of the tunneling process from nanotubes. Before $t < 0$, the wave functions are forced to be strictly inside the nanotube and the self-consistent potential under the external field is obtained. At $t > 0$, the tunneling process of the electrons is described by integrating the time-dependent Schrodinger equation. (Adapted from [7].)

in the carbon nanotubes are orthogonal to the extended states and, therefore, the usual scattering formalism cannot capture the true tunneling rate for the localized states. It would be ideal if the contributions by the traveling states are evaluated within the Laudauer–Bütticker formalism, while the tunneling of the localized states is described from the resonance width or the lifetime. One such approach was proposed in [8], but its application to nanotubes is yet to be performed.

When applied to the field emission of capped (5,5) and (10,10) carbon nanotubes, it was found that the emission currents from the localized states dominate the total current [7, 9]. The simulated image on the screen was dictated by the spatial symmetry of the localized states and displayed a pattern similar to that from the experiment [10] (Figure 4.4).

This approach has been applied to attack a wide range of problems in carbon nanotubes such as oxygen effects [11], BN-nanotube capping [12], double-wall carbon nanotubes [13], doping effects [14], metal nanowires [15], and field emission from alkali-doped BN nanotube [16]. The method was also applied to study the field emission from graphene nanoribbons [17].

4.2.1.2 **Transfer Matrix Method**

In this method [18], the Schrödinger equation is solved using the transfer matrix method. As shown in Figure 4.5, the whole system is divided into three regions: Region I ($z < -aN$) corresponds to a perfect metal. The intermediate region $-aN < z < 0$ contains N periodic repetitions of the unit cell that is not affected by the external fields. By forcing the incident waves travel through this finite length of the nanotube, the band structure effects can also be taken into the consideration. Region II ($0 < z < D$) contains the part of the nanotube subject to the electric field. The potential energy in region II is calculated by employing a pseudopotential for the ion-core potential in which the electronic density associated with the four

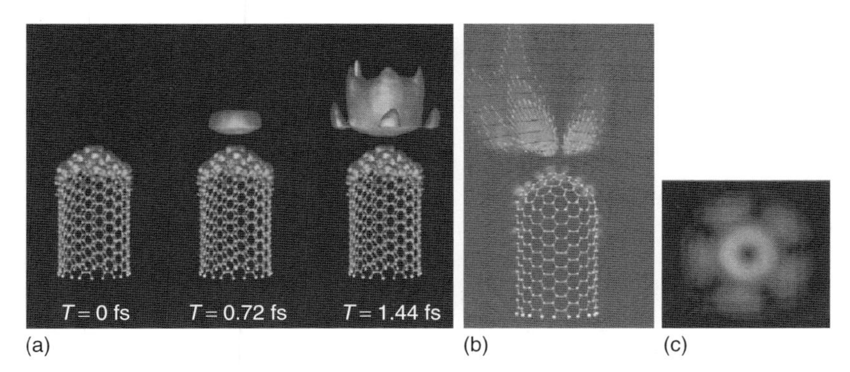

Figure 4.4 (a) Snapshots of the charge distribution of emitted electrons (doubly degenerate localized states) from the capped (10,10) nanotube. The amplitude of the wave function is magnified 30 times in the vacuum region for visual purpose. (b) The current density of one of the localized states at the last instance of (a). The electronic density and the equipotential lines are also displayed. (c) The simulated image on the screen. (Adapted from [9].)

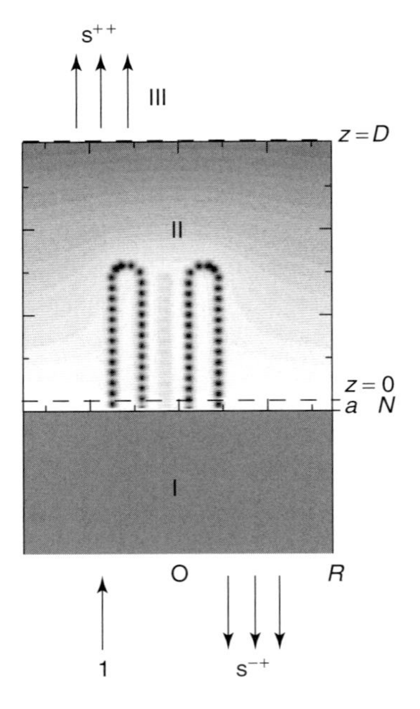

Figure 4.5 Schematic diagram depicting the transfer matrix method to calculate the field emission process. (Adapted from [18].)

valence electrons of each carbon atom is represented by the sum of Gaussian distributions. These electronic densities are displaced from the nuclear positions to reflect the screening of the external field and to produce dipole arrays. The region III ($z > D$) is the field-free vacuum.

By exploiting the cylindrical symmetry of the problem, the traveling states in region I and III can be expressed as follows:

$$\begin{aligned} \psi_{m,j}^{\mathrm{I},\pm} &= A_{m,j} J_m(k_{m,j}\rho) \exp(\mathrm{i}m\phi) \exp\left[\pm \mathrm{i}\sqrt{(2m/\hbar^2)\, E - V_{\mathrm{met}} z}\right] \\ \psi_{m,j}^{\mathrm{III},\pm} &= A_{m,j} J_m(k_{m,j}\rho) \exp(\mathrm{i}m\phi) \exp\left[\pm \mathrm{i}\sqrt{(2m/\hbar^2)\, E z}\right] \end{aligned} \tag{4.5}$$

where $A_{m,j}$ are normalization coefficients, J_m are Bessel functions, $k_{m,j}$ are wave vectors satisfying $J_m(k_{m,j}R) = 0$, E is the electron energy, and V_{met} is the potential energy in the metal in region I. Using Eq. (4.5), the scattering solutions in region I and III are given as follows:

$$\psi_{m,j}^{+} = \begin{cases} \psi_{m,j}^{\mathrm{I},+} + \sum\limits_{m',j'} S_{(m',j'),(m,j)}^{-+} \psi_{m',j'}^{\mathrm{I},-} & (z \leq -aN) \\ \sum\limits_{m',j'} S_{(m',j'),(m,j)}^{++} \psi_{m',j'}^{\mathrm{III},-} & (z \geq D) \end{cases} \tag{4.6}$$

The connection between the coefficients is made through the transfer matrix formalism by dividing the system into thin slabs [19]. It is noted that this procedure is not self-consistent in the sense that the scattering solutions do not affect the potential.

Using this approach, Mayer *et al.* compared the field emission properties of open and closed nanotubes in isolated or bundled configurations and found that the open tubes outperform the closed ones in the emission currents [18]. In addition, they also proposed a formula incorporating the number of tubes and the extraction field and found a deviation from the FN behavior at high electric fields. In [20], the field emission properties of single-wall and multiwall carbon nanotubes were also compared.

A similar approach uses the Lippmann–Schwinger equation [21], which was successfully applied to the field emission from a single-atom electron source [22]. In this case, the reference system is the vacuum–metal junction and the nanotube corresponds to the perturbation. The computational results on the (10,0) nanotube showed an emission distribution that is highly peaked at certain energies, indicating that those currents originate from the localized states at the tip region.

4.2.1.3 **Other Quantum Mechanical Methods**

Although not applied to carbon nanotubes, there have been several attempts to go beyond the conventional FN framework. In [8], Ishida *et al.* proposed an embedding Green function method which can incorporate the field emission current from both extended and localized states. The emission current from the extended states was calculated by the Landauer–Büttiker formula, while that from the localized states was estimated through the lifetime of the resonant states. In [23], a conventional scattering approach based on density functional theory was formulated and applied to study the field emission from the metal represented by the jellium model. Interestingly, the deviation from the FN plot was observed at high fields even for the planar tip geometry. For a realistic system, the method was applied to the field emission from the edges of the graphite ribbon arrays [24]. On the other hand, Ramprasad *et al.* introduced an approach based on the Bardeen transfer Hamiltonian method [25]. In particular, they employed the interpolated local density appoximation (LDA) potential to address the image charge potential. Within this formalism, the left-hand side and the right-hand side (or cathode and anode) are calculated separately. The tunneling rate is then calculated using the wave function tails in the tunneling region and the Fermi golden rule.

4.2.1.4 **Semiclassical Approaches**

In the semiclassical approaches, the electronic structure of the nanotube tip is calculated at the three-dimensional quantum mechanical level but the transmission probability into the vacuum state is evaluated by WKB-style approximations. For example, in [26], the following form was developed for calculating the emission currents:

$$j(F,T) = \frac{1}{C}\sum_{q}\int_{BZ} N\left[E_q(k)\right] D\left[E_q(k), F\right] \mathrm{d}k \tag{4.7}$$

where $N\left[E_q(k)\right]$ is the supply function given as the product of group velocity and the Fermi distribution function. $E_q(k)$ is the dispersion of the nanotube given below (the curvature effect is neglected):

$$E_q(k) = \pm t\sqrt{1 + 4\cos\left(\frac{\sqrt{3}}{2}k_x a\right)\cos\left(\frac{k_y a}{2}\right) + 4\cos^2\left(\frac{k_y a}{2}\right)} \tag{4.8}$$

where t is the hopping parameter, and k and q are quantum numbers running over the Brillouin zone of the nanotube. $D(E,F)$ is the tunneling probability through the one-dimensional potential barrier of the form $U(x, F) = \phi - e\beta Fx - e^2/4x$ and was evaluated by WKB approximation.

This approach has revealed many interesting properties of field emission of the nanotube that depend on the chirality. For example, the chiral effects were examined in [26], and it was found that the emission currents depend more on the energy gap than on the chirality (Figure 4.6). The intrinsic energy spectrum of the emitted electron was investigated and various subpeak features were found with respect to the applied bias and the temperature [27]. The field emission properties of multiwall nanotubes were also discussed in [28]. Recently, considering the energy band structure of the nanotube into account, an analytic expression for the generalized FN formula was also derived for the carbon nanotubes [29].

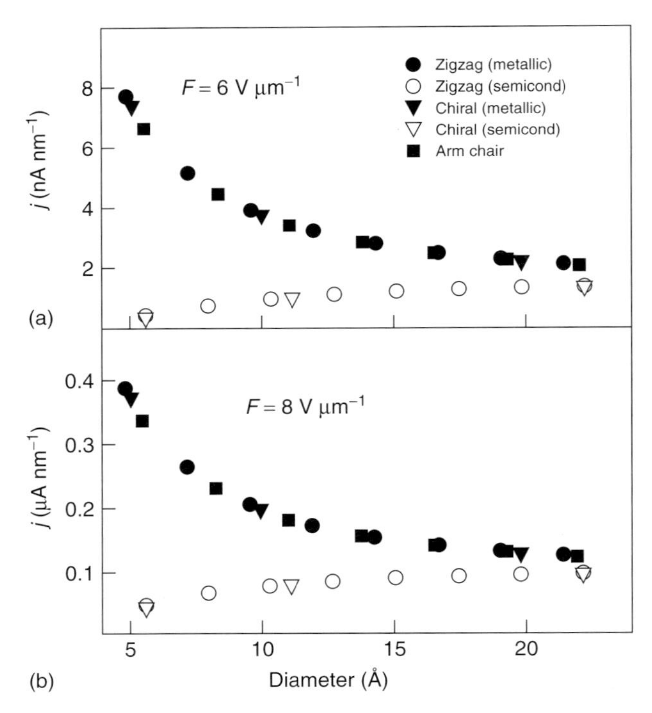

Figure 4.6 Current densities with respect to the diameter of the nanotube. (Adapted from [26].)

In the above method, the field emission process was approximated as one dimensional whereas the full band structure of the nanotube was considered within the tight-binding approximation. In the semiclassical method suggested in [30], the two-dimensional nature of the emitting surface was considered by dividing the supercell into a fine grid and calculating the emission currents along each grid point. This method is capable of evaluating the emission currents with spatial resolution on the plane normal to the emission direction, and therefore the projected image on the screen can be simulated (Figure 4.7).

4.2.2
Current–Voltage Characteristics of Field Emission Currents from Carbon Nanotubes

Several experiments have reported that the field emission currents from carbon nanotubes do not follow the straight FN plot and are rather saturated at high electric fields [31–34]. The nonlinear FN behavior was attributed to effects of space charge or molecules adsorbed at the emission tip. Theoretically, various results indicate that the nonlinear behavior could be intrinsic to the field emission from carbon nanotubes. In [35], the current saturation was attributed to the spatial distribution of electric fields that is specific to the nanotube. A similar result was obtained with more quantum mechanical treatment [36]. In [9], it was noted that the change of the occupation numbers corresponding to the highly emitting localized states slightly deflects the FN plot. On the basis of the explicit consideration of band structures in the nanotube, Liang *et al.* derived an analytical formula for the field emission from the carbon nanotubes (Figure 4.8) [29]. They attributed the non-FN behavior to the Dirac-electron behavior in the small-diameter, low-field region. Although not clarified quantitatively, the field penetration effect at the tip region may play a role in the nonlinear FN behavior, as has been demonstrated in the hybrid quantum mechanical approach on the realistic size of the nanotube [37].

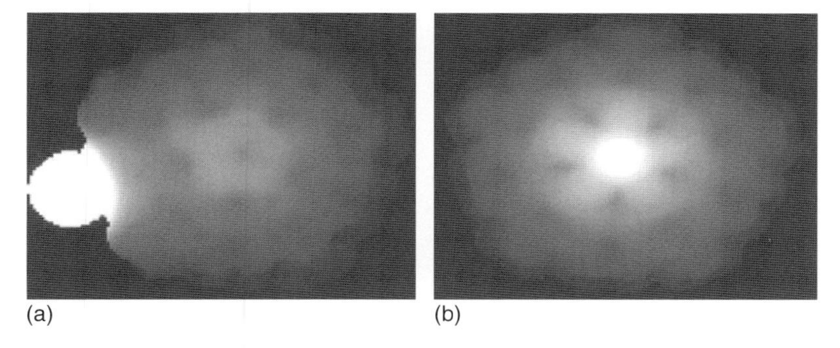

Figure 4.7 The tunneling probability patterns for the Cs adsorbed (a) and trapped (b) at the capped (10,10) nanotube. The external field is 0.3 V Å^{-1}. (Adapted from [30].)

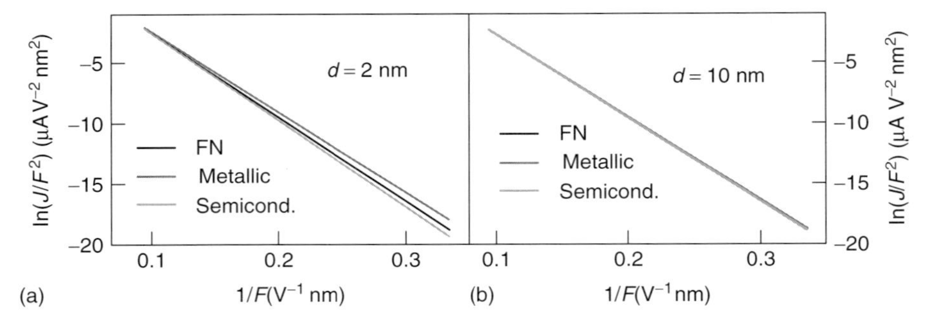

Figure 4.8 The FN plots of the emission currents from single-walled carbon nanotubes with various diameters. For the small-radius carbon nanotube, the deviation from the linear FN behavior is seen. (Adapted from [29].)

4.3
Concluding Remarks

Theoretically, the field emission from carbon nanotubes is a very challenging topic in many respects. Fundamentally, the field emission phenomena are highly nonequilibrium quantum mechanical processes that require the precise description of electronic structures of the nanotubes and the exponentially decaying tail of the tunneling electrons. As examined in this chapter, several approaches have been proposed so far, but none of them seems to provide the complete picture of the field emission of carbon nanotubes. For the refinement of the current methods and further development of new methodologies, feedback from experiments is critical. However, this has been hampered by the difficulties in isolating the emission current from a single carbon nanotube with well-defined geometry such as the length, radius, and chirality. Furthermore, the various possible configurations of pentagons that close the nanotube end will result in different current densities even among nanotubes with the same bulk structure. Therefore, advances in experiments to identify the emission currents from individual nanotubes will greatly assist the development of the theory.

References

1. Folwer, R.H. and Nordheim, L. (1928) Electron emission in intense electric fields. *Proc. R. Soc. London, Ser. A*, **119**, 173–181.
2. Edgcombe, C.J. and Valdrè, U. (2001) Microscopy and computational modeling to elucidate the enhancement factor for field electron emitters. *J. Microsc.*, **203**, 188–194.
3. Gomer, R. (1993) *Field Emission and Field Ionization, American Vacuum Society Classics*, American Institute of Physics, New York.
4. Gadzuk, J.W. and Plummer, E.W. (1973) Field emission energy distribution (FEED). *Rev. Mod. Phys.*, **45**, 487–548.
5. He, J., Cutler, P.H., and Miskovsky, N.M. (1991) Generalization of

Fowler-Nordheim field emission theory for nonplanar metal emitters. *Appl. Phys. Lett.*, **59**, 1644–1646.
6. Han, S. and Ihm, J. (2000) Role of the localized states in field emission of carbon nanotubes. *Phys. Rev. B*, **61**, 9986–9989.
7. Han, S., Lee, M.H., and Ihm, J. (2002) Dynamical simulation of field emission in nanostructures. *Phys. Rev. B*, **65**, 085405.
8. Ishida, H., Wortmann, D., and Ohwaki, T. (2004) First-principles calculations of tunneling conductance. *Phys. Rev. B*, **70**, 085409.
9. Han, S. and Ihm, J. (2002) First-principles study of field emission of carbon nanotubes. *Phys. Rev. B*, **66**, 241402.
10. Saito, Y., Hata, K., and Murata, T. (2000) Field emission patterns originating from pentagons at the tip of a carbon nanotube. *Jpn. J. Appl. Phys.*, **39**, L271–L272.
11. Park, N., Han, S., and Ihm, J. (2001) Effects of oxygen adsorption on carbon nanotube field emitters. *Phys. Rev. B*, **64**, 125401.
12. Park, N., Han, S., and Ihm, J. (2003) Field emission properties of carbon nanotubes coated with boron nitride. *J. Nanosci. Nanotechnol.*, **3**, 179–183.
13. Son, Y.-W., Oh, S., Ihm, J., and Han, S. (2005) Field emission properties of double-wall carbon nanotubes. *Nanotechnology*, **16**, 125–128.
14. Ahn, H.-S., Lee, K.-R., Kim, D.-Y., and Han, S. (2006) Field emission of doped carbon nanotubes. *Appl. Phys. Lett.*, **88**, 093122.
15. Lee, C.-K., Lee, B., Ihm, J., and Han, S. (2007) Field emission of metal nanowires studied by first-principles methods. *Nanotechnology*, **18**, 475706.
16. Yan, B., Park, C., Ihm, J., Zhou, G., Duan, W., and Park, N. (2008) Electron emission originated from free-electron-like states of alkali-doped Boron-nitride nanotubes. *J. Am. Chem. Soc.*, **130**, 17012–17015.
17. Tada, K. and Watanabe, K. (2002) Ab initio study of field emission from graphitic ribbons. *Phys. Rev. Lett.*, **88**, 127601.
18. Mayer, A., Miskovsky, N.M., Cutler, P.H., and Lambin, Ph. (2003) Transfer-matrix simulations of field emission from bundles of open and closed (5,5) carbon nanotubes. *Phys. Rev. B*, **68**, 235401.
19. Mayer, A. and Vigneron, J.-P. (1998) Quantum-mechanical theory of field electron emission under axially symmetric forces. *J. Phys.: Condens. Matter*, **10**, 869–881.
20. Mayer, A., Miskovsky, N.M., and Cutler, P.H. (2002) Theoretical comparison between field emission from single-wall and multi-wall carbon nanotubes. *Phys. Rev. B*, **65**, 155420.
21. Adessi, Ch. and Devel, M. (2000) Theoretical study of field emission by single-wall carbon nanotubes. *Phys. Rev. B*, **62**, 13314.
22. Lang, N.D., Yacoby, A., and Imry, Y. (1989) Theory of a single-atom point source for electrons. *Phys. Rev. Lett.*, **63**, 1499–1502.
23. Gohda, Y., Nakamura, Y., Watanabe, K., and Watanabe, S. (2000) Self-consistent density functional calculation of field emission currents from metals. *Phys. Rev. Lett.*, **85**, 1750–1753.
24. Huang, S.F., Leung, T.C., Li, B., and Chan, C.T. (2005) First-principles study of field-emission properties of nanoscale graphite ribbon arrays. *Phys. Rev. B*, **72**, 035449.
25. Ramprasad, R., Fonseca, L.R.C., and Allmen, P. (2000) Calculation of the field-emission current from a surface using the Bardeen transfer Hamiltonian method. *Phys. Rev. B*, **62**, 5216–5220.
26. Liang, S.-D., Huang, N.Y., Deng, S.Z., and Xu, N.S. (2004) Chiral and quantum size effects of single-wall carbon nanotubes on field emission. *Appl. Phys. Lett.*, **85**, 813–815.
27. Liang, S.-D., Huang, N.Y., Deng, S.Z., and Xu, N.S. (2006) Intrinsic energy spectrum in field emission of carbon nanotubes. *Phys. Rev. B*, **73**, 245301.
28. Liang, S.-D., Deng, S.Z., and Xu, N.S. (2006) Seeking optimal performance of multiwall carbon nanotubes in field emission: tight-binding approach. *Phys. Rev. B*, **74**, 155413.

29. Liang, S.-D. and Chen, L. (2008) Generalized Fowler-Nordheim theory of field emission of carbon nanotubes. *Phys. Rev. Lett.*, **101**, 027602.

30. Khazaei, M., Farajian, A.A., and Kawazoe, Y. (2005) Field emission patterns from first-principles electronic structures: application to pristine and cesium-doped carbon nanotubes. *Phys. Rev. Lett.*, **95**, 177602.

31. Bonard, J.-M., Salvetat, J.-P., Stöckli, T., Forró, L., and Châtelain, A. (1999) Field emission from carbon nanotubes: perspectives for applications and clues to the emission mechanism. *Appl. Phys. A*, **69**, 245–254.

32. Collins, P.G. and Zettl, A. (1996) A simple and robust electron beam source from carbon nanotubes. *Appl. Phys. Lett.*, **69**, 1969–1971.

33. Collins, P.G., and Zettl, A. (1997) Unique characteristics of cold cathode carbon-nanotube-matrix field emitters. *Phys. Rev. B*, **55**, 9391–9399.

34. Xu, X. and Brandes, G.R. (1999) A method for fabricating large-area, patterned, carbon nanotube field emitters. *Appl. Phys. Lett.*, **74**, 2549–2551.

35. Buldum, A. and Lu, J.P. (2003) Electron field emission properties of closed carbon nanotubes. *Phys. Rev. Lett.*, **91**, 236801.

36. Peng, J., Li, Z., He, C., Chen, G., Wang, W., Deng, S., Xu, N., Zheng, X., Chen, G., Edgcombe, C.J., and Forbes, R.G. (2008) The roles of apex dipoles and field penetration in the physics of charged, field emitting, single-walled carbon nanotubes. *J. Appl. Phys.*, **104**, 014310.

37. Zheng, X., Chen, G., Li, Z., Deng, S., and Xu, N. (2004) Quantum-mechanical investigation of field-emission mechanism of a micrometer-long single-walled carbon nanotube. *Phys. Rev. Lett.*, **92**, 106803.

5
Field Emission from Graphitic Nanostructures

Kazuyuki Watanabe and Masaaki Araidai

5.1
Introduction

There are other graphitic nanostructures besides carbon nanotubes that are of great interest for use as field emitters. Such structures consist of graphene, the elementary building block of graphite. Very recently, elaborate techniques have been used to fabricate stable graphitic films that are a few atoms thick [1]. Since then, numerous studies on the electronic [2–4], magnetic, thermal [5], and optical properties [6] of graphene and graphitic ribbons have been carried out, revealing new features unique to two-dimensional carbon networks that could be exploited in potential future electronic applications.

Chen *et al.* performed a field emission (FE) experiment using graphite platelet nanofibers (GPNs) [7] consisting of several thousand graphitic ribbons stacked together like a deck of cards [8]. Very recently, vertically aligned few-layer graphene (FLG), only four to six atomic layers thick and up to several micrometers wide, has been synthesized and used in FE experiments [9]. GPNs and FLGs have suitable mechanical and electronic properties for field emitters: inertness, stiffness, and thin structure of graphene.

A quantitative understanding of the FE mechanism of such graphitic nanostructures requires a first-principles theory because the conventional method, the Fowler–Nordheim (FN) theory, which is based on the Wentzel–Kramers–Brillouin approximation [10], is not suitable for analyzing the highly inhomogeneous electric fields generated by the sharp tips of these graphitic nanostructures.

This aspect has been emphasized and several modern computational methods have been described in Chapter 4 (Part II). In this chapter, we review the time-dependent density functional theory (TD-DFT) [11] studies that have been performed on the FE from graphitic ribbons [12] and graphene [13], focusing on the quantitative aspects of hydrogen-termination effect, electric field direction dependence, and vacancy effects on FE properties. Recent FE studies on FLG [9] and graphite ribbon arrays (GRAs) [14] in other groups are also reviewed. Finally, we briefly describe our numerical results on FE of diamond surfaces [15], illustrating the effects of hydrogen impurities in the subsurface on FE properties.

Carbon Nanotube and Related Field Emitters: Fundamentals and Applications. Edited by Yahachi Saito

ISBN: 978-3-527-32734-8

Detailed description of FE from diamond surfaces are given in Chapter 16 (Part III) in this book.

5.2 Method and Model

We performed conventional DFT and TD-DFT calculations to investigate the electronic states and electron emission properties. We used the norm-conserving pseudopotentials of NCPS97 [16] based on the Troullier–Martins scheme [17] and the generalized gradient approximation by Perdew *et al.* for the exchange-correlation potential [18]. The electronic wave functions were expanded in plane waves.

First, we determine the ground states of atomic and electronic structures of graphitic nanostructures using the conventional DFT in the absence of an electric field. Second, we apply an electric field of $5 \sim 10$ V nm^{-1} and calculate the time evolution of the wave functions using the TD-DFT scheme. Since the Kohn–Sham Hamiltonian is updated at each time step, a part of the dynamical screening effect and some electron correlation effects are automatically taken into account. In this calculation of the time evolution of wave functions, we use the seventh-order Taylor expansion method [19]. The Taylor expansion method guarantees that the normalization condition for the electron number is sufficiently accurate to calculate the emitted current for simulation times $t \leq 150$ a.u. (3.6 fs), although the Suzuki–Trotter split operator method [20] would be more accurate and reliable for long time simulations [21]. Finally, the value of the FE current is evaluated from the gradient of the curve of the number of emitted electrons as a function of time.

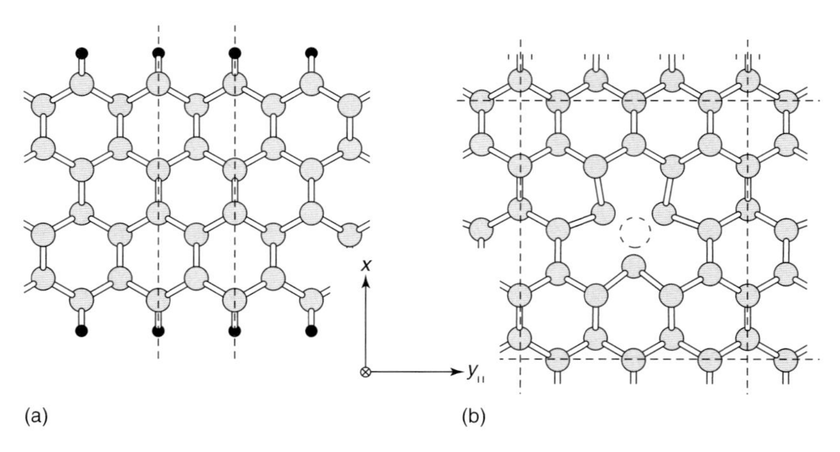

Figure 5.1 Top views of (a) graphitic ribbons with hydrogen (H) termination and (b) graphene sheets with a vacancy defect in the unit cell. Gray circles are carbon atoms and the small black dots in (a) are H atoms. Dashed straight lines denote boundaries of the unit cells and the dashed circle in (b) denotes a vacancy defect. (Reprinted with permission from [13].)

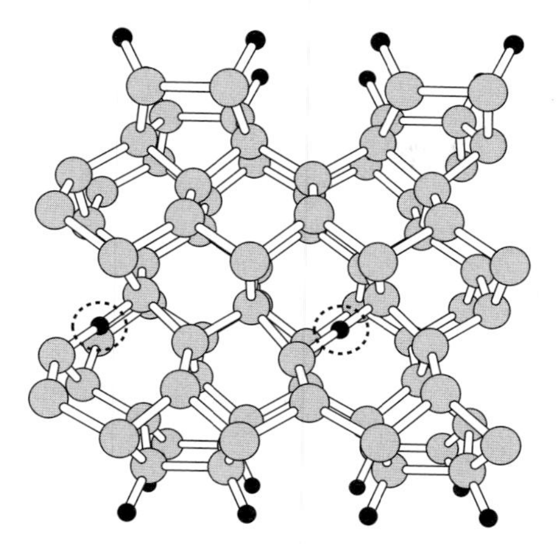

Figure 5.2 Side view of H-terminated diamond C(100) surface with hydrogen impurities of bond center (BC) site. (Reprinted with permission from [15].)

The systems we investigated in this study are zigzag graphitic ribbons with and without H termination (Figure 5.1a), graphene sheets with and without a vacancy defect (Figure 5.1b), and hydrogen (H) terminated (monohydride) diamond C(100) 2×2 surfaces (Figure 5.2). Since the effect of the edge states on the FE is interesting, we selected zigzag ribbons for the present calculations. The x and y axes are defined as shown in Figure 5.1. The atomic structure near the vacancy defect in the graphene sheet was noticeably deformed, as seen in Figure 5.1b. Since the atomic positions of diamond surfaces optimized in our calculations are in good agreement with those obtained using the Vienna *ab initio* simulation package (VASP) [22], our results for the energy band structures and the effective potential in the Kohn–Sham Hamiltonian are reliable. The computational details are given in previous papers by us [12, 13, 15].

5.3 Results

5.3.1 Graphitic Ribbons: H Termination and Field Direction

The band structures as a function of the wave number ($\boldsymbol{k}_y$) are shown for the clean zigzag ribbon (without H termination) in Figure 5.3a and for the H-terminated zigzag ribbon in Figure 5.3b [12]. The characteristic features are the same as those obtained in another study based on the DFT, except for some fine structures caused by the different sizes of the unit cell and/or types of stacking of the graphitic ribbons [23, 24]. The edge state, which has been discussed in detail by Nakada *et al.* [25], is seen above the Fermi level in Figure 5.3a and on the Fermi level in Figure 5.3b

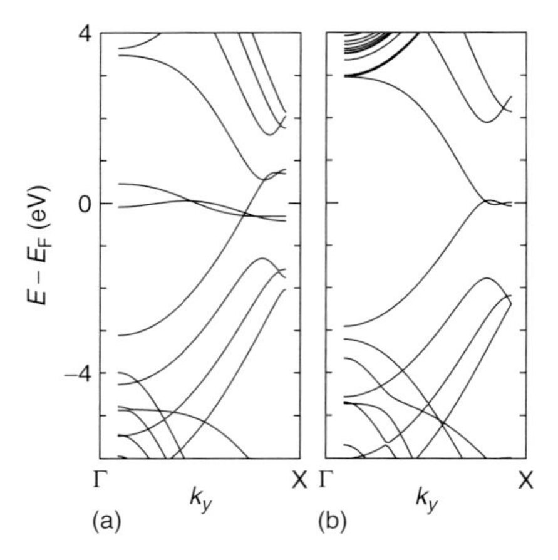

Figure 5.3 Band structures of (a) a clean zigzag ribbon and (b) an H-terminated zigzag ribbon as a function of wave number $\boldsymbol{k}_y$. (Reprinted with permission from [12].)

near X points. We first applied a parallel electric field, $E_{//}$ along the x direction in Figure 5.1 to the zigzag graphitic ribbons with and without H termination [12]. The calculated energy distributions for the emission currents are indicated by broken lines and the electronic states responsible for the peaks are shown in the right panels in Figure 5.4 [13]. For the H-terminated ribbon (Figure 5.4a), there are two peaks originating from σ states in the energy distribution. The electronic distribution of emitted electrons resulting in the higher peak is shown in the upper right panel in Figure 5.4a. The edge state (π state) does not contribute to the FE of the H-terminated ribbon, even though the edge state is near the Fermi level for the H-terminated ribbon. On the other hand, a prominent peak appears in the energy distribution for the clean ribbon (Figure 5.4b). The upper right panel in Figure 5.4 reveals that the electronic states origin of the sharp peak is the dangling bond (DB) state. The reason why electrons are emitted from the σ state (H-terminated) or the DB state (clean) and not from the π states can be elucidated from the directional angle between the local electronic distributions and the applied field, $E_{//}$. Typical electronic orbitals, π, σ, and DB near an edge of the clean ribbon, are schematically shown in Figure 5.5. As seen in Figure 5.5, σ or DB are parallel to the parallel field, $E_{//}$, and tend to interact with the field.

To verify that the electronic states contribute to the FE provided that the electronic orbitals are aligned parallel to the electric field, we investigated whether π orbitals interact with an electric field ($E_{\perp}$, z axis in Figure 5.1) that is perpendicular to the ribbon sheet. The energy distributions of the FE current and the corresponding electronic distributions are depicted by solid lines in Figure 5.4. For the H-terminated ribbon (Figure 5.4a), FE currents are emitted from the π states at the edge, as expected, although no prominent peaks appear, in contrast with the

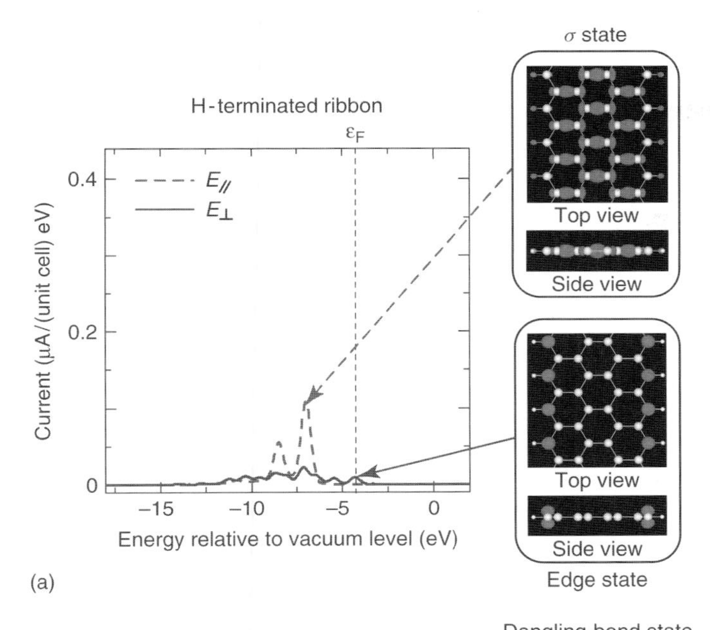

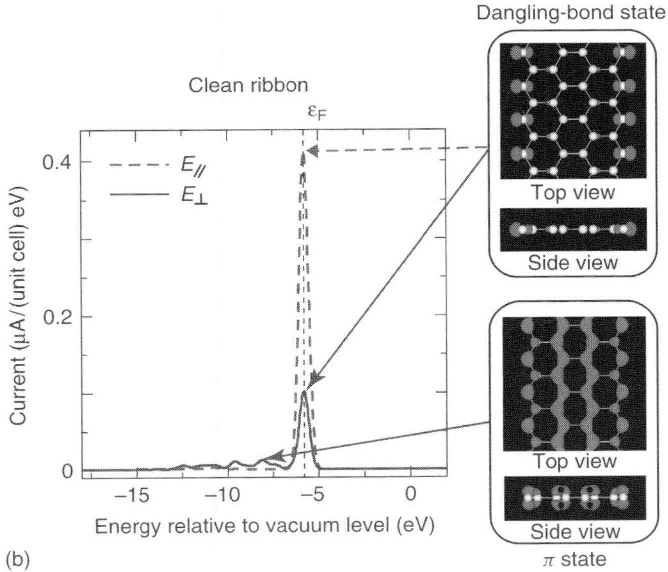

Figure 5.4 Energy distribution of the FE current for (a) a H-terminated ribbon and (b) a clean ribbon. The vacuum level is selected as the energy origin. On the right side of each panel, electronic distributions causing the peaks are shown by gray clouds with carbon atoms (white spheres) and H atoms (white dots). The broken and solid lines denote the results for a parallel ($E_{//}$) and perpendicular ($E_{\perp}$) electric field, respectively. (Reprinted with permission from [13].)

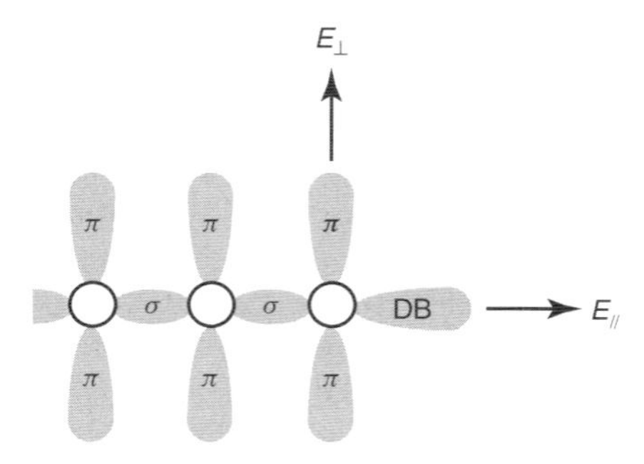

Figure 5.5 Schematic diagram of electron orbitals (gray clouds) of a clean graphitic ribbon around the right edge (side view). $E_{//}$ and $E_\perp$ denote applied electric fields parallel and perpendicular to the ribbon, respectively. DB represents a dangling bond. (Reprinted with permission from [13].)

energy distribution curve for the parallel field, $E_{//}$ (broken line in Figure 5.4a). For the clean ribbon (Figure 5.4b), however, a sharp peak originating from the DB state is observed, similar to the case for the parallel field, $E_{//}$. Negligibly small contributions of the π states to the total FE current are found in the energy distributions. The direction of the electronic distribution of the DB state in the upper right panel in Figure 5.4b is *not* parallel to the perpendicular field, $E_\perp$. It follows from this result that the DB state contributes to the FE even when the direction of the electronic distribution is *not* parallel to the direction of the electric field.

It is important to clarify why π states cannot be the main source of the FE even when π orbitals are aligned parallel to the applied electric field, $E_\perp$ (see Figure 5.5). The detailed analysis of the local electric field around the ribbon edge indicates that the electric field is remarkably enhanced at the edge where the DB orbitals exist. In contrast, the electric field is clearly reduced inside the ribbon compared with that at the edge. This reduction in the electric field is due to the screening effect. A similar feature is also observed in the H-terminated ribbons. Consequently, the π orbitals cannot be the main contributor to FE even when the π orbitals are aligned parallel to the applied field. Detailed discussions about this aspect are given in a previous paper by us [13]. Here, we summarize the FE properties of zigzag graphitic ribbons. The FE currents from clean and H-terminated graphitic ribbons in a parallel electric field of $10\,\text{V nm}^{-1}$ are 0.37 and 0.14 μA per unit cell, respectively. These FE currents are respectively reduced to 0.10 and 0.06 μA per unit cell in a perpendicular field of $10\,\text{V nm}^{-1}$. The DB orbital is the main source of the FE from the clean ribbon in both electric field directions. The DB state disappears on H termination and thus the σ orbitals and the π orbitals become emitting sources in the parallel and perpendicular electric fields, although the contribution of the π orbitals is negligible.

5.3.2
Graphene Arrays: Interlayer Interaction

Let us briefly review two recent studies on FE from graphene arrays by other groups. One is a first-principles theoretical study on GRAs [14] and the other is an experimental study on FLGs [9]. The theoretical study, which is based on a self-consistent method by a DFT similar to that used in a previous study by us [26], demonstrated that the FE properties of pristine and H-terminated zigzag (armchair) ribbons vary remarkably depending on the interribbon separation D. In particular, the emission current from pristine zigzag ribbons changes by 5 orders of magnitude when D increases from 5 to 25 Å. This was interpreted in terms of a reduction in the work function caused by an increase in the edge dipole and by enhancement of the electric field at the edges. The study also emphasizes that the FN analysis fails to evaluate the work function of H-terminated zigzag ribbons. This is because the emitting levels are shifted below the Fermi energy on H termination (see Figure 5.3b), while the FN theory assumes that the emitting levels are near the Fermi energy. The reader is referred to [14] for more details. The conclusion is consistent with that of our previous study [12].

The recent experimental study reported the FE characteristics of vertically aligned FLG (four to six atomic layers) on silicon and titanium substrates [9]. FLG was found to be a good field emitter, characterized by turn-on fields as low as $1\,\mathrm{V\,\mu m^{-1}}$ and field amplification factors of up to 7500, and hydrogen was identified as an efficient etchant for improving FE. A simple model calculation in the study, however, indicated that, as a result of screening of the electric field, only a small fraction of FLG contributes to FE, namely the fraction that is at least 10% higher than the average height and that is positioned several micrometers from other graphenes [9].

5.3.3
Graphene Sheet: Defects

Having clarified the important role of the DB state in the FE from graphitic ribbons, we further explored the FE properties of a graphene sheet with an atomic vacancy, because DB states are generated in the vacancy [13]. The graphene sheets investigated are shown in Figure 5.1b. The electric field was applied perpendicular ($E_{\perp}$) to the sheet.

We investigated the FE current-density images of the graphene sheets to clarify the role of the vacancy on the FE characteristics. Figure 5.6 shows the FE images for the graphene sheet with vacancy defects. The plane on which the FE current distribution is plotted is 4.9 Å above the graphene sheet. The FE current density around the vacancy is about 22 times as large as that of the other regions. The total FE currents obtained from graphene sheets without and with vacancy defects are 0.054 and 0.106 μA per unit cell, respectively. The FE images thus obtained reflect a dramatic change in the electronic distribution around the vacancy defect due to the removal of a carbon atom. The FE images reveal that vacancy defects significantly

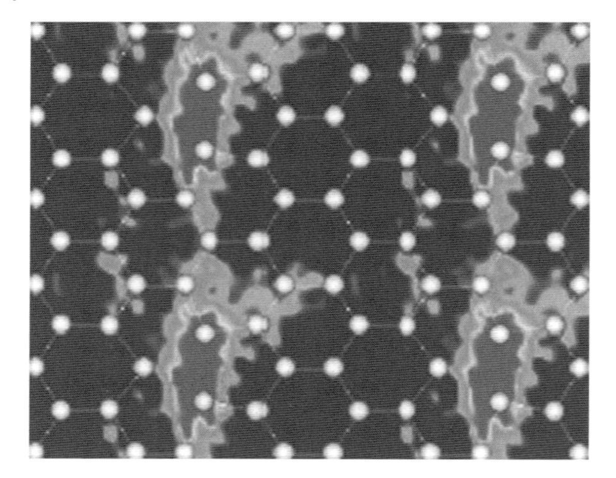

Figure 5.6 FE image of the graphene with vacancy defects. The distance between the FE image plane and the graphene is 4.9 Å. The FE current density around the vacancy is about 22 times larger than that from the other regions. (Reprinted with permission from [13].)

enhance the FE current density of the graphene sheet due to the appearance of the DB states.

5.3.4
Diamond Surfaces: Impurities

Having described the effects of H termination, field direction, interlayer interaction, and vacancy defects on the FE properties of graphitic nanostructures in the previous sections, we finally present the results for the FE emission of diamond with H impurities to understand how the impurity-induced surface electronic states modify the FE properties [15]. First, we performed self-consistent total-energy calculations and obtained a configuration of hydrogen complex and the hydrogen-impurity-induced electronic states in the diamond subsurface. The configuration of hydrogen impurities located in the subsurface is shown in Figure 5.2. This is a side view of H-terminated diamond C(100) 2×2 surfaces (super slab model). Figure 5.2 shows a hydrogen complex at the bond center (BC) site. The BC site has been found to be the most stable one in the subsurface from the *ab initio* calculation [27].

The electronic band structure of the diamond surfaces with hydrogen impurities in the subsurface is shown in Figure 5.7 [15]. The effect of hydrogen impurities appears in the energy band gap. The hydrogen at the BC site generates weakly extended states in the energy band gap, which is partially occupied by electrons (as seen in Figure 5.7). It has been experimentally reported that most hydrogen atoms in diamond are localized at a depth of 1 nm from the surface [28].

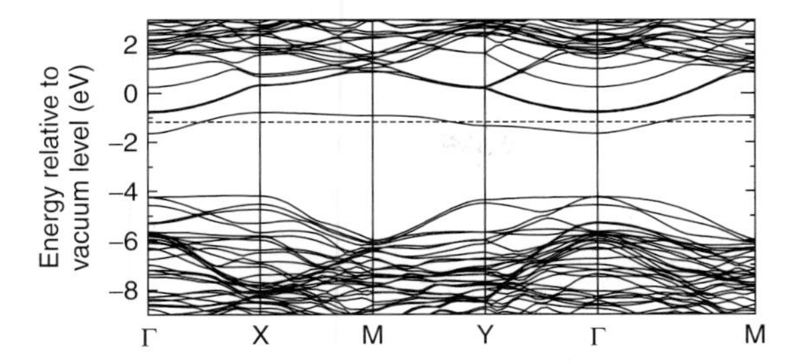

Figure 5.7 Energy band structures of diamond surfaces with BC impurities in the subsurfaces. The hydrogen-impurity-induced states are generated in the energy band gap. The vacuum level is chosen for the origin of the energy. The horizontal broken line denotes the Fermi level. The Fermi level is −1.18 eV. (Reprinted with permission from [15].)

First, we investigate the FE of an ideal H-terminated diamond surface without hydrogen impurities in the subsurface. The minimum ionization energy (IE) of the valence band maximum is 4.07 eV and the total FE current is 423 nA nm^{-2}. The energy distribution of the FE current density is given by the broken line in Figure 5.8 [15]. The peak with an IE of 4.07 eV is found to originate from the bulk or inner state. Thus, even inner states can contribute to the FE when the IE is sufficiently small for electrons to tunnel into the vacuum. Moreover, detailed analysis of the electronic states reveals that the major contributors to the current peaks are the states at the Γ point. Next, we evaluate the FE current from an H-terminated diamond surface with a single hydrogen interstitial at a BC site in the subsurface. The energy distribution of the FE current density (solid line) and the corresponding electronic states are given in Figure 5.8. The highest peak formed by electrons emitted from the electronic state of the BC site (right panel in Figure 5.8) is about 4 times as intense as that formed by electrons emitted from the clean subsurface without any hydrogen impurities. The minimum IE is 1.34 eV and the FE current is 1.99 μA nm^{-2}. The total FE current of this surface is about 5 times that of the pristine diamond surface. The reason for this remarkable increase in the emission current is that the IE of the emitting level of the surface (1.34 eV) is much smaller than that of the pristine surface (4.07 eV).

To summarize, doping hydrogen impurities in the subsurface increases the FE current from 423 nA nm^{-2} for an impurity-free subsurface to 1.99 μA nm^{-2} when there are BC impurities in the subsurface at an electric field of 5 V nm^{-1}. The enhancement in the emission current caused by the hydrogen impurities is revealed by the appearance of impurity-induced states in the energy band gap, leading to a significant reduction in the IEs. Consequently, first-principles calculations reveal the microscopic reason why high emission currents at low electric fields have been experimentally observed from diamond surfaces with many hydrogen atoms in the subsurface [29, 30].

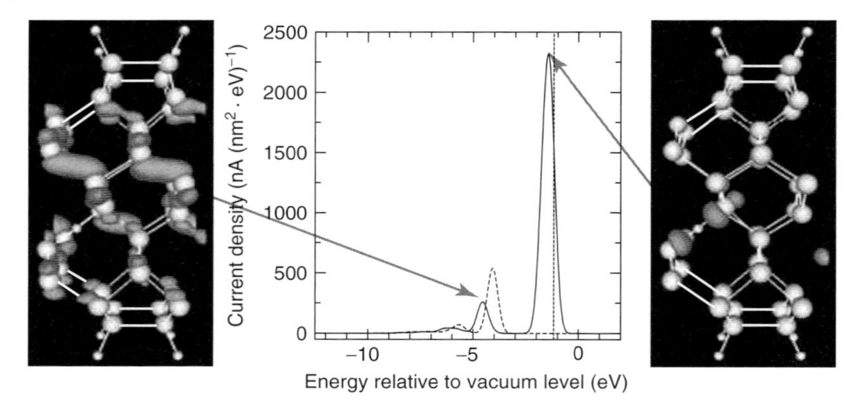

Figure 5.8 Energy distributions of FE current densities of diamond surfaces without impurities (broken line) and with BC impurities (solid line) in the subsurface. The vertical broken line denotes the position of the Fermi level. On the both sides of the panel, the electron distributions causing the peaks are indicated by gray clouds with carbon atoms (white spheres) and hydrogen atoms (white dots) in the diamond slabs. (Reprinted with permission from [15].)

5.4 Conclusion

We have presented the first-principles calculation (TD-DFT) results of FE of various graphitic nanostructures, focusing on the effects of local electronic states, defects (such as vacancies and impurities), and the direction of the electric field on the FE characteristics. DB states are the main sources of FE because their electronic orbitals tend to be concentrated at the edges and are aligned with the electric field in vacuum. The substantially enhanced emission current from the atomic vacancy sites in graphene confirms the essential contribution of the DB states to the FE of graphitic nanostructures. The π states, however, contribute little to the FE even when the electronic orbitals are aligned parallel to the electric field because of the screening effect. Hydrogen impurities in the subsurface also significantly enhance the FE current from diamond surfaces, because the impurity levels generated in the energy band gap have low IEs. The TD-DFT studies have demonstrated that local electronic states, in addition to work functions (or IEs), are crucial for understanding the FE mechanism of graphitic nanostructures. This contrasts markedly with the FE properties of usual micron-sized metallic tips that are interpreted by the FN theory.

Acknowledgments

The authors thank S. Watanabe for fruitful discussions on the method of first-principles calculation of FE. K.W. acknowledges partial financial support from MEXT through a Grant-in-Aid (No. **19540411**) and from the Holistic Computational

Science Research Center (HOLCS) of Tokyo University of Science. Some of the numerical calculations were performed on the Hitachi SR11000 at the Institute for Solid State Physics, the University of Tokyo.

References

1. Novoselov, K.S., Geim, A.K., Morozov, S.V., Jiang, D., Zhang, Y., Dubonos, S.V., Grigorieva, I.V., and Firsov, A.A. (2004) Electric field effect in atomically thin carbon films. *Science*, **306**, 666–669.
2. Novoselov, K.S., Geim, A.K., Morozov, S.V., Jiang, D., Katsnelson, M.I., Grigorieva, I.V., Dubonos, S.V., and Firsov, A.A. (2005) Two-dimensional gas of massless Dirac fermion in graphene. *Nature*, **438**, 197–200.
3. Son, Y.-W., Cohen, M.L., and Louie, S.G. (2006) Half-metallic graphene nanoribbons. *Nature*, **444**, 347–349.
4. Han, M.Y., Özyilmaz, B., Zhang, Y., and Kim, P. (2006) Energy band-gap engineering of graphene nanoribbons. *Phys. Rev. Lett.*, **98**, 206805.
5. Yamamoto, T., Watanabe, K., and Mii, K. (2004) Empirical-potential study of phonon transport in graphitic ribbons. *Phys. Rev. B*, **70**, 245402.
6. Yamamoto, T., Noguchi, T., and Watanabe, K. (2006) Edge-state signature in optical absorption of nanographenes: tight-binding method and time-dependent density functional theory calculations. *Phys. Rev. B*, **74**, 121409(R).
7. Chen, X., Ruoff, R., Edwards, E., and Feinerman, A. (2003) in *Field Emission from Graphite Platelet Nanofibers (GPNs), Technical Digest of the 16th International Vacuum Microelectronics Conference, Osaka* (eds M. Takai, Y. Gotoh, and J. Ishikawa), The 158th Committee on Vacuum Nanoelectronics, Japan Society for the Promotion of Science, p. 215.
8. Rodriguez, N.M., Chambers, A., and Baker, R.T.K. (1995) Catalytic engineering of carbon nanostructures. *Langmuir*, **11**, 3862–3866.
9. Malesevic, A., Kemps, R., Vanhulsel, A., Chowdhury, M.P., Volodin, A., and Haesendonck, C.V. (2008) Field emission from vertically alligned few-layer graphene. *J. Appl. Phys.*, **104**, 084301.
10. Gomer, R. (1993) *Field Emission and Field Ionization*, American Institute of Physics, New York.
11. Runge, E. and Gross, E.K.U. (1984) Density-functional theory for time-dependent systems. *Phys. Rev. Lett.*, **52**, 997–1000.
12. Tada, K. and Watanabe, K. (2002) Ab initio study of field emission from graphitic ribbons. *Phys. Rev. Lett.*, **88**, 127601.
13. Araidai, M., Nakamura, Y., and Watanabe, K. (2004) Field emission mechanisms of graphtitic nanostructures. *Phys. Rev. B*, **70**, 245410.
14. Huang, S.F., Leung, T.C., Li, B., and Chan, C.T. (2005) First-principles study of field-emission properties of nanoscale graphite ribbon arrays. *Phys. Rev. B*, **72**, 035449.
15. Araidai, M. and Watanabe, K. (2004) Ab initio study of field emission from hydrogen defects in diamond subsurfaces. *Appl. Surf. Sci.*, **237**, 483–488.
16. Kobayahsi, K. (1999) Norm-conserving pseudopotential database (NCPS97). *Comput. Mater. Sci.*, **14**, 72–76.
17. Troullier, N. and Martins, J.L. (1991) Efficient pseudopotentials for plane-wave calculations. *Phys. Rev. B*, **43**, 1993–2006.
18. Perdew, J.P., Chevary, J.A., Vosko, S.H., Jackson, K.A., Pederson, M.R., Singh, D.J., and Fiolhais, C. (1992) Atoms, molecules, solids, and surfaces: applications of the generalized gradient approximation for exchange and correlation. *Phys. Rev. B*, **46**, 6671–6687.
19. Yabana, K. and Bertsch, G.F. (1996) Time-dependent local density approximation in real time. *Phys. Rev. B*, **54**, 4484–4487.

20. Suzuki, M. (1992) General nonsymmetric higher-order decomposition of exponential operators and symplectic integrators. *J. Phys. Soc. Jpn.*, **61**, 3015–3019.
21. Sugino, O. and Miyamoto, Y. (1999) Density-functional approach to electron dynamics: stable simulation under a self-consistent field. *Phys. Rev. B*, **59**, 2579–2586.
22. Furthmüller, J., Hafner, J., and Kresse, G. (1996) Dimer reconstruction and electronic surface states on clean and hydrogenated diamond (100) surfaces. *Phys. Rev. B*, **53**, 7334–7351.
23. Miyamoto, Y., Nakada, K., and Fujita, M. (1999) First-principles study of edge states of H-terminated graphitic ribbons. *Phys. Rev. B*, **59**, 9858–9861.
24. Kawai, T., Miyamoto, Y., Sugino, O., and Koga, Y. (2000) Graphitic ribbons without hydrogen-termination: electronic structures and stabilities. *Phys. Rev. B*, **62**, R16349–R16352.
25. Nakada, K., Fujita, M., Dresselhaus, G., and Dresselhaus, M.S. (1996) Edge state in graphene ribbons: nanometer size effect and edge shape dependence. *Phys. Rev. B*, **54**, 17954–17961.
26. Gohda, Y., Nakamura, Y., Watanabe, K., and Watanabe, S. (2000) Self-consistent density functional calculation of field emission currents from metals. *Phys. Rev. Lett.*, **85**, 1750–1753.
27. Kanai, C., Shichibu, Y., Watanabe, K., and Takakuwa, Y. (2002) Ab initio study on surface segregation of hydrogen from diamond C(100) surfaces. *Phys. Rev. B*, **65**, 153312.
28. Kimura, K., Nakajima, K., Yamanaka, S., Hasegawa, M., and Okushi, H. (2001) Hydrogen depth-profiling in chemical-vapor-deposited diamond films by high-resolution elastic recoil detection. *Appl. Phys. Lett.*, **78**, 1679–1681.
29. Wang, C., Garcia, A., Ingram, D.C., Lake, M., and Kordesch, M.E. (1991) Cold field emission from CVD diamond films observed in emission electron microscopy. *Electron. Lett.*, **27**, 1459–1461.
30. Xu, N.S., Latham, R.V., and Tzeng, Y. (1993) Field-dependence of the area-density of 'cold' electron emission sites on broad-area CVD diamond films. *Electron. Lett.*, **29**, 1596–1597.

6
The Optical Performance of Carbon Nanotube Field Emitters

Niels de Jonge

6.1
Introduction

Carbon nanotubes can be used as high-brightness quality electron point sources for electron microscopes [1, 2]. The carbon nanotube electron source could possibly replace the Schottky emitter for high-resolution imaging and analysis in scanning electron microscopes (SEMs) and transmission electron microscopes (TEMs). It is expected that carbon nanotubes will have several advantages over field emission sources made from sharp metal tips (typically tungsten or molybdenum), which have been used for about four decades in electron microscopy [3]. A carbon nanotube with a closed cap and an emitting surface cleaned of adsorbed molecules exhibits a highly stable emission [4] on account of its extremely rigid structure and high melting point [5]. In this chapter, an overview is given of the theory and experimental results describing the optical properties of the carbon nanotube electron sources. The measurements were performed on emitters made of individual carbon nanotubes with closed caps and carefully cleaned surfaces. This sample preparation is necessary to be able to investigate the emission mechanism of a single carbon nanotube, ensuring that the measured emission properties are not from an ensemble of carbon nanotubes and not dominated by impurities on the emission surface. First, the electron emission process will be evaluated in terms of the emitted current density and energy spread of the electron beam. Secondly, the measurements and the theory of the brightness will be presented. The figure of merit of the electron source is the reduced brightness versus the energy spread, which defines the optical performance of the source. Several simplified and practical equations for the optical performance of the carbon nanotube electron source will be provided, which are assumed to have a more general validity for nanometer-sized electron sources.

Carbon Nanotube and Related Field Emitters: Fundamentals and Applications. Edited by Yahachi Saito

ISBN: 978-3-527-32734-8

6.2
Making an Electron Source from an Individual Carbon Nanotube

A series of measurements were performed on several multiwall carbon nanotubes obtained from two different growth techniques, namely, arc discharge [6] containing carbon nanotubes with a wide range of radii and chemical vapor deposition (CVD) [7] with radii mainly between 0.5 and 3 nm. For a precise characterization, individual carbon nanotubes were mounted on tungsten support tips using a nanomanipulator system in an SEM, thereby avoiding the (undesired) mounting of bundles of carbon nanotubes or multiple carbon nanotubes. The method is described in detail elsewhere [8, 9]. A brief summary is shown in Figure 6.1. A sample with carbon nanotubes protruding from a sharp edge placed in the SEM was searched for a long, straight, thin, and freestanding nanotube. This nanotube

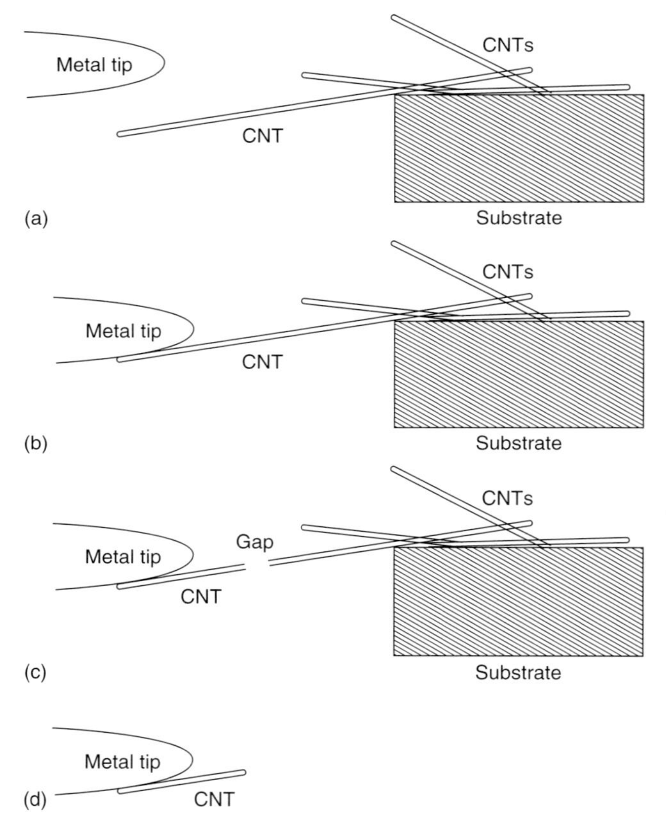

Figure 6.1 Mounting procedure of a carbon nanotube (carbon nanotube) electron source as performed inside a scanning electron microscope. (a) A carbon nanotube protruding from a thin substrate containing many nanotubes is selected and (b) attached to a tungsten support tip. (c) The carbon nanotube is broken by joule heating. (d) The open tube end is finally closed. (Source: Taken from [9].)

was brought into contact with a tungsten tip to which it adhered. Then, the single nanotube was broken off the nanotube sample by applying a voltage difference over the nanotube, leading to joule heating on account of a current of more than 20 μA (corresponding to a current density higher than $6 \times 10^{10} A\, m^{-2}$ for a tube with a radius of 10 nm!). The length and the diameter can be selected with this method to a typical precision of 200 and 5 nm, respectively. Moreover, the contact length and the angle between the nanotube and the support tip can be set.

For optimal use as electron source, carbon nanotubes with open caps are not desired, as the manifold of dangling bonds from an open cap may lead to current fluctuations [10] and may even lead to a quick destruction of the nanotube under the presence of the strong electric field needed for electron emission. However, one would expect that the breaking action in the described mounting procedure would result in an open cap. Considerable research effort was invested in finding a procedure to close the cap, which was predicted to be possible [10, 11]. It appeared indeed possible to close the cap in the mounting process for carbon nanotubes with small radii, as observed directly by *in situ* TEM measurements on one carbon nanotube and on several samples using the indirect method of field emission microscopy [9]. The final step in the preparation the carbon nanotube electron source was its cleaning by heating in an ultrahigh vacuum system (10^{-10} Torr). Each nanotube was first heated to the carbonization temperature [5] of 1000 K for 10 min to remove volatile species from the tube. It would have been advantageous to heat at temperatures of 1500 K at which recrystallization of the graphene planes takes place, but we found that the carbon nanotube was usually removed from the tungsten support tip at these temperatures. The nanotube was then operated at a temperature of typically 800 K during the emission experiments to continuously keep the emitter clean. A filament on which the tungsten wire containing the carbon nanotube was mounted applied the heating. Additional heating was provided by joule heating at higher emission currents [12]. Each carbon nanotube was verified as having a closed cap and being cleaned by recording its field emission pattern [9, 11].

6.3 The Emission Process

6.3.1 The Fowler–Nordheim Model

The electron emission from a cleaned carbon nanotube with a closed cap is caused by field emission with a work function of 5.1 ± 0.1 eV [13, 14]. The current density J follows the Fowler–Nordheim (FN) theory [15], which can be expressed as [13, 16]

$$J = \frac{c_1 F^2}{b_1^2 \phi} \exp\left\{a_2 c_2 c_3 \frac{1}{\sqrt{\phi}}\right\} \exp\left\{-a_1 c_2 \frac{\phi^{3/2}}{F}\right\} \tag{6.1}$$

with work function ϕ and electric field F. The constants were determined elsewhere [14], and are defined as

$$a_1 = 0.958$$
$$a_2 = 1.05$$
$$b_1 = 1.05$$
$$c_1 = e^3/8\pi h$$
$$c_2 = 8\pi\sqrt{(2m)}/3he$$
$$c_3 = e^3/4\pi\varepsilon_0$$

with the electron charge e, Planck's constant h, electron mass m, and the permittivity of free space ε_0 with the values

$$e = 1.6022 \times 10^{-19}\ \text{C}$$
$$h = 6.6262 \times 10^{-34}\ \text{Js}$$
$$m = 9.1098 \times 10^{-31}\ \text{kg}$$
$$\varepsilon_0 = 8.8542 \times 10^{-12}\ \text{Fm}^{-1}$$

The total current is given by

$$I = 2\pi R^2 J \tag{6.2}$$

where a hemispherical emitting surface with radius of curvature R was assumed. We neglect the possibility that the emitting surface is smaller, for example, being a flat end with the area πR^2, or even smaller for the case in which the emission comes from the edges only, or nonhomogeneous emission from localized states. The local field at the apex of a sharp electrically conducting tip equals the product of the extraction voltage U and the field enhancement factor β:

$$F = \beta U \tag{6.3}$$

The value of β can be computed numerically knowing the geometry of the emitter [14]. A graph of log I/U^2 versus $1/U$, called the FN plot, is thus a linear curve. Note that joule heating may occur at larger currents, leading to a deviation from linearity [12]. The FN model can be modified to account for the highly curved emitting surface, resulting in a better fit of the experimental data with the theoretical FN curve [17]. However, the original FN model is sufficient within about 10% accuracy.

6.3.2 Measurement of the Fowler–Nordheim Plot

The current–voltage characteristics were recorded for eight carbon nanotubes. The emission pattern of carbon nanotube 1 (CVD grown) is shown in Figure 6.2a and is typical for a carbon nanotube with a closed cap [9]. The measured I–U characteristic of carbon nanotube 1 is shown in Figure 6.1b. All FN plots were linear, indicating that field emission was occurring for all carbon nanotubes. Values of β were

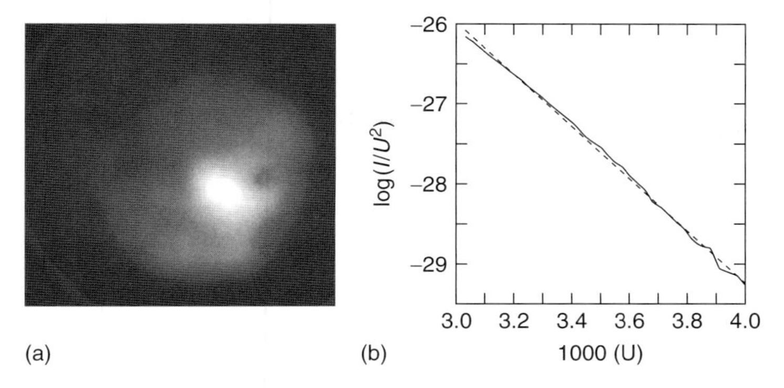

Figure 6.2 Emission measurements on electron source carbon nanotube 1 for $T = 800\,\mathrm{K}$ and a vacuum of 2×10^{-10} Torr. (a) Field emission pattern recorded with a microchannel plate and a phosphor screen. (b) FN plot and a linear fit with a slope of −3244 (dashed).

obtained from the slopes of the FN plots (*slope*) and values of R were derived from the point at which $1/U$ was zero (*zero*), using

$$\beta = \frac{-a_1 c_2 \phi^{3/2}}{slope} \tag{6.4}$$

$$R = \frac{b_1}{\beta}\sqrt{\frac{\phi \exp\left(zero - a_2 c_2 c_3/\phi\right)}{2\pi c_1}} \tag{6.5}$$

The results are shown in Table 6.1. TEM images of the carbon nanotube batch sample and of several mounted carbon nanotubes revealed that most carbon nanotubes had a few walls and a radius between 1 and 3 nm, consistent with the values obtained from the FN plots. A TEM image of carbon nanotube 1 taken after the emission measurements confirmed a radius of 1.1 nm. This was the only sample of this series that was successfully imaged in the TEM after the emission measurements. The values of the field enhancement factor are as expected, on the basis of numerical calculations performed with Munro's software (EMECH package) [14].

6.3.3
The Energy Spread

The current density as a function of the energy E is approximately proportional to [16]

$$J(E) \propto \frac{\exp\left(E/d\right)}{1 + \exp\left(E/k_b T\right)} \tag{6.6}$$

Table 6.1 Emission data of 10 carbon nanotubes.

Carbon nanotube	Type	β (× 10^7 m^{-1})	R (nm)	ϕ (eV)	R_v (nm)	I_r' (nA (sr V)$^{-1}$)
1	Arc	2.3	0.98	5.1	–	3.4
2	CVD	1.7	0.59	5.1	–	2.1
3	CVD	0.81	4.6	5.1	–	2.1
4	CVD	2.0	0.81	5.1	–	1.9
5	CVD	1.2	1.5	5.4	–	1.4
6	CVD	0.74	3.0	–	1.8	5.3
7	CVD	1.6	1.7	–	1.6	2.5
8	CVD	2.1	1.7	–	1.7	4.0
10	Arc	–	–	–	3.0	3.7

Measurements of $I-U$ curves revealed values for β and R (derived from the Fowler–Nordheim plots). The angular current density I_r' was measured at 100 nA with a Faraday cup, and the values of the virtual source size r_v were measured with a point projection microscope. The carbon nanotubes were either obtained by arc discharge or by CVD, as indicated.

Here, k_b is the Boltzmann constant, T is the temperature, and d is the tunneling parameter, given by

$$d = \frac{c_4 F}{b_1 \sqrt{\phi}} \tag{6.7}$$

The constants are

$$c_4 = eh/4\sqrt{(2m)}$$
$$k_b = 1.3806 \times 10^{-23}\ \mathrm{JK}^{-1}$$

The energy spread of the emitted electron beam is often expressed by the full width at half maximum of the energy spectrum ΔE. The width of the low-energy side of the energy spectrum is determined by d, whereas T sets the width of the high-energy side. ΔE can be approximated in the parameter range $100 < T < 1000$ K and $0.2 < d < 0.5$ eV as

$$\Delta E \cong 3.1 \times 10^{-4} T + 0.72d \tag{6.8}$$

The validity of the model is demonstrated in Figure 6.3.

6.3.4 Measurement of Energy Spectra

Energy spectra were recorded with a hemispherical energy analyzer for five carbon nanotubes at different currents (10–500 nA) and temperatures (500–900 K). A typical energy spectrum is shown in Figure 6.4a. These spectra were fitted to Eq. (6.6) to obtain values for d and T. For Figure 6.4a, the values are 0.25 ± 0.01 eV and 588 ± 50 K, respectively. Several spectra measured in the higher current regime contained significant shoulders at the low-energy side of the main peak. From

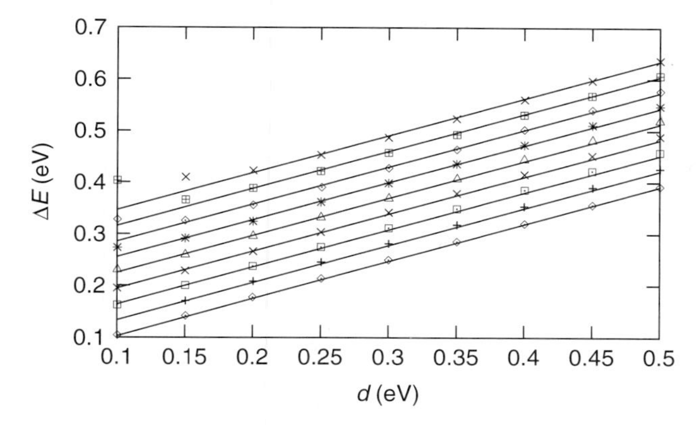

Figure 6.3 Energy spread ΔE as a function of the tunneling parameter d for temperatures of 100–900 K (bottom to top). The lines were generated with Eq. (6.8), while the points present numerically generated energy spectra using Eq. (6.6) from which the values of ΔE were determined by using a program written in LabView (National Instruments).

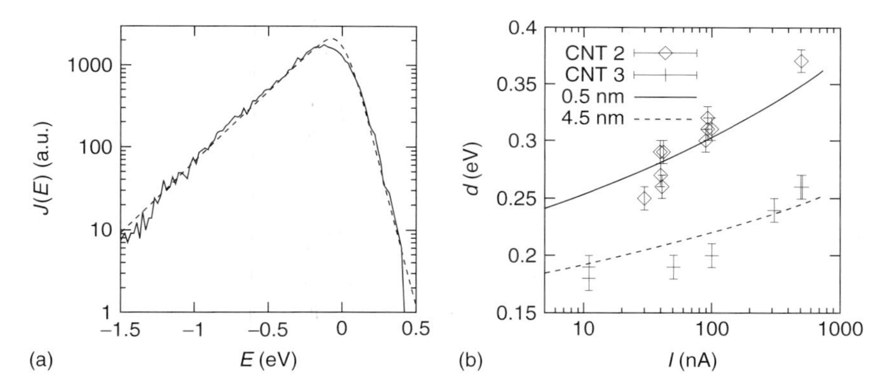

Figure 6.4 (a) Energy spectrum of carbon nanotube 2 recorded at 350 V and 30 nA, and fitted with the FN theory (dashed). (b) Tunneling parameter d as a function of I for two carbon nanotubes, compared with numerically calculated curves. (Source: Taken form [18].)

these spectra, values of d and T could still be determined in most cases, but values of ΔE were not extracted from these spectra. The values of ΔE were corrected for the resolution of the spectrometer by subtracting 50 meV from the obtained values (the validity of this correction was tested by numerically generating energy spectra that were convoluted with a Gaussian with a sigma of 50 meV).

6.3.5
Comparing the Measured Tunneling Parameter with Theory

The plot of d versus I of two carbon nanotubes with respective diameters of 0.6 and 4.6 nm is shown in Figure 6.4b. Numerically calculated curves (using Eqs. (6.1–6.3) and (6.7)) of d versus I for both radii overlap well with the measurements, showing that the data is consistent with the FN model. We can thus express d as a function of R and I only. Note that the value of d increases with decreasing R for a given I. For carbon nanotubes with radii smaller than 1 nm, it would not be possible to obtain sufficient beam current for imaging applications and a small energy spread as well; carbon nanotubes with radii larger than 1 nm are needed.

6.3.6
Determining the Work Function

The data of the FN plot and the energy spectrum can be combined, in order to determine ϕ, using

$$\phi = -1.64\, slope \times d/U \tag{6.9}$$

with the slope of the FN plot and U the extraction voltage at which the energy spectrum was measured [13, 14]. The obtained values are indicated in Table 6.1 alongside each carbon nanotube, and are consistent with both the expected value of 5.1 eV and the value of 5.0 eV obtained by others [19].

6.4
The Brightness

6.4.1
Measuring the Brightness

The most important parameter of an electron source is its brightness. Usually, the brightness is normalized on U, the electron energy at which the brightness is measured, thus obtaining the reduced brightness B_r [16]:

$$B_r = \frac{dI}{d\Omega}\frac{1}{\pi r_v^2}\frac{1}{U} = \frac{I'_r}{\pi r_v^2} \tag{6.10}$$

with the reduced angular current density I'_r and the radius of the virtual source r_v. For a focusing system with ideal lenses, the reduced brightness indicates the amount of current that can be focused into a spot of certain size and electron beam energy. Parts of the following text have been published previously in [18].

I'_r and r_v were measured with a Faraday cup and a point projection microscope, respectively, for several carbon nanotubes to obtain B_r. The method has been described elsewhere [2, 20]. I'_r was obtained for all carbon nanotubes. The average value was 2.8 nA (sr V)$^{-1}$ at $I = 100$ nA (Table 6.1). Measurements at other current

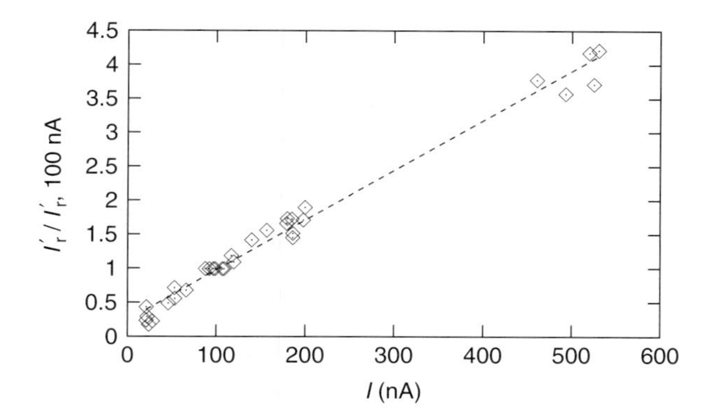

Figure 6.5 Reduced angular current density I_r normalized on I_r at 100 nA for carbon nanotubes 1–8 and a linear fit (dashed). For each carbon nanotube, data points in the full current range were recorded. (Source: Taken from [18].)

levels revealed approximately a linear relationship between I'_r and I. As can be seen from Figure 6.5, the following relationship applies:

$$\frac{I'_r}{I'_{r,100\ \text{nA}}} = 0.25 + 0.0073 \times I\ (\text{nA}) \tag{6.11}$$

Within a factor of 2 accuracy, I'_r can now be expressed by

$$I'_r \cong 0.0073 \times 2.8 \times I\ (\text{sr V}) = 0.02 \times I\ (\text{sr V}) \tag{6.12}$$

The virtual source sizes of three carbon nanotubes were measured; two were found to be almost equal to R and one was smaller by a factor of 1.7, see Table 6.1. Thus for carbon nanotubes with small radii, the following relation applies:

$$r_v \cong R \tag{6.13}$$

This is consistent with previous conclusions based on point projection microscopy and TEM investigations of arc-discharge carbon nanotube electron sources [20].

6.4.2 New Model for the Brightness

Using relations (6.12) and (6.13), and recognizing that d is a function of F and ϕ, the brightness can be rewritten within a factor of 3 accuracy as

$$B_r \cong \frac{0.02I}{\pi R^2} = 0.04J(F,\phi) = 0.04J(d,\phi) \tag{6.14}$$

Since ϕ is constant for a certain material, Eq. (6.14) shows that B_r is a function of d only.

We can now compare this model with our data since we have obtained both B_r and d for a total of five carbon nanotubes. Note that the presentation of just one

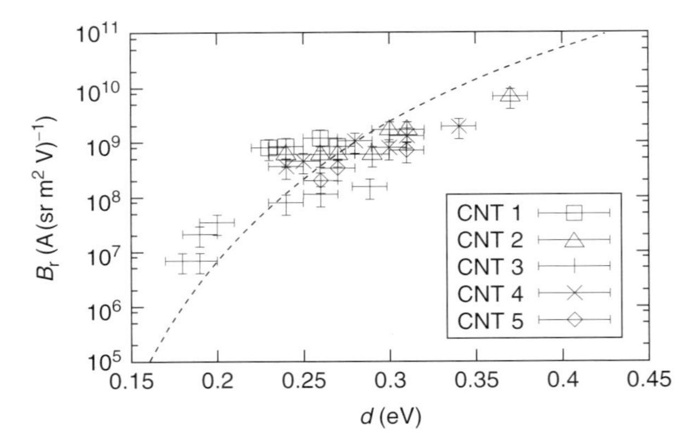

Figure 6.6 Reduced brightness B_r as a function of the tunneling parameter d measured for five carbon nanotubes, compared with the calculated curve using $B_r = 0.04 \times J(d, \phi = 3.0 \text{ eV})$ (dashed), within its error range of a factor of 3 as indicated by the dotted lines. (Source: Taken from [18].)

of these two parameters has a limited use, as high brightness is usually obtained at the expense of a low-energy spread, affecting the performance of the source in an electron optical system. Figure 6.6 shows both the data obtained for the carbon nanotubes and a plot of Eq. (6.14). The data correspond well with the theory. An important aspect of Eq. (6.14) is that it is independent of the actual shape of the carbon nanotube, expressed in R, its length l, and other geometrical factors such as the geometry of the surrounding electrodes. This can be explained as follows: A thin carbon nanotube will result in a relatively large value of d (as shown in Figure 6.4). On the other hand, it has a small value of r_v and a corresponding large brightness, and vice versa for a thicker carbon nanotube.

6.4.3
Discussion of the New Model

This study shows that it is now possible to select the optimal carbon nanotube electron point source from carbon nanotubes with a range of diameters and lengths, as the relation between B_r and d, which determines the optical performance of the source, does not depend on these parameters. It is expected that the result of Eq. (6.14) is more widely applicable to field emitters of nanometer size, for example, nanometer-sized tungsten tips and metallic nanowires. The upper and lower limits are set by the minimal current required for sufficient signal to noise in the application (e.g., imaging) and the maximum current at which the emitter can still operate properly. A further restriction is that the (thermal) vibration amplitude of the emitting end of the carbon nanotube, which is proportional to the ratio R^4/l^3, should be kept sufficiently small [21]. It is important to notice that the determination of the brightness, which is often a complicated experimental procedure, is reduced to recording an FN plot and measuring I'_r.

In the 1960s, an expression was derived for the maximum theoretical brightness of a field emitter [22]:

$$B_r = \frac{Je}{\pi d} \tag{6.15}$$

This equation predicts the dependence on d and ϕ as well, but its value is often a factor of 15–30 larger than obtained from Eq. (6.14). This difference is explained as follows: Typical field emitters consisting of a metal tip with a hemispherical emitting surface with $R > 50\,\text{nm}$ have a much smaller value of r_v than R [16]. For carbon nanotubes, on the contrary, $r_v = R$. Others found $r_v = 0.5R$ for nanometer-sized tungsten electron sources [23]. Thus, r_v approaches R when the emitting surface is reduced to a few nanometers and deviates in shape from hemispherical. Furthermore, the broadening effect of Coulomb interactions [24] was not accounted for in Eq. (6.15).

6.4.4 The Total Figure of Merit for Carbon Nanotube Electron Sources

At room temperature and at low currents, d is almost equal to ΔE and therefore Figure 6.6 presents the figure of merit of the carbon nanotube electron source. However, some heating (to 600–800 K) was required to obtain stable emission, which broadens the energy spectrum. A secondary effect is that joule heating at larger currents also leads to a broadening of the energy spectrum [12]. The figure of merit for electron sources under realistic conditions, B_r versus ΔE, is presented in Figure 6.7. The curve was calculated using Eqs. (6.8) and (6.14) for $T = 600\,\text{K}$. For comparison, data for state-of-the-art electron sources, for example, the tungsten cold field emission gun (CFEG) [3] and the Schottky emitter [25], are included. Because they are cold field emitters, the carbon nanotube and the CFEG have almost identical ΔEs. However, the CFEG often lacks emission stability [3]. In contrast, the carbon nanotube exhibits a highly stable electron beam if it has a closed cap and if it is heat-treated for cleaning of the emitting surface [4]. Note also that the carbon nanotube has a much higher B_r than the CFEG at high currents. The value of the carbon nanotube presents a significant improvement over the Schottky emitter, either a lower ΔE at similar B_r or a much higher B_r at similar ΔE. Future research might aim to find nanomaterials with similar strength as carbon nanotubes, but with an intrinsically lower ϕ, which directly leads to an improvement of the figure of merit. Carbon nanotubes with larger radii than used in this experiment and with hemispherical caps may possibly improve the figure of merit as well.

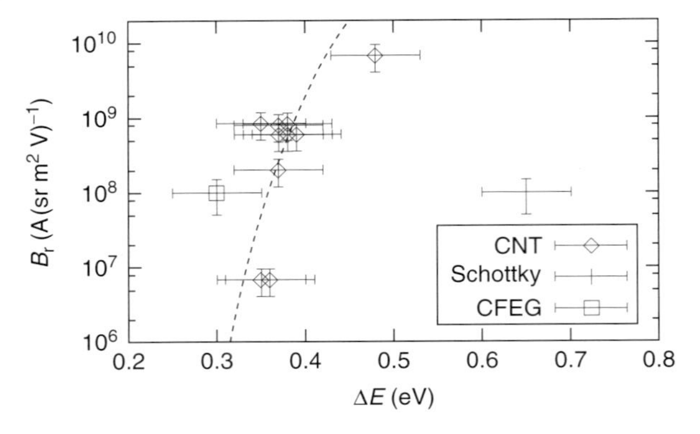

Figure 6.7 Reduced brightness B_r as a function of the energy spread ΔE measured for several carbon nanotubes at currents 10–500 nA and temperatures 600–700 K, calculated curve (dashed) and error range (dotted). The data of the Schottky emitter and the CFEG are also included. (Source: Taken from [18].)

6.5 Conclusions

A carbon nanotube electron source exhibits emission that follows the FN theory of field emission, thus predicting the energy spread for a given current, radius, and temperature. The optical performance of the carbon nanotube under realistic conditions is a significant improvement compared to present state-of-the-art electron point sources. The reduced brightness is a function of the tunneling parameter, which is a measure of the energy spread at low temperatures, but independent of the geometry of the emitter.

Acknowledgments

The authors thank Kenneth B.K. Teo for the CVD carbon nanotube sample; A.G. Rinzler for the arc-discharge carbon nanotube sample; Myriam Allioux, Theo van Rooij, and many others for experimental help; Monja Kaiser for TEM imaging; and Erwin C. Heeres, Sjoerd A.M. Mentink, Greg Schwind, and Gerard van Veen for discussions. This work was supported by FEI company, the EC, the EPSRC, and the Dutch Ministry of Economic Affairs.

References

1. de Jonge, N. and Bonard, J.M. (2004) Carbon nanotube electron sources and applications. *Philos. Trans. R. Soc. London A*, **362**, 2239–2266.
2. de Jonge, N., Lamy, Y., Schoots, K., and Oosterkamp, T.H. (2002) High brightness electron beam from a multi-walled carbon nanotube. *Nature*, **420**, 393–395.

3. Hainfeld, J.F. (1977) Understanding and using field emission sources. *Scan. Electron. Microsc.*, **1**, 591–604.
4. de Jonge, N., Allioux, M., Oostveen, J.T., Teo, K.B.K., and Milne, W.I. (2005) Low noise and stable emission from carbon nanotube electron sources. *Appl. Phys. Lett. (USA)*, **87**, 133118-1–133118-3.
5. Saito, R., Dresselhaus, G., and Dresselhaus, M.S. (1998) *Physical Properties of Carbon Nanotubes*, Imperial College Press, London.
6. Colbert, D.T., Zhang, J., McClure, S.M., Nikolaev, P., Chen, Z., Hafner, J.H., Owens, D.W., Kotula, P.G., Carter, C.B., Weaver, J.H., Rinzler, A.G., and Smalley, R. (1994) Growth and sintering of fullerene nanotubes. *Science*, **266**, 1218–1222.
7. Lacerda, R.G., Teh, A.S., Yang, M.H., Teo, K.B.K., Rupesinghe, N.L., Dalal, S.H., Koziol, K.K.K., Roy, D., Amaratunga, G.A.J., Milne, W.I., Chowalla, M., Hasko, D.G., Wyczisk, F., and Legagneux, P. (2004) Growth of high-quality single-wall carbon nanotubes without amorphous carbon formation. *Appl. Phys. Lett. (USA)*, **84**, 269.
8. de Jonge, N., Lamy, Y., and Kaiser, M. (2003) Controlled mounting of individual multi-walled carbon nanotubes on support tips. *Nano Lett. (USA)*, **3**, 1621–1624.
9. de Jonge, N., Doytcheva, M., Allioux, M., Kaiser, M., Mentink, S.A.M., Teo, K.B.K., Lacerda, R.G., and Milne, W.I. (2005) Cap closing of thin carbon nanotubes. *Adv. Mater. (Germany)*, **17**, 451–455.
10. Rinzler, A.G., Hafner, J.H., Nikolaev, P., Lou, L., Kim, S.G., Tomanek, D., Nordlander, P., Colbert, D.T., and Smalley, R.E. (1995) Unraveling nanotubes: field emission from an atomic wire. *Science*, **269**, 1550–1553.
11. Dean, K.A. and Chalamala, B.R. (2003) Experimental studies of the cap structures of single-walled carbon nanotubes. *J. Vac. Sci. Technol. B*, **21**, 868–871.
12. Purcell, S.T., Vincent, P., Journet, C., and Binh, V.T. (2002) Hot nanotubes: stable heating of individual multiwall carbon nanotubes to 2000 K induced by the field-emission current. *Phys. Revi. Lett. (USA)*, **88**, 105502 - 1–105502-4.
13. Groening, O., Kuettel, O.M., Emmenegger, C., Groening, P., and Schlapbach, L. (2000) Field emission properties of carbon nanotubes. *J. Vac. Sci. Technol. B*, **18**, 665.
14. de Jonge, N., Allioux, M., Doytcheva, M., Kaiser, M., Teo, K.B.K., Lacerda, R.G., and Milne, W.I. (2004) Characterization of the field emission properties of individual thin carbon nanotubes. *Appl. Phys. Lett. (USA)*, **85**, 1607–1609.
15. Fowler, R.H. and Nordheim, L. (1928) Electron emission in intense electric fields. *Proc. R. Soc. Lond. A*, **119**, 173–181.
16. Hawkes, P.W. and Kasper, E. (1996) *Principles of Electron Optics II: Applied Geometrical Optics*, Academic Press, London.
17. Edgcombe, C.J. and de Jonge, N. (2007) Deduction of work function of carbon nanotube field emitter by use of curved-surface theory. *J. Phys. D: Appl. Phys.*, **40**, 4123–4128.
18. de Jonge, N., Allioux, M., Oostveen, J.T., Teo, K.B.K., and Milne, W.I. (2005) The optical performance of carbon nanotube electron sources. *Phys. Rev. Lett.*, **94**, 186807-1–186807-4.
19. Gao, R., Pan, Z., and Wang, Z.L. (2001) Work function at the tips of multiwalled carbon nanotubes. *Appl. Phys. Lett. (USA)*, **78**, 1757.
20. de Jonge, N. (2004) The brightness of carbon nanotube electron emitters. *J. Appl. Phys. (USA)*, **95**, 673–681.
21. Hafner, J.H., Cheung, C.L., Oosterkamp, T.H., and Lieber, C.M. (2001) High-yield assembly of individual single-walled carbon nanotube tips for scanning probe microscopes. *J. Phys. Chem. B (USA)*, **105**, 743–746.
22. Worster, J. (1969) The brightness of electron beams. *Brit. J. Appl. Phys. (J. Phys. D)*, **2**, 457.
23. Qian, W., Scheinfein, M.R., and Spence, J.C.H. (1993) Brightness measurements of nanometer-sized field-emission-electron sources. *J. Appl. Phys. (USA)*, **73**, 7041.

24. Kruit, P., Jansen, G.H., and Orloff, J. (1997) *Handbook of Charged Particle Optics*, CRC Press, p. 275.

25. Swanson, L.W., Schwind, G.A., and Jon, O. (1997) *Handbook of Charged Particle Optics*, CRC Press, New York, p. 77.

7
Heat Generation and Losses in Carbon Nanotubes during Field Emission

Stephen T. Purcell, Pascal Vincent, and Anthony Ayari

7.1
Introduction

Carbon nanotubes (CNTs) have several advantages as field emission (FE) electron sources: chemical stability, high-current carrying capacity, high aspect ratios for low extraction voltages, and low-cost mass production. Since the first experiments [1–3], many authors have shown the extraction of both stable and very large FE currents, I [4–8]. This opens up numerous commercial applications which often demand the highest possible currents. It is now clear that this is controlled by the interplay between Joule heating effects either along the nanotube length [9, 10] or at the contact [11, 12] and the stabilizing heat evacuation mechanisms.

The peculiarity of CNT field emitters is that they emit stably at high temperatures induced by Joule heating, for example, up to ~1600 K. The stable heating state is not often observed for emitters of other materials and has far-ranging implications. It increases sustainable high currents, permits self-cleaning by desorption which improves dramatically the current stability, causes CNT shortening instead of sudden breakdown at extreme currents, and, finally, allows measurements of several intrinsic parameters of the carbon nanotube [13]. In contrast, metal tip emitters suddenly melt into large balls near the apex at high-current densities because of a catastrophic runaway phenomenon that quickly follows induced heating [14, 15]. Three positive feedback mechanisms accentuate temperature and current increases and thus breakdown: (i) strong increase in the metallic resistance with temperature, (ii) increase in FE with temperature, and (iii) rapid diffusion of metal surface atoms to the high-field regions, particularly as temperature increases, which sharpens tips and thus further increases the current. Here, we review the present state of experimental studies and theoretical understanding of the generation and evacuation of heat in carbon nanotubes during FE. Experimental measurements of the temperature by field electron emission spectroscopy (FEES) and measurements of total energy distributions (TEDs) [16, 17] are emphasized, but other pertinent results are also surveyed. The basic theoretical framework is given for heat generation and evacuation accompanied by simulations. An effort

Carbon Nanotube and Related Field Emitters: Fundamentals and Applications. Edited by Yahachi Saito

ISBN: 978-3-527-32734-8

is made to show how this is a very open and eventually complex subject with little quantitative agreement between experiment and theory to date.

Heating effects were reported in one of the very first articles on FE from CNTs [1], though the heating was induced by a focused laser. Light emission ascribed to black body radiation was also observed. We limit our discussion here to current-induced heating. The stability of the FE current from carbon nanotubes was studied early on and discussed within the framework of adsorption and ion retro bombardment [18, 19]. Observations of a gradual destruction of arc-electric multiwall nanotubes (MWNTs) at currents as high as 200 μA during FE were made inside a transmission electron microscope (TEM) [20], though the probable cause by Joule heating was not yet evoked. Heating was originally hypothesized to explain the high-current behavior and degradation of CNTs [21] due to local heating near the apex by Nottingham effects [16] through resonant tunneling states. As well, the existence of characteristic rings in the FE patterns observed previously [19] was used to estimate that field-assisted evaporation of single wall nanotubes (SWNTs) starts at ~1600 °C. The origin of the rings was unclear at that time, but we have recently shown that they can be explained by the self-focusing of thermal-field electrons from the shank just below the cap of a hot nanotube [22], thus confirming the determination of the temperature in [21].

The role of Joule heating was clearly established by FEES on an individual MWNT [9]. This was used to (i) measure the temperature at the emission zone as a function of emission current, $T_L(I)$ (L = CNT length); (ii) show that emission currents (I) on the order of microamperes induced high, stable temperatures reaching 2000 K; (iii) measure the electrical resistance of an individual MWNT, $R(I)$ (equivalently $R(T_L)$), in this case on the order of megohms; (iv) show that the high temperatures were accompanied by light emission from the MWNTs, whose intensity was consistent with Planck's law; (v) show that the high induced temperatures can lead to excellent emission stability by self-cleaning the surfaces of the nanotubes; and (vi) show that even higher currents, and thus higher temperatures, lead to a gradual destruction of the nanotubes. These studies were accompanied by confirming simulations of a 1D model [10] that incorporated Joule heating, thermal conduction to the support, radiation losses, and Nottingham effects. Nottingham effects were estimated to be small compared to Joule heating, thermal conduction, and radiation. This work also showed that FE becomes a new tool for making simultaneous and, therefore, correlated measurements of several physical properties of CNTs.

Since this work, optical spectroscopy measurements of light emission from an MWNT layer from which a small number of emitters were active confirmed that the light has a black body spectrum [23]. Experiments and extension of the modeling of [10] on emission from vertical CNTs on Si substrates were used to compare and predict the critical field, current density, and currents for thermal runaway in CNTs when the electrical resistivity $\rho(T)$ increases at higher temperature, consistent with phonon-controlled mean free paths [12]. 1D modeling was carried out that particularly addressed the Nottingham effect [24], predicting that in certain conditions they can be stronger than radiation effects. A fuller account of the data of [9] and somewhat refined modeling of [10] was presented [25] with a better function for $\rho(T)$, and for the first time an estimate of the temperature dependence of the thermal conductivity $\kappa(T)$. Another aspect was the integration of the TEDs to quantify the Nottingham energy exchange at different temperatures and emission zones. More recently, the 1D model was used to study the influence of Nottingham effects, particularly at high currents and temperatures [26]. In this region, it may cause cooling and the temperature at the CNT cap can be lower than in the near-cap region. Finally, 3D simulations have been made [27] that compare Joule and Nottingham heating without radiation and allow for temperature increases in a W support tip. It was concluded that Joule heating is orders of magnitude larger for currents above picoamperes. Thus, different authors have come to different conclusions for the influence of Nottingham effects, and more work is needed. This may be because the various mechanisms have different scaling (see below) and each author simulated the problem with different parameters, or because more complex averaging of Nottingham effects over the emission surface is necessary for it to be well quantified [25].

7.2 Heat Diffusion Equation for Nanotubes

The measurement and calculation of temperature profiles in nanotubes (and nanowires) is actually a very challenging and open problem that can be treated at many levels of complexity. Our approach has been to find the simplest approximations that contain the essential physics. The nanotube is treated as a simple resistance, which is justified for many nanotubes, for example, MWNTs produced by chemical vapor deposition (CVD), because mesoscopic behavior such as ballistic transport of electrons and phonons observable in high-quality CNTs produced by arc discharge is quickly masked by defects and phonon scattering, particularly at room temperature and above. The simplest model is to treat the CNT as a one-dimensional object of length L in contact with a heat sink fixed at temperature $T = T_0$ at $x = 0$, and include heat generation and losses by (i) Joule effects, (ii) thermal conduction, (iii) radiation, and (iv) Nottingham exchanges at the CNT cap.

The first two mechanisms should be well-enough described by classical expressions. However, we and others [10, 12, 24, 26] have modeled the radiation losses by a differential surface area and the Stefan–Boltzmann law with a constant emissivity. This is clearly not correct [25] because the diameter of most nanotubes is less than the wavelength and the extinction length of photons, and thus a differential element must radiate by its volume and not its surface. Photon emission should be considered within the framework of Rayleigh scattering, where absorptivity a and, hence, emissivity e depend on the particle dimensions, thus adding considerable complexity to the problem. For example, for a small enough dielectric sphere $a(r, \lambda) = (r/\lambda)f(n)$, where $f(n)$ is a dimensionless function of the complex index of refraction n [28]. Evoking Kirchoff's law and inserting this into the Stefan–Boltzmann law makes the emitted power proportional to volume as expected. Unfortunately, this is not sufficient here because antenna and polarized light emission effects must be treated and most nanotubes are not dielectrics. It is not even evident that a local differential expression for radiation losses is appropriate. Doing this correctly may completely change the balance between radiation, thermal conduction, and Nottingham effects. Nottingham effects can be included as a boundary condition for heat flow at the cap $J_{\mathrm{Th}}(L)$ [10, 24–27]. Until now, the maximal current density and field at the cap apex have been used to calculate energy exchanges, but this is a rough approximation that favors heating [25]. The expression should be integrated over the cap apex where lower fields favor cooling as opposed to heating. In the absence of better expressions, we proceed as before with the idea that including the dominating T^4 factor and Nottingham effects at the apex allow discerning general trends.

Under the simplest assumptions, the time-independent heat diffusion equation will be as follows:

$$\rho(T) I_{\mathrm{FE}}^2 \frac{\mathrm{d}x}{A} + \kappa(T) A \frac{\mathrm{d}^2 T}{\mathrm{d}x^2}\mathrm{d}x - 2\pi r e(r) \sigma (T^4 - T_0^4)\mathrm{d}x = 0 \tag{7.1}$$

where $\rho(T)$ and $\kappa(T)$ are the temperature-dependent resistivity and thermal conductivity, A is the cross section, r is the exterior tube radius, σ is the Stefan–Boltzmann constant, and T_0 is the ambient temperature at the support/CNT contact. For all simulations below, we assume $e(r) = 1$ and $A = \pi r^2$, which is a good approximation for nanotubes with the more common central tube diameters. The Nottingham boundary condition at the cap is

$$J_{\mathrm{Th}}(L) = I_{\mathrm{FE}} <E> = -\kappa(T) A \frac{\mathrm{d}T}{\mathrm{d}x} \tag{7.2}$$

where $<E>$ (in electronvolts) is the average energy of emitted electrons with respect to the Fermi level ($<E> < 0$, heating, $<E> > 0$, cooling). $<E>$ can be estimated with the TEDs.

A useful analytic solution can be given when radiation is neglected and κ and ρ are constant, which is reasonable for low temperatures:

$$T(x) = T_0 + \left(\frac{\rho}{\kappa}\frac{I^2}{A^2} - \frac{I <E>}{\kappa A}\right) x - \frac{\rho}{2\kappa}\frac{I^2}{A^2} x^2 \tag{7.3}$$

$$\text{At } x = L,\ T_L = T_0 + \frac{\rho}{2\kappa}\frac{I^2}{A^2} L^2 - \frac{I <E>}{\kappa A} L \tag{7.4}$$

The Nottingham/Joule ratio scales with $\eta_{\mathrm{NJ}} = 2 <E> A/\rho I L$. This varies widely for nanotubes, perhaps explaining why different results have been reported.

To get a feeling of the heating effects, consider a reasonable parameter set (in SI units): $\rho = 10^{-5}$, $\kappa = 100$, $L = 10^{-6}$, $I = 10^{-5}$, $A = 10^{-16}$, and $<E> = -0.1$ (heating). One gets $\Delta T = T_L - T_0 = 500 + 100 = 600^{\circ}$. However, the formula gives widely varying estimates for physically possible nanotubes. A peculiarity of carbon is that both $\rho(T)$ and $\kappa(T)$ vary by orders of magnitude for different forms of graphite [29] and for nanotubes, and they are particularly sensitive to different heat treatments. Obviously, the dimensions of nanotubes also vary by orders of magnitude. Though we assume here ideal structures, in reality the overall radius, the ratio of the inner/outer tube radii, and the structural quality are often not uniform. $\rho(T)$ and $\kappa(T)$ may also vary with the position along the tube because the CNTs may be subjected to temperature gradients. Finally, as stated above, η_{NJ} can vary greatly. This means that each nanotube must be specifically analyzed and one needs the maximum information to understand the high-current and high-temperature behavior.

7.3 Simulations

Solving Eq. (7.1) numerically by iteration is simple if all the CNT parameters are known. Typical examples with and without radiation included are shown in Figure 7.1a and b for $T(x)$ and $T_L(I)$. Nottingham effects have been neglected. One sees that the difference in temperatures for calculations that include either a decreasing $\rho(T)$ or radiation losses progressively increases with temperature. The analytical form is fairly accurate for $\Delta T < 500$ for this choice of parameters.

Though r, A, and L can be determined by electron microscopy, generally one does not know *a priori*, $\rho(T)$, $\kappa(T)$, $e(r,T)$, and $<E>$. Thus, one of the goals of this work is to use FEES experiments and simulations to extract these functions from the data.

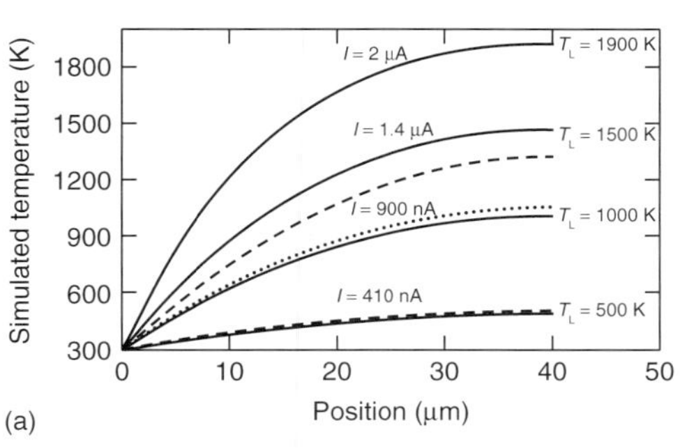

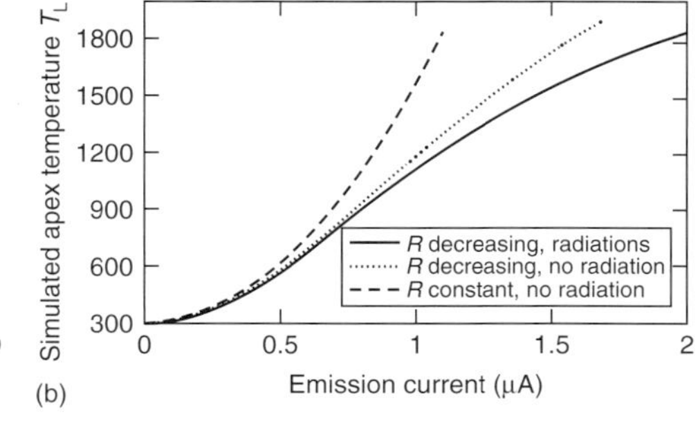

Figure 7.1 (a and b) Temperature profiles along a CNT and $T_L(I)$ for different cases simulated using Eqs. (7.1), or (7.3) and (7.4) ($L = 40$ μm, $r = 10$ nm). Dashed curves: $\kappa = 100$ and $\rho = 1.6 \times 10^{-5}$, no radiation (Eqs. (7.3) and (7.4)). Dotted curves: $\rho(T) = 1.6 \times 10^{-5}$ to $6.4 \times 10^{-9}(T - 300)$, no radiation. Solid curves: same with radiation losses included (see [10]).

In experiments, one generally imposes an applied voltage V and measures the emitted I, which adds an additional degree of complexity to the problem. Using the measured I is sufficient for solving Eq. 7.1 but does not address questions such as the existence of a solution for an elevated stable temperature and the high-current degradation where current changes in time. Actually, I increases with T_L and this creates a positive feedback. A self-consistent solution between the Fowler–Nordheim (FN) equation and the heat equation is needed to find the equilibrium points of V, I, and T_L (and $T(x)$). For this, it is convenient to use a graphic representation where $T_L(I)$ and the thermal $I(V,T_L)$ evolution are plotted on the same graph (Figure 7.2). We use Eq. (7.1) with ρ and κ constant to obtain $T_L(I)$ without loss of generality. Consider a CNT (SI units) with $L = 10^{-5}$, $r = 1.5 \times 10^{-9}$, $\rho = 2.6 \times 10^{-5}$, $\kappa = 100$, and $T_0 = 300$ K. Furthermore, consider a constant field enhancement factor β (field $= \beta V$) and no effect of an IR drop along the CNT; the thermal dependence of the FE current given by $I(V, T_L) \cong I(V, T_L = 0) \times (1 + \text{const.} T_L^2)$ [16]. For a given voltage V_1, $I(V_1, T_L = 0\text{ K})$ defines a point on the x-axis (Figure 7.2a). The emission would start slightly above at $I(V_1, 300\text{ K})$ if the voltage is applied much faster than the heating occurs. $I(V_1, T_L)$ is given by the parabolic solid curve (the inverse function $T_L(V_1, I)$) starting from this point. The self-consistent solution of the thermal and FE equations is the intersection of the dashed curve and the solid curve to give the set V_1, T_{L1} and I_1. In time, I would move along the $I(V_1, T)$ curve. For increasing voltages, we obtain other curves and new stability solutions. Depending on the $T_L(I)$ curves, there is a maximum voltage above which no stability point exists (voltage V_2 in the figure). This defines the maximum temperature T_M and current for which the system is stable. This analysis explains in a simple way the critical temperature, current, and field calculated previously [12] versus the CNT length. At a higher voltage, that is V_3, there is no stationary solution and I and T_L should

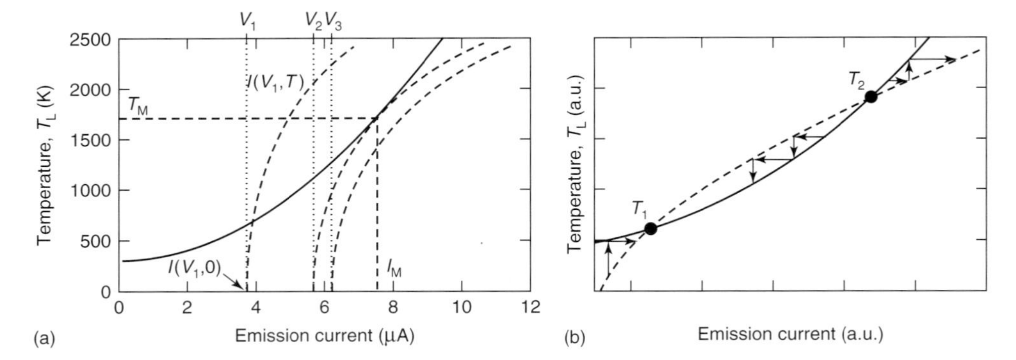

Figure 7.2 (a) Plots of $T_L(I)$ from simulations using Eq. (7.1) and $I(V,T_L)$ (inverted axis). The crossing points define the equilibrium parameter set (V,I,T_L). (b) Zoom near an equilibrium point showing that there are two solutions. The upper solution is unstable (see text) and simulations will follow the arrows.

increase without bounds. T_L and I must then be calculated by the nonstationary diffusion equation. Note that the CNT may degrade at a temperature lower than the maximum mathematical stability point because carbon leaves the CNT cap.

From Figure 7.2b, one sees that two solutions are expected: T_1 and T_2, for a voltage slightly lower than V_2, one more and one less than T_M. These two solutions exist mathematically, but T_1 is the only structurally stable and physical solution. T_2 is structurally unstable and cannot be a physical solution. However, this unstable point could be found by numerical calculations and must be recognized as such.

This is the analysis of the simplest situation but many different cases can be envisaged. For example, a decrease in $\rho(T)$, increase in $\kappa(T)$, and Nottingham effects may cause the $T_L(I)$ curves to bend downward further, thereby pushing the intersection points to higher currents. Also, a $\rho(T)$ that increases with T at higher temperatures [12] could cause $T_L(I)$ to increase again at higher currents, perhaps creating more equilibrium points and thus hysteresis effects. For certain parameter sets, $T_L(I)$ and $I(V, T_L)$ may run roughly parallel, leading to slow responses of the system. In fact, we once observed the current from a nanotube at a fixed voltage slowly increasing by over 100% during roughly a 10-s time frame, while its incandescence also increased greatly, until it suddenly broke down at 40 μA. This is a very long time for a heating phenomenon in a nanometric object.

The cap region usually has the highest temperature and field during current-induced heating, and therefore CNT degradation occurs there preferentially. Regular thermal removal of material at the nano or atomic scale should roughly follow an Arrhenius relation with an average activation energy. Thus time scales are controllable if the temperature is controlled. Through current-induced heating, it is possible to observe and control CNT length reduction on a lab time scale in both simulations and experiments. The gradual degradation phenomenon was first studied in a TEM [20] and then by following I and field electron microscopy (FEM) patterns in time [21] (see also [25, 30]). Notably, in [21] the carbon atoms could leave the SWNTs ring by ring. Striking results from recent detailed TEM studies [31, 32] are shown in Figure 7.3a and b, which demonstrate the gradual and controlled shortening of an MWNT and accompanying $I(V)$ characteristics. The series of $I(V)$ characteristics has a well-defined envelope. This effect has recently been used to precisely tune the resonance frequencies of CNTs in a TEM [33]. Simulations must now include self-consistent solutions between the FN equation and the heat equation in which the length of the CNT also changes. This can be carried out to different degrees of sophistication. Here, we simply limit the length by imposing a maximum temperature at the CNT cap. Changes on a lab time scale (tens of seconds) is known to occur in the 1600–2000 K [9, 21] range. The envelope of the calculated curves in Figure 7.3c is similar to the experimental curves of Figure 7.3b and [30], showing that the regular and reproducible material loss at the apex region can be well simulated and thus understood.

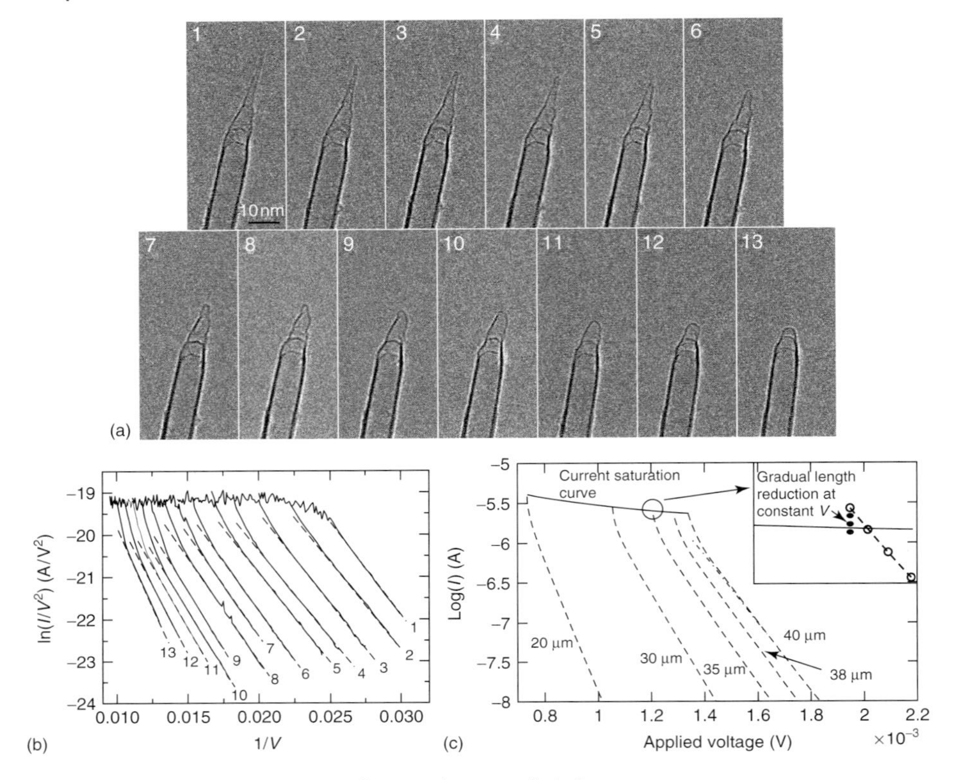

Figure 7.3 (a) TEM images showing the controlled shortening of conical CNT tips by current-induced heating. (b) Corresponding FN plots. (c) Simulated $I(V)$ curves for the shortening of CNT at a set temperature [32]. ((a and b) with permission from [30].)

7.4 Experiments

Our group has studied the high-current behavior particularly with FEES to characterize individual MWNTs at various stages of their thermal and field cleaning. The results were then analyzed by running specific simulations with the heat diffusion equation (Eq. 7.1). The FEES permits simultaneous measurements of both $T_L(I)$ and the voltage drop without which it is difficult to imagine providing meaningful comparisons between theory and experiment, particularly without *a priori* knowledge of $\rho(T)$, $\kappa(T)$, $e(r,T)$, and $<E>$. Unfortunately, there is only one published set of data [9, 25] that measured simultaneously $T_L(I)$ and $R(T_L)$ to guide calculations. The result is that, in our opinion, though the first-order description of the problem is now in place, quantitative comparisons with theory and experiment are largely lacking. This effectively excludes examining many interesting second-order effects.

The details of the experimental procedure are described in the original articles. The experiments were carried out in an ultrahigh vacuum (UHV) system fully equipped for measuring FEM and field ion microscopy (FIM) and FEES. Oriented MWNTs were grown by CVD directly on large Ni support tips. Scanning electron microscopy (SEM) showed that the apex MWNTs were quite straight, with diameters in the range of 20–50 nm and lengths up to ~40 μm. The multiwall character, diameter range, and high number of defects of the MWNTs were confirmed by TEM on samples fabricated by exactly the same procedure. The Ni tip was held in a W spiral to allow *in situ* cleaning by standard Joule heating to ~1300 and up to 1600 K by electron bombardment. The TEDs were measured with a hemispherical electron energy analyzer through a probe hole in the same UHV system. Though many MWNTs are present on the Ni tip, the FE experiments are specific to an individual one because of the selectivity of FE to the highest field emitters.

Two distinct emission regimes before and after the highest temperature cleaning were observed. The first regime is when the surface is rough and consists of disordered nanostructures formed either by adsorbates or the carbon itself. These objects give rise to strong effects in the TEDs similar to those measured from the deposited molecules [34]. The resonant tunneling model [35] has been often invoked to explain most of the effects [25]. The second regime, the "intrinsic MWNT emitter," is when the surface is heated to temperatures reaching 1600 K to produce a "smoother-cleaner" surface without nanometric protrusions. I/V characteristics and TEDs then followed FN theory and excellent current stability was achieved.

The formula for the TED of FE from a free electron gas [16] is the product of a field-dependent transmission probability and the Fermi–Dirac distribution:

$$J(E)\alpha \frac{\exp\left((E-E_{\mathrm{F}})/d\right)}{1+\exp\left((E-E_{\mathrm{F}})/(k_{\mathrm{B}}T_{L})\right)} \tag{7.5}$$

Here, E_{F} is the Fermi energy and $d(\mathrm{eV}) \sim F/\sqrt{(\phi)} \sim 0.2\mathrm{eV}.F$ is the applied field ($\sim$3–7 V nm^{-1}) and ϕ the work function in electronvolts. The TEDs are asymmetric peaks of width $\sim$ 0.3 eV at room temperature. The slope of $\ln(J(E))$ on the low-energy side is $1/d$ and on the high-energy side $(1/d - 1/k_{\mathrm{B}}T) \sim -1/k_{\mathrm{B}}T$. The peak is positioned close to E_{F}. In general, the experimental measurements of TEDs from metallic emitters deviate somewhat from this formula [16]. However, the key to these experiments is that they permit an excellent measure of E_{F} and the temperature at the emission zone. We have found agreement within 20 K between optical pyrometry measurements and fits to Eq. (7.5) in the 1000–1300 K range when the temperature was controlled by the support heating loop, and better than 0.05 eV for E_{F} for emission from W and Pt emitters in the same setup. Three series of TEDs were measured through the probe hole. TEDs from the brightest FE zone (series 1) at various I_{FE} values are shown in Figure 7.4a. As the voltage and I were increased, the TEDs widened on the low-energy side as expected and also shifted to lower energy because of a resistance drop along the MWNT, and widened on the high-energy side because of significant heating effects. We have fitted the data to

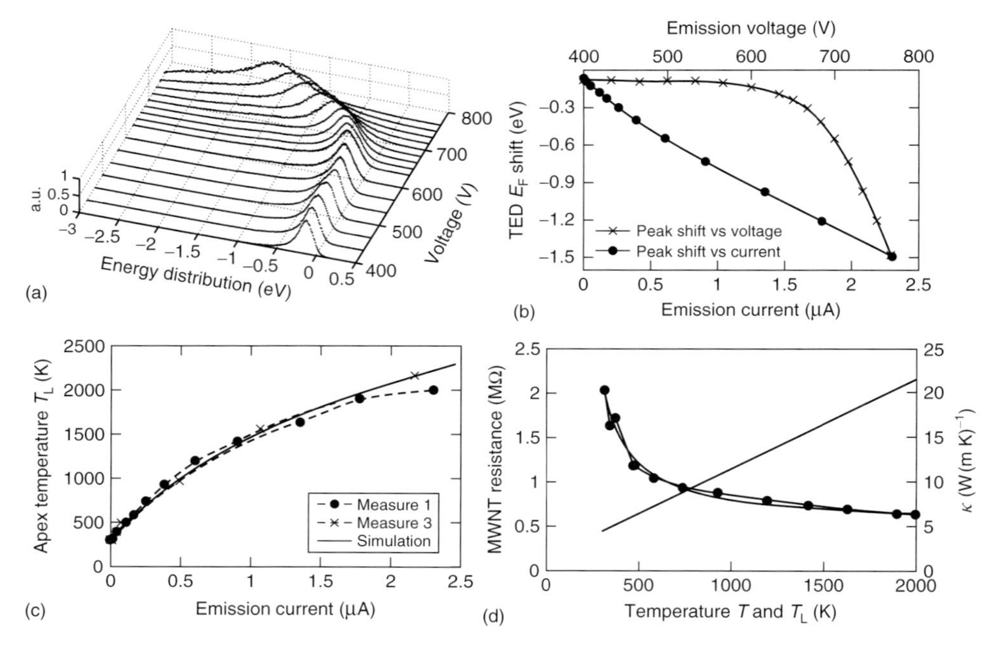

Figure 7.4 (a) Series of TEDs for the MWNT after electron bombardment heating to 1600 K. (b) Shifting of the TEDs with current and voltage showing that the displacement is due to an IR drop. (c) Temperature at the MWNT cap, T_L, measured by fitting the TEDs for two runs. The solid line is a fit to the $T_L(I)$ curve using simulations of the heat equation that determines $\rho(T)$ and $\kappa(T)$. (d) $R(T_L) \equiv E_F/I$ from fitting the TEDs. $\kappa(T)$ is found through self-consistent simulations with Eq. (7.1) ($A = 314\,\text{nm}^2$ and $L = 40\,\mu\text{m}$). (With permission from [25].)

Eq. (7.5) to extract the dependence of the parameters E_F, T_L, and d on the voltage and current.

In Figure 7.4b, we show the fits to the measured TEDs with Eq. (7.5) for E_F against applied voltage and current. They show that E_F displaces to lower energy roughly linearly with I because of a resistive IR drop along the MWNT. The shift gives resistances in the megohm range. In Figure 7.4c, we show that T_L increases from 300 K ($I < 1$ nA) to 2000 K ($I = 2.3$ μA). The results for two runs on different emission zones are shown. As a direct consequence of these results, we proposed that heating increases because of Joule heating along the MWNT. The simultaneous direct measurement of temperature and resistance gives the necessary inputs for simulation of the heat diffusion problem [10].

The combination of independent measures of temperature and resistance allows us to determine the nanotube resistance $R(T_L) \equiv E_F/I_{FE}$, which was in the megohm range. It decreased as T_L increased by ~70% (Figure 7.4d), showing that this MWNT did not have metallic behavior. This $R(T_L)$ was used by all those who have made simulations of the FE current-induced heating problem in

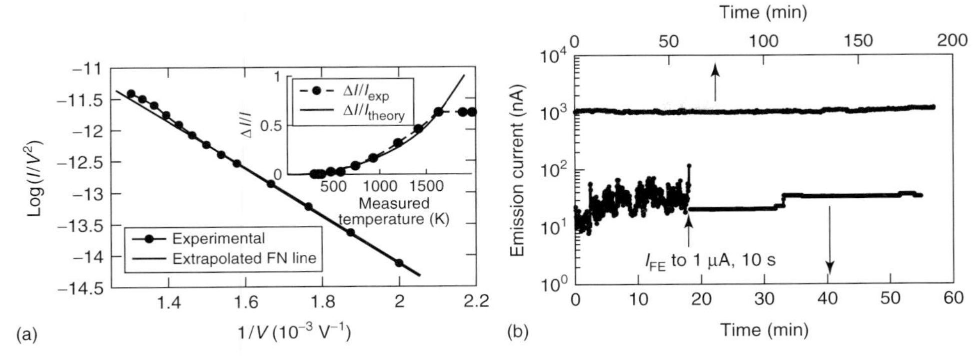

Figure 7.5 (a) FN plot corresponding to Figure 7.4a. The current rises above the FN line because of the current-induced high temperature. Inset: Difference between the measured current and the low current fit to the FN plot as a function of T_L. (b) Current stability measured before and after a current-induced heat flash of 10 s and a 3-h current stability run at 1 μA [25].

CNTs [10, 12, 24, 26, 27]. Also shown is the fit to this data using the heat equation allowing for variable electrical and thermal conductivity. $R(T_L)$ (and hence $\rho(T)$) is an exponentially decreasing function, and $\kappa(T)$ increases linearly with T (Figure 7.4d). The average values are both in the range of those for disordered graphite [29], which can be understood from the TEM images showing many defects. The length and diameter of the MWNT are rather approximate, so the absolute values are crude. However, the ratio $\rho(T)/\kappa(T)$ and the general temperature dependence do not vary strongly with the nanotube dimensions and thus these simulations provide a proof of concept for the technique.

One of the consequences of heating is that the current rises above the FN line at high currents and temperatures above ~1000 K [9, 21, 24], as shown in Figure 7.5a. This is the direct consequence of the increase of $I \alpha T^2$ in FN theory [16], which we used above. In the inset in Figure 7.5a, we plot the difference between the FN fit at low voltage and the measured data against the measured T_L. The fit to the T^2 law is rather good except for the last two points which mark the temperature at which the MWNT undergoes length reduction by partial destruction.

A final consequence of the heating that we treat here is an increase in FE current stability. In Figure 7.5b, we show that the current-induced heating can be used to thermally remove the adsorbates that accumulate by exposure to a poor vacuum. This was demonstrated by stopping the pumping to allow the vacuum to degrade to the 10^{-7} Torr range, leading to an extremely unstable current (Figure 7.5b). After a 10-s flash of 1 μA FE emission, which according to the TEDs raised the temperature to ~1000 K, I became as stable as before. I is even more stable at higher current because the hot nanotube prevents readsorption, as is well known for Schottky emitters. Figure 7.5b shows an almost perfect stability obtained at 1 μA during 3 h. The treatment at 1 μA thus has an effect comparable to a flash surface cleaning, but without any external heat source.

7.5 Conclusion

In this article, we have tried to bring out the advances in our understanding of temperature effects in FE from nanotubes and point out where more work is needed. For the cleaned emitters, what distinguishes CNTs is their ability to function over long times in a condition of current-induced high temperature. This is due to the intrinsic decrease in the MWNT resistance with temperature and the low surface diffusion of carbon which prevents them from falling immediately into a current runaway and explosive breakdown common to metal emitters. The resistive heating of the nanotube is a tool to preferentially clean the principal emitters in an ensemble without external heating. This may be the reason why such excellent stability has been achieved in CNT flat screens where the vacuum conditions are far from UHV. The gradual length reduction at higher currents provides a tool for more uniform emission from a multinanotube emitter at the price of higher extraction voltage. This is useful in applications such as displays where uniformity is critical.

Modeling of the heat transport problem is essential for exploiting the FEES data to extract simultaneous estimates of the physical parameters of CNTs. It permitted the estimation of $\rho(T)$ and $\kappa(T)$ together for the first time. Measurements on different types of CNTs with better characterization of their crystalline structure and dimensions are now needed for more quantitative calculations. The theory can then be extended to include advanced models of the Nottingham effects and radiation cooling.

In conclusion, the combination of $\rho(T)$ and $\kappa(T)$ estimates, the optical emissions, and the *in situ* excitation of mechanical resonances [36] now gives FE access to four fundamental characteristics of a single nanotube or nanowire.

References

1. Rinzler, A.G., Hahner, J.H., Nikolaev, P., Lou, L., Kim, S.G., Tomanek, D., Nordlander, P., Colbert, D.T., and Smalley, R.E. (1995) Unraveling nanotubes: field emission from an atomic wire. *Science*, **269**, 1550–1553.
2. de Heer, W., Châtelain, A., and Ugarte, D. (1995) A carbon nanotube field emission electron source. *Science*, **270**, 1179–1180.
3. Chernozatonskii, L.A., Gulyaev, Yu.V., Kosakovskaja, Z.Ja., Sinitsyn, N.I., Torgashov, G.V., Zakharchenko, Yu.F., Fedorov, E.A., and Val'chuk, V.P. (1995) Electron field emission from nanofilament carbon films. *Chem. Phys. Lett.*, **233**, 63–68.
4. Fransen, M.J., van Rooy, T.L., and Kruit, P. (1999) Field emission energy distributions from individual multiwalled carbon nanotubes. *Appl. Surf. Sci.*, **146**, 312–327.
5. Lovall, D., Buss, M., Graugnard, E., Andres, R.P., and Reifenberger, R. (2000) Electron emission and structural characterization of a rope of single-walled carbon nanotubes. *Phys. Rev. B*, **61**, 5683–5691.
6. Bonard, J.-M., Kind, H., Stöckli, T., and Nilsson, L.-O. (2001) Field emission from carbon nanotubes: the first five years. *Solid-State Electron.*, **45**, 893–914.
7. de Jonge, N. and Bonard, J.-M. (2004) Carbon nanotube electron sources

and applications. *Philos. Trans. R. Soc. London, Ser. A*, **362**, 2239–2266.
8. Minoux, E., Groening, O., Teo, K.B.K., Dalal, S.H., Gangloff, L., Schnell, J.-P., Hudanski, L., Bu, I.Y.Y., Vincent, P., Legagneux, P., Amaratunga, G.A.J., and Milne, W.I. (2005) Achieving high-current carbon nanotube emitters. *Nano Lett.*, **5**, 2135–2138.
9. Purcell, S.T., Vincent, P., Journet, C., and Binh, V.T. (2002) Hot nanotubes: stable heating of individual multiwall carbon nanotubes to 2000 K induced by the field-emission current. *Phys. Rev. Lett.*, **88**, 105502.
10. Vincent, P., Purcell, S.T., Journet, C., and Binh, V.T. (2002) Modelization of resistive heating of carbon nanotubes during field emission. *Phys. Rev. B*, **66**, 075406.
11. Bonard, J.-M., Klinke, C., Dean, K.A., and Coll, B.F. (2003) Degradation and failure of carbon nanotube field emitters. *Phys. Rev. B*, **67**, 115406.
12. Huang, N.Y., She, J.C., Chen, Jun Deng, S.Z., Xu, N.S., Bishop, H., Huq, S.E., Wang, L., Zhong, D.Y., Wang, E.G., and Chen, D.M. (2004) Mechanism responsible for initiating carbon nanotube vacuum breakdown. *Phys. Rev. Lett.*, **93**, 075501.
13. Purcell, S.T., Vincent, P., and Journet, C. (2006) Measuring the physical properties of nanostructures and nanowires by field emission. *Europhys. News*, **37** (4), 26–28.
14. Dyke, W.P., Trolan, J.K., Martin, E.E., and Barbour, J.P. (1953) The field emission initiated vacuum arc. I. experiments on arc initiation. *Phys. Rev.*, **91**, 1043–1054.
15. Dolan, W.W., Dyke, W.P., and Trolan, J.K. (1953) The field emission initiated vacuum arc. II. The resistively heated emitter. *Phys. Rev.*, **91**, 1054–1057.
16. Swanson, L.W. and Bell, A.E. (1973) *Advances in Electronics, Electron Physics*, vol. **32** (ed. L. Marton), Academic Press, New York.
17. Gadzuk, J.W. and Plummer, E.W. (1973) Field emission energy distribution (FEED). *Rev. Mod. Phys.*, **45**, 487–548.
18. Bonard, J.-M., Salvetat, J.P., Stöckli, T., Forro, L., and Chatelain, A. (1999) Field emission from carbon nanotubes: perspectives for applications and clues to the emission mechanism. *Appl. Phys. A*, **69**, 1–10.
19. Dean, K.A., von Allmen, P., and Chalamala, B.R. (1999) Three behavioral states observed in field emission from single-walled carbon nanotubes. *J. Vac. Sci. Technol., B*, **17**, 1959–1969.
20. Wang, Z.L., Poncharal, P., and de Heer, W.A. (2000) Nanomeasurements in transmission electron microscopy. *Microsc. Microanal.*, **6**, 224–230.
21. Dean, K.A., Burgin, T.P., and Chalamala, B.R. (2001) Evaporation of carbon nanotubes during electron field emission. *Appl. Phys. Lett.*, **79**, 1873–1875.
22. Marchand, M., Journet, C., Adessi, C., and Purcell, S.T. in preparation.
23. Sveningsson, M., Jonsson, M., Nerushev, O.A., Rohmund, F., and Campbell, E.E.B. (2002) Blackbody radiation from resistively heated multiwalled carbon nanotubes during field emission. *Appl. Phys. Lett.*, **81**, 1095–1097.
24. Sveningsson, M., Hansen, K., Svensson, K., Olsson, E., Jonsson, M., and Campbell, E.E.B. (2005) Quantifying temperature-enhanced electron field emission from individual carbon nanotubes. *Phys. Rev. B*, **72**, 085429.
25. Purcell, S.T., Vincent, P., Rodriguez, M., Journet, C., Vignoli, S., Guillot, D., and Ayari, A. (2006) Evolution of the field-emission properties of individual multiwalled carbon nanotubes submitted to temperature and field treatments. *Chem. Vap. Deposition*, **12**, 331–344.
26. Wei, W., Liu, Y., Wei, Y., Jiang, K.L., Peng, L.-M., and Fan, S.S. (2007) Tip cooling effect and failure mechanism of field-emitting carbon nanotubes. *Nanoletters*, **7** (1), 64–68.
27. (a) Sanchez, J.A., Mengüç, M.P., and Hii, K.-F. Heat transfer within carbon nanotubes during electron field emission, *J. Thermophys Heat Transf.* **22**, 2008, 281–289; (b) Sanchez, J.A. and Mengüç, M.P. (2008) Geometry dependence of the electrostatic and thermal response of a carbon nanotube during field emission. *Nanotechnology*, **19**, 075702.

28. Bohren, C.F. and Huffman, D.R. (1983) *Absorption and Scattering of Light by Small Particles*, John Wiley and Sons, Inc., New York.
29. Kelly, B.T. (1981) *Physics of Graphite*, Applied Science, London.
30. Vincent, P. (2002) Synthèse, caractérisation, et études des propriétés d'émission de champ de nanotubes de carbone, thesis (in french), Université Claude Bernard Lyon, 1.
31. Wang, M.S., Wang, J.Y., and Peng, L.-M. (2006) Engineering the cap structure of individual carbon nanotubes and corresponding electron field emission characteristics. *Appl. Phys. Lett.*, **88**, 243108.
32. Wang, M.S., Qing, Chen., and Peng, L.-M. (2008) Field emission characteristics of individual carbon nanotubes with a conical tip: the validity of the F-N theory and maximum emission current. *Small*, **4** (11), 1907–1912.
33. Jensen, K., Weldon, J., Garcia, H., and Zettl, A. (2007) Nanotube Radio. *Nano Lett.*, **7** (11), 3508–3511.
34. Swanson, L.W. and Crouser, L.C. (1970) Effect of polyatomic adsorbates on total energy distribution of field emitted electrons. *Surf. Sci.*, **23**, 1–29.
35. Duke, C.B. and Alferieff, M.E. (1967) Field emission through atoms adsorbed on a metal surface. *J. Chem. Phys.*, **46**, 923–937.
36. Purcell, S.T., Vincent, P., Journet, C., and Binh, V.T. (2002) Tuning of nanotube mechanical resonances by electric field pulling. *Phys. Rev. Lett.*, **89**, 276103.

8
Field Emission Microscopy of Multiwall CNTs

Yahachi Saito

8.1
Introduction

Field emission microscopy (FEM) enables imaging of the spatial distribution of the emitted electron current from an electron emitter by using a phosphor screen as anode. The emitter is usually mounted on a heating wire for cleaning the emitter surface. The electron emission is strongly affected by surface structures, local states, and the presence of adsorbates, and is reflected in the FEM patterns. Thus, FEM provides important information on the emission mechanism and surface phenomena on a carbon nanotube (CNT) [1, 2].

CNTs are one of the ideal materials as field emitters because they possess (i) needlelike shape with a sharp tip, (ii) high chemical stability, (iii) high mechanical strength, (iv) low carbon atom mobility, and (v) high electrical and thermal conductivity. The needlelike morphology with an extremely small radius of curvature at the tip is the most prominent advantage of CNTs as an electron emitter. When an electric field is applied to a conductor with a sharp tip, the field concentrates at the sharp point. The field strength at the tip surface is inversely proportional to the radius of curvature r of the tip [3]. The surface of CNTs is inert and stable against residual gas molecules in a vacuum vessel because of the chemical stability of graphite material which constitutes CNTs. The high mechanical strength (tensile strength $\sim$100 GPa [4]) of CNT emitters enables them to endure the high stress caused by electrostatic forces (Maxwell tension). Together with this robustness, the low mobility of carbon atoms in CNTs helps them retain their original shape even under a high electric field. Finally, since CNTs, especially multiwall carbon nanotubes (MWNTs) formed by arc discharge, have high electrical and thermal conductivity, CNTs can transport and emit electrons at high current density (about 10^7 A cm^{-2}) through their tubular walls [5].

This chapter focuses on FEM studies carried out mainly on MWNTs so far. In addition to fundamental properties of CNT field emitters, the molecular images and dynamics of adsorbates on CNT electron emitters revealed by FEM are presented. FEM on single-wall carbon nanotubes (SWNTs) is dealt with in Chapter 10.

Carbon Nanotube and Related Field Emitters: Fundamentals and Applications. Edited by Yahachi Saito

ISBN: 978-3-527-32734-8

8.2
FEM of Carbon Nanotubes

8.2.1
FEM Measurement

For FEM studies, employment of a single, isolated CNT fixed to the apex of a metal needle may be ideal. A few techniques to fabricate such pointed emitters are described in Chapter 2. However, the single CNT emitters prepared under a scanning electron microscope (SEM) [6] are usually covered with contaminants (hydrocarbon deposit) during the SEM observation. Electrophoresis is an alternative method to attach a thin bundle of CNTs to the tip of a metal needle [7], though the control of the number of CNTs in an attached bundle is difficult. The simple and easy method is to glue a bulk bundle of as-grown CNTs to the tip of a heating loop (e.g., a tungsten filament with diameter of 0.15 mm) by using a conductive paste [8]. The last method keeps the tips of the CNTs clean, but an enormous number of CNTs protrude out at the end of the bundle.

A schematic of an FEM apparatus is shown in Figure 8.1. The emitter tip of the CNTs is placed at about 30 mm distance from a phosphor screen, on which field emission (FE) patterns are observed. The base pressure of the FEM vacuum chamber is typically 10^{-7}–10^{-8} Pa. A negative voltage of 0.6–1.6 kV is applied to the emitter relative to the screen.

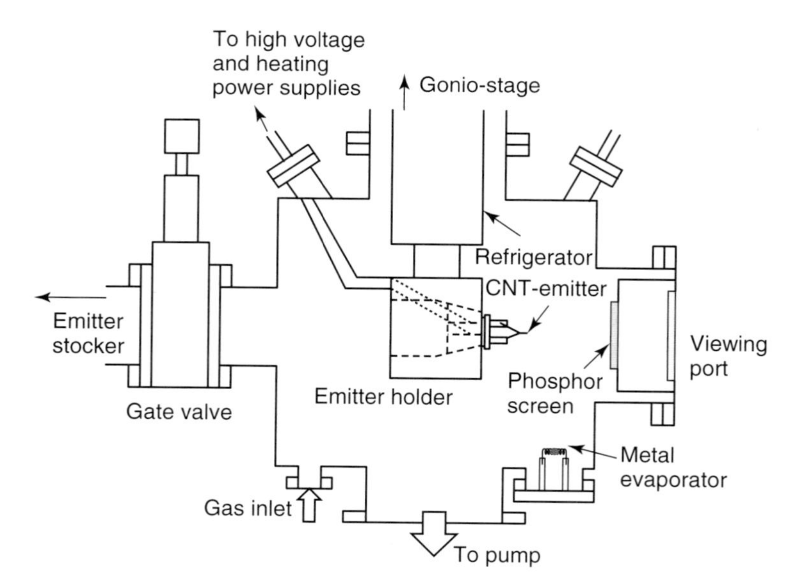

Figure 8.1 Schematic of an FEM apparatus.

8.2.2 MWNTs with Clean Surfaces

MWNTs produced by the arc-discharge technique (Chapter 1) are highly graphitized (i.e., composed of well-developed graphene layers) and thus have high structural perfection. The ends of arc-grown MWNTs are capped by graphite layers with polyhedral shapes (Chapter 1). In order to give a positive curvature to a hexagonal sheet, pentagons (five-membered carbon ring) have to be introduced to the sheet; six pentagons are required to have a curvature of 2π steradians (i.e., a hemispherical cap). The portion where a pentagon is located extrudes like the vertex of a polyhedron, while the other flat regions are made of hexagons.

Typical FEM images of MWNT emitters with clean surfaces are shown in Figure 8.2. Clean CNT tips are obtained by heating at about 1000 °C for a few minutes in an ultrahigh vacuum chamber (e.g., 10^{-8} Pa) during which the

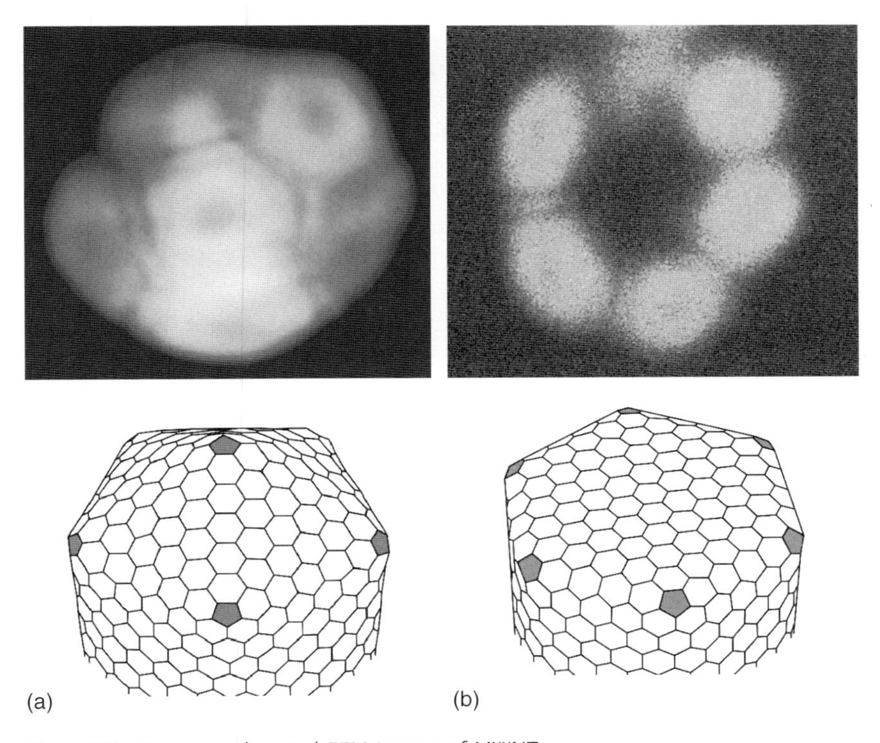

Figure 8.2 Upper panel: typical FEM images of MWNT emitters with clean surfaces. Six pentagonal rings are arranged in fivefold (a) and sixfold symmetry (b). Lower panel: structural models of CNT tips giving the FEM patterns in the upper panel.

adsorbates on CNT surfaces desorb [9]. Six pentagonal rings arranged in fivefold (Figure 8.2a) and sixfold symmetry (Figure 8.2b) can be observed. Each pentagonal region contains a small dark spot at its center. It should also be noted that interference fringes are observed between the neighboring pentagons. Structural models of CNT tips that would give the FEM patterns are shown in the lower panel in Figure 8.2.

Since the pentagons are locally extruding like vertices, the electric field around them would be stronger than that on other flat regions. In addition, it is theoretically indicated that the pentagon site has a higher density of states (DOS) of electrons near the Fermi level E_F than the normal hexagon site [10]. Therefore, the electron tunneling through the pentagons are expected to occur dominantly.

Such patterns (called *pentagon patterns* hereafter) change when gas molecules adsorb on the CNT cap. An adsorbed molecule is usually imaged as a bright spot in the FEM picture, giving rise to an abrupt increase in the emission current [9]. Similar stepwise fluctuations of emission current are also frequently observed in FE sources made of other materials [11, 12]. The origin of the stepwise changes is the adsorption and desorption of molecules on the surface of the emitter. Though most adsorbed molecules in the FEM images appear as simple, bright spots (structureless), some molecules exhibit characteristic shapes reflecting the molecular structure as described in the next section.

Emission patterns from open-ended MWNTs, which were prepared by the oxidation processes, showed bright "doughnutlike" annular rings, reflecting the geometry of the CNT tip [13].

8.2.3
FEM Patterns Depending on Tip Radius

FEM of MWNTs with closed caps shows clear pentagon images, but that of SWNTs, as described in Chapter 10, does not show pentagonal rings but dim (blurred) patterns which resemble scanning tunneling microscope (STM) images of C_{60} fullerenes. Experimental study using CNTs with different apex radii [14] suggested that the difference in the FEM images originates from the difference in the radius of curvature of CNT tips; the "pentagon" patterns are observed for CNTs with apex radii larger than about 2 nm, whereas the "dim" patterns correspond to smaller apex radii.

According to the argument on the spatial resolution of FEM [15, 16], resolutions of 0.2 and 0.35 nm are possible for emitters with tip radius of 1 and 4 nm, respectively. However, it is not enough to resolve individual atoms on the CNT caps. Since the pentagon–pentagon separation s_{p-p} on a CNT cap is roughly the same as the radius of curvature of the tip [14], the s_{p-p} which differentiates the patterns is presumed to be approximately 2 nm.

8.3 Field Emission from Adsorbates on an MWNT

8.3.1 Molecules

8.3.1.1 Hydrogen

Figure 8.3 shows a time-sequential series of FEM patterns from an MWNT exposed to hydrogen gas at a pressure of 1.3×10^{-6} Pa and the corresponding changes in the emission current [17]. The pentagon pattern characteristic of a clean MWNT cap just after flashing (Figure 8.3a) changed to FEM patterns in which one or two small bright spots appeared on the pentagon pattern as shown in Figure 8.3b and c, and a slight increase in the emission current occurred concurrently with the appearance of a bright spot. The number and the position of bright spots changed randomly, indicating frequent adsorption and desorption of hydrogen molecules on the CNT cap preferentially on pentagon sites where the electric field is locally the strongest. After the FE measurement for 11 min with an emission current of 50–100 nA, the hydrogen gas was evacuated and the CNT emitter was subjected to flashing. By this cleaning process, the emission pattern recovered to the original clean pattern, suggesting that hydrogen molecules are inert for the surfaces of MWNTs under this FE condition.

8.3.1.2 Nitrogen

Figure 8.4 shows two types of FEM image of an adsorbate in an atmosphere of nitrogen gas of 8×10^{-7} Pa [18]. A bright spot that appears on a pentagon site changes its brightness and shape. The image is "cocoon" shaped (Figure 8.4a and b)

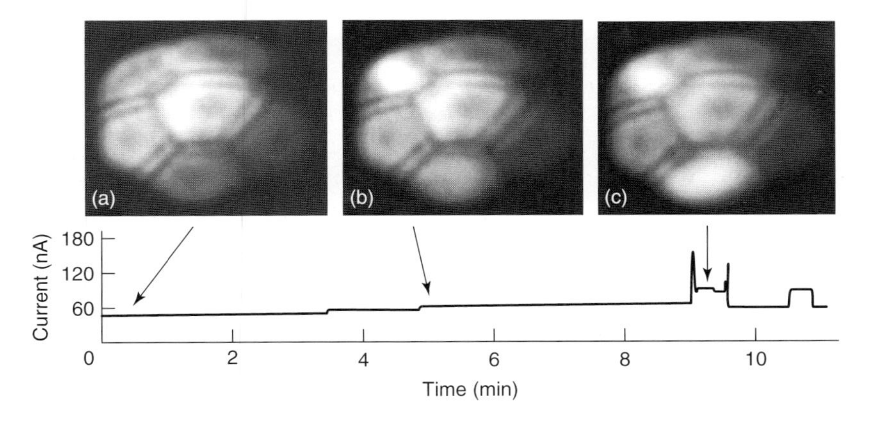

Figure 8.3 Time-sequential series of FEM patterns from an MWNT exposed to hydrogen gas at pressure of 1×10^{-8} Torr and the corresponding changes in the emission current. (Reprinted with permission from K. Hata, A. Takakura, and Y. Saito, *Ultramicroscopy* **95** (2003) 107. Copyright 2003, Elsevier.)

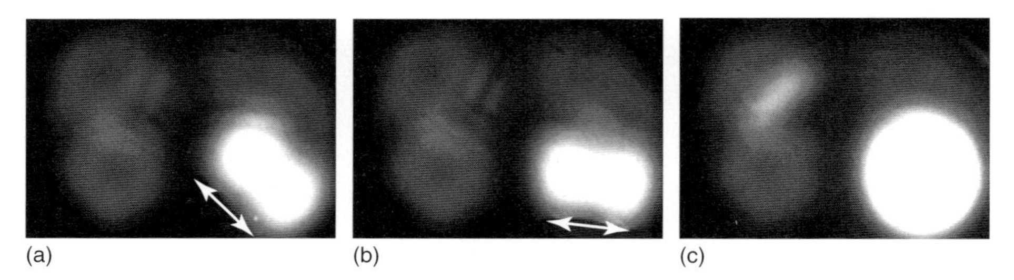

Figure 8.4 Two types of FEM pattern of a nitrogen molecule. (a) and (b) "cocoon"-shaped images with different orientations, and (c) bright circular spot. (Reprinted with permission from S. Waki, K. Hata, H. Sato, and Y. Saito, *J. Vac. Sci. Technol., B* **25** (2007) 517. Copyright 2007, American Vacuum Society.)

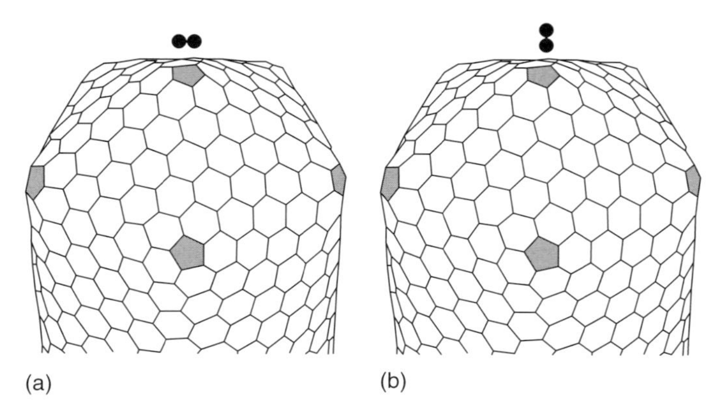

Figure 8.5 Model explaining two different adsorption states of a single nitrogen molecule. The molecular axis (a) parallel and (b) perpendicular to the substrate.

when the current is small, whereas it is a bright circular spot (Figure 8.4c) when the emission current is large. A model explaining two different adsorption states of a single nitrogen molecule is shown in Figure 8.5. If the molecular axis of nitrogen is parallel to the substrate, as illustrated in Figure 8.5a, the cocoon shape reflecting the shape of the molecule (interatomic distance of N_2 is 0.1094 nm) would be expected. When the molecular axis is perpendicular to the substrate (Figure 8.5b), on the other hand, the emission pattern should be a circular bright spot. In the latter configuration, the extended protrusion enhances the field concentration and thus brings about the enhanced emission current, that is, a brighter spot. The perpendicular configuration of the molecule is expected to gain larger adsorption energy than the parallel configuration because a larger polarization force is induced in the perpendicular configuration.

8.3.1.3 Oxygen

Exposure of an MWNT emitter to oxygen (1×10^{-8} Torr) brings about frequent adsorption and desorption phenomena at the pentagon sites [17], which is similar to the case of hydrogen and nitrogen, but the pentagonal rings do not recover after the disappearance of the bright spots. This suggests that pentagons are damaged during electron emission in the oxygen atmosphere. After the FE experiment in the oxygen atmosphere for 10 min, the MWNT emitter was heated (flashing at 1000 °C for 1 minute) again in an ultrahigh vacuum in order to desorb the oxygen molecules. All the pentagons were damaged, and the original pattern was no longer reproduced, exhibiting the very high reactivity of oxygen with the CNT surfaces.

8.3.1.4 Carbon Monoxide

When carbon monoxide was introduced into the FEM chamber up to pressure of 1.3×10^{-6} Pa, large bright spots corresponding to adsorption of the molecules appeared [17]. Even though the common flash-cleaning was applied to the emitter in ultrahigh vacuum after the electron emission for 600 s, the bright spot on the top pentagon did not disappear, and the clean pentagon pattern was not recovered. This suggests that the carbon monoxide molecule was strongly bonding to the pentagon, or it damaged the pentagon during the FE or flashing process.

8.3.1.5 Carbon Dioxide

FEM images of a single CO_2 molecule adsorbed on a MWNT also exhibited a cocoon-shaped bright spot as in Figure 8.6 [19]. Even though the CO_2 molecule is triatomic, it appears diatomic similar to a nitrogen molecule shown in the previous section. The reason for exhibiting the cocoon shape is probably the electric charge distribution within a CO_2 molecule, in which valence electrons lean toward the outer oxygen atoms from the central carbon atom.

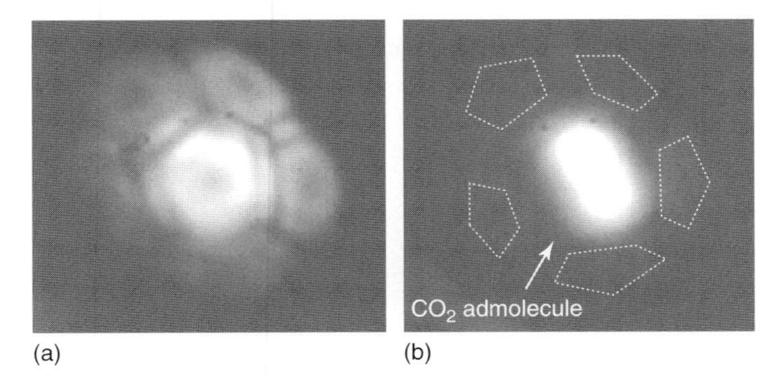

Figure 8.6 FEM images of (a) a clean MWNT tip and (b) a single CO_2 adsorbed on it. The CO_2 molecule exhibits a cocoon shape whose orientation changes randomly among the five discrete directions. (Reprinted with permission from Y. Kishimoto and K. Hata, *Surf. Interface Anal.* **40** (2008) 1669. Copyright 2008, Wiley.)

Though the CO_2 admolecule in Figure 8.6 moved randomly and discretely on the substrate pentagon with a time interval on the order of 100 ms, the orientation of the long axis of the admolecule is found to be arranged statistically equally into five groups, suggesting the presence of five equivalent, stable adsorption sites (orientations) for a CO_2 molecule on the pentagon. The angles between the molecular axes of the adjacent orientations are on average 36° with a slight deviation of only a few degrees, which is in good agreement with the angle expected for the symmetry of the pentagon. The length of the diagonal lines for the carbon pentagon with a side length of 0.142 nm, that is, C–C bond length of graphite, is 0.230 nm. This length is in good agreement with the distance between the two end oxygen atoms in the CO_2 molecule, that is, 0.233 nm. Details of the adsorption sites and the motion of CO_2 on the carbon pentagon are discussed in [19], in which the rotation angle of 72°, instead of 36°, is estimated from an analysis of the motion in the video file from frame to frame (time interval of 1/30 s) and also the rotation of the molecules around the central carbon atom is suggested.

8.3.1.6 Methane

Investigation of the effect of electric field on methane adsorption has revealed that methane adsorption occurs only when a negative electric voltage is applied to the emitter (i.e., being biased to emit electrons), whereas no methane adsorption is observed when a positive voltage or no voltage is applied in an atmosphere of 1.0×10^{-7} Pa methane gas [20].

FEM images of adsorbed methane molecules are usually simple, bright spots like in the case of inorganic molecules such as H_2, O_2, and CO mentioned above. Occasionally, however, a cross-shaped image is observed as shown in Figure 8.7. Since a CH_4 molecule has the tetrahedral structure, it looks like a "cross" when its twofold symmetry axis is normal to the substrate. The size of the cross image is roughly measured to be 0.23 nm on the basis of the size of a carbon pentagon, which was observed under the admolecule. Compared with the size of CH_4 (0.21 nm, the distance between two hydrogen atoms of a methane molecule), the FEM gives

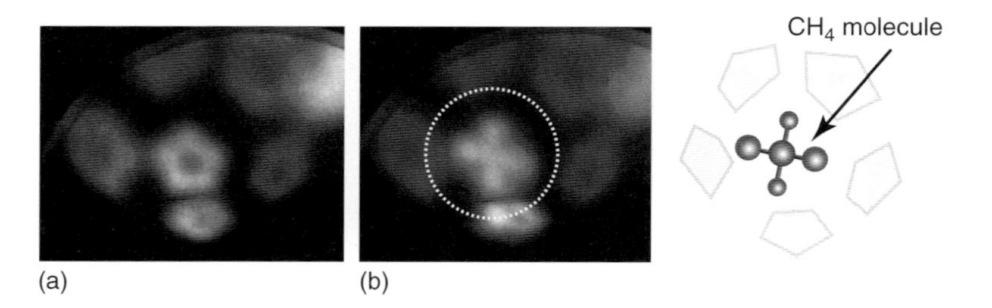

Figure 8.7 FEM images of (a) before and (b) after adsorption of a methane molecule on a pentagon at the MWNT tip. A cross-shaped image, reminiscent of a CH_4 molecule looked along the twofold rotational symmetry axis, is observed in (b).

a little larger image than the real size. This is presumably due to the enhanced magnification of a small protrusion on the round emitter surface [15]. From the shape and the size of the image, we may assume that the pattern corresponds to a single molecule of methane.

8.3.1.7 Comparison with Related Theoretical Studies

Theoretical calculations of FEM images of clean MWNT tips have been reported by Han and Ihm [21] and Khazaei *et al.* [22]. A brief explanation of the theories is given in Chapter 4. However, any theoretical studies that can be compared with FEM images of admolecules presented in this article or explain them are not available, whereas the variations of emission current due to adsorption of some molecules are discussed theoretically. According to Wadhawan *et al.* [23], adsorption of common electronegative gases, for example, O_2 and NH_3, can decrease the current, and inert gases such as He and Ar hardly affect the FE current and its stability. This theoretical prediction contradicts the experimental observations at least for the O_2 adsorption. Park *et al.* [24] ascribed the experimentally observed emission increase by oxygen adsorption to the local enhancement of electric field and the creation of new electronic states.

Li and Wang [25] predicted that the adsorption of CO and CH_4 decrease the emission current because of an increase in the work function. Sheng *et al.* [26], on the other hand, predicted that CO and CO_2 decrease the current, but CH_4 increases it. Contradictions between the theoretical works themselves are found. Experimentally, all kinds of molecules have been shown to bring about emission enhancement upon their adsorption on a CNT.

8.3.2 Aluminum Clusters

Figure 8.8a shows a transmission electron microscope (TEM) picture of Al with a mean thickness of 2.5 nm deposited on MWNTs before the FE experiment. The deposited Al formed a discontinuous film consisting of isolated islands with the diameter of a few nanometers. After the FE experiment, diameter of the Al clusters increased to about 10 nm as revealed in Figure 8.8b.

During the study on the effect of Al deposition on FE properties, intriguing FEM images suggestive of an Al cluster with atomic resolution were observed [27]. Figure 8.9a and b shows the FEM images of an MWNT emitter before and after Al deposition, respectively. A spotty pattern with high symmetry (fourfold symmetry in this case) appeared on the pentagon patterns characteristic of the clean caps of MWNTs (two MWNTs are visible in Figure 8.9a) after the Al deposition, as shown in Figure 8.9b. The contrast of the spotty pattern is reminiscent of the structure of an atom cluster with the shape of cubo-octahedron, which is a crystal form characteristic of face-centered cubic (fcc) metals [28]. A model of the structure consisting of 38 Al atoms is illustrated in Figure 8.10. The fourfold symmetry of the Al image suggests that the Al cluster is oriented with its [100] direction normal to the nanotube surface. Four bright spots observed in the central part of the Al

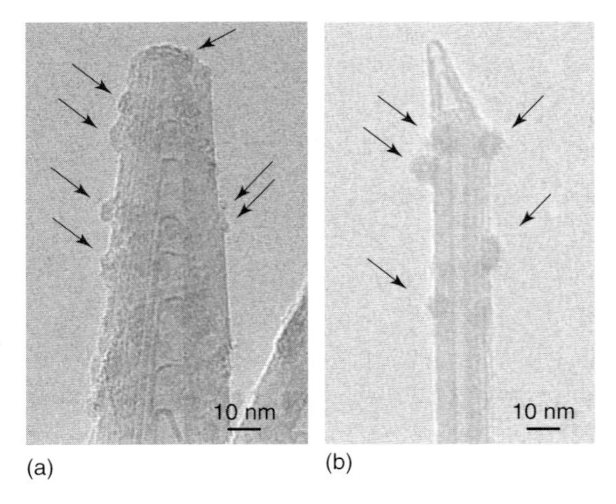

Figure 8.8 TEM images of Al-deposited MWNTs (a) before and (b) after field emission. Mean thickness of the deposited Al is 2.5 nm. Arrows indicate Al clusters. Different MWCNTs are shown in (a) and (b).

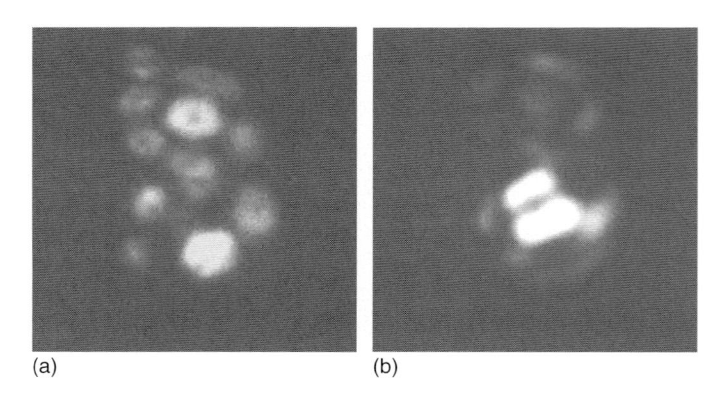

Figure 8.9 FEM images of (a) clean MWNT caps and (b) an Al cluster on a MWNT tip.

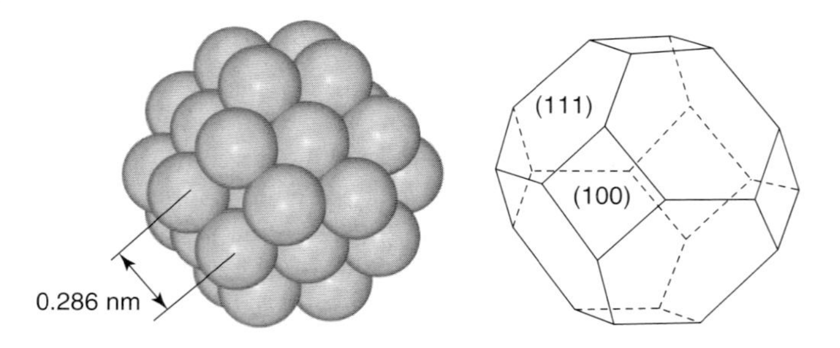

Figure 8.10 Cubo-octahedron of an Al_{38} cluster.

image correspond to the four corners of the top (100) surface. Four dark regions surrounding the central (100) face correspond to (111) faces, which are outlined by bright edges and corners.

The distance between neighboring atoms along the edge of the (100) surface is 0.286 nm when the lattice constant of the cluster is the same as that of bulk Al. Using the size of the carbon pentagon (approximately 0.25 nm in diameter) as a measure of the magnification of FEM images (under the assumption that the pentagon image originates from five carbon atoms comprising a pentagon), the distance between the bright spots at the corners of the (100) face is roughly estimated to be in the range of 0.30 nm, which is a little larger than the nearest neighbor distance on the Al (100) surface due to the local magnification enhancement by a small protrusion on a round tip.

Metal clusters or nanoparticles often exhibit atomic structures different from crystal structures in bulk, for example, icosahedral or multiply twinned structures for elements that form fcc structures in bulk. For Al, however, icosahedral structures have never been observed even for small particles by electron microscopy [28]. Theoretical calculations also suggest that the structural transition from the fcc to the icosahedral structures lies in a range of size between 13 and 55 atom clusters [29]. The present Al cluster falls in this transition range in size. Thus, it is highly probable that the Al cluster exhibits the same structure as the bulk.

The polyhedral Al cluster (Figure 8.9b), which exhibits rotation and migration, disappeared in several seconds from the field of view after its appearance, and finally the original clean cap was recovered. The migration and diffusion of Al clusters on MWNTs are responsible for the increased diameter of Al clusters observed by TEM after the FE experiment, as shown in Figure 8.8b.

8.4 Resolution in FEM and Possible Observation of Atomic Detail

In 1956, Rose [15] gave the equation of FEM resolution δ, which consists of two principal components, namely, the momentum uncertainty and the effect of the transverse velocities of the electrons near the top of Fermi level in the emitter:

$$\delta = \left(\frac{2\hbar\tau}{mM}\right)^{1/2}\left(1 + \frac{2m\tau v_0^2}{\hbar M}\right)^{1/2} \tag{8.1}$$

where M is the magnification, τ is the time of flight of an electron from the emission tip to the screen, v_0 is the average transverse velocity, $\hbar$ is the Planck constant/2π, and m is the electron mass. When M/τ is large enough to assume $2m\tau v_0^2/\hbar M \ll 1$, the term containing v_0 becomes negligible and the resolution is limited by the uncertainty principle. Under such a condition, say $M/\tau \approx 2.5 \times 10^{15}$, he suggested that small protrusions on the surface of the tip can provide resolutions on the order of 0.3 nm so that some of their atomic detail should be observable. M is always reduced by a factor β from that expected for a spherically symmetric geometry where the tip and screen are assumed to be concentric spheres of radii R and z, that is,

$$M = z/\beta R \tag{8.2}$$

Using the approximation $\beta \approx 1.9$, $\tau \approx z(2eV/m)^{-1/2}$, and $v_0 \approx 2 \times 10^5 \text{ms}^{-1}$ ($= 0.11$ eV), the following practical form of solution equation [15, 16] is obtained:

$$\delta = 0.860 \left(\frac{R}{\sqrt{V}}\right)^{\frac{1}{2}} \left(\frac{1 + 2.22R}{\sqrt{V}}\right)^{\frac{1}{2}} \tag{8.3}$$

where δ is in nanometers, R is the tip radius in nanometers, and V is the applied potential in volts between the tip and the screen.

From Eq. (8.3), we see that atomic resolution is attainable for $R/\sqrt{V} \prec 1$. In an experiment using the MWNT as the emitter, R and V are about 5 nm and 1.5 kV, respectively. These parameters give a resolution on the order of 0.3 nm, indicating that some of atomic detail is observable in the present experimental condition.

8.5
Concluding Remarks

CNTs possess unique structural and physicochemical properties distinct from traditional metal emitters (e.g., tungsten and molybdenum): extremely small tip radius (1–10 nm); well-defined, stable surface structures (composed of carbon hexagons and pentagons) made of strong C–C bonds; no oxide formation and no surface diffusion of carbon atoms. This chapter reviews recent FEM studies that suggest near-atomic resolution images of molecules and metal clusters adsorbed CNT emitters. The high resolution is probably due to the small tip radius of CNT emitters as suggested by the Rose's estimation of resolution [15] – though it is rather old. Appearance of pentagons in FEM images as a measure of magnification of the images is another merit of CNT emitters. Adsorbed molecules shown in this chapter are common, small molecules such as N_2 and CO. When metal (W or Mo) needles were employed as the FEM emitter, such molecular images reflecting their structures were never observed because of the presumable reactions between the admolecule and the metal surface. Chemical inertness of CNT surfaces is responsible for the stable observation of these small molecules.

In the 1950s, there was controversy as to whether objects of atomic dimensions can be resolved by FEM. The most interesting and yet controversial FEM patterns are quadruplet or doublet patterns originating from organic dye molecules such as phthalocyanine or flavanthrene reported first by Müller [30]. The advent of CNTs as field emitters will revive the discussion on FEM resolution and open a new scene in the FEM technique for direct observation of adatoms and admolecules. For developing the FEM technique, a realistic theory applicable to CNT emitters is highly required since the simple FN theory is inadequate for more sophisticated analyses of FEM observations.

References

1. Gadzuk, J.W. and Plummer, E.W. (1973) *Rev. Mod. Phys.*, **45**, 487.
2. Melmed, A.J. and Müller, E. (1958) *J. Chem. Phys.*, **29**, 1037.
3. Gomer, R. (1961) *Field Emission and Field Ionization*, Harvard University Press, Cambridge.
4. Demczyk, B.G., Wang, Y.M., Cumings, J., Hetman, M., Han, W., Zettl, A., and Ritchie, R.O. (2002) *Mater. Sci. Eng. A*, **334**, 173.
5. Wei, B.Q., Vajtai, R., and Ajayan, P.M. (2001) *Appl. Phys. Lett.*, **79**, 1172.
6. Nakayama, Y., Nishijima, H., Akita, S., Hohmura, K.I., Yosimura, S.H., and Takeyasu, K. (2000) *J. Vac. Sci. Technol., B*, **18**, 611.
7. Saito, Y., Seko, K., and Kinoshita, J. (2005) *Diamond Relat. Mater.*, **14**, 1843.
8. Saito, Y., Hamaguchi, K., Hata, K., Tohji, K., Kasuya, A., Nishina, Y., Uchida, K., Tasaka, Y., Ikazaki, F., and Yumura, M. (1998) *Ultramicroscopy*, **73**, 1.
9. Hata, K., Takakura, A., and Saito, Y. (2001) *Surf. Sci.*, **490**, 296.
10. Tamura, R. and Tsukada, M. (1995) *Phys. Rev. B*, **52**, 6015.
11. Yamamoto, S., Hosoki, S., Fukuhara, S., and Futamoto, M. (1979) *Surf. Sci.*, **86**, 734.
12. Ishizawa, Y., Aizawa, T., and Otani, S. (1993) *Appl. Surf. Sci.*, **67**, 36.
13. Saito, Y., Hamaguchi, K., Hata, K., Uchida, K., Tasaka, Y., Ikazaki, F., Yumura, M., Kasuya, A., and Nishina, Y. (1997) *Nature*, **389**, 554.
14. Saito, Y., Tsujimoto, Y., Koshio, A., and Kokai, F. (2007) *Appl. Phys. Lett.*, **90**, 213108.
15. Rose, D.R. (1956) *J. Appl. Phys.*, **27**, 215.
16. Brodie, I. (1978) *Surf. Sci.*, **70**, 186.
17. Hata, K., Takakura, A., and Saito, Y. (2003) *Ultramicroscopy*, **95**, 107.
18. Waki, S., Hata, K., Sato, H., and Saito, Y. (2007) *J. Vac. Sci. Technol., B*, **25**, 517.
19. Kishimoto, Y. and Hata, K. (2008) *Surf. Interface Anal.*, **40**, 1669.
20. Yamashita, T., Asaka, K. Nakahara, H., and Saito, Y. (2008) Presented at the 7th International Vacuum Electron Sources Conference (IVESC 2008), Queen Mary, University of London, August 4-6, 2008, London .
21. Han, S. and Ihm, J. (2002) *Phys. Rev. B*, **66**, 241402 (241(R)).
22. Khazaei, M., Farajian, A.A., and Kawazoe, Y. (2005) *Phys. Rev. Lett.*, **95**, 177602.
23. Wadhawan, A., Stallcup, R.E.II, Stephens, K.F.II, Perez, J.M., and Akwani, I.A. (2001) *Appl. Phys. Lett.*, **79**, 1867.
24. Park, N., Han, S., and Ihm, J. (2001) *Phys. Rev. B*, **64**, 125401.
25. Li, Z. and Wang, C.-Y. (2006) *Chem. Phys.*, **330**, 417.
26. Sheng, L.M., Liu, P., Liu, Y.M., Qian, L., Huang, Y.S., Liu, L., and Fan, S.S. (2003) *J. Vac. Sci. Technol., A*, **21**, 1202.
27. Saito, Y., Matsukawa, T., Yamashita, T., Asaka, K., Nakahara, H., and Uemura, S. (2007) Presented at the 14th International Display Workshops (IDW'07), Sapporo Convention Center Sapporo Japan, December 5-7, 2007.
28. Kimoto, K. and Nishida, I. (1977) *Jpn. J. Appl. Phys.*, **16**, 941.
29. Cheng, H.P., Berry, R.S., and Whetten, R.L. (1991) *Phys. Rev. B*, **43**, 10647.
30. Müller, E.W. (1950) *Z. Naturforsh.*, **5a**, 473.

9
In situ Transmission Electron Microscopy of CNT Emitters

Koji Asaka and Yahachi Saito

9.1
Introduction

Since the first experiments demonstrating the excellent electron field emission properties of carbon nanotubes [1, 2], various emitter devices such as flat panel displays, cathode ray tubes, miniature X-ray tubes, microwave devices, and electron sources for electron microscopes have been developed. For the optimization of their performance, it is important to examine directly the behavior of individual nanotube emitters in electric fields and to understand the electron field emission properties since the emission properties depend on the structural and electronic features that are intrinsic to the individual nanotubes as well as configurations between the emitter and a counter electrode. The structural behavior and emission properties of individual multiwall carbon nanotubes (MWNTs) in electric fields have been investigated by scanning electron microscopy (SEM). Wei *et al.* demonstrated that MWNTs with a curved shape, prepared by thermal decomposition of CH_4 and H_2, were oriented parallel to the electric field line under an applied electric field, and then recovered the initial shape after the electric field was removed [3]. A decrease of the length of the MWNTs after field emission at currents of 50–120 nA for 30 min was also observed. Bonard *et al.* observed the degradation and failure of the individual MWNTs after field emission [4]. They also suggested that only a small number of exceptionally long and/or narrow nanotubes contribute to the emission current in large area measurements using nanotube films [5]. However, SEM is not suitable for *in situ* observations of the structural behavior of the individual nanotubes during field emission because the large number of electrons emitted form the nanotubes surpasses the number of signal electrons and saturates the electron detector of the microscope. Alternatively, experimental methods to manipulate the individual nanotubes in a transmission electron microscope and to measure the *in situ* field emission properties with simultaneous imaging have been developed. *In situ* transmission electron microscopy (TEM) possesses the advantage that it is possible to observe the structural dynamics of the individual nanotubes during field emission at a higher spatial resolution. Here, *in situ* TEM studies of the carbon nanotube emitters are described.

Carbon Nanotube and Related Field Emitters: Fundamentals and Applications. Edited by Yahachi Saito

ISBN: 978-3-527-32734-8

9.2
Degradation and Failure of Nanotubes at Large Emission Current Conditions

Pioneering studies of the *in situ* observations of nanotubes in electric fields by TEM have been reported by Wang *et al.* [6, 7]. They observed two types of structural damages: that is, "splitting" and "stripping" of MWNTs under applied high electric field conditions sufficient to extract large emission currents. Figure 9.1 shows a time-sequence series of images of the splitting process of an MWNT during field emission. The diameter of the MWNT, which was produced by the arc-discharge technique, is ~20 nm. The gap from the MWNT tip to the counter electrode used as anode is ~2 μm. The emission current increases from 10 to 250 μA as the applied voltage between the MWNT tip and the anode increases from 80 to 130 V. At 90 V, the emission current increases to 40 μA and the walls of the MWNT split, as shown in Figure 9.1b. They suggested that the mechanism of the damage is the electrostatic force acting on the MWNT tip. The splitting damage is accompanied by an abrupt increase in the emission current during field emission [8, 9]. A stripping structural damage of MWNTs during field emission was observed at a gap width of ~2 μm at applied voltages from 100 to 200 V. The wall of the MWNT was degraded and decreases in diameter and length occurred in the MWNT. During

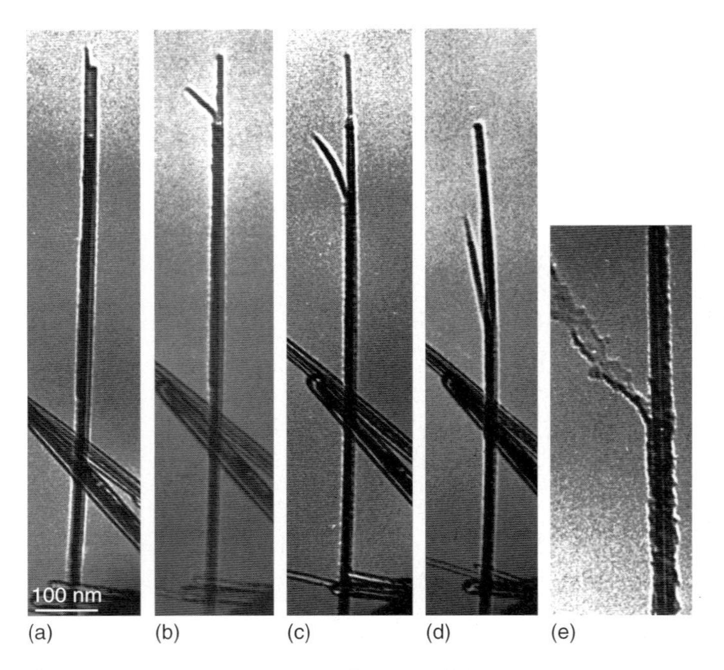

Figure 9.1 Time-sequence series of images of the "splitting" process of an MWNT during field emission. (Reproduced with permission from Z.L. Wang *et al.*, *Appl. Phys. Lett.*, **80**, 856 (2002).)

the degradation process, a decrease of the emission current was also observed [9]. Similar degradation processes of doublewall carbon nanotubes (DWNTs) during field emission were reported by Saito *et al.* [10]. Figure 9.2 shows a time-sequence series of images of the degradation process of DWNT bundles during field emission at applied voltages from 50 to 100 V. In Figure 9.2a, the dark regions at the top and bottom are the surfaces of a tungsten needle attached to the DWNTs and of a copper plate used as anode, respectively. The gap width before the field emission experiment is ~2 μm. The most protruding DWNT bundle A is split at an applied

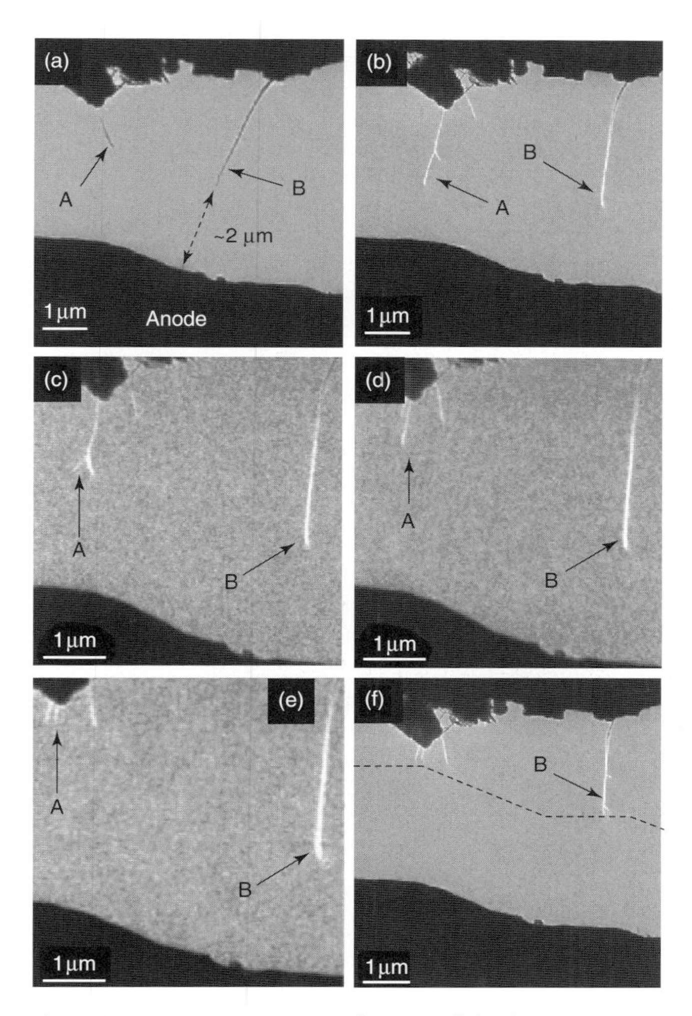

Figure 9.2 Time-sequence series of images of the degradation process of DWNT bundles during field emission. (Reproduced with permission from Y. Saito *et al.*, *Diamond Relat. Mater.*, **14**, 1843 (2005).)

voltage of 60 V (Figure 9.2b) and gradually sublimated between 60 and 85 V, which results in a decrease in the length (Figure 9.2c–e). After the decrease in length of the bundle, the degradation proceeds to the next protruding bundle B. Finally, the degradation ceases when the distances between the bundle tips and the anode surface become uniform (Figure 9.2f). As one cause of the degradation, such as the decrease in length at large emission currents, sublimation by Joule resistive heating was proposed.

Wei *et al.* proposed that the temperature during field emission is highest at the interior rather than the tip of a carbon nanotube on the basis of the one-dimensional heat equation considering the cooling effect due to electron emission and showed by *in situ* TEM that the nanotubes during field emission collapsed at the point close to the highest temperature [11].

Jin *et al.* demonstrated that a single MWNT with 15 nm diameter was capable of emitting large emission currents up to ~26 μA, but after 10 min the emission current suddenly reduced to zero, resulting from the fatal structure damage [12].

9.3
Effect of Tip Structure of Nanotubes on Field Emission

The structures and the surface conditions of the nanotube tip influence field emission properties [12–16]. Wang *et al.* modified the tip structure of MWNTs by controlled field-induced evaporation and measured the field emission properties *in situ* [15]. Figure 9.3a–d shows images of an MWNT tip structure after the modification by field evaporation. The diameter and length of the MWNT are ~10 and 315 nm, respectively. The MWNT in Figure 9.3a has a sharp tip structure. After applying a few voltage scans with a sweeping time of 200 ms, the sharp tip is modified into a blunt one, as shown in Figure 9.3b–d. The emission current leading to evaporation is several tens of microamperes. Figure 9.3e and f shows the current–voltage curves during field emission and their Fowler–Nordheim plots for the MWNT in Figure 9.3a–d, respectively. The gap width is 380 nm. The applied voltage required for start of the emission of the current increases from 19 to 42 V (a–d in Figure 9.3e) as the sharp tip is transformed into a hemispherical one (Figure 9.3a–d). From the slope of the Fowler–Nordheim plots, the corresponding field conversion factors of the MWNTs in Figure 9.3a–d were estimated to be 1/4.6, 1/5.6, 1/6.5, and 1/9.5 nm^{-1}, respectively. The *in situ* observation shows experimentally that the factors decrease with the increase in the radius of curvature of the nanotube tip using the same MWNT. In addition, they observed that the cap of an MWNT with ~37 nm diameter was opened by field evaporation, and examined *in situ* the field emission properties at a gap width of 660 nm. They showed that the field emission easily occurs in an open-ended MWNT rather than in a capped one at a low applied voltage.

Xu *et al.* measured the work function of MWNTs with various tip structures by *in situ* TEM and demonstrated that the work function at the MWNT tips sensitively changed depending on its tip structures and surface conditions [13]. The field

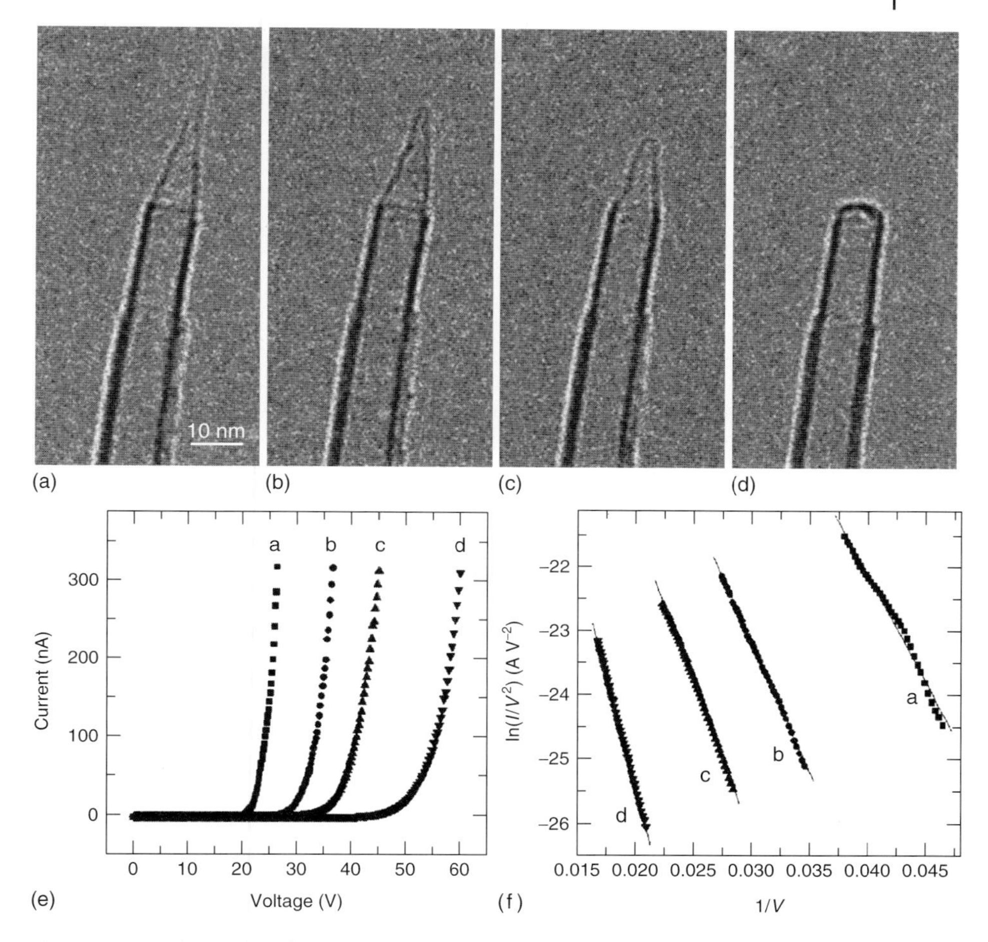

Figure 9.3 (a–d) Images of a MWNT tip structure after the modification by field evaporation. (e) and (f) Current–voltage curves during field emission and their Fowler–Nordheim plots for the MWNT in (a–d), respectively. (Reproduced with permission from M.S. Wang *et al.*, *Appl. Phys. Lett.*, **88**, 243108 (2006).)

conversion factor and the work function are essential parameters that determine the field emission properties. Their precise quantification is important for the evaluation of the field emission properties of nanotube emitters.

9.4 Relationship between Field Emission and Gap Width

The gap width between the nanotube tip and the counter anode is one of the crucial factors to evaluate field emission characteristics. Asaka *et al.* fabricated a

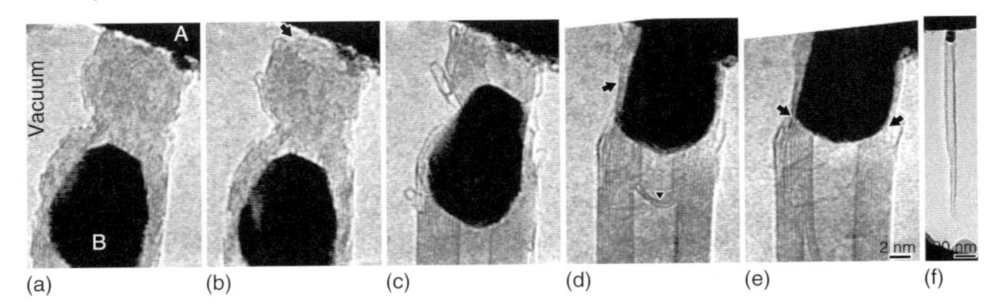

Figure 9.4 Time-sequential series of high-resolution images of the welding process of a MWNT to the metal surface. The tip diameter and length of the MWNT emitter are 5 and 176 nm, respectively, as shown in (f). The MWNT tip is closed. (Reproduced with permission from K. Asaka *et al.*, *Appl. Phys. Lett.*, **92**, 023114 (2008).)

single MWNT emitter on a metal surface by welding in a transmission electron microscope and examined *in situ* the field emission properties at various gap widths [17]. Figure 9.4a–e shows a time-sequential series of high-resolution images of the welding process of an MWNT to the metal surface. In Figure 9.4a, the dark regions at the top A and bottom B are the platinum surface and a platinum particle encapsulated in the MWNT, respectively. In this welding process, each layer consisting of the MWNT is directly connected to the particle at the junction between the MWNT and the particle, as shown by two small arrows in Figure 9.4e, suggesting that an ohmic contact forms at the junction. After welding, the single MWNT emitter is formed on the platinum surface, as shown in Figure 9.4f. The tip diameter and length of the MWNT are 5 and 176 nm, respectively. The nanotube tip is closed. In Figure 9.4f, the dark region at the bottom is a copper plate used as the anode. Figure 9.5a shows the current–voltage curves during field emission at gap widths in the range from 27 to 442 nm. The applied voltage required for an emission current of 100 nA increases with the increase of the gap from 27 to 442 nm. At 27 nm, the emission current starts to be observed at only 29 V and increases up to 14 μA at 42 V. Figure 9.5b shows the Fowler–Nordheim plots obtained from the current–voltage curves in Figure 9.5a. From the slope of these plots, the field enhancement factors (βs) are estimated and presented as a function of the gap width (d) in Figure 9.5c. The field enhancement factors decrease with decreasing gap width, and it is well approximated by $\beta = 1 + d^{0.79}$. The decrease results from an approximately parallel plate configuration of the nanotube tip and the anode, as reported previously [18].

9.5
Other Studies by *In situ* TEM of CNT Emitters

Comings *et al.* performed electron holography experiments of individual field-emitting MWNTs inside a transmission electron microscope and examined the

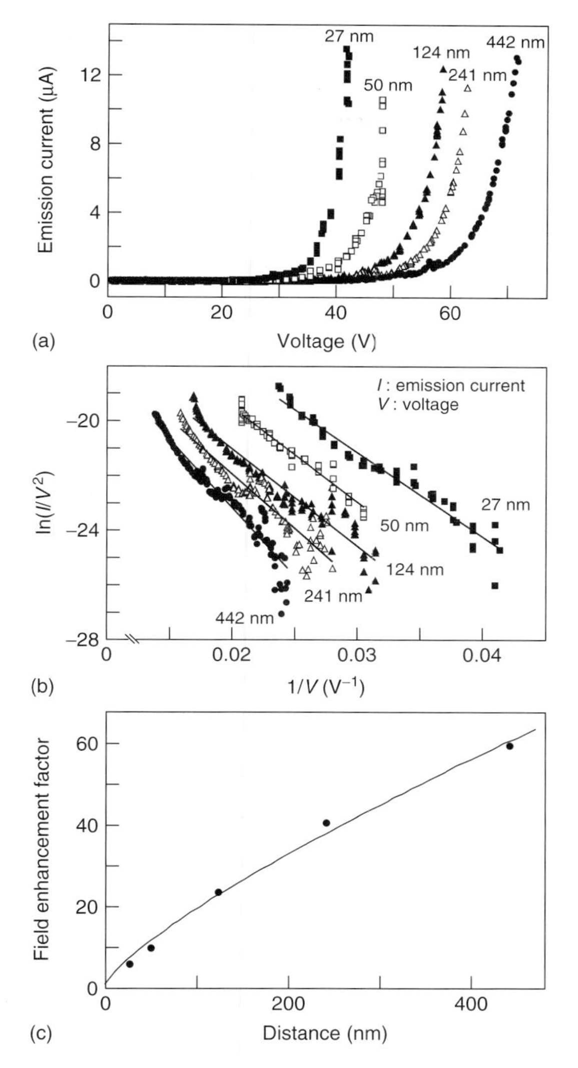

Figure 9.5 (a) Current–voltage curves during field emission at gap widths of 27–442 nm. (b) Fowler–Nordheim plots obtained from the current–voltage curves in (a). (c) Field enhancement factors as a function of the gap distance. Dots represent experimental data, and the solid curve shows a fitting curve, $\beta = 1 + d^{0.79}$. (Reproduced with permission from K. Asaka *et al.*, *Appl. Phys. Lett.*, **92**, 023114 (2008).)

magnitude and spatial distribution of the electric field surrounding the MWNT [19]. They revealed that the field strength was highest at the tip of the MWNT and not at the sidewall defects. They also showed that individual MWNTs can be used as a nanoscale electrostatic biprism for electron holography [20].

Gao *et al.* measured by *in situ* TEM the work function at the tips of individual MWNTs with diameters of 14–55 nm [21]. The majority (70%) of investigated

MWNTs gave work functions in the range 4.6–4.8 eV, which falls in the range of previously reported values for graphite materials [22]. The other MWNTs had a work function of ~6.5 eV. They suggested that the discrepancy might be due to the metallic and semiconductive characteristics of the MWNTs.

References

1. de Heer, W.A., Châtelain, A., and Ugarte, D. (1995) A carbon nanotube field-emission electron source. *Nature*, **270**, 1179–1180.
2. Rinzler, A.G., Hafner, J.H., Nilolaev, P., Lou, L., Kim, S.G., Tománek, D., Nordlander, P., Colbert, D.T., and Smalley, R.E. (1995) Unraveling nanotubes: field emission from an atomic wire. *Science*, **269**, 1550–1553.
3. Wei, Y., Xie, C., Dean, K.A., and Coll, B.F. (2001) Stability of carbon nanotubes under electric field studied by scanning electron microscopy. *Appl. Phys. Lett.*, **79**, 4527–4529.
4. Bonard, J.-M., Klinke, C., Dean, K.A., and Coll, B.F. (2003) Degradation and failure of carbon nanotube field emitters. *Phys. Rev. B*, **67**, 115406-1–115406-10.
5. Bonard, J.-M., Dean, K.A., Coll, B.F., and Klinke, C. (2002) Field emission of individual carbon nanotubes in the scanning electron microscope. *Phys. Rev. Lett.*, **89**, 197602-1–197602-4.
6. Wang, Z.L., Poncharal, P., and de Heer, W.A. (2000) Nanomeasurements in transmission electron microscopy. *Microsc. Microanal.*, **6**, 224–230.
7. Wang, Z.L., Gao, R.P., de Heer, W.A., and Poncharal, P. (2002) In situ imaging of field emission from individual carbon nanotubes and their structural damage. *Appl. Phys. Lett.*, **80**, 856–858.
8. Fujieda, T., Hidaka, K., Hayashibara, M., Kamino, T., Matsumoto, H., Ose, Y., Abe, H., Shimizu, T., and Tokumoto, H. (2004) In situ observation of field emissions from an individual carbon nanotube by Lorentz microscopy. *Appl. Phys. Lett.*, **85**, 5739–5741.
9. Doytcheva, M., Kaiser, M., and de Jonge, N. (2006) In situ transmission electron microscopy investigation of the structural changes in carbon nanotubes during electron emission at high currents. *Nanotechnology*, **17**, 3226–3233.
10. Saito, Y., Seko, K., and Kinoshita, J. (2005) Dynamic behavior of carbon nanotube field emitters observed by in situ transmission electron microscopy. *Diamond Relat. Mater.*, **14**, 1843–1847.
11. Wei, W., Liu, Y., Wei, Y., Jiang, K., Peng, L.-M., and Fan, S. (2007) Tip cooling effect and failure mechanism of field-emitting carbon nanotubes. *Nano Lett.*, **7**, 64–68.
12. Jin, C., Wang, J., Wang, M., Su, J., and Peng, L.-M. (2005) In-situ studies of electron field emission of single carbon nanotubes inside the TEM. *Carbon*, **43**, 1026–1031.
13. Xu, Z., Bai, X.D., Wang, E.G., and Wang, Z.L. (2005) Field emission of individual carbon nanotube with in situ tip image and real work function. *Appl. Phys. Lett.*, **87**, 163106-1–163106-3.
14. Wang, M.S., Peng, L.-M., Wang, J.Y., and Chen, Q. (2005) Electron field emission characteristics and field evaporation of a single carbon nanotube. *J. Phys. Chem. B*, **109**, 110–113.
15. Wang, M.S., Wang, J.Y., and Peng, L.-M. (2006) Engineering the cap structure of individual carbon nanotubes and corresponding electron field emission characteristics. *Appl. Phys. Lett.*, **88**, 243108-1–243108-3.
16. Kaiser, M., Doytcheva, M., Verheijen, M., and de Jonge, N. (2006) In situ transmission electron microscopy observation of individually selected freestanding carbon nanotubes during field emission. *Ultramicroscopy*, **106**, 902–908.
17. Asaka, K., Nakahara, H., and Saito, Y. (2008) Nanowelding of a multi-walled carbon nanotube to metal

surface and its electron field emission properties. *Appl. Phys. Lett.*, **92**, 023114-1–023114-3.

18. Smith, R.C., Cox, D.C., and Silva, S.R.P. (2005) Electron field emission from a single carbon nanotube: effects of anode location. *Appl. Phys. Lett.*, **87**, 103112-1–103112-3.
19. Cumings, J., Zettl, A., McCartney, M.R., and Spence, J.C.H. (2002) Electron holography of field-emitting carbon nanotubes. *Phys. Rev. Lett.*, **88**, 056804-1–056804-4.
20. Cumings, J., Zettl, A., and McCartney, M.R. (2004) Carbon nanotube electrostatic biprism: principle of operation and proof of concept. *Microsc. Microanal.*, **10**, 420–424.
21. Gao, R., Pan, Z., and Wang, Z.L. (2001) Work function at the tips of multiwalled carbon nanotubes. *Appl. Phys. Lett.*, **78**, 1757–1759.
22. Weast, R.C. (1976–1977) *Handbook of Chemistry and Physics*, 5th edn, CRC Press, p. E-81.

10 Field Emission from Single-Wall Nanotubes

Kenneth A. Dean

10.1 Introduction

Single-wall nanotubes (SWNTs) provide the best field-emitting geometry known to humankind. They are routinely formed with diameters on the order of 1 nm and aspect ratios greater than 1000. From a chemical standpoint, their covalently bonded structures make them more robust than traditional metallic structures and immune from electromigration. Simply put, SWNTs make excellent electron emitters.

While SWNT emitters do follow the Fowler–Nordheim tunneling theory under the right conditions, the details of their emission behavior are far richer. The Fowler–Nordheim description was developed for metals, and it is applicable only for very specific surface conditions [1, 2]. SWNTs do not generally present these special surface conditions, and they readily operate outside the Fowler–Nordheim assumptions. The field emission behavior of SWNTs is dominated by surface states arising from both the nanotubes structures themselves and external molecular interactions. Moreover, the stability of nanotube field emitters provides readily observable phenomena that occur at temperature, current, and electric field conditions that are well beyond the destruction limit of other types of emitters.

Many of these behaviors affect the engineering designs of devices incorporating SWNT electron emitters. To promote the design and construction of SWNT field emission devices, this chapter summarizes the current understanding of SWNT field emission behavior over a broad range of operating conditions.

10.2 Single-Wall Nanotubes and Field Emission

SWNTs are typically grown from a seed catalyst nanoparticle by arc evaporation, laser evaporation, or chemical vapor deposition. Each technique produces a diverse

Carbon Nanotube and Related Field Emitters: Fundamentals and Applications. Edited by Yahachi Saito

ISBN: 978-3-527-32734-8

population of SWNTs, and the mixture often includes multiwalled nanotubes (MWNTs) and other carbon forms as well. Each technique produces a population of nanotubes with differences in diameter, length, chirality, and conductivity. Synthesized SWNTs range in diameter from approximately 0.5 to greater than 6 nm, with typical techniques producing nanotubes between 1 and 2 nm [3]. They typically range in length from a few nanometers to many micrometers. The chiral twist of SWNT graphitic sheet walls relative to the tube axis gives rise to a band structure such that approximately one-third of possible SWNT structures are metallic tubes while two-thirds are semiconducting tubes [4]. Of course, the population fractions of these tubes vary with the synthesis technique. SWNTs are also terminated by a cap, and each cap structure presents different electronic states for field emission. In short, SWNTs are not a homogenous material, but rather a population of structures with widely varying field emission properties.

10.3
Measuring the Properties of a Single SWNT

The majority of publications that characterize the field emission properties of nanotubes begin with nanotube films prepared on a flat surface. To use experimentally convenient voltages, the extraction electrode, which is typically a phosphor screen or small-diameter metallic probe, is positioned within 50–500 μm of the emitter surface [5, 6]. While these methods provide relevant information on current density and emitter uniformity, they lack the means to provide a detailed understanding of underlying emitter physics. For example, measuring a large ensemble of nanotubes with different cap structures, diameters, lengths, and conductivities produces a composite current–voltage curve heavily weighted to the behavior of emitters with the highest aspect ratio. Temporal behaviors are also washed out, as the behaviors of many nanotubes are averaged together. It is difficult to understand nanotube physics by studying samples with large numbers of emitters. Moreover, a parallel plate electrode geometry provides poor gas conductance, creating a local environment at the field emitter that is dominated by outgassing from electron-bombarded surfaces, rather than by the pressure of the measurement chamber. Under these conditions, the cleanliness of the field-emitting surface can neither be measured nor controlled, and this undermines the interpretation of any experimental results.

These experimental shortcomings can be overcome by employing traditional field emission microscopy techniques [7]. The electrodes are spaced far apart, providing controllable vacuum conditions. Moreover, the geometry magnifies the electron beam as much as a million times, allowing for real-time observation of the changes in the spatial variations of the electron current density across the emitting surface. The emitters themselves can be mounted on a heating element, allowing for surface modification and cleaning. The field emission microscope technique can control the relevant variables.

While the field emission microscopy technique is powerful, obtaining results from a single nanotube requires some technique in sample preparation. However, the exponential dependence of field emission current on nanotube geometry works in the scientist's favor. Given a population of nanotubes, only the sharpest few will emit, so a sample can contain many more nanotubes than emit. In addition, the field emission microscope geometry can image the emission from several nanotubes, and in many cases, the images are completely nonoverlapping. With these advantages, a number of investigators have been successfully preparing nanotube field emitters on a standard scanning electron microscope filament assembly by applying carbon paint to the filament as an adhesive, and gently touching the wet adhesive to powdered nanotubes. The adhesive picks up a sparse enough population of nanotubes to produce a sample with one emitting nanotube.

An experimental challenge with SWNTs is measuring the structure of the nanotube that is field emitting. Typically, individual SWNTs (as opposed to bundles of SWNTs) are too small to be observed and accurately measured by scanning electron microscopy, although their presence can be detected by transport measurements between two probes. While SWNTs are readily observed via transmission electron microscopy, the primary field emitters are typically micrometer-long, free-ended nanotubes. These vibrate in the microscope, causing blurring of the tube everywhere but at its anchor point. Distinct field-emitting SWNTs have rarely been structurally characterized.

Recently, Arnold *et al.* demonstrated a means for purifying and separating nanotubes by size and chirality [8]. This opens the possibility of measuring a known type of nanotube, based on pure starting material. To date, however, no such measurements have been reported.

10.4
Field Emission from a Clean SWNT Surface

Researchers have observed that nanotubes, like metals, emit electrons from adsorbate states unless the surface is cleaned properly [9]. In the vast majority of experimental situations, the adsorbate states dominate SWNT behavior, and this will be covered in Section 10.5. The pure behavior of the nanotube cap is obtained when the nanotube surface is cleaned properly.

Dean and Chalamala reported that SWNTs can be cleaned in an ultrahigh vacuum (UHV) field emission microscope by thermal desorption at 900 K under an applied field [10]. During the cleaning process, the field emission microscope image changed rapidly as adsorbates were thermally excited, but the image became stable and static once the surface was clean. When the sample temperature was returned to room temperature, the clean image remained the same, as long as the surface remained clean (minutes to hours). It should also be noted that the adsorbate states emitted 10–100 times more current than the clean nanotube surface, so the cleaning processes results in a significant decrease in electron current. This cleaning process and the resulting field emission behavior were confirmed by Liu *et al.* [11] for SWNTs and Hata *et al.* for MWNTs [12].

10.4.1
Clean SWNT Field Emission Microscope Images

Field emission microscopy has been used to study metal surfaces for many decades [7]. For clean metal and semiconductor surfaces, the field-emitted electron beam produces a spatial distribution or "pattern" on the phosphor screen, which reflects the crystal symmetry of the emitting surface. Contrast arises in the image as a result of differences between the electron work functions of the crystal planes bisected by the surface and the differences between atomic steps in the crystal planes. Consequently, the detailed structure in the field emission pattern is representative of the electronic structure of the emitting surface.

Carbon nanotubes do not have a plane crystal structure as metals do; their electronic structure is very different. SWNTs are most often terminated with cap structures, and theorists have predicted that local states exist on the nanotube caps. Tamura and Tsukada computed that the several pentagon rings of carbon atoms on the cap enhance tunneling at each these locations [13]. The existence of these end states has been verified by Carroll *et al.* using scanning tunneling spectroscopy [14]. Thus, one would expect the field emission microscopy images of nanotubes to produce patterns corresponding to the spatial distribution of these local cap states.

There are numerous potential SWNT electronic structures. The typical nanotube cap illustrated in the literature is the fivefold-symmetry cap of the (10,0) tube with a C_{60}-type structure. However, Astakhova *et al.* identified seven caps that could theoretically exist at the end of this nanotube. While each cap contained six pentagons, most structures showed either two- or fourfold symmetry [15]. Thus, an FEM image, which is a map of the nanotube caps, can be much more complicated. Furthermore, Osawa *et al.* predict thousands of different cap configurations for the 90 SWNT structures with diameters smaller than 1.6 nm [16]. Thus, if a nanotube FEM image is a map of local electronic structure, and perhaps of the pentagonal carbon rings, one would expect to encounter a wide variety of patterns in a given population of nanotube structures. Further variety is expected because, for chiral nanotubes, the axis of symmetry of many nanotube caps does not lie along the axis of symmetry of the nanotube. This means that the side of the nanotube cap sits at the very end of the nanotube where the applied electric field is largest during field emission. As a result, the FEM image will be most intense off the axis of symmetry, producing a lopsided projection of the cap.

Dean *et al.* first reported experimentally obtained clean field emission microscope images showing more complicated structures anticipated from SWNT caps (Figure 10.1) [10, 17, 18]. They reported a clean nanotube image that was stable over time at room temperature for minutes to hours, until readsorption of an adsorbate. However, the same clean image was obtained again after thermally removing the adsorbate, over several months. Thus, they reported that the clean images were

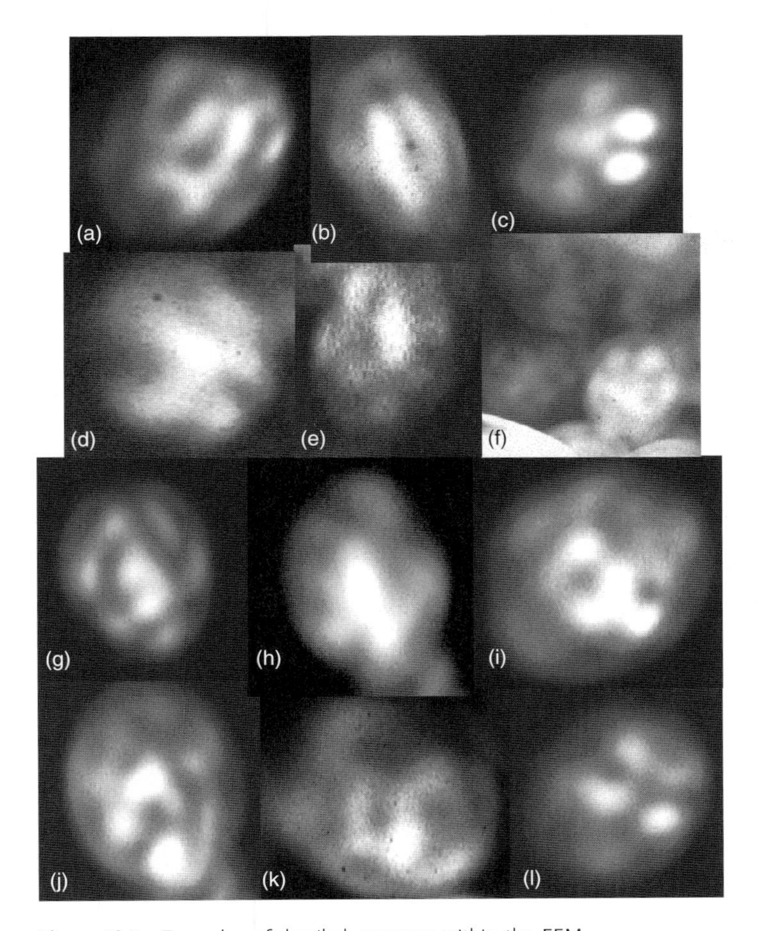

Figure 10.1 Examples of detailed structure within the FEM images of single-wall nanotubes. In (f), a group of closely spaced SWNTs all show fine structure. Images (g), (h), and (j) were obtained from Rice SWNTs. The rest of the images were obtained from Material and Electrochemical Research's arc-grown SWNTs.

repeatable, stable, and consistent with a clean nanotube surface. A fine-structure image with stable behavior was also reported by Sun *et al.* [19].

Dean *et al.* reported a multitude of different patterns, consistent with the large distribution of nanotube cap configurations occurring naturally. Some SWNT images show five- and sixfold symmetry, which they compared to scanning tunneling microscopy images of C_{60} molecules. These symmetries are expected for commonly proposed nanotube cap configurations of achiral nanotubes. More commonly, however, they reported that the clean nanotube images showed less symmetry, suggesting that other cap structures (especially chiral ones) were not as symmetrical [17].

Several authors have performed theoretical computations of the electronic structure of nanotube caps during field emission. They have created spatial maps of the local density of states in their efforts to explain the observed images [20, 21]. This work appears promising, but there is still an opportunity for additional understanding and measurement in this space.

Dean and Chalamala reported evidence for capped SWNTs, with all images having a clear structure in the middle (using several nanotube samples with narrow population distributions between 0.8 and 1.4 nm). They reported never observing an annular ring at room temperature. Even immediately following field evaporation (Section 10.1), they reported no evidence of open or uncapped nanotubes. However, Lui *et al.* [11] reported an annular SWNT field emission pattern with discrete dots in the ring from a SWNT bundle. They interpreted the image as an open (16, 0) nanotube. Consequently, more work is needed in this area to reach a general understanding.

An important question in interpreting these SWNT field emission microscope images is whether the FEM instrument is capable of resolving features with atomic resolution or nearly atomic resolution. Traditional field emission microscopy work on metals was performed on metal tips with large radii of curvature, thus limiting the resolution of the FEM to well short of atomic resolution. However, calculations by Ashworth show that with a tip radius of 5 Å, a resolution of 2 Å is possible [22], and Rose calculates that objects ~3 Å apart can be resolved with the FEM [23]. Binh *et al.* have resolved atomic scale images with the FEM [24]. In addition, Brodie presents strong evidence that extremely sharp objects and, particularly, whiskers with very high aspect ratios can and do show atomic resolution [25]. Finally, Dou *et al.* reported atomic resolution from sharpened W tips [26]. Both theoretical computations and a growing body of experimental work support atomic or near-atomic resolution for extremely sharp objects like nanotubes.

10.4.2
Clean SWNT *I–V*s

Current–voltage ($I-V$) measurements of clean SWNTs show stable, repeatable tunneling behavior and enormous obtainable current density. Dean *et al.* reported measurements for several samples, each containing a single emitting SWNT [27]. At 300 K, each clean nanotube surface produced a stable, reproducible $I-V$ curve, providing an excellent fit to the Fowler–Nordheim equation up to an emitted current ranging from 0.3 to 1 μA depending on the sample (Figure 10.2). The field emission image was also stable and reproducible.

The high overall current infers a high-current density exceeding 10^8 A cm^{-2} DC, using the mean diameter of the Rice University source nanotubes to estimate the area. This field emission current density is likely to be two orders of magnitude larger than can be obtained from typical metal field emitters without pulsing the bias voltage. A similarly high nanotube current density was measured under

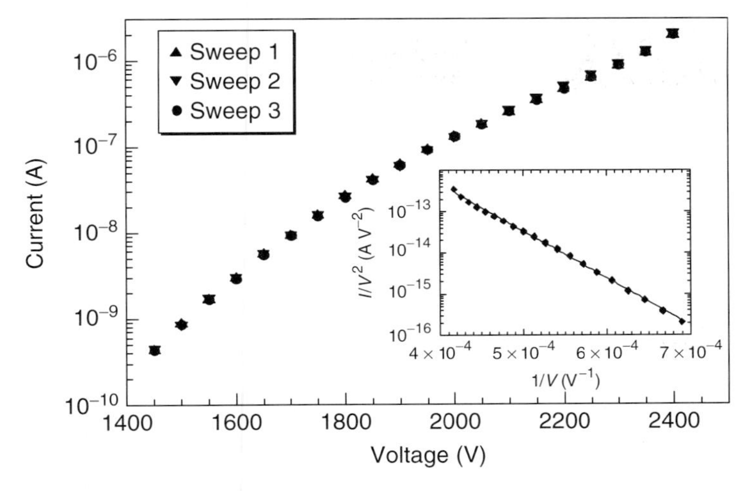

Figure 10.2 The current–voltage characteristic of a single, clean SWNT. The data fits the Fowler–Nordheim theory well (inset). The three symbols denote three consecutive *I–V* sweeps. (Reprinted with permission, Dean and Chalamala, *Appl. Phys. Lett.* (2002), American Institute of Physics [27].)

transport (non-field-emission conditions) on an individual SWNT by Yao and Dekker [28].

Dean *et al.* reported additional reproducible *I–V* behavior at electric fields up to 5% higher than discussed above, but the *I–V* curve in this region increased much faster than predicted by the Fowler–Nordheim equation, with maximum current limit reaching 2 μA. The field emission image also became blurred and a ring formed around the field emission image (Figure 10.3) [18, 29]. Ring formation has been observed by others in diode-type field emission configurations [30].

The increased-current blurred images and encircling ring suggest a local increase in temperature at the nanotube tip, resulting in thermally assisted field emission, and similar observations have been made for metals at high-current densities [31, 32]. In addition, nanometer-sized metal protrusions exhibit a local temperature increase exceeding 200 °C with only 4 nA of field emission current, as measured in the energy distribution of emitted electrons [33]. The metal protrusions were destroyed by melting at currents larger than 100 nA.

Dean *et al.* estimate that the local tip temperature reaches 1600 K based on the change in the nanotube's high current limit versus temperature [29]. Purcell *et al.* measured a similar increase in MWNT tip temperature at high current (from the slope of the electron energy distribution curves), demonstrating a temperature of 2000 K for a current of 1.3 μA [34]. Nanotubes can apparently handle an enormous current density without damage. However, currents exceeding the 1–2 μA range have been shown to cause permanent changes in the nanotubes, which is discussed in Section 10.4.4.

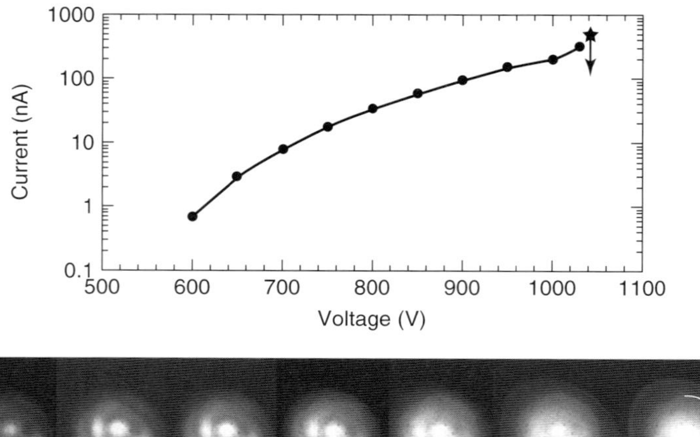

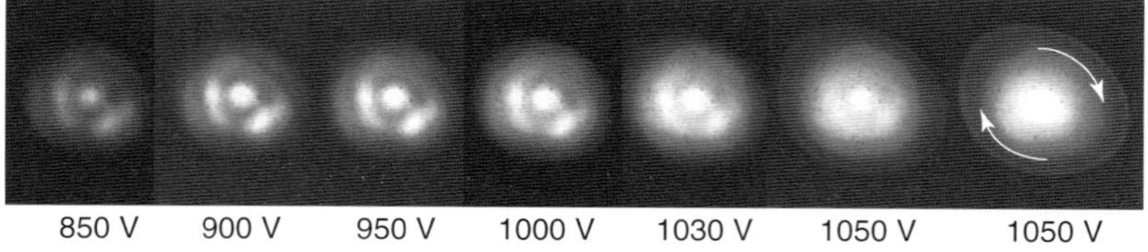

Figure 10.3 An $I-V$ curve and corresponding field emission microscope images of a single SWNT. At extremely high currents, the slope of the ln $(I)-V$ behavior begins to increase, the field emission image blurs, a ring forms around the image, and the image begins to rotate. (Reprinted with permission, Dean *et al.*, *Appl. Phys. Lett.* (2001), American Institute of Physics [29].)

10.4.3
Thermal Field Emission

The previous section presented results concluding that nanotubes tips heat up at high currents producing thermally assisted field emission behavior without applied external heat. Thermal field emission behavior can also be examined by measuring both the change in current with temperature and the energy distribution of emitted electrons.

Of particular interest is that typical nanotube samples contain a mix of chiral (semiconducting) and nonchiral (metallic) nanotubes, with approximately 2 : 1 ratio. Dean *et al.* reported a significant difference in the temperature dependence of current from among nanotubes [35]. In a sample set containing 12 nanotubes, 4 increased in current by ~2 times between 300 and 1450 K (expected for metals with ~5 eV work function), while 8 increased by orders of magnitude (Figure 10.4a). The nanotubes with the highest current density at 1450 K were undetectable at 300 K under the applied voltage. The ratio of low thermally dependent nanotubes to high thermally dependent nanotubes in that experiment is the same as the anticipated ratio of conducting to semiconducting nanotubes, suggesting an explanation. Additional thermal field emission measurements also suggest a strong influence of the nanotube cap structure on the thermal dependence [17].

The energy distribution of field-emitted electrons provides some additional information. Dean *et al.* reported electron energy distribution measurements for

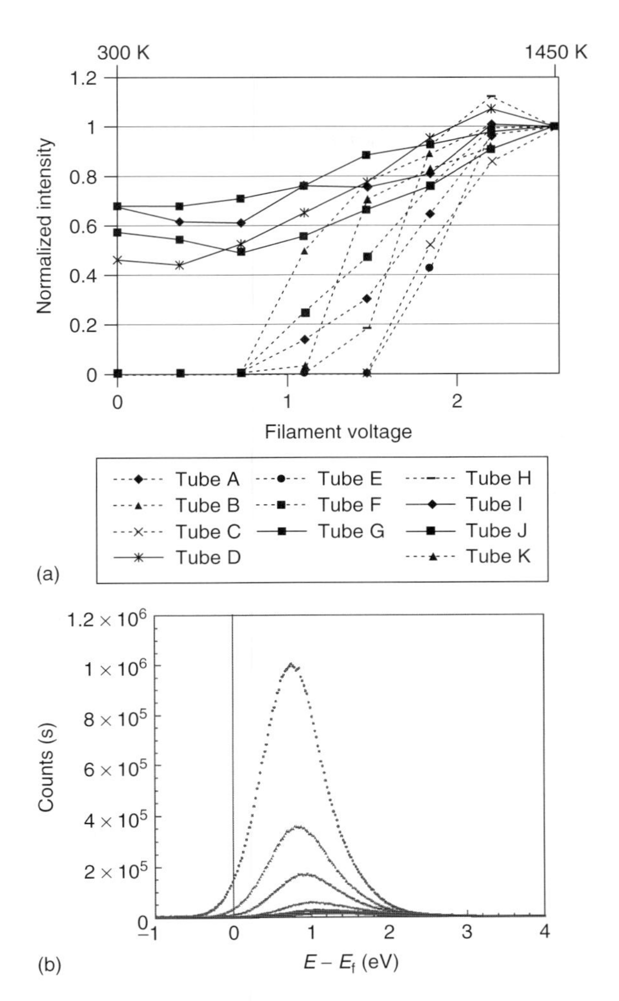

Figure 10.4 Current versus temperature behavior of 12 nanotubes mounted on a heater filament (a). Four of 12 nanotubes show metal-like temperature dependence, while the rest are highly temperature dependent. The current was computed from the measured integrated intensity of light on the phosphor screen. (Reprinted with permission, Dean *et al.*, New Diamond and Frontier Carbon Technology (2002), MYU-KK [35].) (b) Electron energy distribution spectra of field-emitted electrons from SWNTs at 930 K showing the emission state well above the Fermi level. (Reprinted with permission, Dean *et al.*, *Appl. Phys. Lett.* (1999), American Institute of Physics [36].)

a nanotube sample showing strong temperature dependence in the clean state. They report observing electron tunneling from broad states 1–2.8 eV above the Fermi level between 930 and 1160 K (Figure 10.4b) [36]. While it is unclear whether the multiple states were made from one or from a few nanotubes, the field emission behavior was clearly nonmetallic, and there was negligible current density coming from the Fermi level region. The total emitted current at 930 K (emission from a single, broad peak, suggesting one CNT) was 120 nA, so the electric field was presumably large. The peak positions shifted linearly with applied voltage, characteristic of resonant tunneling states, rather than linearly with current, suggesting a conductivity limitation. These measurements suggest that cap states are primarily responsible for the temperature dependence of field emission from nanotubes showing a strong temperature dependency.

Collazo *et al.* reported measuring the electron energy distribution from a population of clean nanotubes [37]. Room temperature electron emission peaked at energies just below the Fermi level. Dean *et al.* also observed peaks only near the Fermi level at room temperature. Collazo *et al.* did not report measurements at elevated temperature.

10.4.4 High Current and Field Evaporation

Section 10.4.3 discussed the stable and reproducible field emission from clean SWNTs, although at high currents researchers found that nanotubes were getting hot. Dean *et al.* reported that above a threshold current of 1–2 μA per SWNT, the individual nanotubes irreversibly changed [29]. Movement commenced within the blurred center of the field emission image. They observed that the motion was an infrequent "pop" for currents just above the threshold, but the speed of motion increased rapidly with increasing current. At currents two to three times above the threshold, the movement was so frequent that it appeared as the continuous spinning of a blurred image with the frequency of rotation exceeding 5 Hz. The intensity of the outer ring also fluctuated as the blurred central image spun (Figure 10.5).

Reducing the current below the threshold stopped the motion, apparently freezing the field emission image in place. The field emission image was generally not the same as the original. The new field emission image and associated I–V characteristics were stable and reversible under all conditions below the threshold current. In short, movement in the field emission images at high current coincided

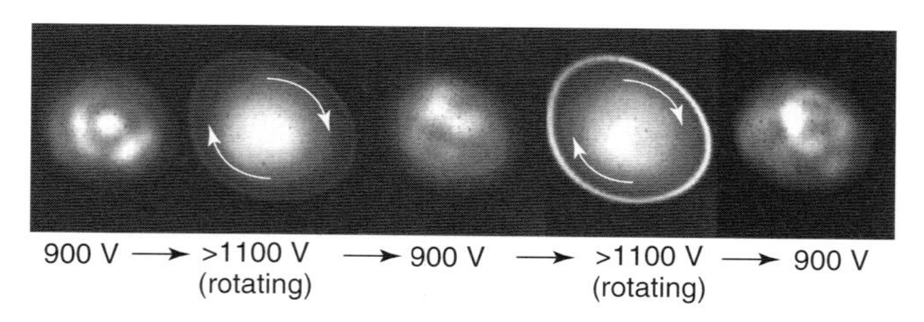

Figure 10.5 The field emission image of an SWNT showing the blurring at high currents. A ring forms around the image and the nodes in the central image spin rapidly. When the current is reduced, a different field emission image is produced, indicating that the cap structure has changed. (Reprinted with permission, Dean *et al.*, *Appl. Phys. Lett.* (2001), American Institute of Physics [29].)

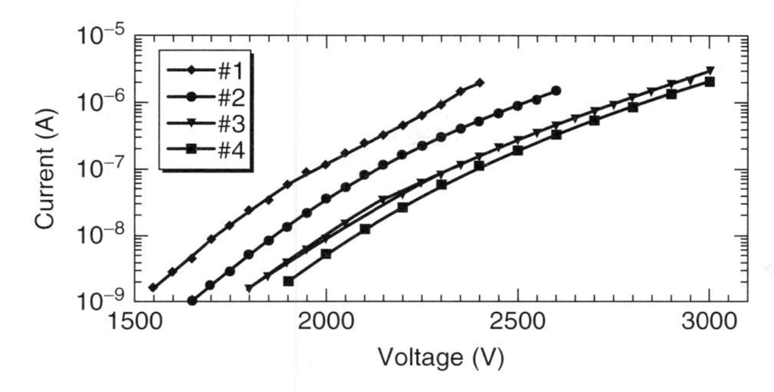

Figure 10.6 The *I*–*V* behavior of an individual SWNT degrading during periods of emission at high currents, accompanied by rapid movement in the FEM image. Ten minutes of high-current emission separate curves 1 and 4. (Reprinted with permission, Dean *et al.*, *Appl. Phys. Lett.* (2001), American Institute of Physics [29].)

with a permanent change in the spatial distribution of field emission current from SWNTs (Figure 10.5).

For small periods of movement in the FEM image, the current–voltage behavior showed only minor changes (as will be discussed later). Allowing the spinning motion to proceed for several minutes under continuous bias conditions resulted in a new FEM image with degraded $I-V$ characteristics. The shift in the $I-V$ curve is consistent with a reduction in the field enhancement of the nanotube by shortening as much as 25% (Figure 10.6).

Dean *et al.* explained the observed phenomenon as thermally assisted field evaporation [29]. Traditional field evaporation is measured under the opposite bias conditions of field emission, and requires substantially higher electric fields. Hata *et al.* experimentally measured room temperature field evaporation of ionized carbon clusters from both MWNTs and SWNTs at an electric field of $\sim$10 V nm^{-1} under the opposite bias conditions (field ion microscopy) used for field emission [38]. The increase in temperature of the nanotube at high current lowers the evaporation barrier, allowing field evaporation under electric fields compatible with field emission, and with the same polarity. Interestingly, thermally assisted field evaporation of metals under field emission conditions catastrophically destroyed emitters in great arcs. In contrast, SWNT thermal field evaporation appeared to be gentle, and generally stable.

Dean *et al.* reported no evidence for open-ended SWNT FEM images similar to the images Saito *et al.* observed for MWNTs [39]. The new FEM patterns that appeared when the voltage was reduced were consistent with an immediate reformation of a new end structure.

These reported changes in field emission images upon thermal evaporation opened up the possibility of measuring the variety of cap structures that could occur on just one SWNT. Investigators had proposed that there were a finite number of atomic structures that an SWNT would support. If the electronic

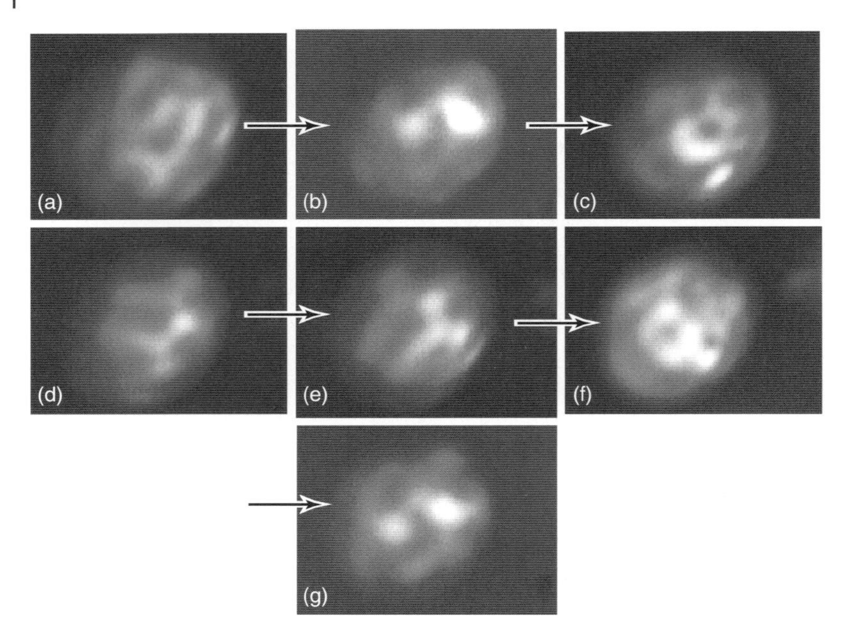

Figure 10.7 A subset of field emission images obtained from a single nanotube showing repeating patterns. Patterns (b) and (g) are identical, (d) and (e) are rotated, and (c) and (f) are closely related. Patterns presented are in the order they were obtained, but several intermediate patterns have not been included. (Reprinted with permission, Dean *et al.*, *J. Vac. Sci. Technol.* (2002), American Institute of Physics [17].)

structure and the atomic structures were closely linked, there might be only a finite number of nanotube patterns available. By repeatedly field-evaporating the end of a nanotube, Dean *et al.* reported that some nanotubes produced more than 50 unique patterns without any pattern repetition, while others showed repetition of exact or very similar patterns (Figure 10.7) [17].

10.4.5
Anomalous High-Temperature Behavior

Dean *et al.* repeatedly observed instability in SWNT field emission which has not been adequately explained [18]. Field emission current was reported to be stable for clean SWNTs at room temperature. This current was also stable for temperatures above 900 K, but only for higher currents (and voltages/electric fields). Below a threshold voltage, which was a found to be a linearly increasing function of temperature, the emission current dropped rapidly over time, typically decreasing by two orders of magnitude over 100 s. The entire $I-V$ curve was then shifted to higher voltages and continued to shift over time during field emission. However, application of a voltage much higher than the degradation threshold caused an increase in current over time (at that fixed voltage) which accelerated rapidly over nominally 20 s. This current increase terminated abruptly when the original current (at that voltage) was reached, restoring completely the original $I-V$ curve. Again, this behavior was found to be extremely reproducible on each sample, among samples, and among nanotubes from different sources and production techniques; yet, no mechanism for this behavior has yet been proposed.

10.5 SWNT-Adsorbate Field Emission

Unless the field-emitting surfaces of nanotubes are specifically cleaned, field emission behavior of nanotubes is dominated by adsorbate–nanotube electronic states. These states produce distinctive field emission patterns, current fluctuations, and temperature behaviors, which are described in the following section.

10.5.1 Field Emission Microscopy

Early SWNT field emission experiments reported lobed field emission images on phosphor-coated anodes [40, 41]. In fact, these images have been commonly observed by researchers using electrode geometries that provide magnification [30, 42, 43]. Dean and Chalamala proposed that these images resulted from chemisorbed adsorbates states [10]. First, they documented that SWNTs showed field emission patterns with one, two, and four lobes (Figure 10.8). However, a large percentage of these lobes (>50%) appeared to be lopsided versions of the three basic patterns, which would occur if the axis of symmetry of the field-emitting surface was not along the axis of symmetry of the nanotube. This structure typically occurs in chiral nanotubes, where the apex is not along the axis of symmetry of the nanotube. Most of the time, the images were not constant with time; they flickered and changed patterns.

In the literature, lobe-type field emission images were readily observed on metal surfaces, and were found to originate from preferential tunneling through specific adsorbate molecules [7, 44–46]. For example, the emission patterns of large chemisorbed molecules on metals produce bright one-, two-, and four-lobed images superimposed over the metal images. These surface states create a resonant tunneling condition for electrons, greatly increasing the local tunneling current at the molecule [47, 48]. While the mechanism is tunneling, and often produces a good linear fit on a $\ln(I/V^2)$ versus $1/V$ plot, it is not described by the Fowler–Nordheim model, and the slope is not proportional to $\phi^{3/2}$ (where ϕ is the emitter work function) [45].

The behavior of adsorbates on metal surfaces had been studied extensively, and several other behaviors had been identified from field emission images: (i) the image nodes changed with time, (ii) the rate of change of the image increased with temperature, (iii) the images were removed by desorbing the molecule, leaving the image of a clean metal surface, (iv) the adsorbates readsorbed onto a clean nanotube surface by supplying adsorbate molecules, and (v) the adsorbate substantially increases the local field emission current. Dean and Chalamala verified

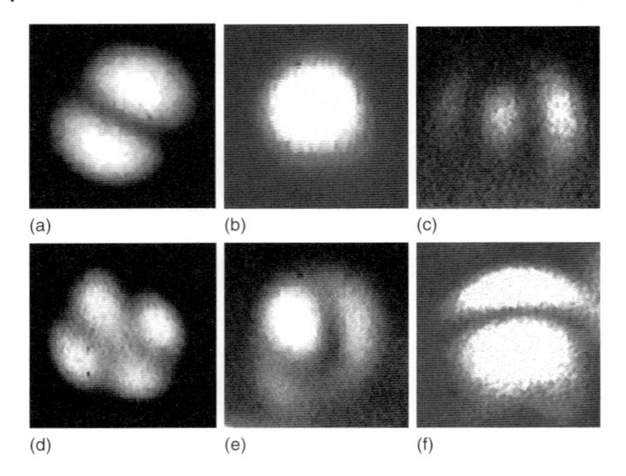

Figure 10.8 Field emission patterns with one to four lobes observed at room temperature (a–d). Tilted versions of the symmetrical lobed patterns that probably occur when electrons are emitted from surfaces not parallel to the anode (e–f). (Reprinted with permission, Dean *et al.*, *J. Appl. Phys.* (1999), American Institute of Physics [10].)

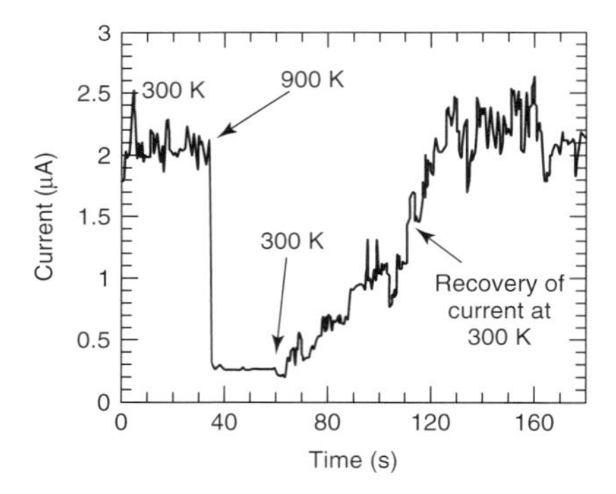

Figure 10.9 Current versus time showing a drop in current as the temperature is raised to 900 K, and a recovery of current when the temperature is returned to 300 K. (Reprinted with permission, Dean *et al.*, *J. Vac. Sci. Technol.* (1999), American Institute of Physics [18].)

all of these behaviors in the SWNT FEM images (Figure 10.9) [10]. In addition, they reported that adsorbates on SWNTs increased the total tunneling current by up to two orders of magnitude, an observation verified by Collazo *et al.* [37].

Several groups have sought the identity of the adsorbate species responsible for enhanced SWNT emission. Collazo *et al.*, Dean *et al.*, and Nilsson *et al.* each reported that in a baked out vacuum system under UHV conditions, the time to re-absorb adsorbates on a clean surface and recover the full current was approximately 1 h [18, 37, 42]. However, Dean *et al.* observed that, if the phosphor was omitted

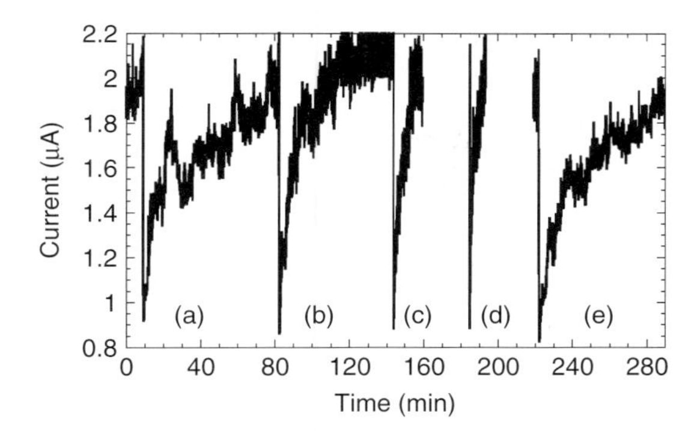

Figure 10.10 The effect of water partial pressure on the recovery time of field emission current after heating the emitter to 900 K. (a) 5×10^{-9} Torr H_2O, (b) 5×10^{-8} Torr H_2O, (c) 4×10^{-7} Torr H_2O, (d) 5×10^{-6} Torr H_2O, and (e) 2×10^{-9} Torr H_2O after 3 h of pumping after measurement of (d). (Reprinted with permission, Dean *et al.*, *J. Vac. Sci. Technol.* (1999), American Institute of Physics [18].)

from the bakeout, the recovery time was only 3 min, suggesting that the adsorbate responsible was present in the phosphor and could be outgassed below 60 °C. Nilsson *et al.* verified this observation, also reporting that the adsorbate was present in the phosphor. Dean *et al.* performed gas introduction experiments in a field emission microscope with H_2, O_2, H_2O, and Ar. The effects of CO and CO_2 were also screened. They reported that addition of a partial pressure of water produced a strong response on the current recovery time (<60 s), whereas the other gases showed no effect (Figure 10.10). Consequently, Dean *et al.* concluded that water or a related species (OH^- or H_3O^+) was responsible for the adsorbate states. However, Nilsson *et al.* reported no change in emission with the introduction of H_2 or water. Consequently, additional investigation is needed in this area.

Finally, it should be noted that very similar adsorbate field emission patterns have been reported from field-emitting MWNTs, diamond-like carbon, nanodiamond, microcrystalline graphite, and nanocrystalline graphite [12, 35]. (In the case of MWNTs, the surface is large enough to accommodate multiple, spatially isolated adsorbates.) Clearly, this chemical interaction is a property of carbon materials in general.

10.5.2 Electron Energy Distributions

The mechanisms proposed for adsorbate-state tunneling incorporate a resonant electronic state. These models predict that the energy distribution of emitted electrons will show a sharp peak at the energy level of the resonant state. Both Dean *et al.* and Collazo *et al.* reported electron energy distribution measurements of SWNTs emitting from adsorbate states in which the energy peak is located just below the Fermi level [36, 37]. Collazo *et al.* measured a clear difference between the peak of the clean nanotubes and the peak of adsorbate states on nanotubes. The

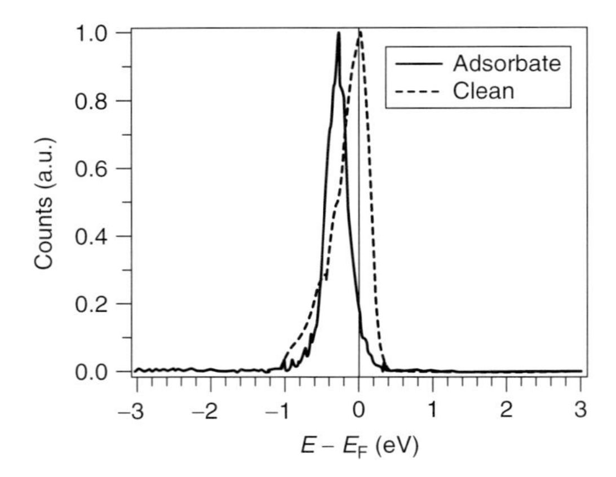

Figure 10.11 Field emission electron energy distribution spectra of nanotubes at an applied bias of 757 V for the clean and adsorbate state. (Reprinted with permission, Collazo *et al.*, *Diamond Relat. Mater.* (2002), Elsevier [37].)

clean nanotube energy distribution peak was located at the Fermi level and showed a typical asymmetric peak shape characteristic of metallic emitters. In contrast, the adsorbate peak was shifted approximately 0.4 eV below the Fermi level and had a symmetrical shape (Figure 10.11). The adsorbate peak position was less dependent on voltage than the Fermi level peak. Collazo *et al.* concludes that the adsorbates emit through a resonant state located close to the Fermi level.

10.5.3 Current Saturation and Field-Emission-Induced Surface Cleaning

Several investigators, including Collins and Zettl [49, 50], Bonard *et al.* [51], and Xu and Brandes [52], have shown that both SWNTs and MWNTs exhibit field emission current saturation at high fields. This property is of particular interest for field emission displays (FEDs) because existing prototypes require external current-limiting resistors in series with each emitter. These resistors are necessary with refractory metal and silicon emitters because they eliminate current runaway and arcing and improve current uniformity among emitters [53]. The intrinsic emission current saturation characteristics of carbon nanotubes may make them a better emitter choice for display applications.

The emission current saturation reported in the above references was observed in measurements of ensembles of emitting nanotubes; so, it was unclear whether saturation was an ensemble behavior or an individual SWNT behavior. However, current saturation had been observed in all test geometries including parallel plate, ball-plane, and point-plane (FEM). It had also been reported under vacuum conditions ranging from 10^{-5} to 10^{-9} Torr.

Using several different samples, each with a single, isolated SWNT emitter, investigators reported that current saturation was a property of a single SWNT

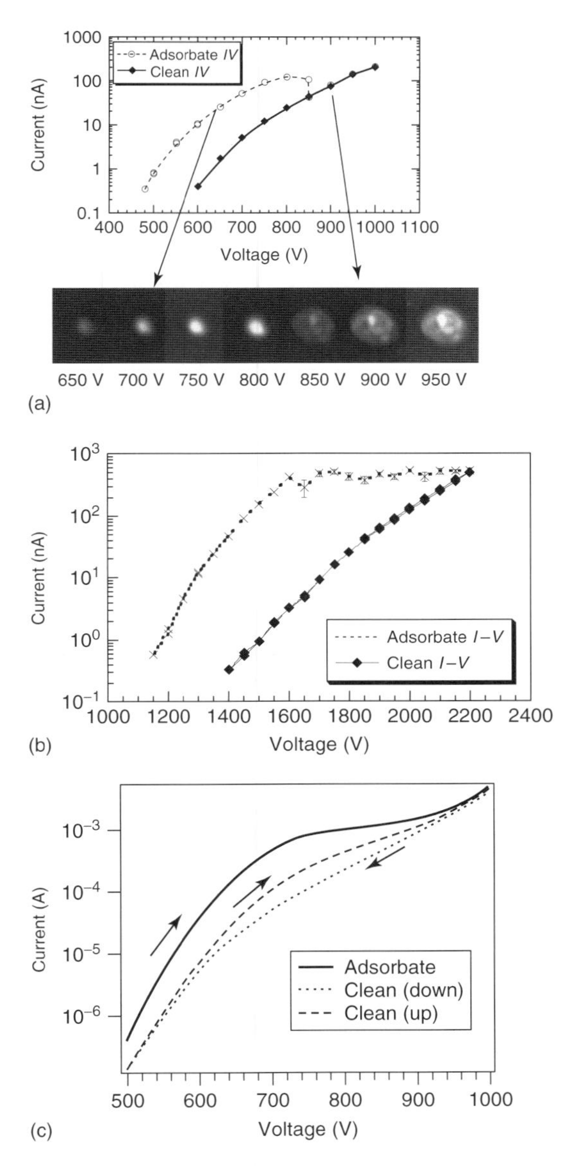

Figure 10.12 The field emission current of an individual SWNT saturating at approximately 100 nA because the adsorbate molecule desorbs. The *I*–*V* curve follows the dashed line until desorption, whence it follows the solid line for all voltages (a). In a typical case, the desorption event occurs through multiple states leading to a broad, saturated region (b). In (b) the error bars represent the standard deviation of current fluctuation during the current measurement at each point. (Reprinted with permission, Dean *et al.*, New Diamond and Frontier Carbon Technology (2002), MYU-KK [35].) (c) Current–voltage (*I*–*V*) curve for the SWNT thin film. (Reprinted with permission from Collazo *et al.*, *Diamond Relat. Mater.* (2002), Elsevier [37].)

[27, 35]. They found that a clean nanotube emitter under UHV conditions showed Fowler–Nordheim tunneling behavior with no saturation whatsoever, while the same nanotube under nonideal vacuum conditions demonstrated strong $I-V$ current saturation. The low-current data increased approximately exponentially with voltage over >50% increase in electric field, as expected. However, above approximately 100–300 nA per nanotube (varying by sample), the current increased only minimally over the next 30% increase in field.

After considering a host of possible mechanisms, Dean *et al.* concluded that, as the current increased, the end of the nanotube heated up, causing the adsorbate to occupy less favorable tunneling states until it finally desorbed. The result was an $I-V$ curve that looked saturated (Figure 10.12a and b). As part of their description, they reported that, as the current increased, both the motion of the adsorbate in the FEM image and the current fluctuation at constant extraction voltage increased substantially, peaking at about 100 nA [35]. With the application of sufficient electric field in the saturated current region, the adsorbate was eventually removed, and the $I-V$ curve transitioned into the clean-emitter $I-V$ behavior. Moreover, the FEM image motion, current fluctuation behavior, and desorption event could all be achieved at constant current simply by applying heat from an external source.

This self-cleaning behavior led to an $I-V$ hysteresis under proper UHV conditions with a high-current "adsorbate-state" upsweep and a low-current "clean-nanotube" downsweep. Collazo *et al.* went further to demonstrate this result with a film of SWNTs containing a great number of emitting nanotubes under UHV conditions (Figure 10.12c) [37].

However, under poor vacuum (10^{-7} Torr, unbaked chamber) Dean *et al.* reported that readsorption of adsorbates was so rapid that the downsweep tracked back along the adsorbate-state upsweep curve, even for single SWNTs. No hysteresis but just a repeatable saturated $I-V$ curve was observed.

Cleaner conditions than those above, which are required to eliminate the adsorbates, are not met in most nanotube investigations reported in the literature. Moreover, it is not clear whether sufficient cleanliness conditions will be met in practical devices to eliminate adsorbates. Consequently, field emission from adsorbates is an important aspect of both nanotube characterization and nanotube-based devices.

10.6 Field Emission Stability

Recent research on carbon-based field emitters has focused on properties necessary for FEDs and other vacuum microelectronic applications. These devices require thousands of hours of highly stable emitter operation.

The first FEDs employed metal-tip field emitters. Metal field emission tips are modified by oxidation and sputtering during operation. They have a tendency to sharpen over time, providing a positive feedback loop which results in their destruction by a catastrophic arc, often in less than 1 h under UHV conditions [54–56]. To reduce or eliminate device failures due to arcing, many vacuum microelectronic devices employ a series resistance in the emitter circuit to provide negative feedback, thus creating stability.

Surprisingly, SWNTs do not need a ballast resistor in series to prevent current runaway. Stable operation has been demonstrated for hundreds of hours under the same conditions that destroy metal emitters in an hour. It is believed that strong covalent bonding, resistance to sputtering and oxidation, adsorbate current saturation mechanisms, and a gentle field evaporation process all contribute to this stability.

10.6.1 Current Fluctuation

As shown in previous sections, a clean SWNT surface is very stable over time. For applications that support a UHV environment, and tolerate periodic surface cleans of $>600\,^{\circ}C$ (for example, scanning electron microscope electron sources), this stability can be realized. For most other applications, adsorbate molecules populate the surface. Current from a field emitter is very sensitive to the presence of adsorbate molecules. More importantly, adsorbate molecules occasionally move, resulting in large current swings. Current fluctuation (or noise) is a major reason why metal microtip field emission sources were not able to displace solid-state semiconductors in amplifier applications. SWNT adsorbate-state field emission increases SWNT emission current by 10–100 times, meaning that SWNTs are particularly noisy.

Dean *et al.* reported that SWNT fluctuation was a function of nanotube temperature. Current fluctuation of an individual SWNT (with adsorbates) increased by roughly an order of magnitude between room temperature and approximately $225\,^{\circ}C$, which is the peak fluctuation temperature. More importantly, extracting field emission current from nanotubes heated them, thereby increasing their current fluctuation. They found that fluctuations peaked at approximately 100 nA of current per nanotube, which is a fairly typical operating condition (Figure 10.13).

Consequently, SWNTs are noisy field emission sources. Noise can be minimized by employing them in applications requiring great numbers of emitters in parallel, thereby averaging out the fluctuations. FEDs make the best use of nanotube averaging, often employing hundreds or thousands of nanotubes per pixel.

10.6.2 Current Degradation

SWNTs are reported to be more susceptible to degradation than other emitters. Bonard *et al.* reported that SWNT field emitters degraded faster than MWNT emitters [51]. Uemura *et al.* reported that DWNTs (the smallest of MWNTs) provided significantly enhanced stability over SWNTs, without sacrificing the sharp

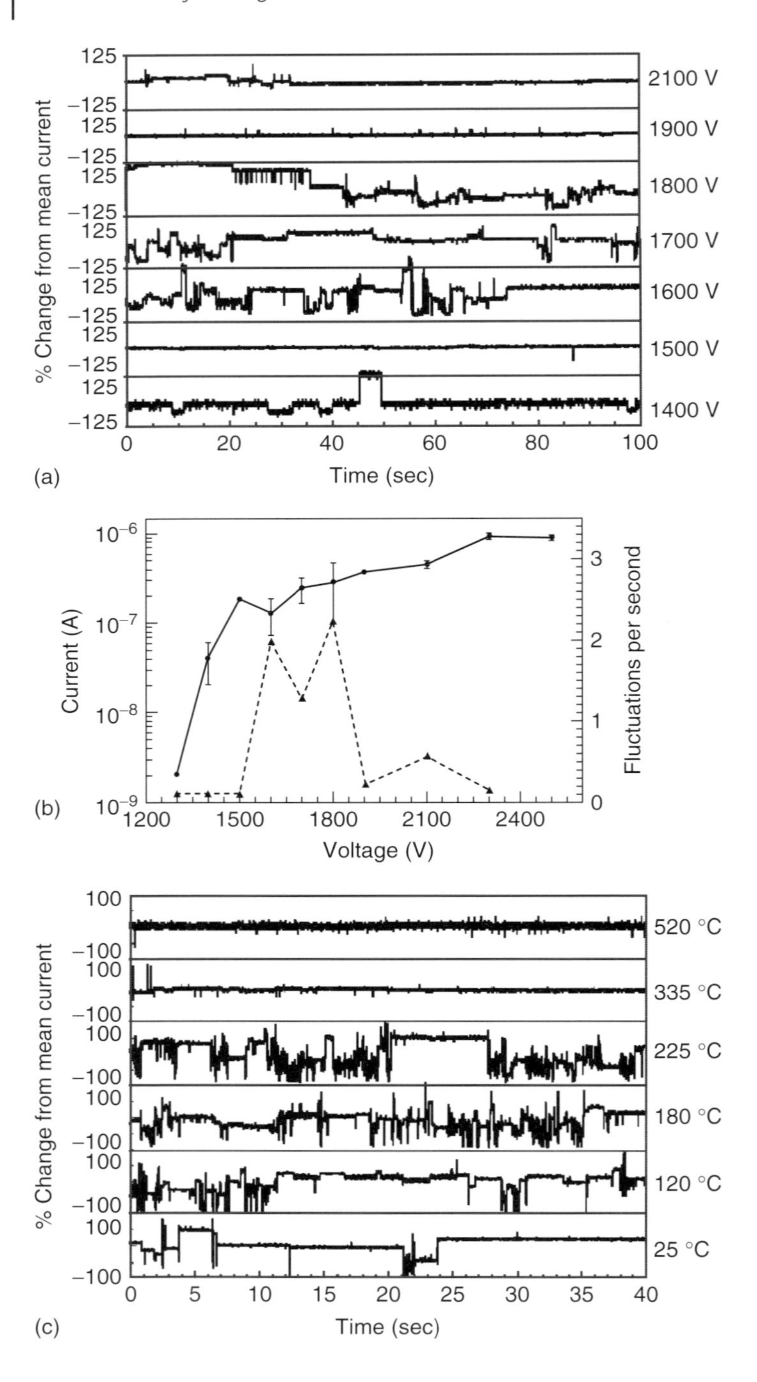
125
−125
% Change from mean current
2100 V
1900 V
1800 V
1700 V
1600 V
1500 V
1400 V
0
20
40
60
80
100
(a)
Time (sec)
Current (A)
Fluctuations per second
10^{-6}
10^{-7}
10^{-8}
10^{-9}
1200
1500
1800
2100
2400
0
1
2
3
(b)
Voltage (V)
100
−100
520 °C
335 °C
225 °C
180 °C
120 °C
25 °C
0
5
10
15
20
25
30
35
40
(c)
Time (sec)

Figure 10.13 The fluctuation in current from an individual SWNT increasing with increasing current through the onset point of current saturation. In (a) and (b), the current saturation onset occurs at a nominal current of 100 nA, corresponding to an applied voltage of 1600 V. The magnitude and frequency of current fluctuation peak near the saturation onset, with a fluctuation rate 100 times that at lower voltages (b). Above the strongly current-saturated region (1600–1800 V), the current fluctuation decreases with increasing current. Errors bars in (b) depict the standard deviation of current magnitude measured at each point. In (c), external thermal excitation greatly increases the current fluctuation frequency, producing a fluctuation behavior similar to that shown in (b), but without increasing the current. The current at room temperature is nominally 300 nA, but it decreases with increasing temperature to 200 nA at 225 °C and 60 nA at 520 °C as adsorbates are removed. The maximum current fluctuation rate occurs at excitation conditions of 225 °C and 200 nA. (Reprinted with permission, Dean *et al.*, *New Diamond Front. Carbon Technol.* (2002), MYU-KK [35].)

geometry [57]. To find out what factors contribute to SWNT degradation, Dean and Chalamala demonstrated 350 h of direct current (100% duty cycle) operation without degradation in UHV field emission microscope experiments [58]. They also reported no degradation when exposed to substantial H_2 and Ar partial pressures, but substantial degradation with 10^{-8} Torr of O_2 present. A small amount of degradation was also observed with 10^{-6} Torr of water present (Figure 10.14).

Dean and Chalamala proposed a reactive sputter-etching mechanism to explain their data, noting that studies on the sputter-etching of graphite show much faster etching in O_2 than in Ar, demonstrating a reactive ion etching effect [59]. In addition, carbon etching studies have also found that water chemically enhances the sputter-etching of carbon [59]. Nilsson verified that H_2 partial pressures had

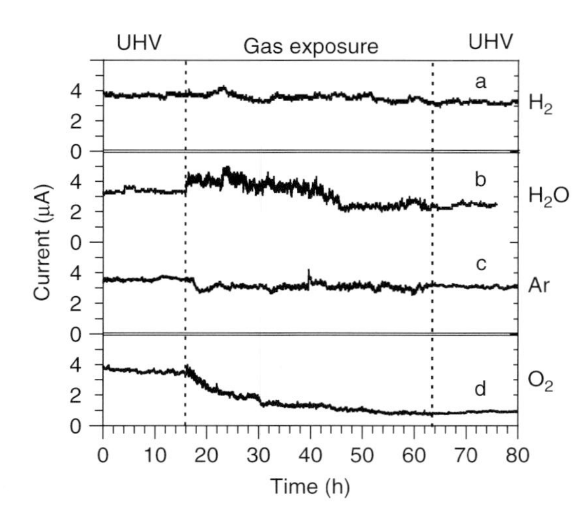

Figure 10.14 Emitter operation under various gas ambients: (a) 0.15 Torr s H_2 (150 000 Langmuir), (b–d) 0.015 Torr s (15 000 Langmuir). (Reprinted with permission, Dean *et al.*, *Appl. Phys. Lett.* (1999), American Institute of Physics [58].)

negligible effect, but O_2 partial pressures degraded emission [42]. Thus, reactive sputter-etching explains the irreversible current decrease observed with both O_2 and H_2O, although O_2 exposure is considerably more severe.

10.7 Conclusions

Clean SWNTs are highly stable and reproducible field emitters under UHV conditions. Scientists have observed structure in field emission microscope images indicative of the local electronic structure of nanotube caps. In addition, electron energy distribution measurements and thermal field emission behavior also suggest that nanotubes emit through special cap electronic states.

However, adsorbate molecules strongly and readily modify SWNT field emission behavior. It is not clear whether sufficient cleanliness conditions will be met in practical devices to eliminate adsorbates. Consequently, field emission from adsorbates is an important (and perhaps dominant) aspect of both nanotube characterization and nanotube-based devices. While enhancing overall current, these adsorbates create current noise, which is objectionable for many applications. Thus, SWNT field emitters are most suited to device designs that support periodic cleaning, or incorporate hundreds of electrically parallel emitters to average out current fluctuations.

On the positive side, SWNTs show incredible stability when compared to metals. They can source orders of magnitude more current density without degradation; they are effectively immune from electromigration; and they degrade gently under high-current, high-temperature, and high-electric-field conditions. Moreover, their physics provides for current self-limiting mechanisms, rather than runaway feedback mechanisms in metals. However, SWNT field emitters are degraded by oxygen in the environment. MWNTs, even very small diameter ones, are more robust to oxygen. It is hoped that this understanding of SWNT field emitters, including both their strengths and their weaknesses, will lead to new and exciting FEDs.

References

1. Fowler, R.H. and Nordheim, L. (1928) *Proc. R. Soc. London*, **A119**, 173.
2. Modinos, A. (1984) *Field, Thermionic, and Secondary Electron Emission Spectroscopy*, Plenum, New York.
3. For example: Rinzler, A.G., Liu, J., Dai, H., Nikolaev, P., Huffman, C.B., Rodriguez-Macias, F.J., Boul, P.J., Lu, A.H., Heymann, D., Colbert, D.T., Lee, R.S., Fischer, J.E., Rao, A.M., Eklund, P.C., and Smalley, R.E. (1998) *Appl. Phys. A*, **67**, 29.
4. Saito, R., Fujita, M., Dresselhaus, G., and Dresselhaus, M.S. (1992) *Appl. Phys. Lett.*, **60**, 2204.
5. Zhu, W., Kochanski, G.P., Jin, S., and Seibles, L. (1996) *J. Vac. Sci. Technol., B*, **14**, 2011.
6. Coll, B.F., Jaskie, J.E., Markham, J.L., Menu, E.P., Talin, A.A., and vonAllmen, P. (1998) *Covalently Bonded Disordered*

Thin Films, MRS Symposium Proceedings, vol. 498 (eds M.P. Siegal, W.I., Milne, and J.E., Jaskie), Materials Research Society, Warrendale, p. 185.

7. Dyke, W.P. and Dolan, W.W. (1956) *Adv. Electron. Electron Phys.*, **8**, 89.
8. Arnold, M.S., Green, A.A., Hulvat, J.F., Stupp, S.I., and Hersam, M.C. (2006) *Nat. Nanotechnol.*, **1**, 61.
9. Gomer, R. (1993) *Field Emission and Field Ionization*, American Institute of Physics, New York, p. 34 and 148.
10. Dean, K.A. and Chalamala, B.R. (1999) Field emission microscopy of carbon nanotube caps. *J. Appl. Phys.*, **85**, 3832.
11. Liu, W., Hou, S., Zhang, Z., Zhang, G., Gu, Z., Luo, J., Zhao, X., and Xue, Z. (2003) Atomically resolved field emission patterns of single-walled carbon nanotubes. *Ultramicroscopy*, **94**, 175.
12. Hata, K., Takakura, A., and Saito, Y. (2001) Field emission microscopy of adsorption and desorption of residual gas molecules on a carbon nanotube tip. *Surf. Sci.*, **490**, 296.
13. Tamura, R. and Tsukada, M. (1995) *Phys. Rev. B*, **52**, 6015.
14. Carroll, D.L., Redlich, P., Ajayan, P.M., Charlier, J.C., Blasé, X., De Vita, A., and Car, R. (1997) *Phys. Rev. Lett.*, **78**, 2811.
15. Astakhova, T.Yu., Buzulukova, N.Yu., Vinogradov, G.A., and Osawa, E. (1999) *Fullerene Sci. Technol.*, **7**, 223.
16. Osawa, E., Yoshida, M., Ueno, H., Sage, S., and Yoshida, E. (1999) *Fullerene Sci. Technol.*, **7**, 239.
17. Dean, K.A. and Chalamala, B.R. (2003) Experimental studies of the cap structure of single-walled carbon nanotubes. *J. Vac. Sci. Technol., B*, **21**, 868.
18. Dean, K.A., von Allmen, P., and Chalamala, B.R. (1999) Three behavioral states observed in field emission from single-walled carbon nanotubes. *J. Vac. Sci. Technol., B*, **17**, 1959.
19. Sun, J.P., Zhang, Z.X., Hou, S.M., Zhang, G.M., Gu, Z.N., Zhao, X.Y., Liu, W.M., and Xue, Z.Q. (2002) *Appl. Phys. A*, **75**, 479–483.
20. Buldum, A. and Lu, J.P. (2003) *Phys. Rev. Lett.*, **91**, 236801.
21. Khazaei, M., Dean, K.A., Farajian, A., and Kawazoe, Y. (2007) *J. Phys. Chem. C*, **111**, 6690.
22. Ashworth, F. (1951) *Adv. Electron.*, **3**, 1.
23. Rose, D.J. (1956) *J. Appl. Phys.*, **27**, 215.
24. Binh, V.T., Garcia, N., and Purcell, S.T. (1996) in *Advances in Imaging and Electron Physics*, vol. 95 (ed. P. Hawkes), Academic, New York, pp. 63–153.
25. Brodie, I. (1978) *Surf. Sci.*, **70**, 186.
26. Dou, J., Chen, E., Zhu, C., and Yang, D. (2000) *J. Vac. Sci. Technol., B*, **18**, 2681.
27. Dean, K.A. and Chalamala, B.R. (2000) Current saturation mechanisms in carbon nanotube field emitters. *Appl. Phys. Lett.*, **76**, 375.
28. Yao, Z., Kane, C.L., and Dekker, C. (2000) *Phys. Rev. Lett.*, **84**, 2941.
29. Dean, K.A., Burgin, T.P., and Chalamala, B.R. (2001) Evaporation of carbon nanotubes during electron field emission. *Appl. Phys. Lett.*, **79**, 1873.
30. Zhu, W., Bower, C., Zhou, O., Kochanski, G.P., and Jin, S. (1999) *Appl. Phys. Lett.*, **75**, 873.
31. Latham, R.V. (ed.) (1995) *High Voltage Vacuum Insulation*, Academic Press, New York, pp. 211–218.
32. Dyke, W.P., Trolan, J.K., Martin, E.E., and Barbour, J.P. (1953) *Phys. Rev.*, **91**, 1043.
33. (a) Binh, V.T., Purcell, S.T., Gardet, G., and Garcia, N. (1992) *Surf. Sci. Lett.* **279**, 197; (b) Also done via TED in review article, Binh, V.T., Garcia, N., and Purcell, S.T. (1996) *Advances in Imaging and Electron Physics*, vol. 95 (ed. P. Hawkes), Academic Press, New York, pp. 63–153.
34. Purcell, S.T., Vincent, P., Journet, C., and Binh, V.T. (2002) *Phys. Rev. Lett.*, **88**, 105502.
35. Dean, K.A., Chalamala, B.R., Coll, B.F., Wei, Y., Xie, C.G., and Jaskie, J.E. (2002) *New Diamond Front. Carbon Technol.*, **12**, 165.
36. Dean, K.A., Groening, O., Kuttel, O.M., and Schlapbach, L. (1999) Nanotube electronic states observed with thermal field emission electron spectroscopy. *Appl. Phys. Lett.*, **75**, 2773.
37. Collazo, R., Schlesser, R., and Sitar, Z. (2002) *Diamond Relat. Mater.*, **11**, 769.
38. Hata, K., Ariff, M., Tohji, K., and Saito, Y. (1999) *Chem. Phys. Lett.*, **308**, 343.

39. Saito, Y., Hamaguchi, K., Hata, K., Unichida, K., Tasaka, Y., Ikazaki, F., Yumura, M., Kasuya, A., and Nishina, Y. (1997) *Nature*, **389**, 554.
40. de Heer, W.A., Chatelain, A., and Ugarte, D. (1995) *Science*, **270**, 1179.
41. Saito, Y., Hata, K., and Murata, T. (2000) *Jpn. J. Appl. Phys.*, **39** (Pt 2), 271.
42. Nilsson, L., Groening, O., Kuettel, O., Groening, P., and Schlapbach, L. (2002) *J. Vac. Sci. Technol., B*, **20**, 326.
43. Bonard, J.-M., Salvetat, J.-P., Stockli, T., de Heer, W.A., Forro, L., and Chatelain, A. (1998) *Appl. Phys. Lett.*, **73**, 918.
44. Melmed, A.J. and Muller, E.W. (1958) *J. Chem. Phys.*, **29**, 1037.
45. Morikawa, H., Okamoto, K., Yoshino, Y., Iwatsu, F., and Terao, T. (1997) *Jpn. J. Appl. Phys.*, **36**, 583.
46. Gomer, R. and Speer, D.A. (1953) Molecular images with the projection microscope. The ionization potential of zinc phthalocyanine. *J. Chem. Phys.*, **21**, 73.
47. Duke, C.B. and Alferieff, M.E. (1967) Field emission through atoms adsorbed on a metal surface. *J. Chem. Phys.*, **46**, 923.
48. Swanson, L.W. and Crouser, L.C. (1970) *Surf. Sci.*, **23**, 1.
49. Collins, P.G. and Zettl, A. (1997) *Phys. Rev. B*, **55**, 9391.
50. Collins, P.G. and Zettl, A. (1996) *Appl. Phys. Lett.*, **69**, 1969.
51. Bonard, J.-M., Maier, F., Stockli, T., Chatelain, A., de Heer, W.A., Salvetat, J.-P., and Forro, L. (1998) *Ultramicroscopy*, **73**, 918.
52. Xu, X. and Brandes, G.R. (1999) *Appl. Phys. Lett.*, **74**, 2549.
53. Ghis, A., Meyer, R., Rambaud, P., Levy, F., and Leroux, T. (1991) *IEEE Trans. Electron Devices*, **38**, 2320.
54. Janssen, A.P. and Jones, J.P. (1971) The sharpening of field emitter tips by ion sputtering. *J. Phys. D*, **4**, 118.
55. Zeitoun-Fakiris, A. and Juttner, B. (1991) On the dose of bombarding residual gas ions for influencing pre-breakdown field emission in vacuum. *J. Phys. D*, **24**, 750.
56. Chalamala, B.R., Reuss, R.H., and Dean, K.A. (2001) *Appl. Phys. Lett.*, **78**, 2375.
57. Yotani, J., Uemura, S., Nagasako, T., Kurachi, H., Yamada, H., Ezaki, T., Maesoba, T., Nakao, T., Saito, Y., and Yumura, M. (2003) *Soc. Inf. Disp. Digest*, **34**, 918.
58. Dean, K.A. and Chalamala, B.R. (1999) *Appl. Phys. Lett.*, **75**, 3017.
59. Holland, L. and Ojha, S.M. (1976) The chemical sputtering of graphite in an oxygen plasma. *Vacuum*, **26**, 233.

11
Simulated Electric Field in an Array of CNTs

Hidekazu Murata and Hiroshi Shimoyama

11.1
Introduction

For practical application of an array of carbon nanotubes (CNTs) to field emission sources, it is very important to make clear the field emission characteristics from such arrays. These characteristics strongly depend on the electric field distribution on the CNT apexes. In this chapter, we describe the computer simulation analysis of the electric field for an array of CNTs, especially for vertically aligned carbon nanotubes (VA-CNTs) [1, 2] and network-structured CNTs [3], by means of an improved 3D boundary charge method (BCM) and make clear the dependence of the electric field strength at the CNT apex on the geometrical parameters of CNTs such as the shape of the CNT apex, the radius of curvature of the CNT apex, the average density of CNTs, and the nonuniformity in the length of the CNTs.

11.2
Simulation Method

First of all, we have to develop a computer simulation method and a computational model for the electric field analysis of a VA-CNT system. If we can establish a proper computational model and a computer simulation method, then computer simulation of electric field analysis for VA-CNTs becomes possible, and we can make clear the field emission characteristics from VA-CNTs.

The simulation method used for the electric field analysis of a VA-CNT system is an improved three-dimensional (3D) BCM developed by us. The method is sometimes called a *surface charge method*. In this section, we briefly introduce and outline some features of our 3D BCM for the sake of the following sections. For a detailed description of the method, the readers are referred to [4, 5].

In this method, we determine the surface charge density on each conductor surface under the assumption that the geometry and potential of each conductor are known in advance. In order to determine the surface charge density, we divide the whole conductor surface into n small surface elements ΔS_i $(i = 1, 2, \ldots, n)$.

Carbon Nanotube and Related Field Emitters: Fundamentals and Applications. Edited by Yahachi Saito

ISBN: 978-3-527-32734-8

We then determine the surface charge density σ_i on each small surface element ΔS_i by solving n-dimensional simultaneous linear equations with σ_i as unknown under the assumption that each σ_i on each ΔS_i is uniform. In this case, the coefficient matrix element F_{ji} of the n-dimensional simultaneous linear equation is expressed by a double integral, the numerical calculation of which is a key point of the 3D BCM in terms of both numerical accuracy and computation time. That is to say, the integral F_{ji} involves a singularity when $i = j$. Numerical calculation of this type of integral (i.e., a singular integral) is one of the most difficult problems because it is inevitably accompanied by a serious loss of accuracy. Moreover, the direct numerical calculation of the double integral has been a serious obstacle to the practical use of the 3D BCM because of an extremely long computation time.

The first problem, that is, a serious loss of accuracy occurring in case of numerical integration of a singular integral, is mainly caused by the following two reasons: one is due to the use of a numerical integration formula which is inadequate for a singular integral; and the other is due to the cancellation of significant figures occurring in case of subtraction around the singular point. These difficulties have been overcome by utilizing a method developed by Uchikawa *et al.* [6], who succeeded in removing numerical errors arising from numerical calculation of a singular integral by using a double exponential formula for numerical integration and also by transforming the integrand to a trigonometric expression that does not involve subtraction causing cancellation of significant figures.

The second problem, that is, an extremely long computation time needed for the direct numerical calculation of the double integral, has been overcome by utilizing a computational model developed by us as follows: Any given conductor geometry is faithfully modeled by a suitable combination of six basic surfaces: that is, a plane surface, a cylindrical surface, a conical surface, a discoidal surface, a spherical surface, and a torus surface, as shown in Figure 11.1. For example, a CNT with a hemispherical cap is modeled by a combination of a hemispherical surface for the cap part and a cylindrical surface for the tube part (Figure 11.2). This procedure greatly simplifies the execution of integration; that is to say, the first integration in the double integral F_{ji} can be carried out analytically for the above-mentioned six basic surfaces, thereby greatly reducing the computation time without any loss of accuracy.

Once each surface charge density σ_i on each surface element ΔS_i is determined, then we can determine both the potential and electric field at an arbitrary point in space. It must be noted that our 3D BCM enables us to calculate the electric field distribution directly from the surface charge distribution on the whole conductor surface without any numerical differentiation of the potential, thus giving us a numerical method for high-accuracy, high-speed calculation of both the potential and the electric field.

The other most remarkable feature of our 3D BCM is that the method can be applied even to a case with a great difference in electrode size without any difficulty. In the case of field emission from CNTs, for example, the radius of curvature of the CNT apex (r) is of the order of several tens of nanometers or even smaller, and the

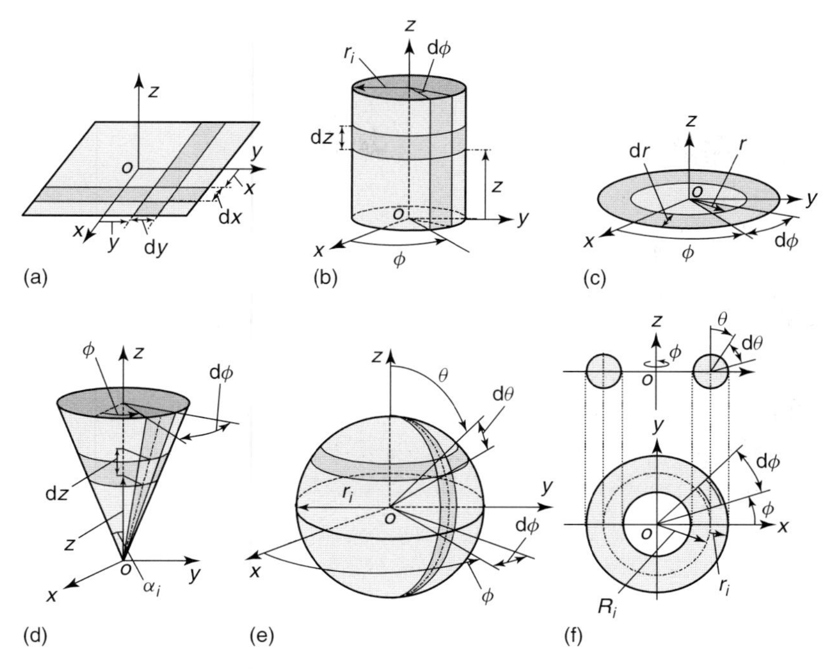

Figure 11.1 Six basic surfaces used in 3D BCM for modeling given conductor geometries: (a) plane, (b) cylindrical, (c) discoidal, (d) conical, (e) spherical, and (f) torus.

distance between the field emitter and the extraction anode (ℓ) is typically 10 mm. Then, the ratio of geometrical dimensions (ℓ/r) becomes 10^5–10^6. Even for such an extreme case, the field analysis is very easy.

11.3
Computational Model

In this section, we describe the computational model for electric field analysis for VA-CNTs [1, 2]. According to the field emission experiment conducted by de Heer *et al.* [7], the average density of CNTs is about $10^9\,\text{cm}^{-2}$, that is, the number of CNTs per 1 mm^2 is as large as 10 millions. Exact modeling of such a large number of VA-CNTs is practically impossible, because an extremely long computation time and a large memory size are required. In our simulation, therefore, the VA-CNT system is modeled by 9×9 CNTs standing vertically on the cathode substrate, as shown in Figure 11.3. We assume that CNTs are conductive materials. Each CNT is modeled by a combination of a hemispherical surface for the cap part and a cylindrical surface for the tube part. The cathode substrate and the anode plate are modeled by plane surfaces. Table 11.1 shows the geometrical parameters of the 9×9 CNT computational model used for simulation. The potentials of the

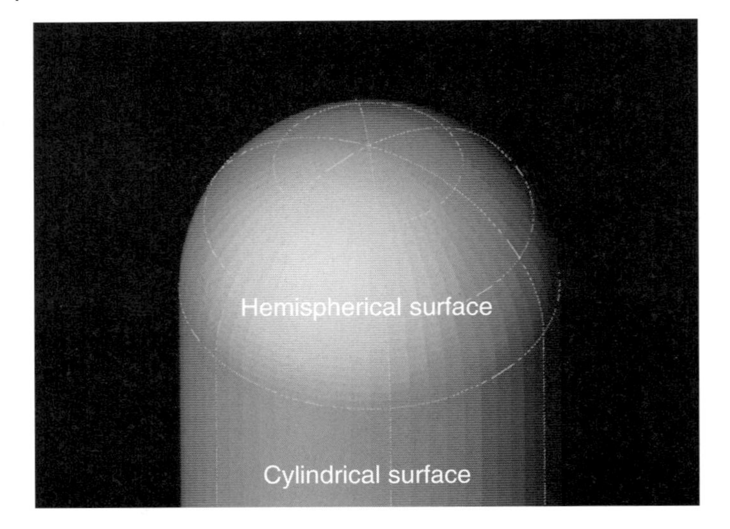

Figure 11.2 CNT with a hemispherical cap modeled by a combination of a hemispherical surface for the cap part and a cylindrical surface for the tube part. This is a typical example of discretization of the central CNT in the 9 × 9 CNTs computational model. Total discretization number is 3602.

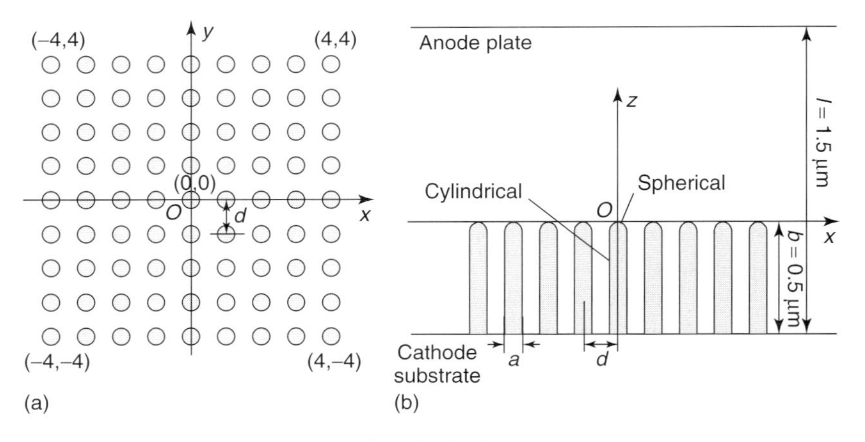

Figure 11.3 9 × 9 CNT computational model for the VA-CNT system: (a) top view and (b) side view.

cathode substrate and the anode plate are set to 0 and 9.75 V, respectively, which corresponds to the condition that an average electric field of 9.75 V μm^{-1} is applied between the anode plate and CNTs.

In this model, the whole conducting surface is divided or discretized into a large number of small surface elements. A typical example of discretization is

Table 11.1 Geometrical parameters of the 9 × 9 CNT computational model.

a	12.5–50 nm
d	200 nm (CNT density: 2.5×10^9 cm^{-2})
b	0.5 μm
ℓ	1.5 μm
Length of one side of cathode (or anode) plate	$16d = 3.2$ μm

a = CNT outer diameter
= twice the radius of curvature of CNT apex.
d = center-to-center distance between neighboring CNTs.
b = length of CNT (uniform length).
ℓ = distance between cathode substrate and anode plate.

Table 11.2 Typical example of discretization of the 9 × 9 CNT computational model.

Number of discretization						
One CNT	Spherical surface	Cylindrical surface	Subtotal number	Cathode substrate	Anode plate	Total number
	18	24	42	100	100	3602
9 × 9 = 81 CNTs			3402			

shown in Table 11.2, where the total number of discretization is 3602. The total number of discretizations determines the accuracy in the electric field calculation. In order to examine the appropriateness of our 9 × 9 CNT computational model, we have to examine two things: The first is to examine what is the discretization number necessary for reasonable accuracy in electric field calculation. For this purpose, we have to carry out an error analysis of the electric field strength calculated as a function of the discretization number. The second is to examine whether the 9 × 9 CNT computational model well simulates a real VA-CNT system where a great number of CNTs exist. For this purpose, we first calculate the electric field strength at the apex of the central CNT, that is, the CNT situated at the origin of the coordinate system. We next calculate the electric field strength at the apex of the CNT adjacent to the central CNT. If the difference between them is sufficiently small, then we can say that the electric field strength at the CNT apex in a real VA-CNT system is well represented by the electric field strength at the apex of the central CNT of the 9 × 9 CNT computational model.

11.4 Field Analysis for the VA-CNT System

11.4.1 Dependence of Numerical Accuracy in Electric Field Calculation on the Discretization Number

In this section, we will examine how many surface elements are needed for discretization for reasonable accuracy in electric field calculation. For this purpose, we carry out the calculation of the electric field strength at the CNT apex for four cases as shown in Table 11.3, where the total discretization number is varied from 3602 (case I) to 5576 (case IV) for two different values of the CNT outer diameter a. Details of discretization of each conducting surface for case I are shown in Figure 11.2 and Table 11.2. For cases II–IV only the central CNT is discretized into smaller surface elements, while the discretization numbers of the rest CNTs, the cathode substrate, and the anode plate are kept same as those given in Table 11.2. The discretization numbers of the hemispherical and cylindrical surfaces of the central CNT are also given in Table 11.3.

The discretization number of the hemispherical surface is most important, because the field emission characteristics strongly depend on the electric field distribution on the CNT apex. Figure 11.4 shows the electric field strength at the central CNT apex calculated as a function of the discretization number of the hemispherical surface for two values of the outer diameter of the CNT. The calculated value of the electric field strength at the CNT apex seems to converge to a certain value as the discretization number increases. We can thus conclude that case IV gives a sufficiently accurate value, that is, a correct value of the electric field strength. The relative errors in the other cases are also listed in Table 11.3. Even in case I, where the total discretization number is 3602 and the

Table 11.3 Error analysis of electric field strength calculated as a function of the discretization number.

Case	Total number of discretization	Number of discretization of central CNT		CNT outer diameter $a = 50$ nm		CNT outer diameter $a = 12.5$ nm	
		Spherical surface	Cylindrical surface	Electric field strength at CNT apex $(\text{V cm}^{-1}) \times 10^5$	Relative error (%)	Electric field strength at CNT apex $(\text{V cm}^{-1}) \times 10^6$	Relative error (%)
I	3602	18	24	5.1354	0.87	2.0034	8.22
II	3728	72	96	5.1069	0.31	1.9095	3.14
III	4316	324	432	5.0972	0.12	1.8791	1.50
IV	5576	868	1152	5.0910	0.00	1.8513	0.00

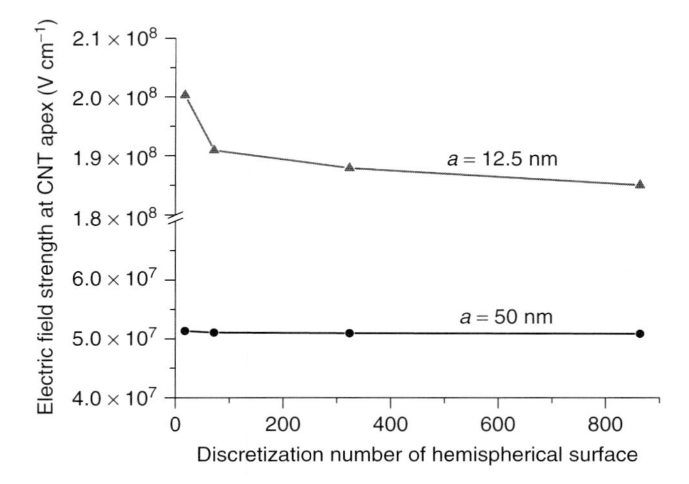

Figure 11.4 Electric field strength at the central CNT apex calculated as a function of the discretization number of the hemispherical surface for the CNT outer diameters $a = 12.5$ and 50 nm.

discretization number of the hemispherical surface of the central CNT is only 18, the relative error is less than 10% for the case of $a = 12.5$ nm. For the case of $a = 50$ nm, the relative error is less than 1%. We can therefore conclude that the total discretization number required for reasonable accuracy in electric field calculation is about 4000.

11.4.2
Appropriateness of 9 x 9 CNT Computational Model

We will examine whether the 9×9 CNT computational model well simulates a real VA-CNT system where a great number of CNTs exist. For this purpose, we calculate the electric field strengths at respective apexes of CNTs aligned on the x axis, and see how uniform they are. The result is shown in Figure 11.5. The horizontal axis shows the location of each CNT, and zero corresponds to the location of the central CNT. The vertical axis is the electric field strength at each CNT apex. It is seen that the electric field strength at the CNT apex in the vicinity of the central CNT is nearly uniform. This situation is also confirmed by Table 11.4, which shows that the relative difference between the electric field strength at the apex of the central CNT and that at the apex of the CNT adjacent to the central CNT is less than 0.5%.

We can therefore conclude that the real VA-CNT system, where a great number of CNTs exist, is well simulated by the present 9×9 CNT model. That is to say, the electric field strength at the CNT apex in a real VA-CNT system is well represented by that at the apex of the central CNT of the present 9×9 CNT computational model.

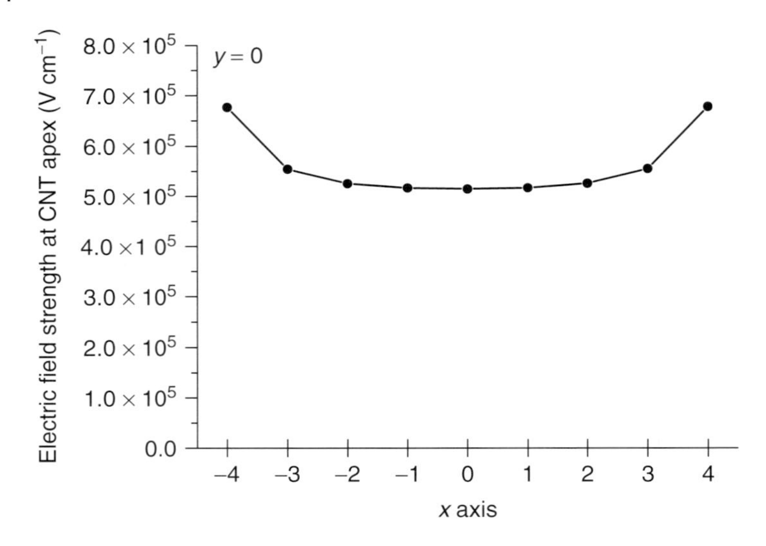

Figure 11.5 Electric field strengths at apex of each CNT aligned on the x axis.

Table 11.4 Relative difference between electric field strength at the apex of central CNT and that at apex of each CNT aligned on the x axis.

x	Electric field strength at CNT apex (V cm^{-1}) x 10^5	Relative difference (%)
−4	6.767	31.78
−3	5.538	7.85
−2	5.250	2.24
−1	5.158	0.45
0	5.135	0.00
1	5.158	0.45
2	5.250	2.24
3	5.538	7.85
4	6.767	31.78

11.5 Field Analysis for VA-CNT System with Uniform Length

In this section, we will describe a rather ideal case, that is, the calculation of the electric field strength at the CNT apex of the hemispherical-capped CNT of uniform length. It will be shown how the electric field strength is affected by the geometrical parameters of the CNTs such as their outer diameter (or the radius of curvature of the CNT apex) and their average density.

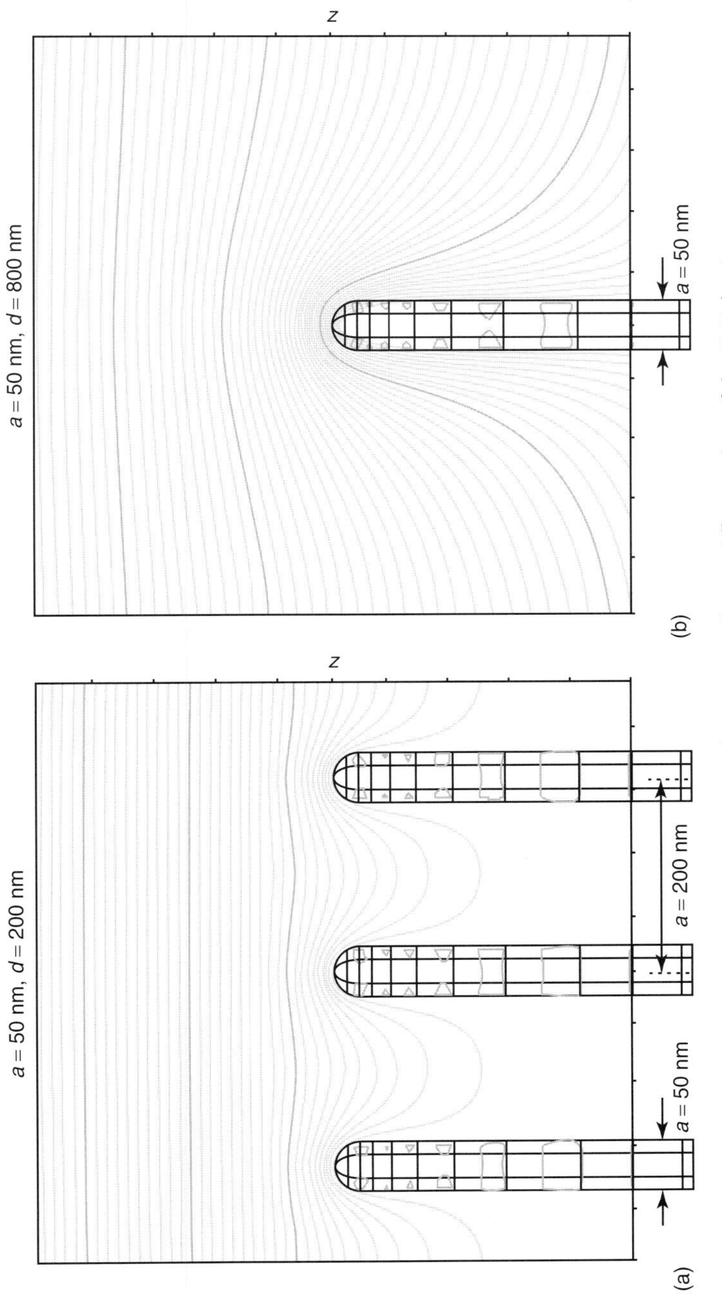

Figure 11.6 Equipotential lines (0.1 V/step) in the vicinity of CNT apexes for two different values of the CNT density.

11.5.1
Dependence of the Electric Field Strength at the CNT Apex on Geometrical Parameters of the CNTs

Figure 11.6 shows a typical example of the simulation for equipotential lines in the vicinity of CNT apexes for two different values of the CNT density, that is, for two different values of the center-to-center distance d between neighboring CNTs. It is clear that the electric field strength at the CNT apex greatly decreases when d becomes small, that is, when the CNT density increases.

Figure 11.7 shows the electric field distribution on the hemispherical surface of the capped CNT for different values of the outer diameter of the CNT under the condition that the average CNT density is $2.5 \times 10^9\,\mathrm{cm}^{-2}$ ($d = 200\,\mathrm{nm}$). It is seen that the smaller the outer diameter of the CNT, the higher the electric field strength.

Figure 11.8 shows the electric field strength at the CNT apex as a function of the outer diameter a for different values of the CNT density. Since the outer diameter a is equal to twice the radius of curvature of the CNT apex, we can say that the electric field strength at the CNT apex is inversely proportional to the radius of curvature of the apex. It is clear that if the CNT density exceeds $10^9\,\mathrm{cm}^{-2}$, then the electric field strength greatly decreases. On the other hand, if the CNT density is less than $10^8\,\mathrm{cm}^{-2}$, then the electric field strength is almost independent of the CNT density.

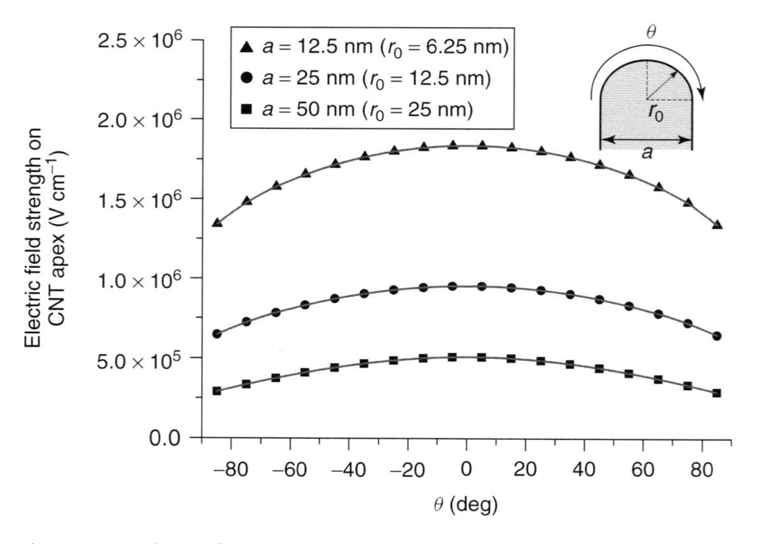

Figure 11.7 Electric field distribution on the hemispherical surface of the capped CNT for different values of the CNT outer diameter under the condition that the average density of CNTs is $2.5 \times 10^9\ \mathrm{cm}^{-2}$ ($d = 200\,\mathrm{nm}$).

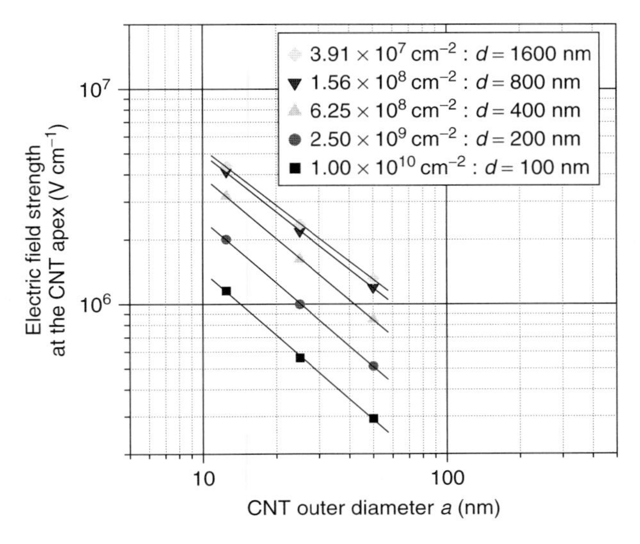

Figure 11.8 Electric field strength at the CNT apex as a function of the CNT outer diameter for different values of the CNT density.

11.5.2
Universal Curve

In order to make a more systematic analysis of the electric field strength at the CNT apex, we introduce a normalized electric field strength. That is, the electric field strength at the CNT apex, under the condition that many CNTs exists on the cathode substrate, is normalized by the one that would be obtained under the condition that only one CNT exists on the cathode substrate, as shown in Figure 11.9. The table in Figure 11.9 shows the electric field strength at the apex of the sole CNT where a voltage of 9.75 V is applied between the cathode substrate and the anode. The electric field strength at the CNT apex in the case of 9×9 CNTs is normalized by the above value. Figure 11.10 shows the normalized electric field strength at the CNT apex as a function of the average CNT density for different values of the CNT outer diameter. The value of d (i.e., center-to-center distance between neighboring CNTs) is also given on the upper horizontal axis for reference. It is clearly seen that the electric field strength at the CNT apex significantly decreases when the CNT density exceeds $10^9\,\text{cm}^{-2}$, and that it is almost independent of the CNT density when the CNT density is less than $10^8\,\text{cm}^{-2}$, which means that electric field strength at the CNT apex for the CNT density less than $10^8\,\text{cm}^{-2}$ is practically equal to that of the sole CNT.

We now introduce a new geometrical parameter p which represents the wall-to-wall distance between neighboring CNTs, as shown in Figure 11.11a. Figure 11.12 shows the normalized electric field strength at the CNT apex for different geometrical parameters of CNTs, where p is used as the horizontal axis

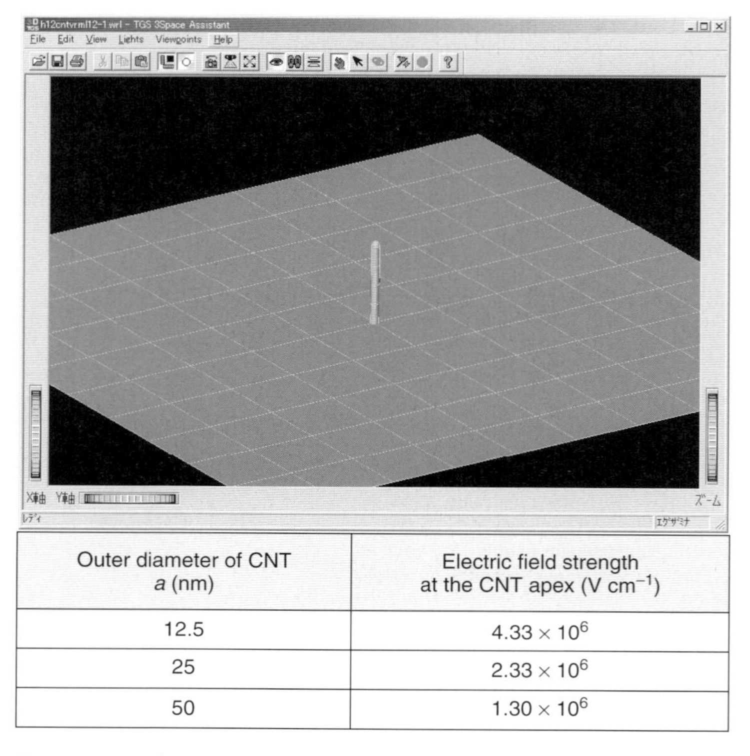

Outer diameter of CNT a (nm)	Electric field strength at the CNT apex (V cm^{-1})
12.5	4.33×10^6
25	2.33×10^6
50	1.30×10^6

Figure 11.9 Sole CNT used for normalization. Electric field strength at the apex of the sole CNT is given in the table, where a voltage of 9.75 V is applied between the cathode substrate and the anode plate.

instead of d. It is clearly seen that the normalized electric field strength can be expressed by a single curve, which means that this curve can be regarded as a universal curve for the various geometrical parameters of CNTs. When we want to know the electric field strength at the CNT apex for a specific geometrical condition of the CNTs, it is not necessary to calculate the electric field strength under the condition that many CNTs exist on the cathode substrate, but it is just sufficient to calculate the electric field strength at the apex of the sole CNT instead. Figure 11.12 also shows that in any VA-CNT system with the value p less than 800 nm, the electric field strength at the apex starts to decrease.

11.6 Field Analysis for VA-CNT System with Nonuniform Length

In this section, we will treat a more realistic case, that is, the effect of nonuniformity of the CNT length on the electric field strength at the CNT apex. Figure 11.11b

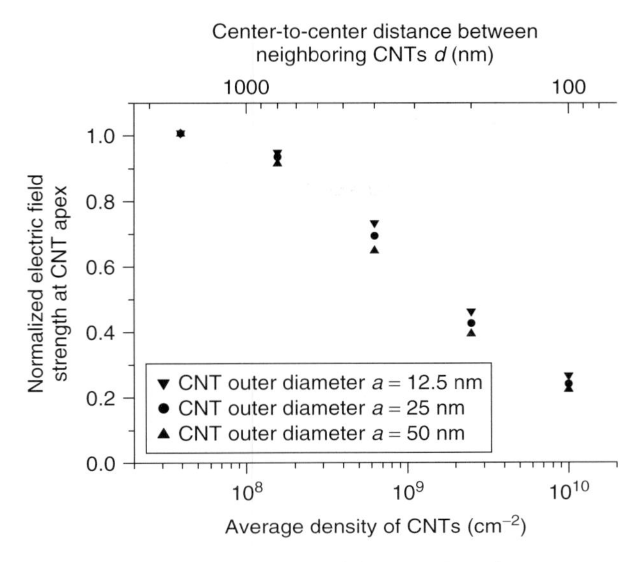

Figure 11.10 Normalized electric field strength at the CNT apex as a function of the average density of CNTs for various geometrical parameters of CNTs.

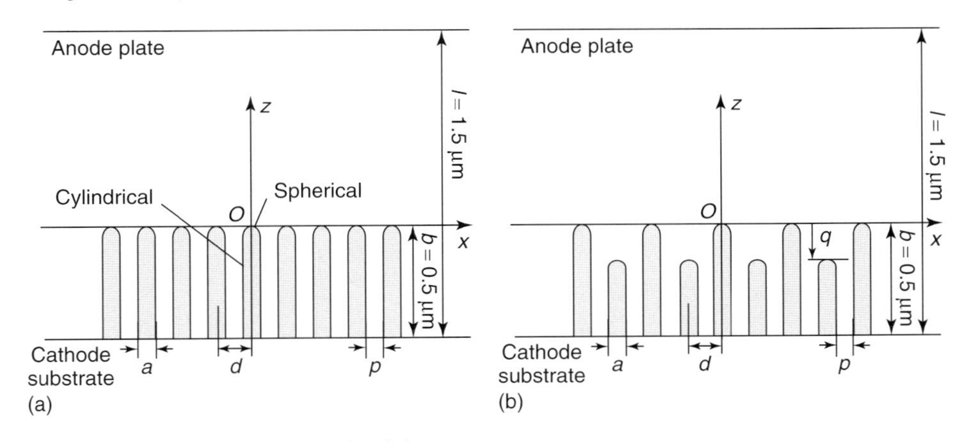

Figure 11.11 Computational models of the VA-CNT system with (a) uniform and (b) nonuniform lengths.

shows the computational model in which the CNTs having two different lengths, that is, higher CNTs and lower CNTs, are arranged one after the other. We introduce a new geometrical parameter q, which represents the difference in length between higher and lower CNTs, as shown in Figure 11.11b. The electric field strength at the apex of the higher CNT or the lower CNT is normalized by that of the CNT with a uniform length.

Figure 11.13 shows the normalized electric field strength at the higher CNT apex for different values of the geometrical parameter, where a new geometrical

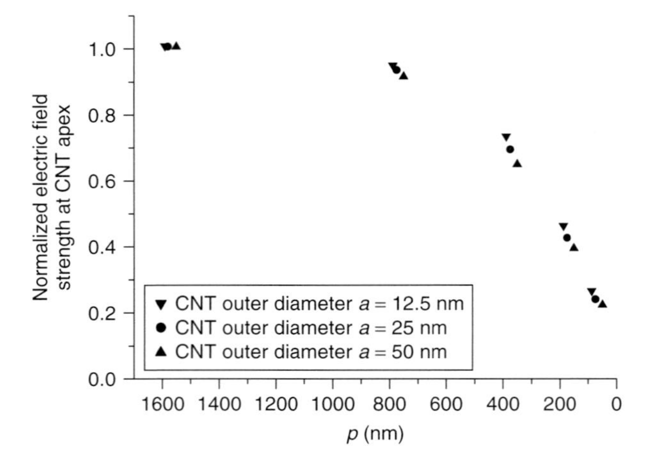

Figure 11.12 Normalized electric field strength at the CNT apex as a function of the wall-to-wall distance between neighboring CNTs for various geometrical parameters of CNTs.

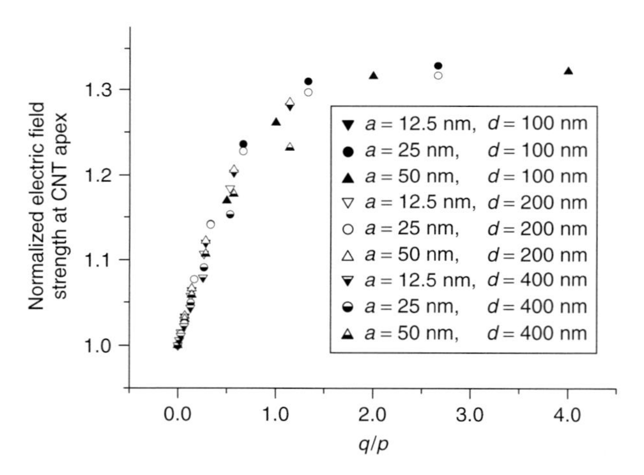

Figure 11.13 Normalized electric field strength at the higher CNT apex as a function of q/p for different geometrical parameters.

parameter q/p is introduced on the horizontal axis. It is seen that the normalized electric field strength at the higher CNT apex is expressed by a single curve, which means that this curve can be regarded as a universal curve of the field strength at the higher CNT apex for various geometrical parameters of CNTs. It is also clear that if the value of q/p is greater than 1.5, then the normalized field strength at the higher CNT apex is independent of the value of q/p, which means that the effect of lower CNTs can be neglected.

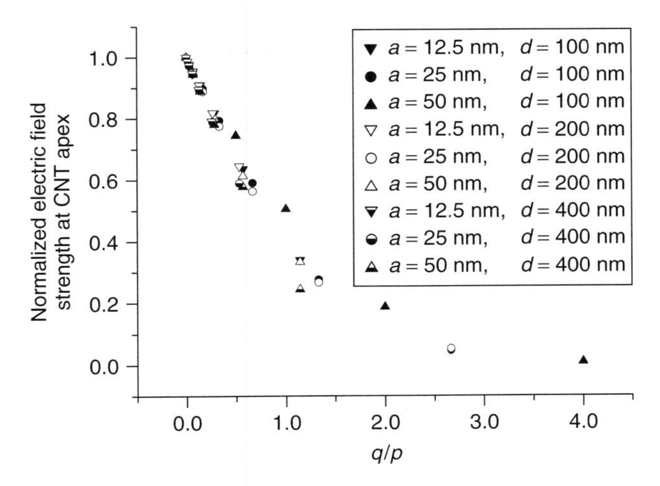

Figure 11.14 Normalized electric field strength at the lower CNT apex as a function of q/p for different geometrical parameters.

Figure 11.14 shows the normalized electric field strength at the lower CNT apex as a function of q/p for different geometrical parameters. The normalized electric field strength is again expressed by a single curve for various geometrical parameters, which means that this curve can also be regarded as a universal curve of the electric field strength at the lower CNT apex. It is also clearly seen that if the value of q/p is greater than 1.5, then the normalized field strength at the lower CNT apex decreases remarkably, which means that if q/p is greater than 1.5, then the field emission from lower CNTs can be neglected.

11.7 Effect of Shape of CNT Apex

In this section, we will examine the dependence of the electric field strength at the CNT apex on various shapes of the CNT apex, as shown in Figure 11.15.

Figure 11.16 shows the highest value of the electric field strength as a function of the radius of curvature r_0 of the CNT apex. The highest electric field strength occurs at the center of the CNT apex for a hemispherical-capped CNT and a horn-capped CNT. For a flat-capped CNT, the highest electric field strength occurs along the circumferential corner between the flat cap and the cylindrical wall, and the radius of curvature of the circumferential corner is indicated as r_0. For a cap-free CNT, the highest electric field strength occurs along the torus region on the top edge of the cylindrical wall, and the radius of curvature of the torus region is indicated as r_0.

As can be seen in Figure 11.16, for any shape of the CNT apex the electric field at the CNT apex is inversely proportional to the radius of curvature of the apex. For the same value of the radius of curvature of the apex, the electric field strength at

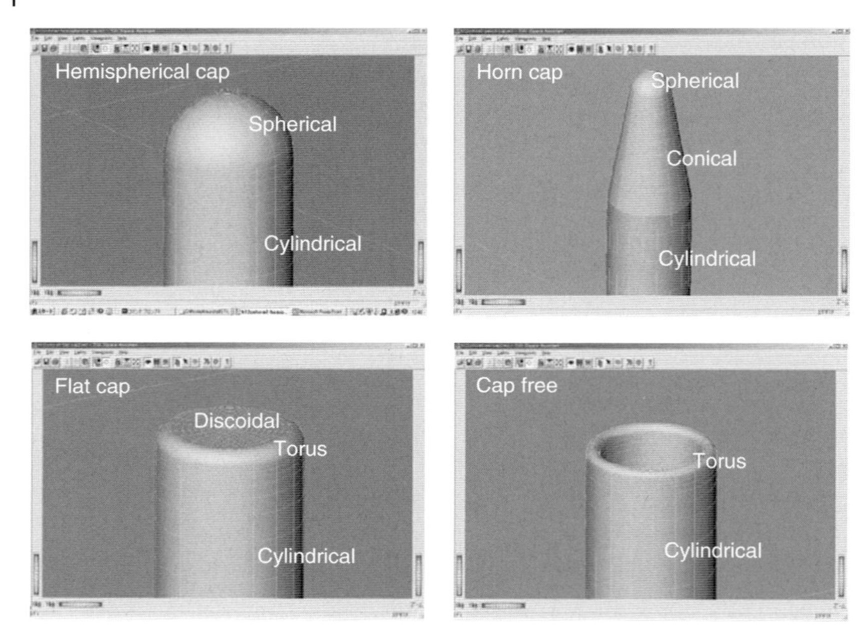

Figure 11.15 Various shapes of the CNT apex.

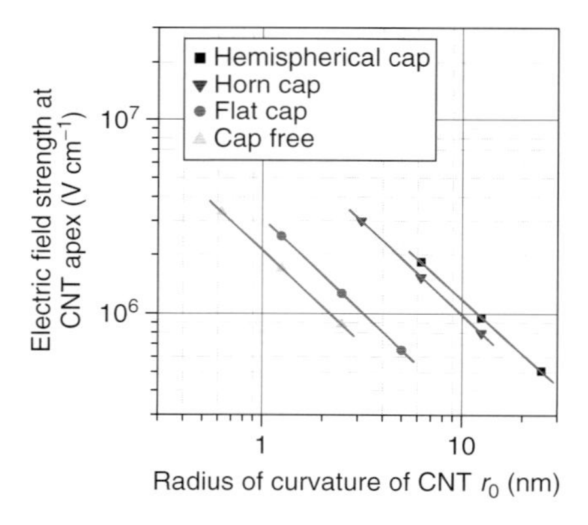

Figure 11.16 Electric field strength at the CNT apex as a function of the radius of curvature of the apex for various shapes of the CNT apex under the condition that the average density of CNTs is 2.5×10^9 cm^{-2} ($d = 200$ nm).

the apex of hemispherical-capped CNT is the highest and becomes about five times as high as that of a cap-free CNT.

11.8 Effect of CNT Length

We will next examine the effect of length of the CNT on the electric field strength at the CNT apex. For this purpose, a more practical VA-CNTs system, in which the sizes of the cathode substrate and the anode plate are 1 mm^2 and the distance

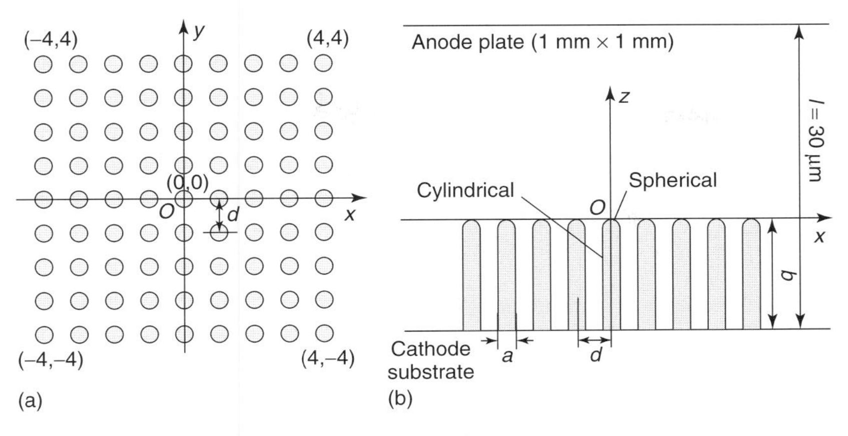

Figure 11.17 Computational model of a practical VA-CNTs system: (a) top view and (b) side view.

between the cathode substrate and the anode plate is 30 μm, is modeled by 9 × 9 CNTs standing vertically on the cathode substrate, as shown in Figure 11.17. Each CNT is modeled by a combination of a hemispherical surface and a cylindrical surface. The potentials of the cathode substrate and the anode plate are set to 0 and 195 V, respectively. The diameter of each CNT is 50 nm. Figure 11.18 shows the electric field strength at the apex of the central CNT as a function of the average CNT density for different values of the CNT height b. It is clearly seen that the electric field strength at the CNT apex significantly decreases as the CNT density increases. It seems that in the case of a short CNT the electric field strength at the

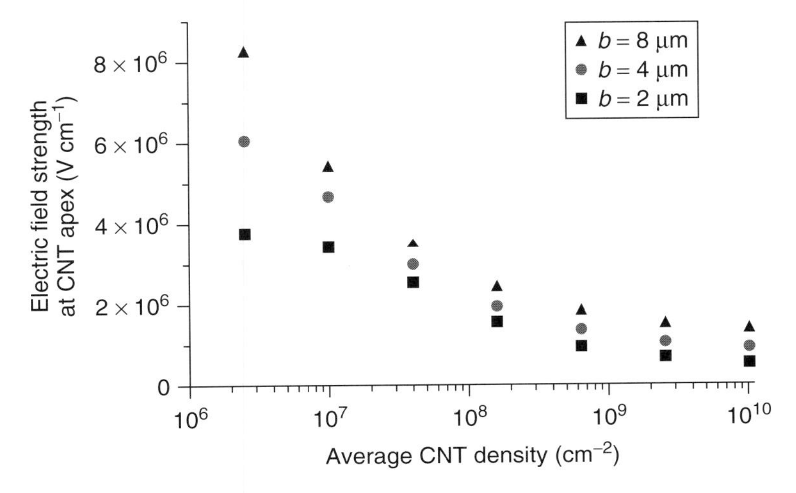

Figure 11.18 Electric field strength at CNT apex as a function of the average density of CNTs for different values of CNT height.

apex is getting saturated as the CNT density decreases. This means that when the CNT density is sufficiently small, the electric field strength at the apex of a short CNT is practically equal to that of the sole CNT standing alone vertically on the cathode substrate. In contrast, in the case of a long CNT the electric field strength at the apex of the CNT seems to keep on increasing even when the CNT density is less than 10^6 cm^{-2}.

11.9
Electric Field Analysis of Network-Structured CNT System

In this section, we will discuss the electric field analysis of network-structured CNTs [3] grown by thermal-chemical vapor deposition, which form an entangled gossamer network structure, and we examine the dependence of the electric field strength at the apexes of CNTs on the geometrical parameters such as CNT length and the average CNT density. The network-structured CNTs are modeled by a horizontal 10×10 square metal mesh grid parallel to the substrate and vertical 9×9 hemispherical-capped CNTs perpendicular to the substrate, as shown in Figure 11.19. For the sake of comparison, the electric field analysis of the 9×9 CNT computational model without the horizontal 10×10 square mesh grid is also performed under the same geometrical condition. Figure 11.20 shows typical examples of the computational model of the network-structured CNTs. The potentials of the cathode substrate including the CNTs and the anode plate are set to 0 and 195 V, respectively. The average CNT density is varied from 6.1×10^5 ($d = 12$

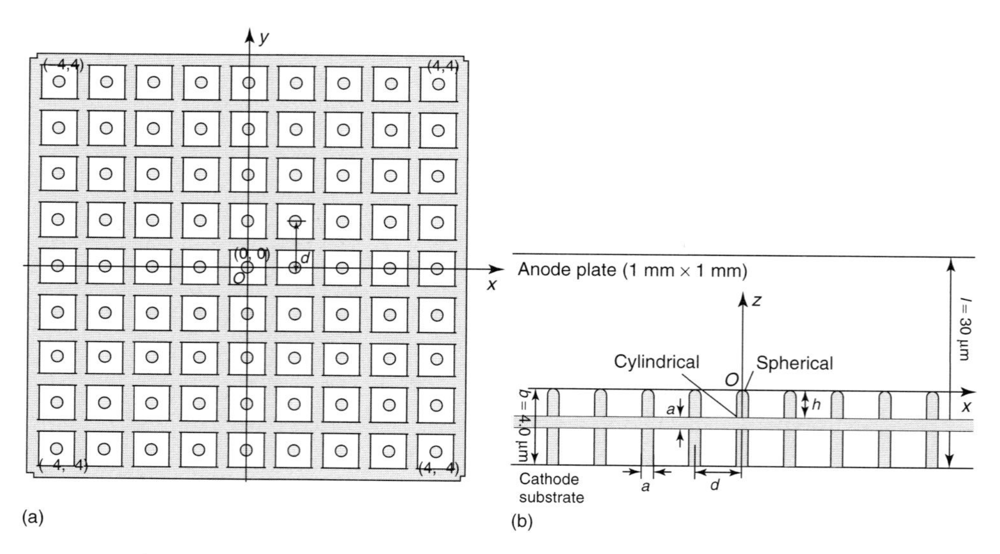

Figure 11.19 Computational model of a network-structured CNT system: (a) top view and (b) side view.

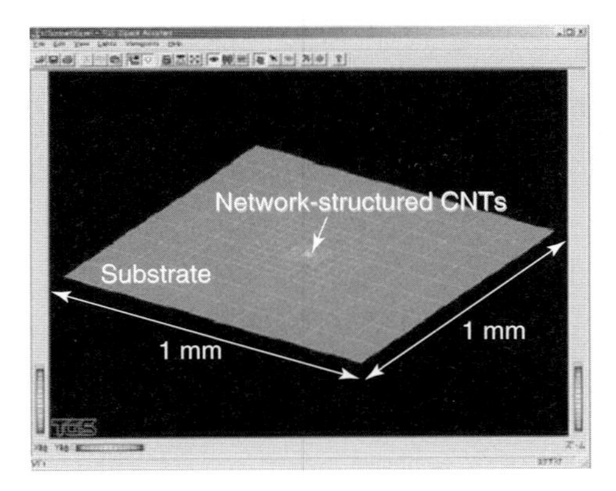

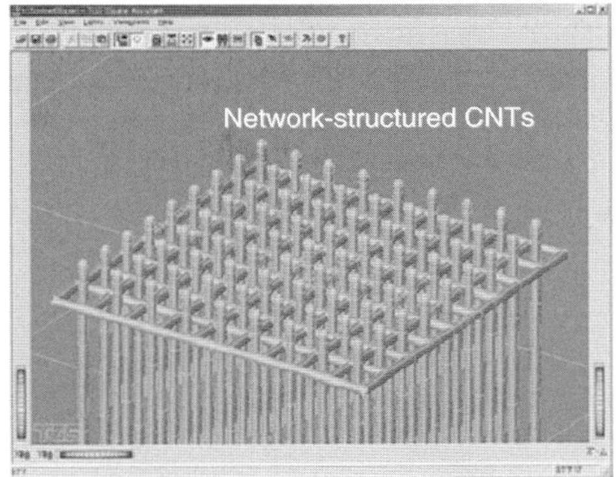

Figure 11.20 Typical examples of the computational model of the network-structured CNT system.

800 nm) to $4.4 \times 10^9\ \text{cm}^{-2}$ ($d = 150$ nm) under the conditions that the CNT outer diameter a is 50 nm and the CNT length b is 4 μm. The electric field strength at the CNT apex is calculated for different values of CNT protrusion h, which represents the height from the horizontal 10×10 square mesh grid to the apex of CNTs. The results for $h = 200$ nm are plotted in Figure 11.21. As can be seen in the figure, the electric field strength at the CNT apex in the case of network-structured CNTs is much weaker than that in the case of VA-CNTs for an average CNT density ranging from 10^6 to $10^7\ \text{cm}^{-2}$. From these results, it is found that the electric field strength at the CNT apex decreases by the existence of the network-structured CNTs.

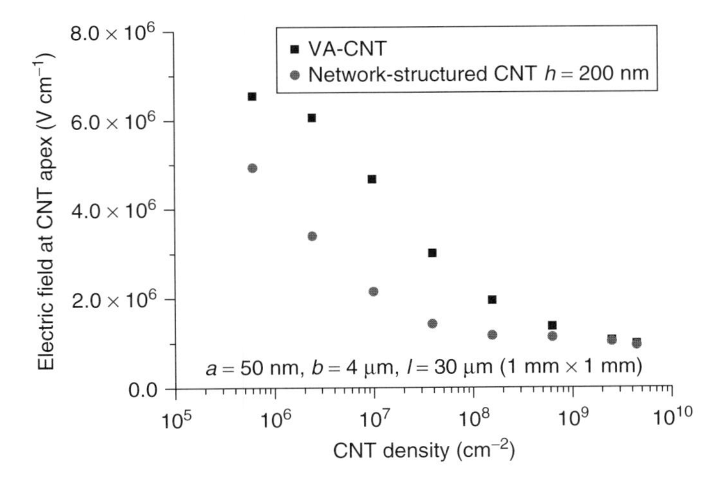

Figure 11.21 Electric field strength at the CNT apex as a function of the average density of CNTs for network-structured CNTs (protrusion $h = 200$ nm) and VA-CNTs.

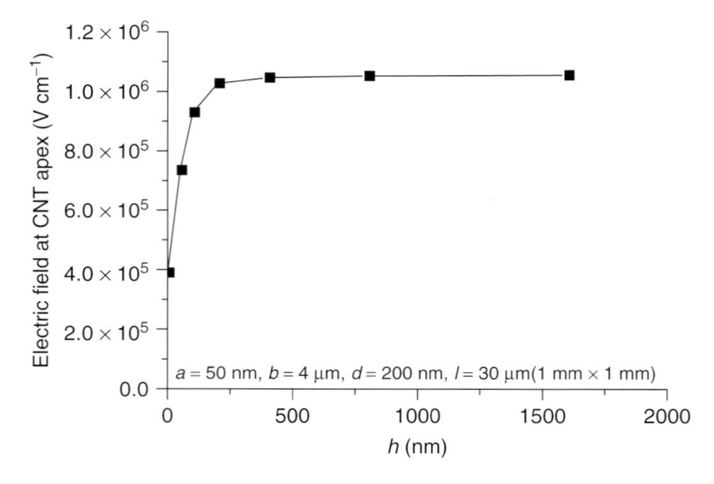

Figure 11.22 Electric field strength at the CNT apex calculated as a function of CNT protrusion h for an average CNT density of 2.5×10^9 cm^{-2} ($d = 200$ nm).

Figure 11.22 shows the electric field strength at the CNT apex calculated as a function of the CNT protrusion h for an average CNT density of 2.5×10^9 cm^{-2} ($d = 200$ nm). From this result, it is found that the electric field strength at the CNT apex decreases rapidly when h is below 200 nm. This means that the electric field strength at the CNT apex is decreased by the screening effect of the network structure when the CNT protrusion h is below the center-to-center distance between neighboring CNTs.

References

1. Murakami, H., Hirakawa, M., Tanaka, C., and Yamakawa, H. (2000) Field emission from well-aligned, patterned, carbon nanotube emitters. *Appl. Phys. Lett.*, **76**, 1776.
2. Kusunoki, M., Suzuki, T., Kaneko, K., and Ito, M. (1999) Formation of self-aligned carbon nanotube films by surface decomposition of silicon carbide. *Philos. Mag. Lett.*, **79**, 153–161.
3. Yotani, J., Uemura, S., Nagasako, T., Kurachi, H., Yamada, H., Ezaki, T., Maesoba, T., Nakao, T., Ito, M., Ishida, T., and Saito, Y. (2004) Emission enhancement by excimer laser irradiation over a web-like CNT layer. *Jpn. J. Appl. Phys.*, **43**, L1459–L1462.
4. Murata, H., Ohye, T., and Shimoyama, H. (1996) in *Charged-Particle Optics II*, Proceedings of SPIE, vol. 2858, (ed. E. Munro), The International Society for Optical Engineering, pp. 103–114.
5. Murata, H., Ohye, T., and Shimoyama, H. (1997) in *Charged Particle Optics III*, Proceedings of SPIE, vol. 3155 (ed. E. Munro), The International Society for Optical Engineering pp. 113–124.
6. Kuno, Y., Yagi, A., Morishima, T., and Uchikawa, Y. (1988) Fundamental study for high accuracy calculation of 3-D electromagnetic field. *IEEE Trans. Magn.*, **MAG-24**, 295–298.
7. de Heer, W.A., Châtelain, A., and Ugarte, D. (1995) A carbon nanotube field-emission electron source. *Science*, **270**, 1179–1180.

12
Surface Coating of CNT Emitters

Yoshikazu Nakayama

The main reasons why carbon nanotubes (CNTs) have attracted attention as field emitter devices are their low operational voltage by virtue of their small tip diameter and their high aspect ratio, in addition to expectations of a long, useful life due to the CNT structure with its strong C–C double bonds. From this perspective, the field emission properties of CNTs and their application to devices have been examined extensively during the past decade, and promising results have been reported [1–10]. However, the performance of CNT field emitters that have been investigated to date remains disappointing; they remain inadequate for practical use in flat panel displays, scanning electron microscopes (SEMs), X-ray sources, and other field-emission-based devices. Further decreases in the operational voltage are necessary for flat panel displays to simplify their driving circuit and device structures and to increase the emission current for electron microscopes and X-ray tubes. The reduction of fluctuation over time of the emission current and the damage of the emission sites are other important issues for all practical applications.

Coating the surface of CNTs with functional materials has been studied as an effective approach to overcome the issues described above [11–23]. Increasing the tip radius of electron emitters by coating them is a negative factor; however, the operating voltage is reduced and the emission current is increased when the coating layer decreases the effective potential barrier through which electrons have to pass to be emitted into vacuum. The coating layer, which has a secondary electron emission phenomenon or a high density of states for electrons, also enhances the emission current. The coating layer has the function of a varistor and can suppress the fluctuation over time of the emission current. Moreover, the layer can prevent deterioration of an electron emitter by protecting the CNT tips from high-energy ions that would impinge during operation, thereby lengthening the emitter lifetime. This chapter presents a summary of the expected effects of the surface coating of CNTs on field emission properties and demonstrates some cases.

Carbon Nanotube and Related Field Emitters: Fundamentals and Applications. Edited by Yahachi Saito

ISBN: 978-3-527-32734-8

12.1
Effects of Surface Coating of CNT Emitters

12.1.1
Parameters Determining Field Emission Properties

The field emission current in a diode structure consisting of two parallel plates is given as

$$I \propto (V/d)^2 \exp\left(-Bd\phi^{3/2}/\beta V\right) \tag{12.1}$$

where V signifies the applied voltage, d stands for the electrode gap, ϕ denotes the work function of the emitter, β represents the compensation term considering the mirror effect, and $B = 6.83 \times 10^7(\mathrm{e}^{-3/2}\mathrm{V}^{-1/2}\,\mathrm{cm}^{-1})$. According to Eq. (12.1), a plot of I/V^2 versus $1/V$ or I/F^2 versus $1/F$ (the so-called Fowler–Nordheim plot) shows a straight line with a slope depending on ϕ, where $F = V/d$.

For field emissions from numerous CNTs deposited on an emitter plate, their properties are usually analyzed using Eq. (12.1), although many types of arrangements of CNTs exist: vertically aligned, partially raised, randomly oriented, and so on. In this case, β has a different meaning. It is called the *field amplification factor*, which is larger when the CNTs are thinner.

However, we cannot apply Eq. (12.1) to the field emitter of a single needle tip because the electric field concentrates locally on the tip. Assuming that the CNT tip is a hemisphere, the local electric field at the tip is given as

$$F_0 = V/kr \tag{12.2}$$

where k is the shape factor with a value of 3–5 and r is the tip apex radius. The local electric field near the tip decreases according to $F = Vr/k(r+x)^2$, with increasing distance x from the tip surface. Considering this electric field, the emission current is written as [5, 6]

$$I(F) \propto F_0^2 \exp\left\{-Cr\left[\frac{eV}{k(eV - K\phi)}\left(\frac{\pi}{2} - \sin^{-1}\sqrt{\frac{eV - k\phi}{eV}} - \sqrt{\phi}\right)\right]\right\} \tag{12.3}$$

where $C = 1.02 \times 10^8(\mathrm{e}^{-3/2}\mathrm{V}^{-1/2}\,\mathrm{cm}^{-1})$ and the mirror effect is not considered. The Fowler–Nordheim plot of the curves calculated using Eq. (12.3) is not straight, which differs from the case with Eq. (12.1). However, in the limited region within which practical data are usually located, the curves calculated using Eq. (12.3) can fit to a straight line whose slope is that for the case with 1.30 times the value of B in Eq. (12.1). Therefore, we can use the modified Eq. (12.1) with $B = B' = 8.0 \times 10^7(\mathrm{e}^{-3/2}\mathrm{V}^{-1/2}\,\mathrm{cm}^{-1})$ and $\beta = 1$ instead of the strict and complex expression of Eq. (12.3) for estimating ϕ or r [5, 6]:

$$I \propto (V/r)^2 \exp\left(-B'r\phi^{3/2}/V\right) \tag{12.4}$$

Equations (12.2) and (12.3) show that the electric field near the tip depends on the applied voltage and the tip apex radius and the emission current is independent of the emitter–collector gap. The turn-on voltage of the emission current from bundled CNTs with about 45 μm length consisting of CNTs with about 10 nm diameter to a collector plate does not depend on the electrode gap when the electrode gap is less than 310 nm. However, the turn-on voltage increases when the electrode distance increases to more than 310 nm [5]. The averaged electric field – the applied voltage divided by the electrode gap – has no meaning for a single CNT emitter.

Consequently, the field emission devices of numerous CNTs deposited on the electrode plate and a single CNT emitter can be estimated using the formulas similar to Eqs. (12.1) and (12.4), respectively. For both cases, the parameters controlling the emission current are the tip apex radius and the work function. A coating on the CNT tip can decrease the effective work function or the potential barrier. However, the tip apex radius increases because of the coating, which is a negative point. A selected combination of the coating layer thickness and the reduction in the potential barrier can reduce the operating voltage and increase the emission current.

12.1.2 Lowering of the Potential Barrier

Values of 4.5–4.85 eV have been used as the work function of the tip of CNTs. These values represent the potential barrier through which electrons are emitted into vacuum by a tunneling effect when CNTs are applied to the electron emitters. The potential barrier can be decreased in two ways by the coating of CNTs.

12.1.2.1 Coating Layer with Low Work Function

Cesium (Cs) is well known as a *conductive material* with a work function as low as 1.93 eV. When a Cs layer is coated onto the CNT tip, the potential barrier for the field emission ideally becomes 1.93 eV, which is much smaller than the work function of CNTs. Estimation using Eq. (12.4) indicates that the emission current from a single CNT emitter with an apex radius of 10 nm coated with a 2-nm-thick Cs layer is more than two orders of magnitude larger at $V = 100$ V than that from an uncoated, single CNT emitter of equal size. An increase of six orders of magnitude by coating Cs onto randomly aligned single-walled CNTs deposited on an electrode has been reported [11]. The Cs layer, however, reacts actively with oxygen to degrade its function. As a material stable to oxygen, TiC, which has excellent electrical and thermal conductance and high melting point, has been examined [12–14]. Its work function is less than 3.0 eV [24], which is smaller than that of CNTs.

12.1.2.2 Coating Layer with Wide Band Gap

For the coating of a monolayer of BN on a CNT, the interaction between the localized states of the CNT and the BN layer was investigated using first-principles pseudopotential calculation [15]. When a capped (5,5) CNT is covered with a capped (10,10) BN tube, the localized states near the Fermi level of the CNT tip are hybridized with the localized states in the conduction band of the BN layer at an electric field higher than 5 V nm^{-1}, thereby forming new states. Electrons in the

CNT cap are transferred into the BN cap through the new states. The new states have an electron affinity (corresponding to the work function in Eqs. (12.1), (12.3), and (12.4)) lower than the work function of the CNT cap. The tunneling probability of the electrons becomes large. The case of the open end of a (4,4) CNT covered with a cap of (8,8) BN nanotube enhances the field emission current over widely varying electric fields.

In the experiments, coating layers that are thicker than the monoatomic layers have been used. In addition, various wide-band-gap materials such as MgO, TiC, and SiO_2 have been applied to achieve a great effect in reducing the operating voltage and increasing the emission current [16–23]. Figure 12.1 shows a model that can explain the experimental results. The hybridized states are formed near the interface region in the wide-band-gap layer (WBGL), as described in the theoretical investigation [15]. Electrons at the CNT cap can enter the WBGL through the newly formed hybridized states connected to the conduction band of the WBGL. The electrons in the conduction band can tunnel into vacuum through the potential barrier or the electron affinity of the WBGL, which is lower than the work function of the CNT. For example, the electron affinity of MgO is 0.85 eV [21]. However, when the WBGL is thick, electrons lose their energy through inelastic scattering in the WBGL and the emission current is decreased. The inelastic scattering provides amplification of the emission current under an optimized condition if the WBGL has a high efficiency of the secondary electron emission [22].

Matching of the coating layer and the CNT is important to achieve the positive effects described above. The coating layer of amorphous carbon formed by electron-beam-induced chemical vapor deposition in SEM – which has a density of less than 1.0 g cm^{-3} [25] – increases the turn-on voltage for the field emission and causes instability of the emission current [26, 27]. On the other hand, the amorphous carbon layer prepared using cathodic arc deposition, which has a density of 2.6 g cm^{-3}, demonstrated an effective decrease in the turn-on voltage [18].

We consider now the difficulty in the two-step tunneling model. In this model, electrons in the CNT tunnel the potential barrier at the interface with the WBGL to reach the conduction band of the WBGL and then tunnel its electron affinity to be

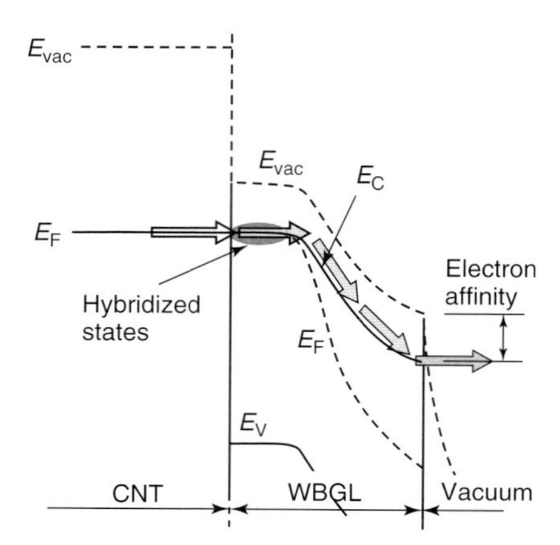

Figure 12.1 Band structure of a CNT coated with a wide-band-gap layer to explain the field emission mechanism at a high electric field, where E_{vac} is the energy level of vacuum, E_F is the Fermi energy, E_C is the conduction band bottom, and E_V is the valence band top.

emitted into vacuum. The probability of the first tunneling is, in general, smaller than that of direct tunneling from the CNT to the vacuum because the relative permittivity of the coating layer is much greater than unity in vacuum. It decreases the electric field in the layer and thereby increases the potential barrier for the first tunneling.

12.1.3 Stabilization of Emission Current

It is well known that the instability of the emission current from CNT emitters is caused mainly by the adsorption and desorption of molecules at the CNT tip [10] and degradation of the CNT tip, in addition to the statistical fluctuation inherent in the quantum mechanical tunneling effect. One can expect that the fluctuation of the emission current described above is suppressed if the layer coated onto CNT tips has a function similar to that of a varistor. We should consider the moment at which the current starts to increase for some reason. The coating layer resistance decreases, thereby weakening the electric field applied to the coating layer. Consequently, the increase in the emission current stops. When the current starts to decrease, the resistance – the electric field of the coating layer – becomes large and the decrease in the current stops. This mechanism serves to keep the current constant [23]. The coating layer also plays a role in protecting the CNT tip from degradation.

12.2 Field Emission from Individual CNT Coated with BN

An individual multiwalled CNT was attached to a W needle tip using electron-beam-induced deposition [28]. Figure 12.2 presents the SEM and transmission electron microscope (TEM) images of the sample. The CNT was produced with an arc discharge, with end caps and with 15 nm outer diameter and 500 nm length. The 3-nm-thick BN layer uniformly covered the CNT; its tip radius was measured as 11 nm. The BN layer was deposited on the individual CNT sample heated at 650 °C using a remote plasma with BC_{13} and N_2. The BN layer was polycrystalline, containing B, N, C, and O atoms in the ratio 39 : 37 : 10 : 14; C and O were incorporated unintentionally.

Field emission properties of CNT and BN-coated CNT samples were measured at room temperature at a vacuum of 2×10^{-8} Pa. The emitter–collector gap was 25 mm. Their BN thicknesses were 3, 6, and 8 nm, and the respective tip radii of the BN-coated CNT samples were 11, 14, and 16 nm.

The turn-on electric field (defined as the electric field at an emission current of 1.0 nA) increases in the order of the samples with BN layers of 3, 6, 0, and 8 nm thickness. Figure 12.3 portrays the Fowler–Nordheim plot of the CNT and BN-coated CNT samples. The data fall roughly on straight lines. The slope is proportional to $r\phi^{3/2}$ according to Eq. (12.4). The effective potential barrier of

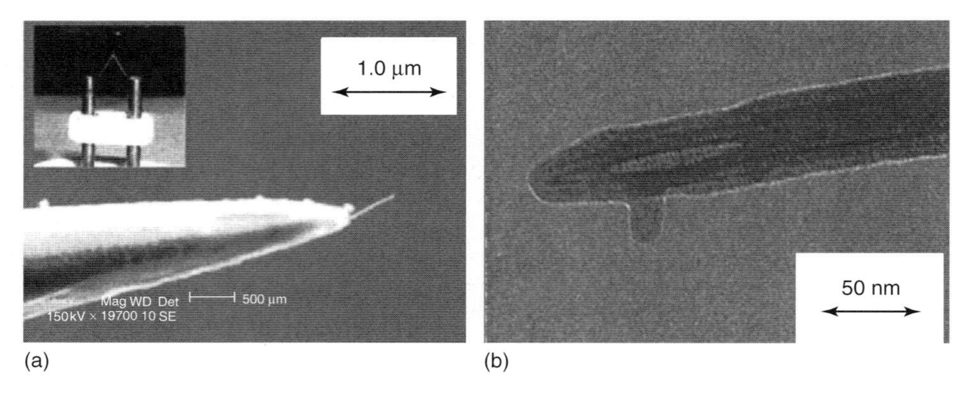

Figure 12.2 SEM image of uncoated CNT attached on a tungsten tip. The inset shows a sample holder with a tungsten tip (a) and TEM image of a BN-coated CNT (b). (Reprinted with permission from [16]. Copyright [2008], AVS: Science & Technology Society.)

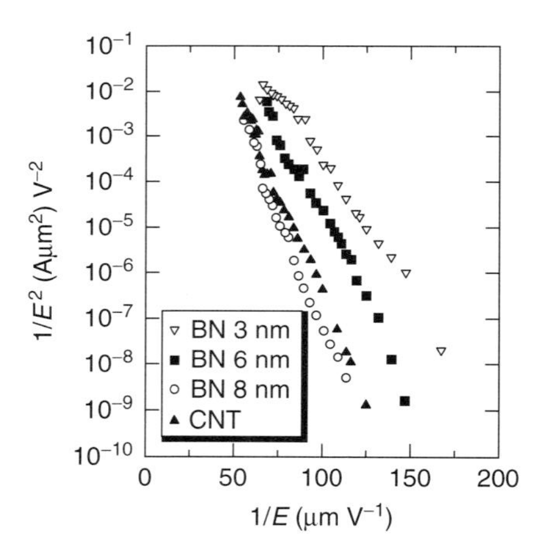

Figure 12.3 Fowler–Nordheim plot of uncoated and BN-coated CNTs with different thicknesses. (Reprinted with permission from [16]. Copyright [2008], AVS: Science & Technology Society.)

BN-coated (3 nm) CNT is estimated as 2.8 eV by assuming that the work function of the CNT tip is 4.85 eV. The thin coating of BN on the CNT tip markedly reduces the effective potential barrier. It thereby decreases the turn-on voltage and increases the emission current, as discussed in Section 12.1.2.

Figure 12.4 depicts the temporal variation of emission current around 10 μA for samples with and without the BN layer with 3 nm thickness. The fluctuation of the emission current is 8.4% for the CNT sample and is reduced to 2.8% for the BN-coated (3 nm) CNT sample. It is demonstrated that the BN coating is effective in achieving stable operation. This fluctuation suppression results from the coating layer, as discussed in Section 12.1.3.

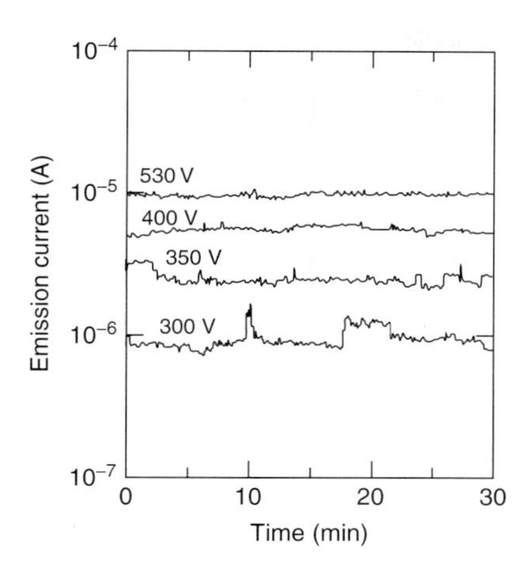

Figure 12.4 Temporal variation of the field emission current of the BN-coated (3 nm) CNT at different operating voltages. (Reprinted with permission from [16]. Copyright [2008], AVS: Science & Technology Society.)

12.3
Field Emission from Brush-Like CNTs Coated with MgO

The selective growth of highly aligned multiwalled CNTs was performed in the CVD chamber, using the patterned Fe catalyst deposited on Si substrates. The pattern is an array of squares of $10 \times 10\,\mu m^2$ with a repetition period of $90\,\mu m$. The CNT growth was performed at $700\,^{\circ}C$ using diluted C_2H_2 with He at 13%. The grown aligned CNTs patterned on Si substrates were subsequently coated with MgO layers of different thicknesses of 2, 5, 10, and 20 nm. The MgO coating was carried out at $350\,^{\circ}C$ using an electron beam evaporation technique with MgO powder. The coated MgO has a crystalline structure, as confirmed through X-ray diffraction.

Figure 12.5a and b respectively present the SEM images of top surface of uncoated and MgO-coated (10 nm) CNTs [23]. The corresponding high-magnification images of the selected area of the top surfaces are shown in the insets of the figures. The insets of Figure 12.5 reveal that the CNT diameter changed from about 20 to 35 nm by the coating of a 10-nm-thick MgO layer. The MgO coating was throughout the total surface of the nanotubes, not only on the tips.

Figure 12.6 presents the field emission current versus voltage plots of uncoated and MgO-coated CNTs with various layer thicknesses. The MgO-coated (10 nm) CNTs show the lowest turn-on voltage (here defined as the applied voltage necessary to obtain an emission current of 0.1 nA). For the field emission for a thin MgO-coated (2 nm) sample, the turn-on voltage becomes high: the value is 200 V, which is higher than that of uncoated CNTs. However, 5- and 10-nm-thick MgO layer coated CNTs show a continuous decrease in the turn-on voltage, indicating improved emission properties. The turn-on voltage, however, increases again for the 20-nm MgO-coated CNTs, more than those of 5- and 10-nm MgO-coated CNTs.

(a)

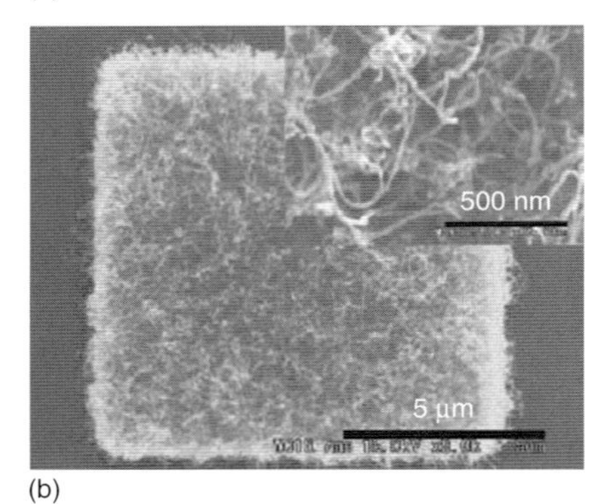

(b)

Figure 12.5 SEM images of the top surfaces of (a) uncoated CNTs and (b) MgO-coated (10 nm) CNTs. Insets show corresponding high-magnification images. (Reprinted with permission from [23]. Copyright [2007], APEX/JJAP.)

Figure 12.7 presents a Fowler–Nordheim plot showing field emissions from uncoated and MgO-coated CNTs. As shown in Eq. (12.1), the slope of this plot is proportional to $\phi^{3/2}/\beta$. The field enhancement factor is related to the CNT diameter. It is reasonable to assume that the well-separated CNT tips are responsible for the field emission because the electric field intensity on the CNT tip is about three times larger than that on the CNT sidewall [29]. Moreover, the randomly oriented CNT tips become aligned easily parallel to the field when an electric field is applied across electrodes [30]. Therefore, assuming $\beta \propto 1/r$, which is related to Eq. (12.2), we can roughly estimate the effective potential barrier of each sample using a work function of 4.5 eV for the CNTs. The estimated value for the MgO-coated (10 nm) CNTs is 2.5 eV, which is much lower than the value of 4.5 eV of the CNTs.

This decrease in the effective potential barrier is explainable using the model described in Section 12.1.2.2, although the potential barrier is higher than the reported electron affinity of 0.85 eV of MgO [21]. The reason why the 2-nm-thick MgO coating is ineffective is not well understood, but it is considered that the thin layer is insufficient to modify the band structure, as presented in Figure 12.1.

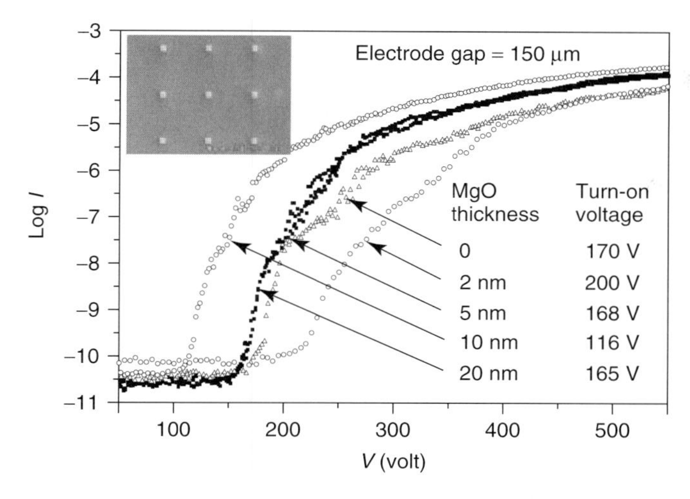

Figure 12.6 I–V characteristics of the field emission from aligned, uncoated CNTs and MgO-coated CNTs with different thicknesses. The inset shows an SEM image of the patterned CNTs. (Reprinted with permission from [23]. Copyright [2007], APEX/JJAP.)

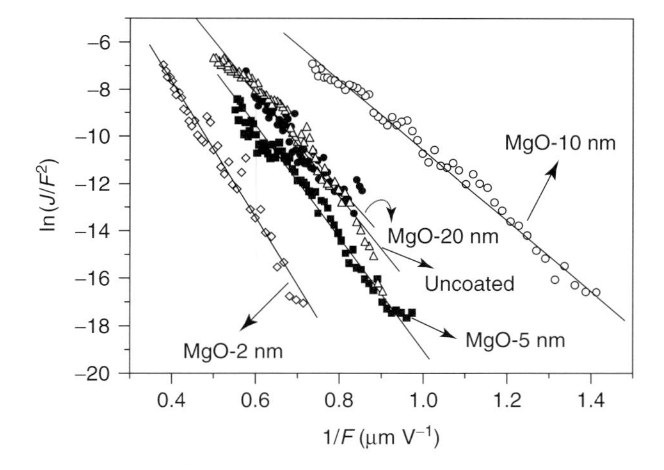

Figure 12.7 Fowler–Nordheim plot of the uncoated and MgO-coated CNTs with different thicknesses. (Reprinted with permission from [23]. Copyright, APEX/JJAP.)

The stability, that is, the emission current fluctuation with time for a fixed applied voltage, was also studied. A gradual increase in the stability is observed for 5- and 10-nm MgO-coated CNTs compared to that of uncoated CNTs. The current fluctuation becomes as low as 9% for MgO-coated (10 nm) CNTs, although it is 31% for uncoated CNTs. For the MgO (2 nm) layer, electrons were transported to be emitted into the vacuum by tunneling, showing less suppression of the current fluctuation. However, thicker layers have diffusive electron transport and function as a varistor to suppress the current fluctuation, as described in Section 12.1.3.

In addition, MgO is well known to have a large coefficient of secondary electron generation. Therefore, one can expect that the kinetic energy of a high-energy electron is transferred to the lattice of MgO to generate secondary electrons in the diffusive electron transport. Its generation coefficient is sensitive to the potential slope inside the MgO layer. This phenomenon also contributes to the reduction of the current fluctuation [23].

12.4 Field Emission from Brush-Like CNTs Coated with TiC

A good candidate for the coating conducting material with a work function (<3.0 eV) [24] lower than that of CNT is TiC. This material also has good thermal conductivity, high melting point, and high hardness [31]. These properties provide low resistive emitters that can operate at high temperatures and that are not vulnerable to ion bombardment during operations.

The patterned brush-like CNTs, as presented in Figure 12.8, which have short lengths of less than 10 μm and diameters of approximately 10–20 nm, were coated

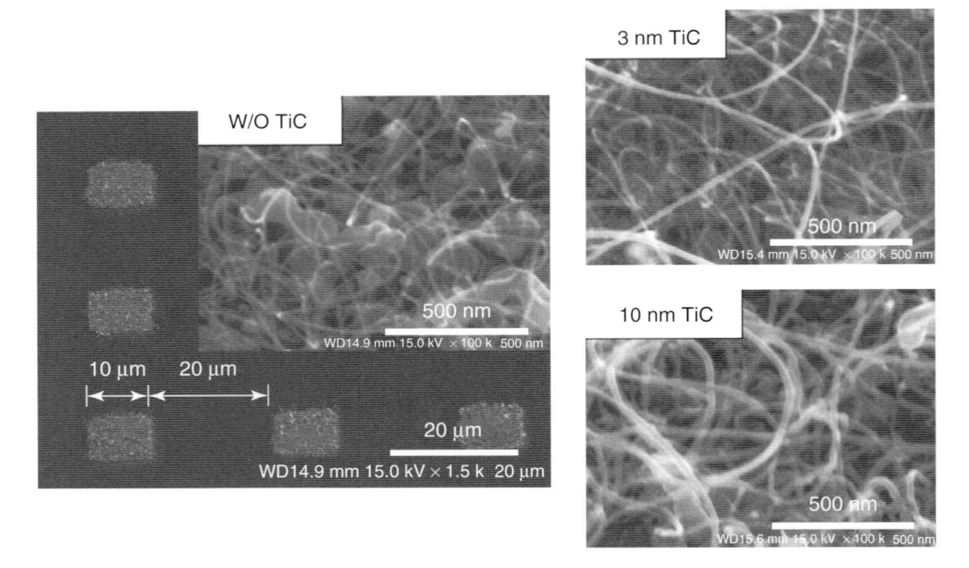

Figure 12.8 SEM image of a CNT array of squares of $10 \times 10\,\text{mm}^2$ with a repetition period of 30 mm. The enlarged images of the uncoated and TiC-coated CNTs with 3 and 10 nm thickness are also shown. (Reprinted with permission from [12]. Copyright, Wiley-VCH Verlag GmbH & Co. KGaA.)

with a carbonized Ti layer [12]. The coated TiC layers have Raman spectral peaks at approximately 260, 420, and 605 cm^{-1}, which are almost identical to those for the commercially available TiC powder. The tube diameter becomes markedly larger with the increase of layer thickness of the TiC coating. Figure 12.8 also shows the SEM images of TiC-coated CNTs.

The voltage–current curves for the as-grown CNT arrays and those after coating with 1, 3, and 10 nm TiC layers were measured. The turn-on voltage (defined as the voltage when the emission current reaches 1.0 nA) for the as-grown CNT sample is 411 V. For CNTs coated with 10-nm-thick TiC, the turn-on voltage increases slightly because of the increase in the tip diameter by the TiC coating, leading to the reduction of the local electric field at the emission sites. However, the turn-on voltage shows a decrease for the 3-nm TiC-coated CNTs and a further decrease for the 1-nm TiC-coated CNT. The reduction of the turn-on voltage cannot be explained simply using the geometric changes of the TiC-coated CNT: the reduction is caused by the changes of band structures caused by the TiC coating.

Figure 12.9 presents the Fowler–Nordheim plot of these data, indicating approximately straight lines with different slopes. Using the slopes for the as-grown CNTs and TiC-coated (1 nm) CNTs, the effective work function of TiC can be estimated under the following two assumptions: (i) The work function of CNT is the same as that of graphite of 4.5 eV. (ii) The coating of 1-nm-thick TiC has little effect on the field amplification factor β in Eq. (12.1). The estimated work function of TiC is 2.8 eV, which is much lower than that of CNTs and is even lower than that reported in the literature. In the system of a TiC-coated CNT, unlike from a dielectric-material-coated CNT, the TiC itself is a good conductor and the electrons are supplied from the CNT to the TiC layer with no barrier. They are then emitted from the surface of the TiC to vacuum, as described in Section 12.1.2.1. Under the same electric field, the tunnel barrier is thinner when the emission site's work function is smaller. Therefore, the electrons are more readily emitted from the

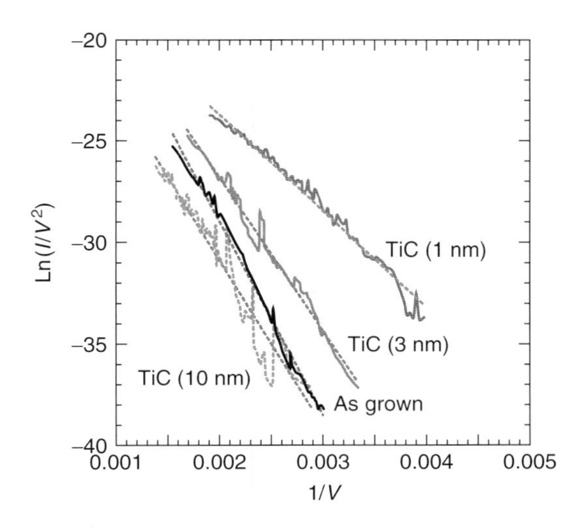

Figure 12.9 Fowler–Nordheim plot of the uncoated and TiC-coated CNTs with different thicknesses. (Reprinted with permission from [12]. Copyright, Wiley-VCH Verlag GmbH & Co. KGaA.)

TiC surfaces than from the CNTs. For that reason, the turn-on voltage for the TiC-coated (3 nm) CNTs is smaller than for as-grown CNTs even though the TiC coating increases their average diameter and sequentially decreases the β value.

References

1. Walt, A., de Heer, W.A., Chatelain, A., and Ugarte, D. (1995) A carbon nanotube field-emission electron source. *Science*, **270**, 1179–1180.
2. Murakami, H., Hirakawa, M., Tanaka, C., and Yamakawa, H. (2000) Field emission from well-aligned, patterned, carbon nanotube emitters. *Appl. Phys. Lett.*, **76** (13), 1776–1778.
3. Saito, Y. and Uemura, S. (2000) Field emission from carbon nanotubes and its application to electron sources. *Carbon*, **38** (2), 169–182.
4. Choi, W.B., Chung, D.S., Kang, J.H., Kim, H.Y., Jin, Y.W., Han, I.T., Lee, Y.H., Jung, J.E., Lee, N.S., Park, G.S., and Kim, J.M. (1999) Fully sealed, high-brightness carbon-nanotube field-emission display. *Appl. Phys. Lett.*, **75** (20), 3129–3131.
5. Akita, S., Matsumoto, S., Murakami, A., and Nakayama, Y. (2000) Field emission properties of carbon nanotubes and its application to flat panel display (in Japanese). *J. Soc. Mat. Sci. Jpn.*, **50** (4), 364–367.
6. Nakayama, Y. and Akita, S. (2001) Field-emission device with carbon nanotubes for a flat panel display. *Synth. Met.*, **117**, 207–210.
7. de Jonge, N., Lamy, Y., Schoots, K., and Oosterkamp, T.H. (2002) High brightness electron beam from a multi-walled carbon nanotube. *Nature*, **420**, 393–395.
8. Moon, J.H., Lim, S.H., Yoon, H.S., Park, K.C., Kang, S., Bae, C., Kim, J.J., and Janga, J. (2005) Spindt tip composed of carbon nanotubes. *J. Vac. Sci. Technol. B*, **23** (5), 1964–1969.
9. Kawakita, K., Hata, K., Sato, H., and Saito, Y. (2006) Development of micro-focused x-ray source by using carbon nanotube field emitter. *J. Vac. Sci. Technol. B*, **24** (2), 950–952.
10. Hata, K., Takakura, A., and Saito, Y. (2003) Field emission from multi-wall carbon nanotubes in controlled ambient gases, H_2, CO, N_2 and O_2. *Ultramicroscopy*, **95**, 107–112.
11. Wadhawan, A., Stallcup, R.E.II , and Perez, J.M. (2001) Effects of Cs deposition on the field-emission properties of single-walled carbon-nanotube bundles. *Appl. Phys. Lett.*, **78** (1), 108–110.
12. Pan, L., Shoji, T., Nagataki, A., and Nakayama, Y. (2007) Field emission properties of titanium carbide coated carbon nanotube arrays. *Adv. Eng. Mat.*, **9** (7), 584–587.
13. Kim, Y.-K., Kim, J.-P., Park, C.-K., Yun, S.-J., Kim, W., Heu, S., and Park, J.-S. (2008) Electron-emission properties of titanium carbide-coated carbon nanotubes grown on a nano-sized tungsten tip. *Thin Solid Films*, **517**, 1156–1160.
14. Qin, Y. and Hu, M. (2008) Characterization and field emission characteristics of carbon nanotubes modified by titanium carbide. *Appl. Surf. Sci.*, **254**, 3313–3317.
15. Park, N., Seungwu Han, S., and Ihm, J. (2003) Field emission properties of carbon nanotubes coated with boron nitride. *J. Nanosci. Nanotech.*, **3** (1/2), 179–183.
16. Morihisa, Y., Kimura, C., Yukawa, M., Aoki, H., Kobayashi, T., Hayashi, S., Akita, S., Nakayama, Y., and Sugino, T. (2008) Improved field emission characteristics of individual carbon nanotube coated with boron nitride nanofilm. *J. Vac. Sci. Technol. B*, **26** (2), 872–875.
17. Yu, K., Zhang, Y.S., Xu, F., Li, Q., Zhu, Z.Q., and Wan, Q. (2006) Significant improvement of field emission by depositing zinc oxide nanostructures on screen-printed carbon nanotube films. *Appl. Phys. Lett.*, **88** (3), 153123.
18. Dimitrijevic, S., Withers, J.C., Mammana, V.P., Monteiro, O.R.,

Ager, J.W. III, and Brown, I.G. (1999) Electron emission from films of carbon nanotubes and ta-C coated nanotubes. *Appl. Phys. Lett.*, **75** (17), 2680–2682.

19. Chen, Y., Sun, Z., Chen, J., Xu, N.S., and Tay, B.K. (2006) Field emission properties from aligned carbon nanotube films with tetrahedral amorphous carbon coatings. *Diamond Relat. Mater.*, **15** (9), 1462–1466.
20. Moon, J.S., Alegaonkar, P.S., Han, J.H., Lee, T.Y., Yoo, J.B., and Kim, J.M. (2006) Enhanced field emission properties of thin-multiwalled carbon nanotubes: Role of SiOx coating. *J. Appl. Phys.*, **100**, 104303–104307.
21. Yi, W., Jeong, T., Yu, S., Heo, J., Lee, C., Lee, J., Kim, W., Yoo, J.-B., and Kim, J. (2002) Field-emission characteristics from wide-bandgap material-coated carbon nanotubes. *Adv. Mater.*, **14** (20), 1464–1468.
22. Son, Y., Han, S., and Ihm, J. (2003) Electronic structure and the field emission mechanism of MgO-coated carbon nanotubes. *New.J. Phys.*, **5**, 152–159.
23. Chakrabarti, S., Pan, L., Tanaka, H., Hokushin, S., and Nakayama, Y. (2007) Stable field emission property of patterned MgO coated carbon nanotube arrays. *Jpn.J. Appl. Phys.*, **46** (7A), 4364–4369.
24. Chaddha, A.K., Parsons, J.D., and Kruaval, G.B. (1995) Thermally stable, low specific resistance ($1.30 \times 10^{-5} \Omega\,cm^{-2}$) TiC Ohmic contacts to n-type 6H $\alpha-$SiC.*Appl. Phys. Lett.*, **66** (6), 760–762.
25. Sawaya, S., Akita, S., and Nakayama, Y. (2006) In situ mass measurement of electron-beam-induced nanometer-sized W-related deposits using a carbon nanotube cantilever. *Appl. Phys. Lett.*, **89**, 193115–193113.
26. Tanaka, H., Akita, S., Pan, L., and Nakayama, Y. (2004) Barrier effect on field emission from stand-alone carbon nanotube. *Jpn.J. Appl. Phys.*, **43** (2), 864–867.
27. Tanaka, H., Akita, S., Pan, L., and Nakayama, Y. (2004) Instability of field emission from a standalone multiwalled carbon nanotube with an insulator barrier. *Jpn.J. Appl. Phys.*, **43** (4A), 1651–1654.
28. Nishijima, H., Kamo, S., Akita, S., Nakayama, Y., Hohmura, K.I., Yoshimura, S.H., and Takeyasu, K. (1999) Carbon-nanotube tips for scanning probe microscopy: preparation by a controlled process and observation of deoxyribonucleic acid. *Appl. Phys. Lett.*, **74** (26), 4061–4063.
29. Konishi, Y., Hokushin, S., Tanaka, H., Pan, L., Akita, S., and Nakayama, Y. (2005) Comparison of field emissions from side wall and tip of an individual carbon nanotube. *Jpn.J. Appl. Phys.*, **44** (4A), 1648–1651.
30. Saito, Y., Seko, K., and Kinoshita, J. (2005) Dynamic behavior of carbon nanotube field emitters observed by in situ transmission electron microscopy. *Diamond Relat. Mater.*, **14** (11-12), 1843–1847.
31. Zaima, S., Adachi, H., and Shibata, Y. (1984) Promising cathode materials for high brightness electron beams. *J. Vac. Sci. Technol. B*, **2** (1), 73–78.

Part III
Field Emission from Related Nanomaterials

Carbon Nanotube and Related Field Emitters: Fundamentals and Applications. Edited by Yahachi Saito

ISBN: 978-3-527-32734-8

13
Graphite Nanoneedle Field Emitter

Takahiro Matsumoto and Hidenori Mimura

13.1
Introduction

In the typical use of a field emitter, a massive vacuum system is required to obtain an extremely low residual pressure of less than 10^{-7} Pa. To realize a novel field emitter which has both stable electron emission and high brightness at the high residual pressure region of 10^{-4} Pa, we have developed a graphite nanoneedle (GRANN) field emitter with a two-dimensional (2D) graphene sheet structure. This structure is promising for the field emission of electrons because the carrier mobility and electron mass were reported to have exceptionally large and small values, respectively, due to the quantum relativistic effect [1–3].

This chapter provides an overview of the fabrication method, structure characterization, and field emission characteristics of the GRANN field emitter. The performance of the GRANN cathode is also overviewed by the demonstration of a time-resolved X-ray radiography system and a field emission scanning electron microscope (FE-SEM) system. To reduce the emission current fluctuation at the high residual pressure region, the thermal operation of a GRANN cathode is shown to be a powerful method. A stochastic model based on the physisorption of atoms and/or molecules on the emission sites is presented to analyze the current fluctuation as a function of temperature. Finally, this model is shown to yield a highly sensitive, precise determination of the physisorption energy of molecules, and we include the physisorption energies for various molecules on the carbon nanostructure.

13.2
Fabrication and Structure Characterization

The GRANN field emitters are fabricated by hydrogen plasma etching of a carbon rod in a microwave (frequency 2.45 GHz) plasma chemical vapor deposition equipment as shown in Figure 13.1. The typical etching condition is as follows: microwave power, 800 W; gas pressure, 1.3 kPa; H_2 gas flow rate, 80 sccm; substrate temperature, 600 °C; substrate bias −200 V; and etching time 30 min [4, 5]. A carbon

Carbon Nanotube and Related Field Emitters: Fundamentals and Applications. Edited by Yahachi Saito

ISBN: 978-3-527-32734-8

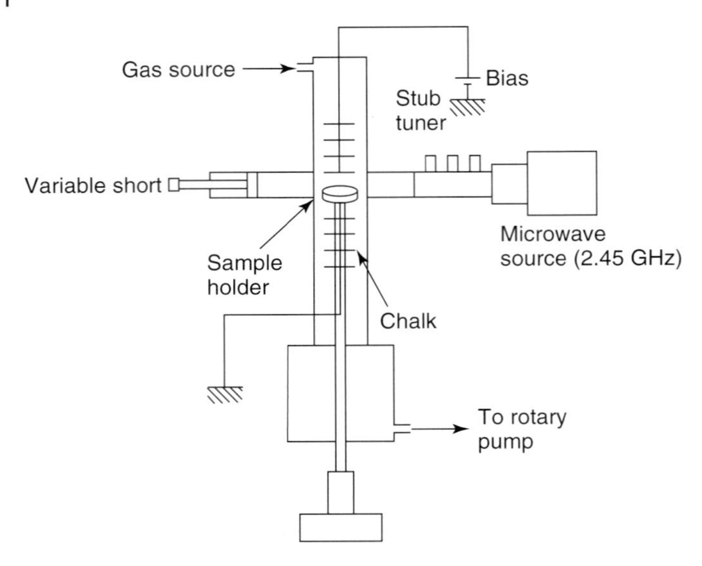

Figure 13.1 Schematic of a microwave plasma CVD equipment for hydrogen plasma etching.

rod of diameter 0.5 mm is mechanically sharpened at one end to less than 10 μm diameter as shown in Figure 13.2a before the plasma etching, and then nanoneedles are fabricated on the tip by hydrogen plasma etching. An SEM image of a carbon rod after hydrogen plasma etching is shown in Figure 13.2b. A number of nanoneedles are fabricated on the spearhead region of the graphite rod [6, 7]. The transmission electron microscope (TEM) image of one nanoneedle is shown in Figure 13.2c. The typical aspect ratio of the nanoneedle is on the order of 1000. The radius of curvature in the top region of the needle is less than 5 nm. This small radius and the high aspect ratio make it suitable for a field electron emission cathode. A high-resolution TEM image of the nanoneedle is shown in Figure 13.2d. A lattice fringe pattern is clearly observed from the bottom to the top of the needle as shown in Figure 13.2d. Based on the lattice fringe and diffraction patterns (*c* axis) shown in the inset of Figure 13.2d, the nanoneedle consists of a two-dimensional graphene sheet with an interplanar spacing of 0.36 nm [7, 8]. This value is larger than that of the hexagonal graphite structure (0.34 nm), which indicates that the *c*-axis lattice is relaxed. Another diffraction pattern (*a* axis) whose direction is orthogonal to the interplanar direction is observed. Based on the distance of *a*-axis diffraction patterns, we can determine the atomic level of spacing, which is 0.21 nm. This value corresponds to the (010) plane spacing of the six-membered ring in the graphene sheet. The 2D graphene sheet structure with the lattice fringes from the bottom to the top is promising for field emission, because the carrier mobility and electron mass are shown to have exceptionally large ($\mu = 15\,000\,\text{cm}^2\ \text{V}^{-1}\ \text{s}^{-1}$) and small ($0.007\ \text{m}_\text{e}$, m_e: free electron mass) values in this 2D system [1–3].

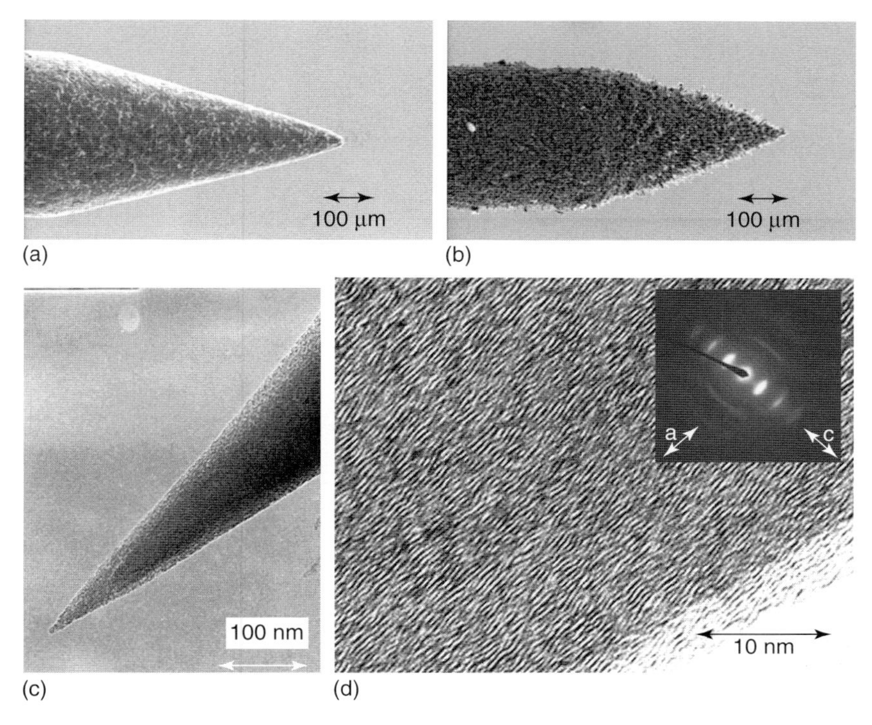

Figure 13.2 SEM image of a carbon rod: (a) before hydrogen plasma etching and (b) after hydrogen plasma etching. (c) TEM image of a single graphene nanoneedle. (d) High-resolution TEM image. Inset shows the selected area electron diffraction pattern of a single graphene nanoneedle.

13.3 Field Emission Characteristics

Figure 13.3 shows a typical logarithmic current–voltage ($I-V$) characteristic of the GRANN field emitter. The figure also shows the $I-V$ characteristic of a sharpened graphite rod without the nanostructure. The current was collected on a 3-mm-diameter aluminum anode, which was located at a distance of 100 μm in front of the cathode. The measurements were carried out in a vacuum chamber with a residual pressure of 1×10^{-6} Pa. The mechanically sharpened graphite rod without the nanostructure showed little field emission current (open circles), while the GRANN emitter starts to emit electrons at an average electric field of about $3\,\mathrm{V\,\mu m^{-1}}$ and the emission current exceeds 2 mA at an applied electric field of $11\,\mathrm{V\,\mu m^{-1}}$ (solid circles) [7]. A Fowler–Nordheim plot of the emission current from the GRANN emitter is shown in the inset of Figure 13.3. Linear dependence of the Fowler–Nordheim plot suggests that the electron emission is dominated by the Fowler–Nordheim tunneling process.

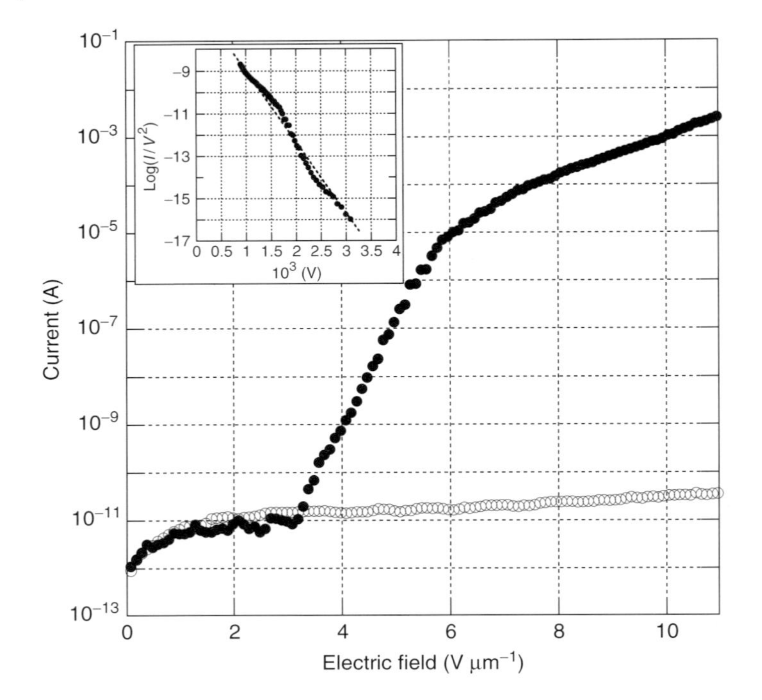

Figure 13.3 Current–voltage characteristics for a mechanically sharpened carbon rod (open circles) and a GRANN cathode (solid circles). Inset shows the Fowler–Nordheim plots ($\ln(I/V^2)$ vs $1/V$) for the I–V characteristics of the GRANN.

13.4 Applications

13.4.1 Pulse X-ray Generation and Time-Resolved X-ray Radiography

We observed the unsaturated behavior of electron emission for the GRANN cathode. This characteristic is desirable for high-power operation of devices such as X-ray and microwave generators. In this section, we show the performance of the GRANN as a cold cathode as demonstrated by high-intensity pulse X-ray generation and time-resolved X-ray radiography [4].

Figure 13.4a shows a schematic of a triode-type field emission X-ray tube composed of a GRANN cathode, a metal grid (100 mesh placed at 0.5 mm from the cathode), and a Cu metal target. The X-ray tube was evacuated by a turbomolecular pump to a base pressure of 10^{-4} Pa. For pulse X-ray generation, a negative pulse voltage with a peak height of 1–10 kV and pulse duration between 1 ms and 10 μs (repetition rate 1–100 Hz) was applied to the cathode. The metal grid was grounded and a constant positive bias of about 20 kV was applied to the anode.

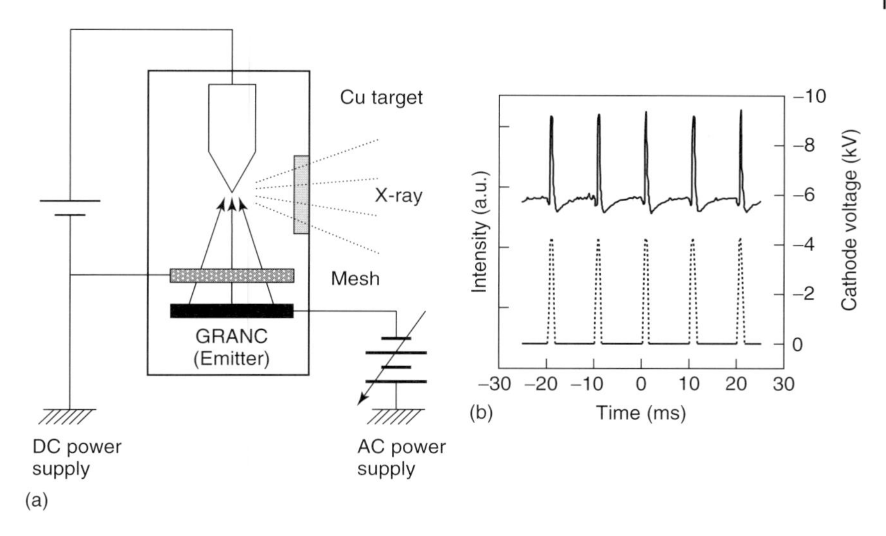

Figure 13.4 (a) Schematic of a triode-type field emission X-ray tube and (b) generated X-ray pulses.

Figure 13.4b shows the pulse voltage applied to the GRANN cathode (pulse duration of 1 ms; dotted line) and the generated X-ray pulses (solid line) detected by a Gd_2O_2S : Eu phosphor with a photomultiplier. The high-intensity X-ray pulse was obtained by applying a negative pulse voltage to the cathode. A pulse duration of 1 ms for the applied voltage was used, because the response of our detection system was limited by the phosphor decay, which was about 500 μs. A much shorter X-ray pulse, on the order of 10 μs, was generated, and by using this pulse we demonstrated in the following two applications in which the GRANN cold cathode can be used for high-speed X-ray radiography.

In the first application, as shown in Figure 13.5a, single-shot X-ray transmission image of a rotating chopper (7500 rpm) was obtained by placing the chopper between the pulse X-ray emission source and a cooled charge-coupled device (CCD) camera with a Gd_2O_2S : Eu phosphor. The image was obtained at an applied anode bias of 25 kV (DC) and a cathode bias of −10 kV. Based on the angular velocity of the rotating chopper and the sharpness of the obtained image, the generated X-ray pulse duration was estimated to be about 10 μs.

In the second application (Figure 13.5b), *in situ* image of a rotating drill and the process of making a hole in a wood plate were obtained using the same condition used in the first application. For this demonstration, we used a 2-mm-diameter drill (2600 rpm) and a 5-mm-thick wood plate. Images were obtained of the rotating drill moving inside the wood plate, and this frame was detected by a single-shot X-ray flash (10 μs duration). The advantage of the single-shot X-ray detection is that clear dynamic transmission images of a fast motion can be obtained without the use of a sophisticated high-speed camera.

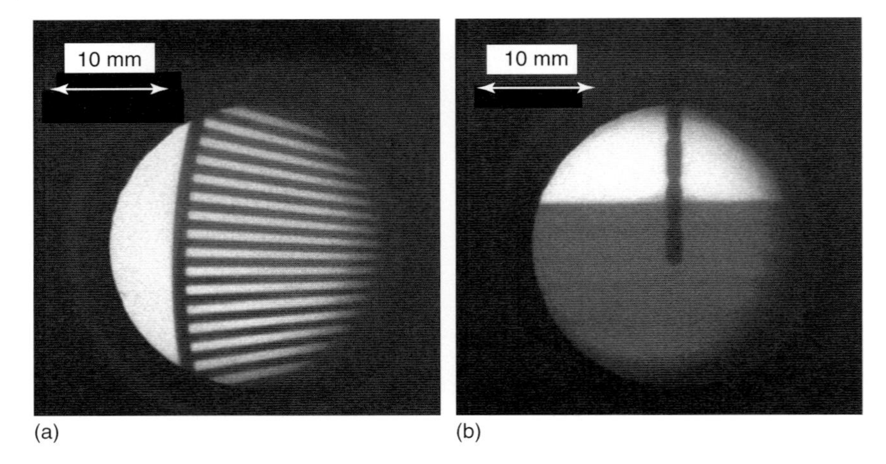

Figure 13.5 Single-shot X-ray transmission images of a (a) rotating chopper (7500 rpm) and a (b) rotating drill (2600 rpm) moving inside of the wood plate.

13.4.2
Construction of a Compact FE Scanning Electron Microscope (FE-SEM) System

Intense electron emission at a high residual pressure of 10^{-4} Pa seems to be promising to construct a high-resolution electron microscope system without the need for a massive ultrahigh vacuum system. In this section, we show the performance of the GRANN as a cold cathode as demonstrated by the construction of a compact FE-SEM system, where the brightness of the GRANN cathode was determined to be on the order of 10^{12} A sr^{-1} m^{-2} [6, 7].

A schematic of an SEM optical system equipped with an etched graphite rod is shown in Figure 13.6. The SEM experiments were performed under a residual pressure of about 10^{-4} Pa. This pressure was much higher than that of typical FE-SEM with a tungsten tip, which generally requires a low residual pressure, below 10^{-7} Pa. Either an etched graphite rod or a tungsten filament thermal emission (TE) cathode was mounted on the SEM system. Electrode 1, shown in Figure 13.6, was used as a wehnelt for the TE cathode and as an extracting gate electrode (0.1–1 kV) for the FE cathode. Other electrodes, such as a suppressor to focus the electron beam, were not included in this SEM system. A single final objective lens was used to focus the crossover on the target (sample holder), where the electron beam diameter and convergence angle were measured. The objective lens was composed of a permanent magnet and a supplementary coil and was designed to focus at 3.0 kV acceleration voltage. The aperture diameter was 0.3 mm. The spatial resolution was evaluated by obtaining the images of a 4-μm-wide copper mesh located on the sample holder. In this lens configuration, the source size was reduced to 0.182 times the sample target. In addition, a Faraday cup on the sample holder was used to collect the probe current.

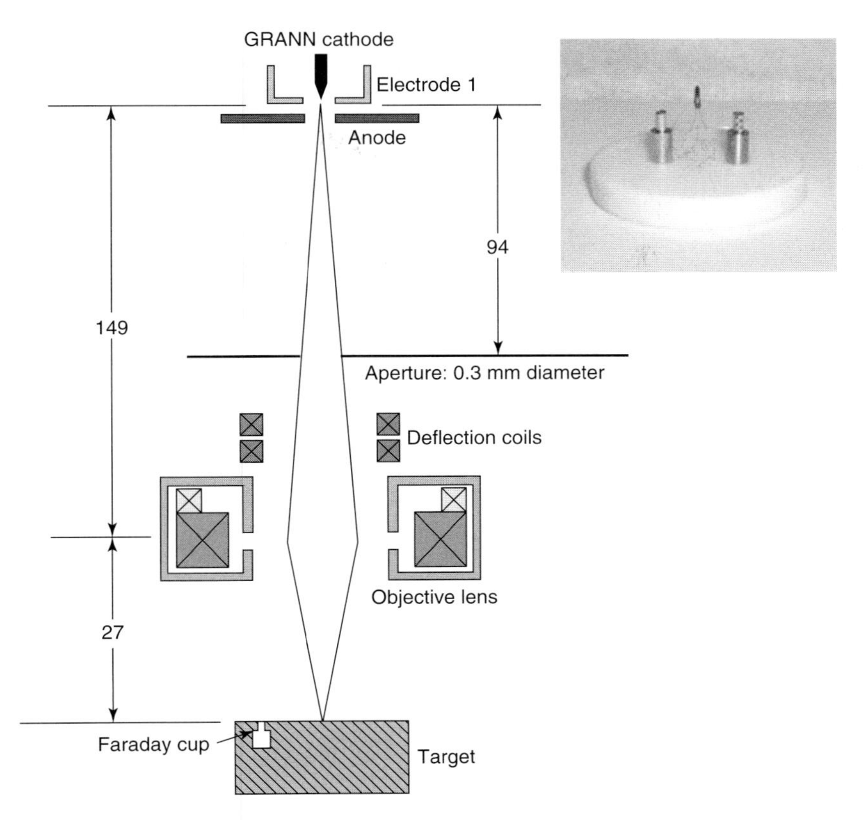

Figure 13.6 FE-SEM system equipped with a GRANN cathode.

Figure 13.7a shows the SEM image of the copper mesh obtained using the TE cathode to compare both the resolution and brightness with those of the GRANN cathode. The image obtained using the TE cathode was smeared because of its large source size. The brightness B was calculated as

$$B = \frac{I}{\mathrm{d}s\,\mathrm{d}\Omega} = \frac{I}{\pi r^2\,\pi\alpha^2} \tag{13.1}$$

where I is the probe current, ds is the source size, dΩ is the solid angle, r is the radius of the beam, and α is the open angle of the electron optics. For the TE cathode, the beam diameter $2r$ was estimated at about 4 μm and the measured I was about 0.6 μA. Therefore, the brightness B of the TE cathode was estimated to be about 2×10^8 A m^{-2} sr^{-1}, which coincides with the brightness reported for a tungsten filament TE cathode [9]. This B value of the TE cathode shows that the method is appropriate for estimating B. Figure 13.7b shows the SEM image of the copper mesh obtained using the GRANN cathode. The spatial resolution was clearly improved compared to that obtained using the tungsten filament TE cathode. This result shows that an SEM image can be obtained using the GRANN cathode despite a high residual pressure on the order of 10^{-4} Pa. However, as shown in

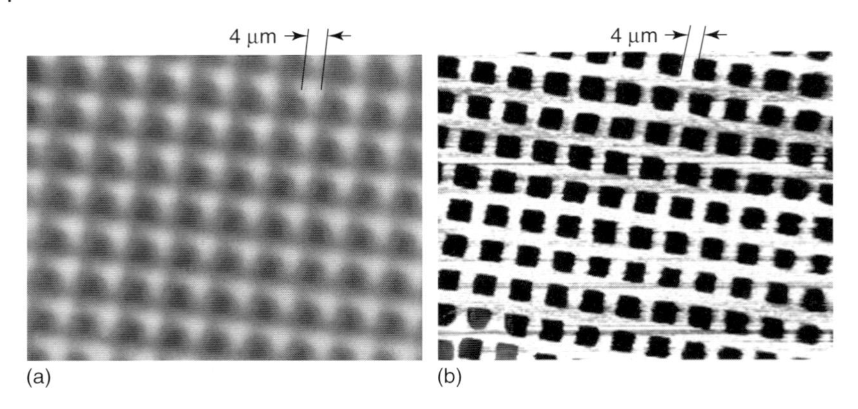

Figure 13.7 SEM images of a copper mesh obtained using a (a) TE cathode and a (b) GRANN cathode.

Figure 13.7b, many horizontal noise lines were observed. This is because beam fluctuations occurred during the scanning of the electron beam to obtain SEM images, which is known to be a step-and-spike noise due to ion bombardment and/or atom adsorption [10–12].

13.4.3 Stabilization of the FE-SEM System by Thermal Field Operation

The fluctuation of the emission current originates from the adsorption and desorption of atoms and/or molecules on the emission sites of the GRANN cathode [7, 8, 11]. To stabilize the emission current, generally the evacuation of the residual gas to the degree of ultrahigh vacuum (10^{-9} Pa) is performed, and this can be achieved only with a massive and costly vacuum system. On the other hand, the number of adsorbed atoms per unit time depends on the temperature of the cathode. Therefore, the fluctuation can be reduced by heating the GRANN cathode. In this section, we show the reduction of the current fluctuation as a function of the GRANN cathode temperature at a residual pressure above 10^{-3} Pa, and show clearer FE-SEM images compared to those obtained with the FE-SEM without heating the GRANN cathode.

The FE-SEM optical system equipped with the designed GRANN gun is the same as shown in Figure 13.6, where the GRANN cathode was attached to the W filament by using a graphite dispersion in order to heat the cathode and the temperature of the GRANN cathode was measured by a pyro- or a radiation thermometer. The stability of the emission current was measured by using a Faraday cup on the sample holder under a residual pressure of about 10^{-3} Pa. The histogram of the emission current intensity (probability vs current) measured during 400 s at room temperature (300 K) shown in Figure 13.8a (solid rhombuses) gives the deviation of the current fluctuation, $\Delta I = 70–80$ pA, where ΔI is the full width at half maximum (FWHM) of the current fluctuation. The histogram of the emission current intensity at 1200 K shown in Figure 13.8a (solid circles) gives the deviation

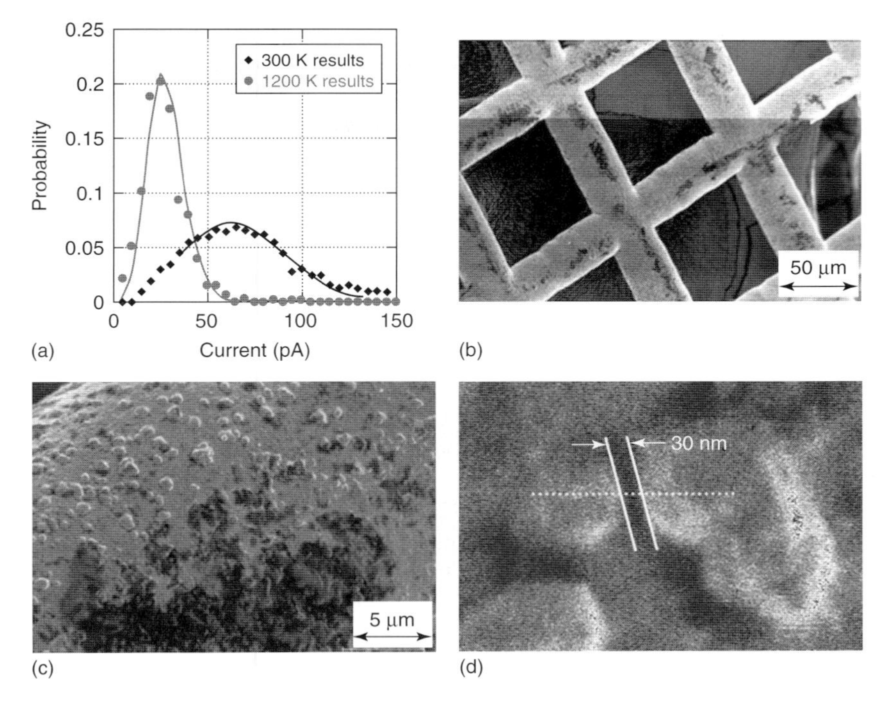

Figure 13.8 (a) Histograms of the emission current intensity at 300 K (solid rhombuses) and 1200 K (solid circles). (b and c) Copper mesh images obtained using the GRANN cathode. (d) Highest resolution image obtained by using the GRANN cathode.

of the current fluctuation, $\Delta I = 20$ pA. The current fluctuation was reduced by a factor of 4 by heating the cathode, thereby showing that heating is an effective way to stabilize the field emission current, especially in the high residual pressure region (10^{-3}–10^{-4} Pa).

By using this stabilized electron emission, it is possible to obtain clearer FE-SEM images compared to those obtained by the GRANN cathode at room temperature shown in Figure 13.7. Figure 13.8b shows the FE-SEM images obtained using the GRANN cathode at 1200 K. The stability of the emission current was improved compared to that of the GRANN cathode without heating, thus leading to clearer images compared to those obtained using the GRANN cathode at room temperature. Figure 13.8c shows a higher resolution image of the copper mesh. Copper grains are clearly evident, which indicates that the emission source size was reduced compared to a TE cathode and that the heated GRANN cathode was stable during the acquisition of the SEM image (60 s). Figure 13.8d shows the highest resolution image obtained by using the GRANN cathode. Based on this result, the maximum spatial resolution of this SEM optical system was analyzed by the distance between Cu grains, which was estimated to be 30 nm and indicated by the solid white lines in

Figure 13.8d. Both the stability and the spatial resolution obtained by the GRANN cathode are promising to construct a compact and high-resolution FE-SEM system, because it is possible to obtain higher resolution images of less than 1 nm by using a commonly used 200× magnification lens at this high residual pressure region.

The source size was estimated to be 160 nm. The maximum emission current measured by the Faraday cup was about 70 nA. Based on these experimentally determined parameters, the brightness of the GRANN cathode was estimated as 5×10^{11} A sr^{-1} m^{-2}, which is similar to that of CNTs [13]. A higher brightness of the order of 10^{13} A sr^{-1} m^{-2} should be attainable if a higher extraction voltage is applied to the cathode.

13.5 Stochastic Model

The electronic properties of nanomaterials are very sensitive to the adsorption of molecules [7, 8, 10–12]. This characteristic is clearly observed when measuring fluctuations of field emission current of carbon nanostructures, where the emission current stability is easily lost under high residual pressure and stability is obtained by thermally desorbing adsorbed molecules from the emission sites. This phenomenon can be explained as the adsorption and desorption of molecules on emission sites of carbon nanostructures with the molecules adsorbing onto the nanostructures affecting the electronic properties, such as charge transfer and tunneling probability [14, 15].

In this section, we show a model in which the field emission current fluctuation originates from the adsorption and desorption of molecules on the emission sites, which is described using a stochastic birth-and-death model [16]. The emission current fluctuation is analyzed as a cathode temperature using differential equations obtained from the model. Finally, we show the method of the determination of the physisorption energy of various molecules using the field emission current fluctuation.

Figure 13.9 shows the fluctuating emission current distribution measured in a 10^{-4} Pa H_2 atmosphere at different temperatures. The current intensity distribution at 300 K shown in Figure 13.9 (solid circles) gives the current fluctuation deviation, $\Delta I/I_p = 0.31$, where ΔI is the FWHM of the fluctuation and I_p is the peak value of the current. The deviation can be reduced by heating the cathode: $\Delta I/I_p = 0.21$ at 700 K (solid triangles), and $\Delta I/I_p = 0.16$ at 925 K (solid squares). This reduction in the deviation suggests that heating is an effective way to stabilize the field emission current. Along with the reduction of the deviation, the peak of the current distribution shifts to the lower current side with increasing cathode temperature. Both the reduction of the deviation and the lower peak shift of the current distribution can be interpreted qualitatively as the adsorption and desorption of atoms or ions on or off the surface of the cathode [11, 12].

Figure 13.10 illustrates the model in which the emission current fluctuation originates from the adsorption and desorption of molecules and/or ions. The

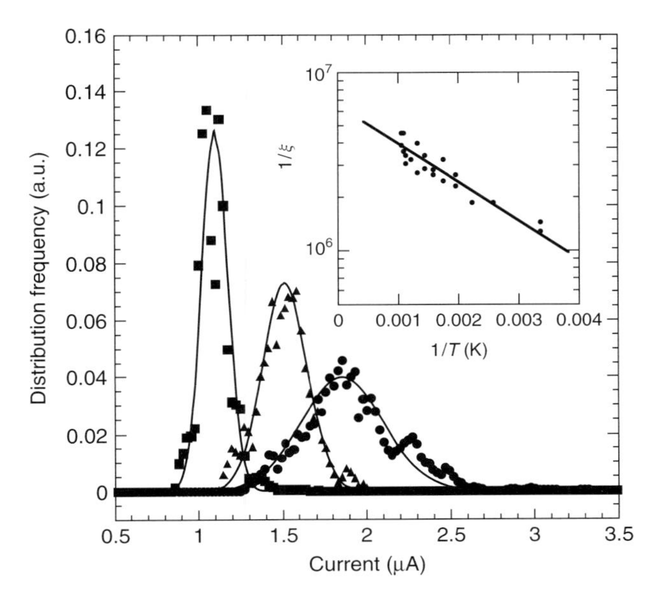

Figure 13.9 Histogram of the emission current intensity at 300 K (solid circles), 700 K (solid triangles), and 925 K (solid squares) in a H_2 atmosphere; the solid lines are the theoretically fitted curves. The inverse of the variance, $1/\xi$, for each temperature obtained by the theoretical fitting is shown as solid circles in the inset. The solid line in the inset is the fitted curve with a physical adsorption energy of 45 meV.

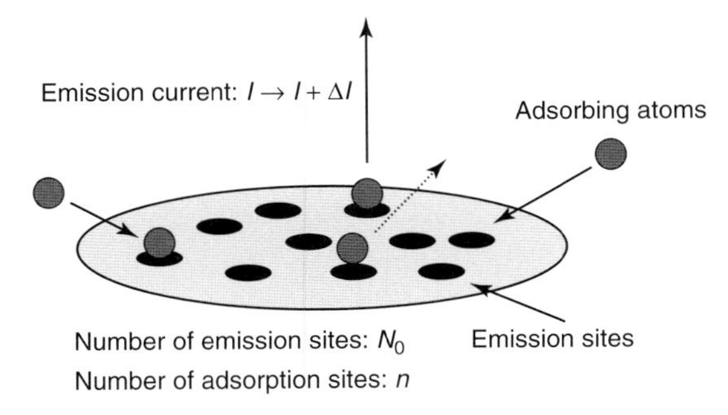

Figure 13.10 Physical desorption model in which fluctuation of the emission current originates from the adsorption and desorption of atoms and/or molecules onto the emission sites.

current fluctuation occurs as a result of the occupation of the emission sites by the adsorbed molecules. Here, we postulate that the magnitude of the current (I) is proportional to the number of occupied states (n) (e.g., $I(n) = I_0 + \eta n$, where I_0 is the emission current of unoccupied state and η is the magnitude of the current hop due to the adsorption of single molecule) to explain both the reduction of the deviation and the lower peak shift of the current distribution as shown in

Figure 13.9. For this model, we apply a stochastic differential equation described by birth-and-death processes [8, 16]. A detailed analysis will give the stationary distribution p_n as the following Poisson distribution:

$$p_n = \frac{\xi^n}{n!} \exp(-\xi) \tag{13.2}$$

where n is the number of adsorbed sites and ξ is the FWHM of the histograms. The solid curves in Figure 13.9 are the theoretically fitted ones given by Eq. (13.2), where the best fits were obtained using the fitted values, $\xi = 2.76 \times 10^{-7}$ for 920 K, $\xi = 3.45 \times 10^{-7}$ for 700 K, and $\xi = 7.72 \times 10^{-7}$ for 300 K. By fitting the histogram of the current fluctuation at various cathode temperatures with Eq. (13.2), and then plotting these theoretically determined values logarithmically [$\ln(1/\xi)$] as a function of the inverse of the temperature, $1/T$, as described by the relation,

$$\ln(1/\xi) \propto -E_{\mathrm{ad}} \cdot (1/kT) \tag{13.3}$$

we can determine the physical adsorption energy, E_{ad}, from the slope, as shown in the inset of Figure 13.9. The solid line shown in the inset of Figure 13.9 is the theoretical line computed with a physisorption energy of $E_{\mathrm{ad}} = 45$ meV, which is similar to the physisorption energy of H_2 onto a graphite surface [17].

Figure 13.11 shows the values of $1/\xi$ obtained by measuring the emission current fluctuations as a function of temperature for CO (solid circles), Ar (solid squares), and He molecules (solid triangles). The slope of the solid line shows

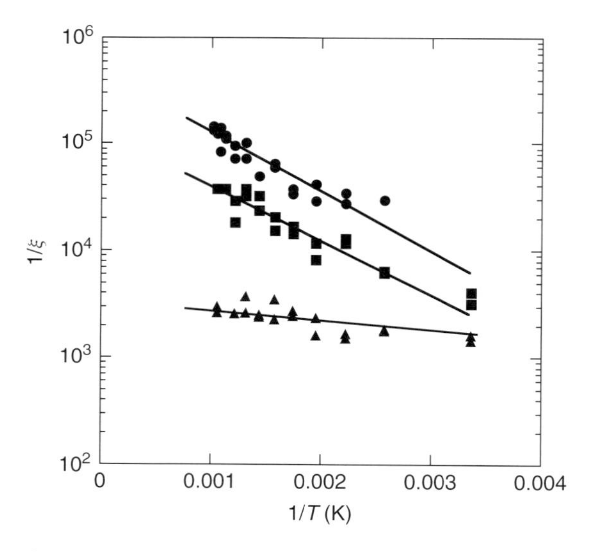

Figure 13.11 Inverse of the variance, $1/\xi$, for CO molecules (solid circles), Ar molecules (solid squares), and He molecules (solid triangles) as a function of GRANN cathode temperature. The solid lines are the fitted curves with physical adsorption energies of 110 meV for CO molecules, 100 meV for Ar, and 15 meV for He.

the theoretically determined physical adsorption energies, E_{ad}: 110 meV for CO molecules, 100 meV for Ar, and 15 meV for He. These physisorption energies for the GRANN emitter are similar to the physisorption energies for the graphite surface [17]. Therefore, we consider that the electron emission occurs not from the edge of the nanoneedle but from the basal plane of the graphene sheet.

13.6 Summary

GRANN field emitters were fabricated by simple hydrogen plasma etching of a carbon rod. The GRANN has a two-dimensional graphene sheet structure with lattice fringes from the bottom to the top, which is promising for FE operation. The GRANN emitter has a high field emission current of 2 mA at an average field of 11 V μm^{-1} as well as a high brightness of more than 10^{12} Asr^{-1} m^{-2}. The performance of the GRANN emitter was demonstrated by the construction of a time-resolved X-ray radiography system and an FE-SEM system. For clear FE-SEM image acquisition at the high residual pressure region (10^{-3}–10^{-4} Pa), stability of the emission current is required. We showed that the thermal field operation of the GRANN emitter is a way to obtain a stable field emission, leading to the acquisition of clear FE-SEM images under the high residual pressure. A stochastic model with the birth-and-death process was applied to describe the emission current fluctuation originating from the adsorption and desorption of molecules on the emission sites. Based on the stochastic model, we have presented a new method to determine the physisorption energies of various molecules on the emission sites of the GRANN cathode. We believe that the model presented here will become a powerful tool to determine the physisorption energies of various atoms and/or molecules on the surface of various carbon nanomaterials.

References

1. Novoselov, K.S., Geim, A.K., Morozov, S.V., Jiang, D., Katsnelson, M.I., Grigorieva, I.V., Dubonos, S.V., and Firsov, A.A. (2005) *Nature*, **438**, 197.
2. Zhang, Y., Tan, Y.W., Stormer, H.L., and Kim, P. (2005) *Nature*, **438**, 201.
3. Geim, A.K. and Novoselov, K.S. (2007) *Nat. Mater.*, **6**, 183.
4. Matsumoto, T. and Mimura, H. (2004) *Appl. Phys. Lett.*, **84**, 1804.
5. Matsumoto, T. and Mimura, H. (2005) *J. Vac. Sci. Technol., B*, **23**, 831.
6. Neo, Y., Mimura, H., and Matsumoto, T. (2006) *Appl. Phys. Lett.*, **88**, 073511.
7. Matsumoto, T., Neo, Y., Mimura, H., Tomita, M., and Minami, N. (2007) *Appl. Phys. Lett.*, **90**, 103516.
8. Matsumoto, T., Neo, Y., Mimura, H., and Tomita, M. (2008) *Phys. Rev. E*, **77**, 031611.
9. Joy, D.C. (1977) *Scan. Electron Microsc.*, **I**, 1.
10. Beitel, G.A. (1972) *J. Vac. Sci. Technol.*, **9**, 370.
11. Yamamoto, S., Hosoki, S., Fukuhara, F., and Futamoto, M. (1979) *Surf. Sci.*, **86**, 734.
12. Hosoki, S., Yamamoto, S., Fukumoto, M., and Fukuhara, S. (1979) *Surf. Sci.*, **86**, 723.

13. de Jonge, N., Lamy, Y., Schoots, K., and Oosterkamp, T. (2002) *Nature*, **420**, 393.
14. Dyke, W.P. and Dolan, W.W. (1956) *Adv. Electron. Electron Phys.*, **8**, 89.
15. Gadzuk, J.W. and Plummer, E.W. (1973) *Rev. Mod. Phys.*, **45**, 487.
16. Feller, W. (1957) *An Introduction to Probability Theory and its Application*, 3rd edn, vol. 1, John Wiley & Sons, Inc., New York, p. 444.
17. Vidali, G., Ihm, G., Kim, H., and Cole, M.W. (1991) *Surf. Sci. Rep.*, **12**, 133.

14
Field Emission from Carbon Nanowalls

Masaru Hori and Mineo Hiramatsu

14.1
General Description of Carbon Nanowalls

Since the report of carbon nanotubes by Iijima [1], the fabrication of carbon nanostructures has been studied intensively. One-dimensional carbon nanostructures, such as carbon nanotubes and carbon nanofibers, have attracted interest for applications such as electrochemical devices, electron field emitters, hydrogen storage materials, and scanning probe microscopy because of their particular physical, chemical, and mechanical characteristics. Carbon nanostructures with different morphologies can now be fabricated on a substrate using several different techniques, such as plasma-enhanced chemical vapor deposition (PECVD) and hot filament chemical vapor deposition (HFCVD).

Carbon nanowalls, that is, two-dimensional carbon nanostructures standing vertically on a substrate, have been recently fabricated [2–4]. Carbon nanowalls can be described as graphite sheet nanostructures with edges that are composed of stacks of planar graphene sheets standing almost vertically on the substrate, which forms a wall structure with a high aspect ratio (Figure 14.1). The thickness of carbon nanowalls ranges from a few nanometers to a few tens of nanometers. The unique structure of carbon nanowalls, such as the high aspect ratio and high surface-to-volume ratio, means that carbon nanowalls can be potentially used as electron field emitters and as catalyst support materials. Furthermore, carbon nanowalls essentially consist of graphene sheets, so they are expected to have high mobility for carriers and large sustainable current densities. Therefore, carbon nanowalls are considered to be one of the most promising carbon materials used for nanoscale electronic devices.

Carbon Nanotube and Related Field Emitters: Fundamentals and Applications. Edited by Yahachi Saito

ISBN: 978-3-527-32734-8

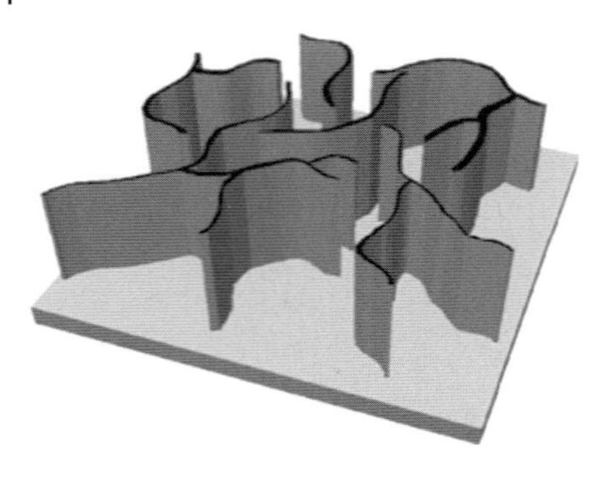

Figure 14.1 Schematic illustration of carbon nanowalls.

14.2 Synthesis of Carbon Nanowall Films

14.2.1 Synthesis Techniques

The first report on the fabrication of two-dimensional carbon nanostructures dates back more than 10 years. Ando *et al.* [5] found petallike "carbon roses" during the fabrication of carbon nanotubes in 1997. In 2002, Wu *et al.* [2] reported the fabrication of two-dimensional carbon nanostructures, "carbon nanowalls," standing vertically on catalyzed substrates. Both cases were found incidentally, during the fabrication of carbon nanotubes. To date, carbon nanowalls have been grown using various chemical vapor deposition (CVD) methods such as microwave plasma [2, 6], radio frequency (rf) inductively coupled plasma [4], rf capacitively coupled plasma assisted by H radical injection [3, 7], helicon-wave plasma [8], electron beam excited PECVD [9], HFCVD [10], and even by sputtering of a graphite target [11].

Synthesis techniques for carbon nanowalls are similar to those used for diamond films and carbon nanotubes. In general, a mixture of hydrocarbon and hydrogen or argon gases, typically CH_4 and H_2, is used as source gases for the synthesis of carbon nanowalls. High-density plasmas such as microwave plasma and inductively coupled plasma are suitable for decomposing H_2 molecules efficiently, because a large number of H atoms are required for the growth of carbon nanowalls, which is similar to that for diamond growth. Metal catalysts such as Fe and Co are required for the growth of carbon nanotubes, whereas carbon nanowalls do not require such catalysts. Consequently, carbon nanowalls have been fabricated on several substrates, including Si, SiO_2, Al_2O_3, Ni, and stainless steel, at substrate temperatures of 500–700 °C without the use of catalysts [12].

Hiramatsu *et al.* have proposed a novel method of plasma processing – radical controlled processing using a radical injection technique [13, 14]. The successful formation of vertically aligned carbon nanowalls was demonstrated using fluorocarbon (typically, C_2F_6) PECVD with H atom injection [3]. The morphology of carbon nanowalls is affected by the ratio of H and CF_3 radicals in the plasma [7].

14.2.2
Characterization

Carbon nanowalls are graphite nanostructures with edges comprised of stacked planar graphene sheets standing vertically on a substrate. The sheets form a wall structure with thicknesses in the range from a few nanometers to a few tens of nanometers, and with a high aspect ratio. The large surface area and sharp edges of carbon nanowalls may prove useful for a number of different applications. In particular, vertically standing carbon nanowalls with a high surface-to-volume ratio would serve as an ideal catalyst support material for fuel cells and gas storage materials.

Scanning electron microscope (SEM) images of a typical carbon nanowall film that was synthesized using electron beam excited PECVD employing a mixture of CH_4 and H_2 [9] are shown in Figure 14.2a and b. These images indicate the vertical growth of the two-dimensional carbon sheets on the substrate.

The low-magnification transmission electron microscope (TEM) image of Figure 14.3a shows a typical carbon nanowall with a micrometer-high planar

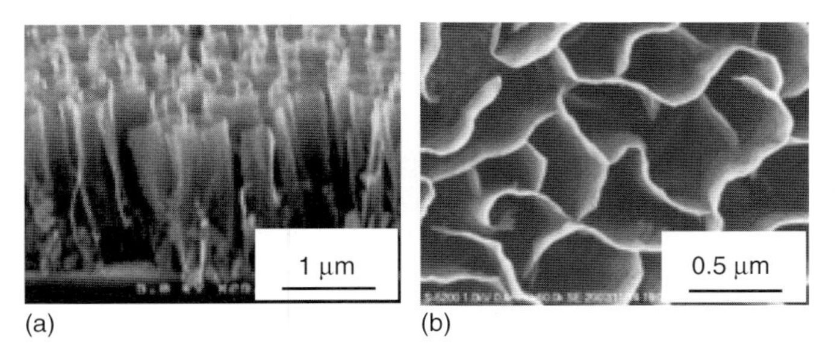

Figure 14.2 SEM images of a typical carbon nanowall film grown using electron beam excited plasma-enhanced CVD employing a CH_4 and H_2 mixture [9].

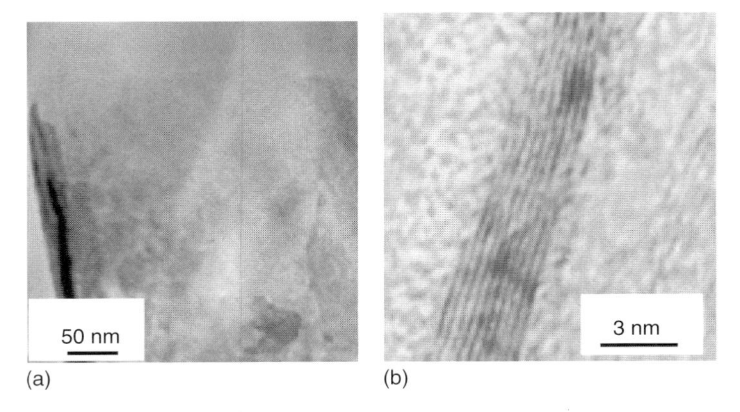

Figure 14.3 TEM images of a typical carbon nanowall [9].

nanosheet structure with a relatively smooth surface. The carbon nanowall is composed of nanodomains of a few tens of nanometers. The high-resolution TEM image of the carbon nanowall shown in Figure 14.3b reveals graphene layers, which indicates the graphitized structure of the carbon nanowalls. The spacing between neighboring graphene layers was measured as approximately 0.34 nm.

Micro-Raman spectra were measured for a carbon nanowall film using the 514.5 nm line of an Ar laser. The size of the laser spot was approximately 1 μm in diameter with a 100× objective lens. Raman spectra of carbon nanowall film recorded using different configurations of laser direction and carbon nanowall orientation are shown in Figure 14.4. In spectrum (i), the laser was normal to the substrate surface, and irradiated the top of the aligned carbon nanowalls. On the other hand, in spectrum (ii), the laser was parallel to the substrate surface and irradiated the cross-section of the carbon nanowall film. Spectrum (i), which is a typical Raman spectrum of carbon nanowalls, is found to have a G band peak at 1590 cm^{-1}, which indicates the formation of a graphitized structure, and a D band peak at 1350 cm^{-1}, which corresponds to the disorder-induced phonon mode. The peak intensity of the D band is twice that of the G band. It is noted that the G band peak is accompanied by a shoulder peak at 1620 cm^{-1} (D′ band). This shoulder peak is associated with finite-size graphite crystals and graphene edges [15, 16]. The strong and sharp D band peak and D′ band peak suggest a more nanocrystalline structure and the presence of graphene edges and defects, which are prevalent features of carbon nanowalls. On the other hand, spectrum (ii) reveals G and D peaks with comparable intensities. Spectrum (i) provides information on the upper part of the carbon nanowalls, including many edges, while spectrum (ii) represents the properties of a carbon nanowall surface with fewer edges. Therefore, the strong D band peak of spectrum (i) is mainly attributed to edges with defects on the top of aligned carbon nanowalls. For spectrum (ii), a D band peak that is comparable to the G band indicates carbon nanowalls composed of nanographite domains, as shown in the TEM image of Figure 14.3b.

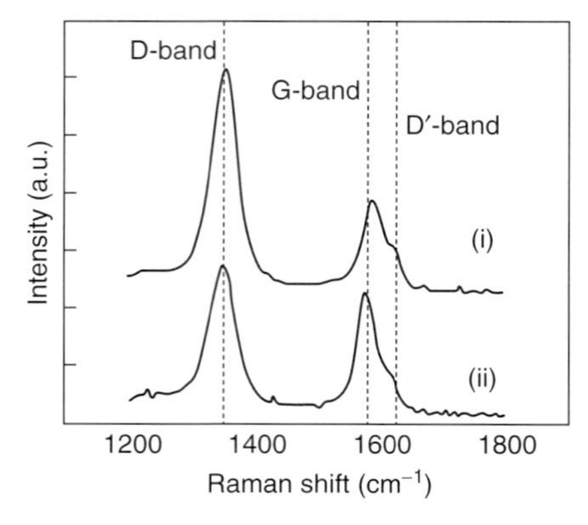

Figure 14.4 Raman spectra obtained for a carbon nanowall film using different configurations of the laser direction and carbon nanowall orientation. Spectrum (i) was obtained with the laser irradiated to the top of the aligned carbon nanowalls. Spectrum (ii) was obtained with the laser irradiated to the cross section of carbon nanowall film.

14.2.3
Morphology of Carbon Nanowall Film

Vertically standing graphite sheets such as carbon nanowalls could serve as an efficient edge emitter for electron field emission. In the case of application to an efficient emitter for electron field emission, thin edges with moderate spacing and uniform height distribution are required. However, the morphology of carbon nanowall films is influenced by the source gases, plasma generation methods, pressure, and substrate temperature.

SEM images of carbon nanowalls grown by rf inductively coupled PECVD employing CH_4/Ar at different gas flow rate ratios are shown in Figure 14.5a–c. With increase of the Ar flow rate, the spacing between the nanowalls decreased (density of nanowalls increased), which suggests that Ar ions contribute to the nucleation of carbon nanowalls.

In the case of carbon nanowall growth by rf capacitively coupled PECVD with H radical injection employing a fluorocarbon/hydrogen system, the morphology of the carbon nanowalls is dependent on the types of carbon source gases [12]. SEM images of carbon nanowall films grown by rf capacitively coupled PECVD with H radical injection employing C_2F_6/H_2, CF_4/H_2, and CHF_3/H_2 systems, respectively, are shown in Figure 14.6a–c. The morphology of the carbon nanowall film grown using the CF_4/H_2 system was similar to that using the C_2F_6/H_2 system shown in Figure 14.6a, except that the spacing between the nanowalls was very large compared to the case of the C_2F_6/H_2 system. The morphology of the carbon nanowall film grown using the CHF_3/H_2 system was also similar to that using the C_2F_6/H_2, and the spacing between nanowalls was slightly larger than that of the nanowall film grown using the C_2F_6/H_2 system. In contrast, definite carbon nanowalls were not grown using the C_4F_8/H_2 system. C_2F_6 is expected to yield CF_3

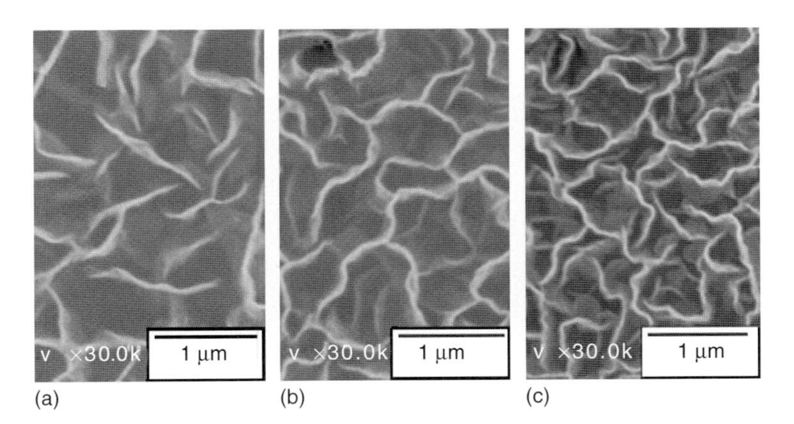

Figure 14.5 SEM images of carbon nanowalls grown by rf inductively coupled plasma-enhanced CVD employing CH_4/Ar at different gas flow rate ratios (CH_4 : Ar) of (a) 7 : 7 sccm, (b) 3 : 12 sccm, and (c) 1.5 : 13.5 sccm.

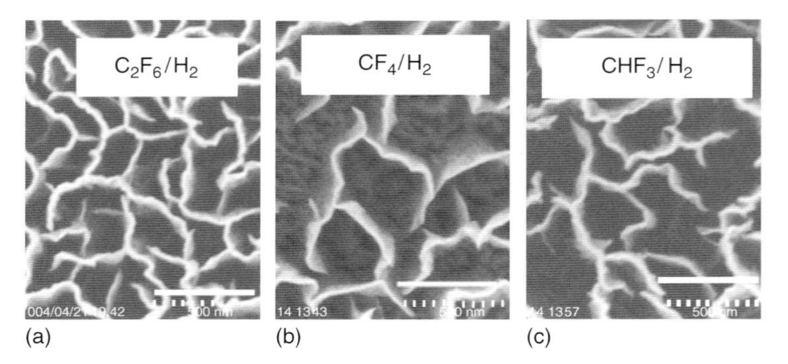

Figure 14.6 SEM images of carbon nanowall films grown by rf capacitively coupled plasma-enhanced CVD with H radical injection employing (a) C_2F_6/H_2, (b) CF_4/H_2, and (c) CHF_3/H_2 mixtures. Scale bar: 500 nm.

radicals most effectively. CF_3 radicals could also be generated in the CF_4 or CHF_3 plasma to some extent. However, the density of CF_3 radicals in the C_4F_8 plasma is considered to be low, although large amounts of CF_2 radicals are generated in a C_4F_8 plasma by electron impact dissociation from C_4F_8, because of the cyclic structure of the C_4F_8 molecule. Therefore, in the case of nanowall growth using a fluorocarbon/hydrogen system, the density ratio of H to CF_3 radicals in the plasma was found to influence the morphology of carbon nanowalls [7].

The crystallinity of carbon nanowalls can be improved by additives during the CVD process. Cross-sectional SEM images of carbon nanowall films grown by very high frequency (VHF) capacitively coupled PECVD with H radical injection employing a C_2F_6/H_2 mixture are shown in Figure 14.7a and b [17]. The image in Figure 14.7a corresponds to a typical carbon nanowall film grown on a Si substrate without O_2, with slightly branching two-dimensional carbon sheets

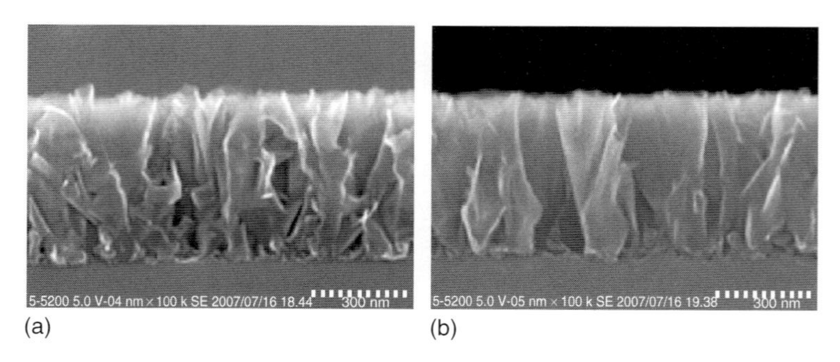

Figure 14.7 Cross-sectional SEM images of carbon nanowall films grown by VHF capacitively coupled plasma-enhanced CVD with H radical injection employing a C_2F_6/H_2 mixture (a) without O_2 addition and (b) with O_2 addition.

standing almost vertically on the substrate. In contrast, a carbon nanowall film grown with the addition of O_2 at a flow rate of 5 sccm is shown in Figure 14.7b. These carbon nanowalls exhibit less branching than those produced without O_2, so that monolithic multilayered graphene structures are obtained; however, the growth rate is reduced by approximately 33%. Compared with the carbon nanowalls grown without O_2, those grown with O_2 had larger planar sheets with a smooth surface, which is effective for improving electrical conduction along the nanowall surface.

14.3 Field Emission Properties of Carbon Nanowalls

Carbon nanowalls have a high density of atomic-scale graphitic edges that are potential sites for electron field emission. The characteristic curve of the electron emission current as a function of the applied field strength is shown in Figure 14.8 for a typical carbon nanowall film fabricated using the C_2F_6/H_2 system (Figure 14.6a). The measurement system used was a diode structure which was comprised of a spherical stainless steel anode (radius: 3 mm) at a distance of 500 μm from carbon nanowall film sample. The sample was biased with a negative voltage and the maximum applied voltage was 10 kV. The electric field is expressed as the applied voltage divided by the anode-to-sample distance. The threshold electric field, which is defined as the field when the emission current attains a value of 0.1 μA, was 4.5 V μm^{-1}, which is similar to the value reported for the carbon nanosheet sample [18].

The electron field emission properties of carbon nanowall samples with different morphologies fabricated using the C_2F_6/H_2, CF_4/H_2, CHF_3/H_2, and CH_4/Ar systems are shown in Figure 14.9. The $I-V$ characteristics of carbon nanowall samples with different morphologies indicate that the threshold electric field is dependent on the morphology; the carbon nanowall sample grown using C_2F_6/H_2 system shown in Figure 14.6a exhibited the lowest threshold electric field.

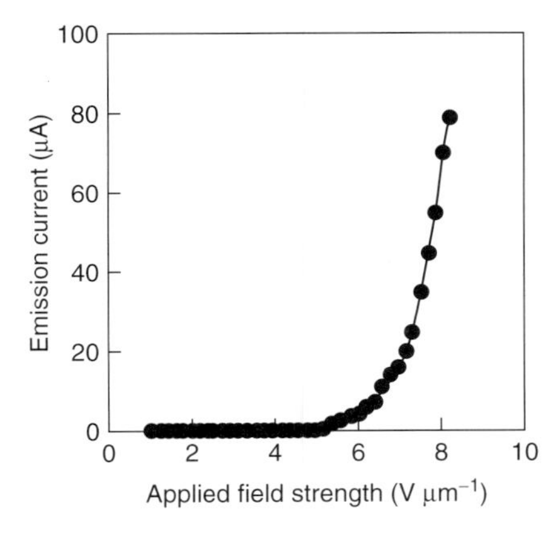

Figure 14.8 Characteristic electron emission current curve as a function of the applied voltage for a typical carbon nanowall film fabricated using the C_2F_6/H_2 system (Figure 14.6a).

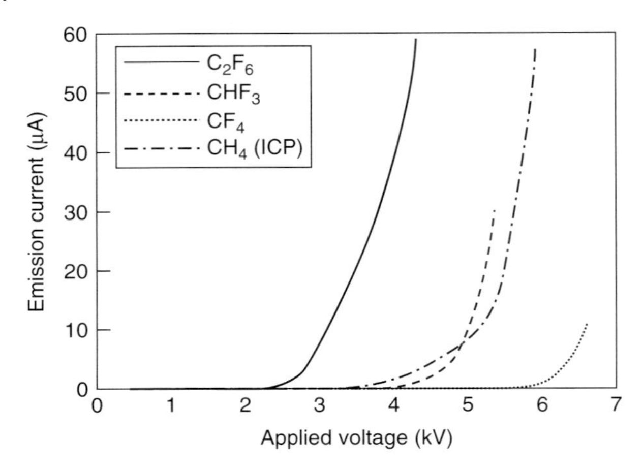

Figure 14.9 Electron field emission properties of carbon nanowall samples with different morphologies, fabricated using the C_2F_6/H_2, CF_4/H_2, CHF_3/H_2, and CH_4/Ar systems.

14.4 Surface Treatment for Improvement of Field Emission Properties

Unlike carbon nanotubes, which have closed networks of sp^2-bonded carbons, the electronic states of carbon materials with graphite open edges are influenced by their shape and composition. It was recently reported that a limited number of emission sites of graphite edges of as-grown carbon nanowalls contribute to the emission current [18]. Furthermore, carbon nanowalls are composed of nanodomains that are a few tens of nanometers in size and each carbon nanowall has many edges and defects, which would limit the emission current. However, the field emission performance can be improved by surface modification or coating of the carbon nanowalls. Recently, it was reported that the field emission properties of carbon nanowalls can be improved by coating with chromium oxide to a thickness of 1.5 nm [18]. Improvement of the field emission properties of carbon nanowalls was investigated by (i) depositing metal nanoparticles on the surface and (ii) plasma treatment for surface modification.

14.4.1 Metal Nanoparticle Deposition

Metal nanoparticles such as Pt and Ti supported on the carbon nanowall surface can enhance the field emission properties of carbon nanowalls. Pt nanoparticles were formed by metal-organic chemical fluid deposition (MOCFD) employing a supercritical fluid (SCF) [19]. SCF-MOCFD using supercritical carbon dioxide ($scCO_2$) as a solvent for the metal-organic compound (trimethyl(methylcyclopentadienyl) platinum: $MeCpPtMe_3$) resulted in highly dispersed Pt nanoparticles deposited on the entire surface of the carbon nanowalls.

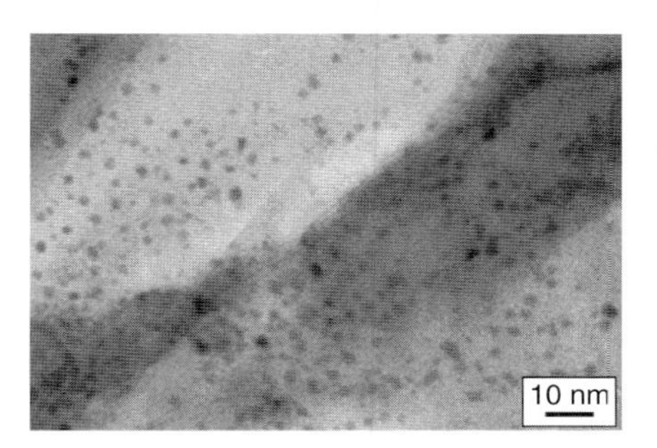

Figure 14.10 TEM image of a carbon nanowall surface supporting Pt nanoparticles.

A TEM image of detached carbon nanowalls after the SCF-MOCFD treatment is shown in Figure 14.10, and indicates the presence of dispersed Pt nanoparticles of approximately 2 nm size supported on the carbon nanowall surface. The electron field emission properties of as-grown carbon nanowalls and Pt-deposited carbon nanowalls after the SCF-MOCFD treatment are shown in Figure 14.11a. The threshold electric field of the Pt-deposited carbon nanowall film decreased from 5 to 4 $V\,\mu m^{-1}$ after the SCF-MOCFD, and its emission current increased by more than 250% compared to that of the as-grown carbon nanowalls. Note that the emission properties of carbon nanowalls exposed to $scCO_2$ without the Pt precursor were the same as those of the as-grown carbon nanowalls; the morphology of carbon nanowalls remained unchanged after exposure to a high-pressure supercritical environment. Figure 14.11b shows these field emission characteristics for typical as-grown and Pt-supported carbon nanowalls plotted on a Fowler–Nordheim plot. The Fowler–Nordheim plots obtained from the electron emission measurements have almost identical slopes, although the intercept for

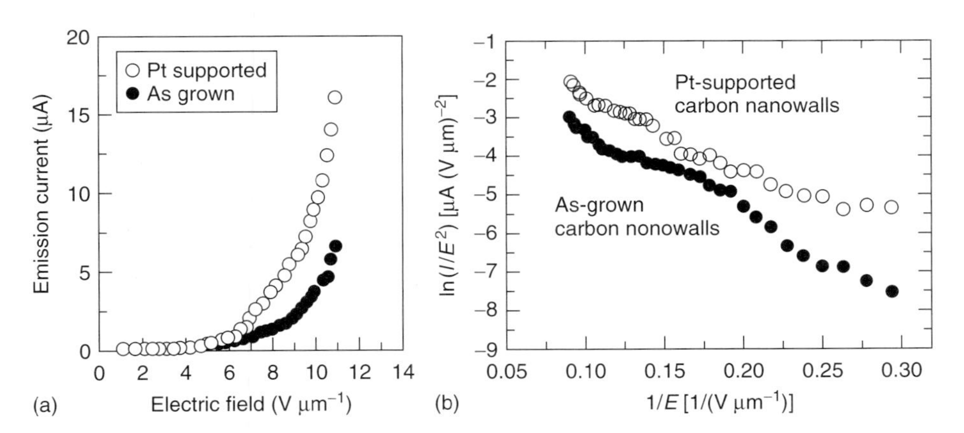

Figure 14.11 (a) Field emission current as a function of the average electric field for typical as-grown and Pt-deposited carbon nanowall films. (b) Corresponding plots according to the Fowler–Nordheim theory.

the Pt-deposited carbon nanowall film was higher than that for the as-grown carbon nanowall film. Therefore, the enhancement factors of typical as-grown and Pt-supported carbon nanowalls are suggested to be almost the same. The enhancement factor is calculated from the Fowler–Nordheim plot and has a value of 5000–9000, assuming a work function of 4.7 eV.

It is considered that there are potentially idle emission points on as-grown carbon nanowalls, and the Pt nanoparticles may activate these emission points, resulting in an increase of the number of emission sites of the Pt-supported carbon nanowalls. However, the carbon nanowalls have been reported to consist of nanodomains that are a few tens of nanometers in size [20], and individual carbon nanowalls were found to have many edges and defects [16]. In the case of SCF-MOCFD, Pt nanoparticles would be preferably formed on the edges of carbon nanowalls as well as at the defects and the grain boundaries of the carbon nanowall surfaces, which would effectively improve the electrical conduction along the carbon nanowall surface.

14.4.2
N_2 Plasma Treatment

In the case of field emission from carbon nanowalls, control of the electronic states of graphite edges that emit electrons would be important for improvement of the field emission properties of carbon nanowalls. Plasma surface treatment would be effective in modifying the electronic states at the surface of carbon nanowall films with many graphite edges, as well as defects or grain boundaries.

A carbon nanowall film was synthesized by VHF capacitively coupled PECVD with H radical injection employing a C_2F_6/H_2 mixture [17]. The morphology of the as-grown carbon nanowall film was similar to that shown in Figure 14.7a. The carbon nanowall film was subsequently exposed to N_2 plasma using the same CVD system, at a substrate temperature of 600 °C for 10 min. As a result of N_2 plasma treatment, the D to G band peak intensity ratio in the Raman spectrum (I_D/I_G) was significantly decreased from 2.24 to 1.37, mainly because of the reduction in the peak intensity of the D band. The D′ band peak intensity also decreased after N_2 plasma treatment. These results suggest that the inclusion of nitrogen in the carbon nanowall surface results in the reduction of defects. However, when compared with the surface morphology of the as-grown carbon nanowall film, the sharp edges and the orientation of the individual carbon nanowalls are unchanged, even after N_2 plasma treatment.

An example of the field emission enhancement of carbon nanowall film as a result of N_2 plasma treatment is shown in Figure 14.12. The electron emission current from the N_2 plasma-treated carbon nanowall film was significantly increased, while the threshold field was almost the same as that of the untreated carbon nanowall film. Therefore, surface modification by N_2 plasma treatment contributes mainly to the improvement of electrical conduction along the carbon nanowall surface.

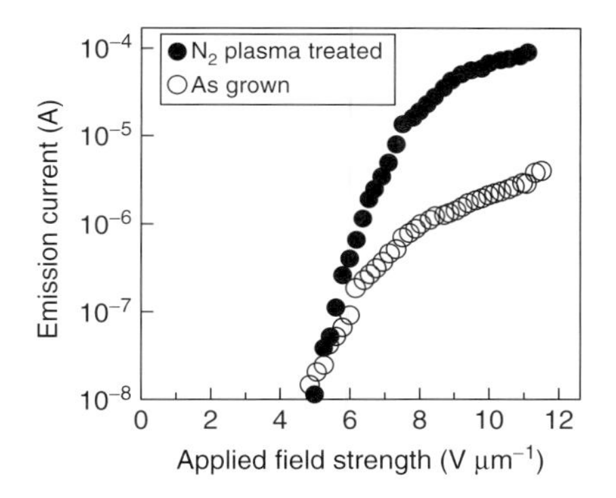

Figure 14.12 Field emission properties of a carbon nanowall film after N_2 plasma treatment.

14.5 Prospects for the Future

Carbon nanowalls are expected to be promising materials for electron field emission because of a high density of atomic-scale graphitic edges and good electrical conductivity. In contrast, the spatial uniformity of field emission sites and long-term stability of the field emission current density must be improved with respect to practical applications. In addition, further elucidation of the field emission mechanism of metal-deposited carbon nanowalls and nitrogen-doped carbon nanowalls is required.

References

1. Iijima, S. (1991) Helical microtubules of graphitic carbon. *Nature (London)*, **354**, 56–58.
2. Wu, Y.H., Qiao, P.W., Chong, T.C., and Shen, Z.X. (2002) Carbon nanowalls grown by microwave plasma enhanced chemical vapor deposition. *Adv. Mater. (Weinheim, Ger.)*, **14** (1), 64–67.
3. Hiramatsu, M., Shiji, K., Amano, H., and Hori, M. (2004) Fabrication of vertically aligned carbon nanowalls using capacitively coupled plasma-enhanced chemical vapor deposition assisted by hydrogen radical injection. *Appl. Phys. Lett.*, **84** (23), 4708–4710.
4. Wang, J.J., Zhu, M.Y., Outlaw, R.A., Zhao, X., Manos, D.M., Holloway, B.C., and Mammana, V.P. (2004) Free-standing subnanometer graphite sheets. *Appl. Phys. Lett.*, **85** (7), 1265–1267.
5. Ando, Y., Zhao, X., and Ohkohchi, M. (1997) Production of petal-like graphite sheets by hydrogen arc discharge. *Carbon*, **35** (1), 153–158.
6. Chuang, A.T.H., Robertson, J., Boskovic, B.O., and Kozio, K.K.K. (2007) Three-dimensional carbon nanowall structures. *Appl. Phys. Lett.*, **90** (12), 123107–123109.
7. Hiramatsu, M. and Hori, M. (2006) Fabrication of carbon nanowalls using novel plasma processing. *Jpn. J. Appl. Phys.*, **45** (6B), 5522–5527.
8. Sato, G., Morio, T., Kato, T., and Hatakeyama, R. (2006) Fast growth of carbon nanowalls from pure methane using helicon plasma-enhanced chemical vapor deposition. *Jpn. J. Appl. Phys.*, **45** (6A), 5210–5212.
9. Mori, T., Hiramatsu, M., Yamakawa, K., Takeda, T., and Hori, M. (2008) Fabrication of carbon nanowalls using electron beam excited plasma-enhanced chemical vapor deposition. *Diamond Relat. Mater.*, **17** (7-10), 1513–1517.

10. Shimabukuro, S., Hatakeyama, Y., Takeuchi, M., Itoh, T., and Nonomura, S. (2008) Preparation of carbon nanowall by hot-wire chemical vapor deposition and effects of substrate heating temperature and filament temperature. *Jpn. J. Appl. Phys.*, **47** (11), 8635–8640.
11. Zhang, H., Yoshimura, I., Kusano, E., Kogure, T., and Kinbara, A. (2004) Formation of carbon nano-flakes by RF magnetron sputtering method. *Shinku (J. Vac. Soc. Jpn.)*, **47** (2), 82–86 (in Japanese).
12. Shiji, K., Hiramatsu, M., Enomoto, A., Nakamura, M., Amano, H., and Hori, M. (2005) Vertical growth of carbon nanowalls using rf plasma-enhanced chemical vapor deposition. *Diamond Relat. Mater.*, **14** (3-7), 831–834.
13. Ikeda, M., Ito, H., Hiramatsu, M., Hori, M., and Goto, T. (1995) Synthesis of diamond using RF magnetron methanol plasma chemical vapor deposition assisted by hydrogen radical injection. *Jpn. J. Appl. Phys.*, **34** (11, Part 1), 2484–2488.
14. Hiramatsu, M., Inayoshi, M., Yamada, K., Mizuno, E., Nawata, M., Ikeda, M., Hori, M., and Goto, T. (1996) Hydrogen-radical-assisted radio-frequency plasma-enhanced chemical vapor deposition system for diamond formation. *Rev. Sci. Instrum.*, **67** (6), 2360–2365.
15. Yu, J., Zhang, Q., Ahn, J., Yoon, S., Rusli, F.Y., Li, J., Gan, B., Chew, K., and Tan, K.H. (2001) Field emission from patterned carbon nanotube emitters produced by microwave plasma chemical vapor deposition. *Diamond Relat. Mater.*, **10** (12), 2157–2160.
16. Kurita, S., Yoshimura, A., Kawamoto, H., Uchida, T., Kojima, K., Tachibana, M., Molina-Morales, P., and Nakai, H. (2005) Raman spectra of carbon nanowalls grown by plasma-enhanced chemical vapor deposition. *J. Appl. Phys.*, **97** (19), 104320.
17. Kondo, S., Hori, M., Yamakawa, K., Den, S., Kano, H., and Hiramatsu, M. (2008) Highly reliable growth process of carbon nanowalls using radical injection plasma-enhanced chemical vapor deposition. *J. Vac. Sci. Technol., B.*, **26** (4), 1294–1300.
18. Hou, K., Outlaw, R.A., Wang, S., Zhu, M., Quinlan, R.A., Manos, D.M., Kordesch, M.E., Arp, U., and Holloway, B.C. (2008) Uniform and enhanced field emission from chromium oxide coated carbon nanosheets. *Appl. Phys. Lett.*, **92** (13), 133112.
19. Machino, T., Takeuchi, W., Kano, H., Hiramatsu, M., and Hori, M. (2009) Synthesis of platinum nanoparticles on two-dimensional carbon nanostructures with an ultrahigh aspect ratio employing supercritical fluid chemical vapor deposition process. *Appl. Phys. Express.*, **2**, 025001.
20. Kobayashi, K., Tanimura, M., Nakai, H., Yoshimura, A., Yoshimura, H., Kojima, K., and Tachibana, M. (2007) Nanographite domains in carbon nanowalls. *J. Appl. Phys.*, **101** (9), 094306.

15
Flexible Field Emitters: Carbon Nanofibers

Masaki Tanemura and Shu-Ping Lau

15.1
Introduction

As was described in the previous chapters, because of their high aspect ratios, small radii of curvature of the tips, and high chemical stability, one-dimensional (1D) nanocarbon materials such as carbon nanotubes (CNTs) [1] and carbon nanofibers (CNFs) are quite promising as field electron sources, for example, for flat panel displays [2, 3]. Arc discharge [1], laser ablation [4], and chemical vapor deposition (CVD) [5–7] have conventionally been employed for their synthesis. In those methods, however, growth temperatures higher than 500 °C are generally required. For their applications to flexible displays using plastic substrates and from a standpoint of an eco-process, however, they should be grown at lower temperatures, ideally at room temperature (RT). In this respect, plasma-enhanced CVD at and below 120 °C has been attempted [8, 9]. In this chapter, we will deal with a new approach to synthesize CNFs at RT and their applications to flexible emission sources.

15.2
Room Temperature Fabrication of Ion-Induced Carbon Nanofibers

Ion irradiation of solid surfaces sometimes induces the formation of nano- to microsized surface structures, such as ripples, pyramids, conical protrusions, and whiskers, even at RT [10, 11]. In addition, surface texturing is sometimes enhanced by a simultaneous supply of so-called seed materials which are different from the constituent of the target surface during ion irradiation [12, 13]. These imply that ion irradiation is promising as a basic technique to fabricate nanomaterials or nanostructures at low temperatures.

Figure 15.1a shows a typical scanning electron microscope (SEM) image of an ion-induced CNF on a glassy carbon surface irradiated by Ar^+ ions of 3 keV with a simultaneous supply of Mo at RT [14]. As seen in Figure 15.1a, a conical protrusion

Carbon Nanotube and Related Field Emitters: Fundamentals and Applications. Edited by Yahachi Saito

ISBN: 978-3-527-32734-8

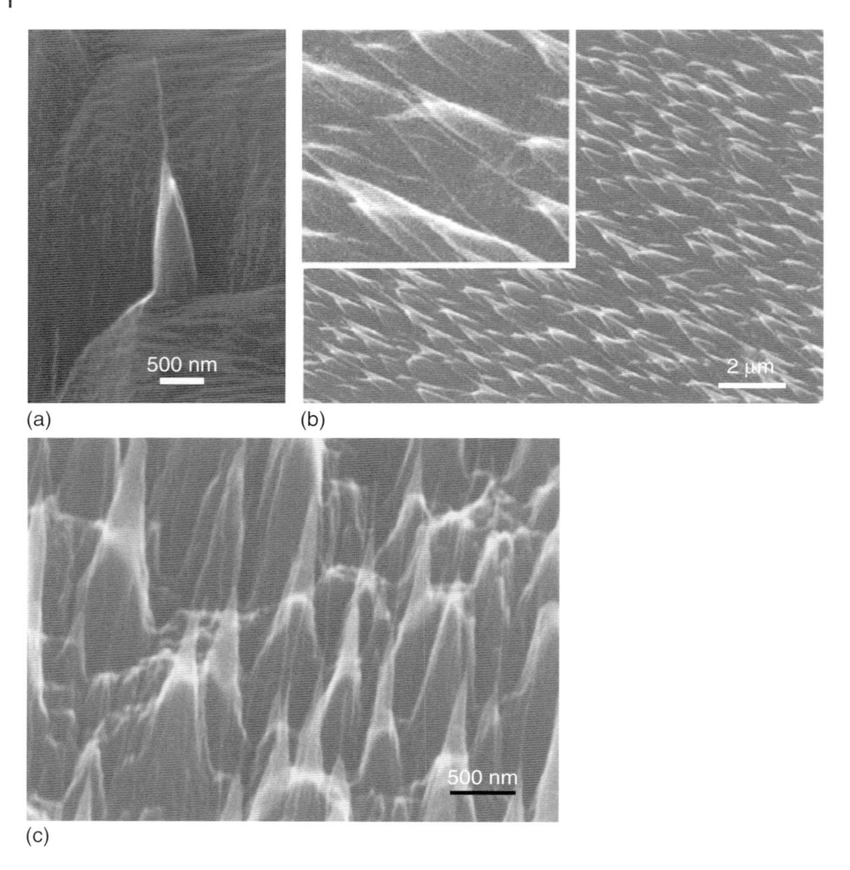

Figure 15.1 SEM images of (a) an isolated CNF formed on glassy carbon, and densely distributed CNFs formed on (b) carbon-coated Si, and (c) polyimide surfaces. Inset in (b) shows an enlarged image.

formed on the surface, and a linear single CNF, ~50 nm in diameter and ~1 μm in length, grew on the tip. Both the basal cone and the CNF were pointing in the ion-beam direction. Although Mo was simultaneously supplied during the ion irradiation for the CNF growth in Figure 15.1a, the supply of Mo or metal particles was not a prerequisite for the ion-induced CNF growth. In other words, the supplied metal particles do not act as a catalyst for the CNF growth, and thus the RT growth of ion-induced CNFs is possible without catalyst on graphite and carbon-coated substrates (metals, semiconductors, and plastics) as well as on glassy carbon [15–17].

The fabrication of densely distributed CNF-tipped cones is also possible. Figure 15.1b shows a typical example of those formed on a carbon-coated Si (C/Si) surface at RT without catalyst [15]. The ion-induced CNFs were linear and aligned in the ion-beam direction. They were 0.3–1 μm in length, and almost uniform in diameter, ~20 nm, in the growth direction independent of the CNF length. CNF size was independent of the cone size. It should be noted that single CNFs were formed on almost all the cones (inset of Figure 15.1b). In the ion-irradiation method, CNFs generally grow only on the cone tips, and more than one CNF do not grow on the respective cone tips. It should be also mentioned the ion-irradiation method requires no heat supply. This allows a greater choice of substrates. In fact, similar to the bulk carbon (see also Figure 15.6a) and Si substrates above exemplified, densely distributed CNF-tipped cones grew also on plastic substrates as shown in Figure 15.1c [15].

Figure 15.2a shows a typical transmission electron microscope (TEM) image of an Ar^+-induced CNF-tipped cone formed on a graphite substrate at RT without catalyst [18]. The CNF is ~20 nm in diameter and longer than 500 nm in length. No clear boundary between the CNF and the conical tip is recognizable. In addition, no hollow structure is observed in CNFs, suggesting that they are not CNTs but CNFs. Figure 15.2b and c shows the electron diffraction patterns (EDPs) taken at a middle part of the CNF and a stem region of the cone, respectively. The EDP from the stem region of the cone was characterized by the spotty rings arising from the graphite lattice (Figure 15.2c), implying that the stem part of the cones maintained the polycrystalline nature of the graphite substrate. By contrast, only hallow-like rings were observed and any graphite-related spot was not recognizable in the EDP from the CNF (Figure 15.2b), suggesting the amorphous-like or very fine-crystallite structure of the CNF. The crystallographic features are similar to those observed for CNFs grown on a carbon-coated Ni mesh [17]. Hofmann *et al.* also demonstrated that CNTs synthesized by plasma-enhanced CVD at 120 °C were low in graphitization quality [9]. Thus, the amorphous-like structure is a feature common to carbon nanomaterials grown at low temperatures.

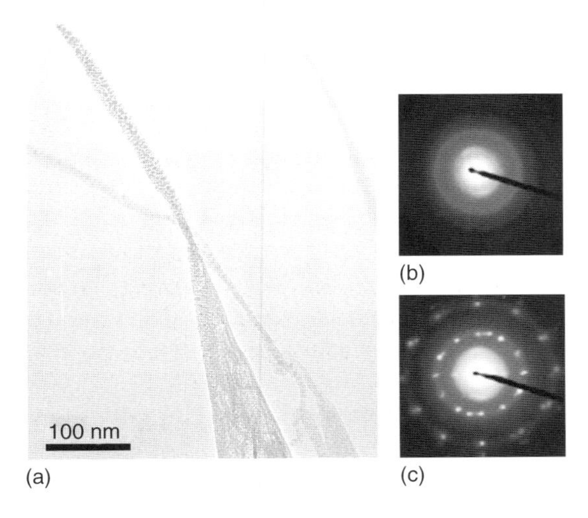

Figure 15.2 (a) TEM image of a typical CNF-tipped cone formed on graphite. (b) shows the EDPs from a middle part of the CNF in (a), and (c) from the stem part of the cone in (a).

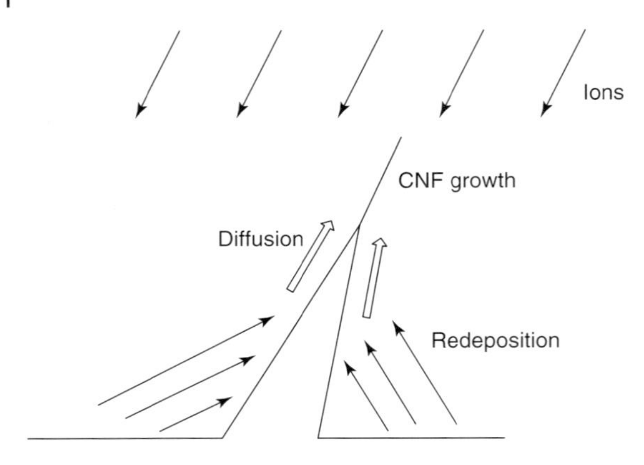

Figure 15.3 Growth model of the ion-induced CNF.

The growth mechanism of ion-induced CNFs is thought to be as follows [14, 17, 19, 20]: (i) formation of conical protrusions, (ii) redeposition of carbon atoms sputter-ejected from the surface onto the sidewall of the conical protrusions, and (iii) the surface diffusion of the redeposited carbon atoms toward the tips during sputtering (Figure 15.3). Because CNFs grown on glassy carbon at elevated temperatures, for example, at 200 °C, were longer than those formed at RT, it is obvious that surface diffusion plays a decisive role in the growth of ion-induced CNFs [14].

15.3
Applications to Field Electron Emission Sources

Ion-induced CNFs cover many of the application fields of CNTs. Thanks to their high aspect ratios, small radius of curvature of the tip, and high chemical stability, their application to field emission (FE) sources is very promising.

15.3.1
Current–Voltage (I–V) Characteristics

Figure 15.4 shows the FE characteristics obtained for the CNFs formed on graphite and carbon-coated Si (C/Si) cathodes [17]. The FE characteristics were measured for an applied voltage range of 0–3500 V under a parallel plate configuration at a working pressure of 5.0×10^{-4} Pa. It should be noted that the FE measurements were carried out under conventional vacuum condition. The tested emission area and gap width between the anode and the cathode were 0.79 cm^2 (10 mm in diameter) and 1 mm, respectively. The corresponding Fowler–Nordheim (FN) plot is also shown in the inset of Figure 15.4. From Figure 15.4, the *threshold fields*, defined as the field at which current density reaches 1 $\mu A\ cm^{-2}$, were estimated to be 2.2 and 1.8 $V\ \mu m^{-1}$ for CNFs on graphite and C/Si, respectively. The field enhancement factors (β) calculated from the FN plot were 1310 and

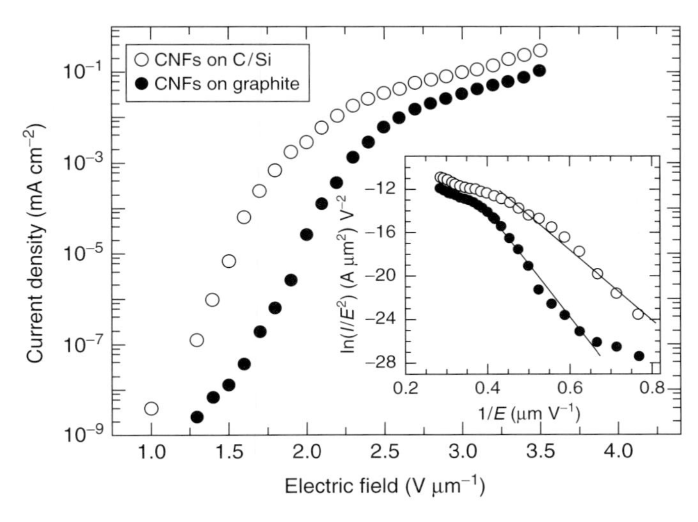

Figure 15.4 *I–V* characteristics of CNFs grown on graphite and C/Si surfaces. Inset shows the corresponding FN plot.

1951, respectively, assuming a work function of 4.6 eV for graphite. It must be stressed that these emission characteristics are comparable to those reported for CNTs prepared by conventional CNT synthesis methods at high temperatures [21–24], and are much better than those obtained for CNFs grown by CVD at low temperatures [25, 26].

As is well known, the FE properties of multiple emitter systems depend on the site density of emitter tips as well as their size. For higher and stable emission currents, a large number of emitter tips are preferable. An excess numerical density of emitter tips, however, leads to the saturation in total emission current as a result of the so-called screening effect of adjacent tips [27]. In this respect, ion-induced CNFs grown on conical bases are advantageous to prevent the excess of emission-site density, and hence reduce the screening effect. In addition, the geometrical structure of CNF-tipped cones is a kind of "multistage emitter structure" leading to the larger field enhancement [28]. Thus, these factors are thought to be responsible for the observed good FE properties for the ion-induced CNFs.

15.3.2 Lifetime

The FE current in general decreases with the operation time, mainly because of the damage of emitter tips induced by the bombardment with ionized residual gas molecules. Once the tips are damaged, they do not contribute any more to the intense electron emission. This is the reason why ultrahigh vacuum (UHV), usually better than 10^{-6} Pa, is required for stable and long-term operation. For

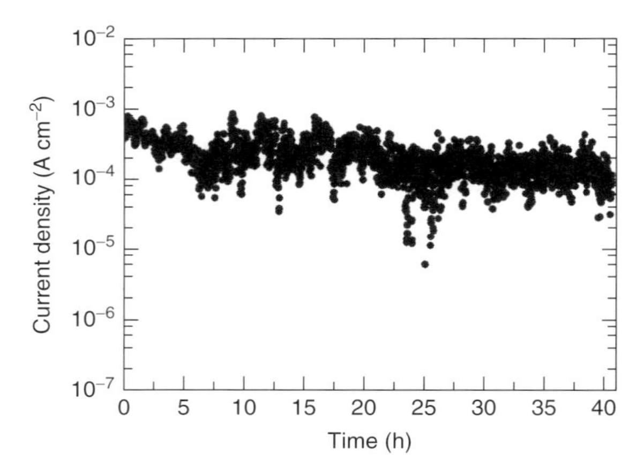

Figure 15.5 Time dependence of the emission-current density measured at an electric field of 10 V μm^{-1}.

Pd-catalyzed CNTs, for example, a stable emission for as long as 600 h under an extreme UHV condition, 3×10^{-9} Pa, has been demonstrated [29]. Such an extreme UHV, however, is unrealistic for practical applications. Thus, the development of tough emitters with a long lifetime operating under low vacuum is an important area of research.

Figure 15.5 represents the result of a lifetime test for CNFs grown on a graphite plate for about 40 h at a constant applied electric field of 10 V μm^{-1} under non-UHV conditions (10^{-4} Pa range) [19]. The emission-current density gradually decreased with time (t) from ~7 to ~1.7×10^{-4} A cm^{-2} at the initial stage ($t = 0$–6 h) of the lifetime test, and then strangely increased slightly up to ~3×10^{-4} A cm^{-2}. This value was sustained for $t = 7$–20 h. For $t > 20$ h, the current density was stable at ~1.7×10^{-4} A cm^{-2}. It must be noted that the emitter operated stably for more than 40 h under such a non-UHV condition.

Figure 15.6 shows a typical SEM image of surface structure before and after the lifetime test. The CNFs on the cones observed before the lifetime test disappeared. Instead, rodlike protrusions and nonaligned, thick CFs, 0.2–0.5 μm in diameter, newly grew at the central (Figure 15.6b) and peripheral regions (Figure 15.6c) of the emission area, respectively. As described above, the basic growth mechanism of the sputter-induced CNFs is the redeposition of sputter-ejected carbon atoms onto the sidewall of conical protrusions and the surface diffusion of the carbon atoms to the tips during Ar^+ sputtering. In the present FE source operating under a non-UHV condition, the cathode surface would be continuously irradiated with ionized residual gas molecules, and hence carbon atoms would be sputter-ejected from the cathode surface. Those carbon atoms would redeposit onto surface projections and diffuse toward the tips, similar to the growth process of Ar^+-induced CNFs. In addition, the surface diffusion would be enhanced as a result of the local resistive heating induced by FE currents. Thus, thick CFs would grow during the FE process. This is a "self-regeneration" of the emitter tips. As the diameter of the CFs

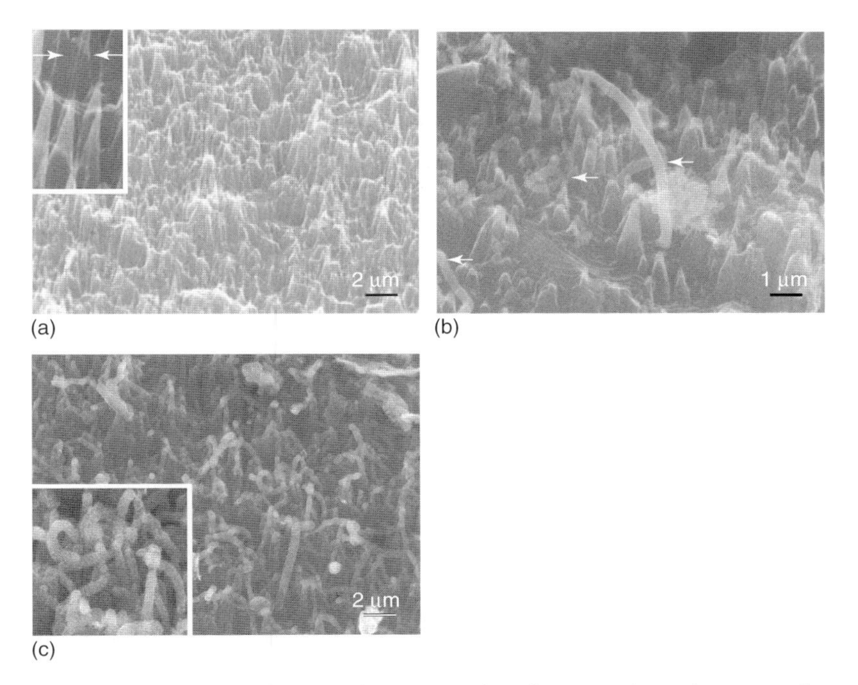

Figure 15.6 Typical SEM images of (a) before and (b and c) after lifetime test for 40 h. (b) Central and (c) peripheral regions of the emission area taken after the lifetime test. Insets in (a) and (c) show the enlarged images of typical CNF-tipped cones and entangled CFs newly grown during the FE process. Arrows in (a) and (b) represent the initial CNFs before the lifetime test and CFs formed after lifetime test, respectively.

newly grown during the life-time test is almost uniform in the growth direction, the electric field around the CF tips is nearly constant. This appears to be the source of the long-sustained emission. The optimization of the self-regenerative condition may lead to novel FE sources with a long lifetime operating under non-UHV conditions.

15.3.3 Flexible CNF Cathode

As described above, ion-induced CNFs grow on any solid substrates, such as plastics, at RT. Is FE possible from CNFs grown on plastic substrates? If FE occurs, flexible plastic-based devices will be realized using the ion-induced CNFs. Figure 15.7a shows an SEM image of the geometrical configuration of an anode microprobe (tungsten) and an isolated single CNF, ~700 nm in length and ~16 nm in diameter, tipped on a nanocone of 300 nm in length formed on a Kapton polyimide foil [30]. Similar to the CNFs grown on graphite and Si substrates, FE comparable to that from CNTs was possible from the CNFs on the polyimide substrates (Figure 15.7b). The maximum current density of the single CNF attainable was in a range of

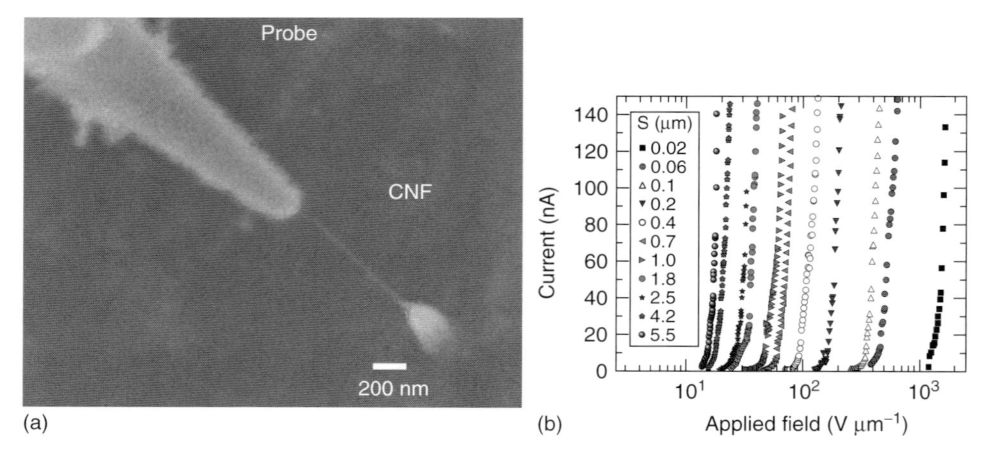

Figure 15.7 (a) SEM micrograph of an isolated single CNF and an anode probe. (b) Emission current against applied field as a function of the gap width (*S*).

10^9 A m^{-2}, which is comparable to that from CNTs [31], and the FE current achievable was as high as 15 μA without causing damage.

As will be seen in a later chapter, nanocarbon is quite promising as FE X-ray sources [32–35]. Taking this point into account, a prototype of a flexible plastic X-ray source consisting of ion-induced CNFs grown on a polyimide (cathode) and Cu target (anode) was developed. The experimental system, which was evacuated by a turbomolecular pump, was an open type X-ray tube with a beryllium window directing the X-rays in a vertically upward direction [36]. Figure 15.8a is a photograph showing a mock-up of the flexible emitter and the copper anode when inside the X-ray tube. The surface of the Cu anode was angled approximately at 45° to

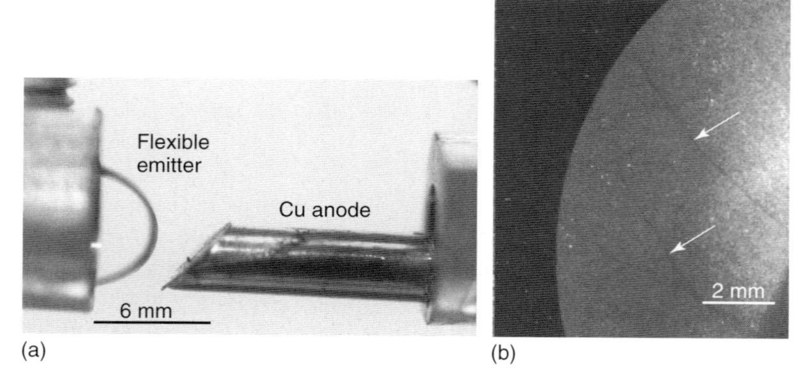

Figure 15.8 (a) Photograph of a flexible emitter and a copper anode for X-ray generation. (b) Radiograph of a bougainvillea flower. Arrows in (b) represent veins.

assist X-ray emission. The radius of the polyimide emitter curl was approximately 3 mm. Similar to the graphite and carbon-coated Si cases, CNF-tipped cones were densely distributed on the polyimide substrate with a number density of around 4.6×10^7 cm^{-2}. The length and diameter of the CNFs measured ~1 μm and 10–15 nm, respectively. The flexible emitter cathode was connected to a pulse voltage generator operating at 600 Hz with output voltage ranging from 10 to 25 kV.

Figure 15.8b shows an X-ray image of a bougainvillea flower obtained by the flexible emitter exposed for about 45 s using 13 kV [36]. It should be noted that the flexible X-ray source produces a high-quality picture with high resolution on the biological sample. Veins (~250 μm in thick) on the petals are clearly seen in Figure 15.8b. This proves that the flexible emitter is capable of emitting sufficient current to generate the X-ray flux necessary for a high signal-to-noise ratio even at such a low voltage. In continuous operation of the flexible X-ray source at 16 kV, stable X-ray generation for longer than 30 min with low fluctuation of the X-ray output was confirmed by a dosimeter. This suggests that the flexible-polyimide-based CNFs can have excellent durability for use as high-power electron emitter sources. The use of such a flexible emitter will allow for greater choices of X-ray tube design previously not possible in radiotherapy. It will be also possible to make an X-ray-transparent emitter as the window with a concave target surface to achieve a focused X-ray beam, thereby reducing the spot size of the X-ray beam. This will allow for more precise radiotherapy than is possible with the usual diverging X-ray source.

For polyimide-based CNFs, FE occurs under compressive stress conditions as well as tensile stress conditions, as shown in Figure 15.8a. In order to check the FE characteristics under various bending conditions, a plastic flexible emitter consisting of a CNF-covered polyimide, 1.0×2.0 cm^2 (cathode), and an electrically conductive copper tape (anode) separated by a 100 μm-thick polytetrafluorethylene (PTFE) insulation spacer was prepared [37]. Similar to the flexible X-ray source in Figure 15.8a, CNF-tipped cones were abundant and almost uniformly distributed on the cathode surface. Basal cones were 0.5–1 μm in length and 200–500 nm in stem diameter, whereas CNFs were ~15 nm in diameter. A hole of 6 mm was punched through the spacer, and hence the test exposed emission area was 0.28 cm^2. This flexible emitter was bent to different curvatures and directions, that is, inward or outward bending for tensile or compressive stress, respectively, by using three cylindrical structures – 10, 25, and 50 mm in diameter. FE characteristics of the flexible emitter thus prepared were measured in a parallel plate configuration under a base pressure of 10^{-6} Pa.

Figure 15.9a illustrates the current–voltage ($I-V$) characteristics measured under various bending conditions [37]. The threshold field at which the current density of 1 μA cm^{-2} is achieved at flat (no bending) condition was ~3.2 V μm^{-1}, whereas that for the emitter under various bending conditions was in the range of 3.1–4.2 V μm^{-1}. Thus, the remarkable change in the threshold field was not observed under different bending curvatures for both tensile and compressive stressed emitters. Lifetime test of the flexible emitter proved the stable emission with average current densities of 7.25 and 7.11×10^{-5} A cm^{-2} for a duration

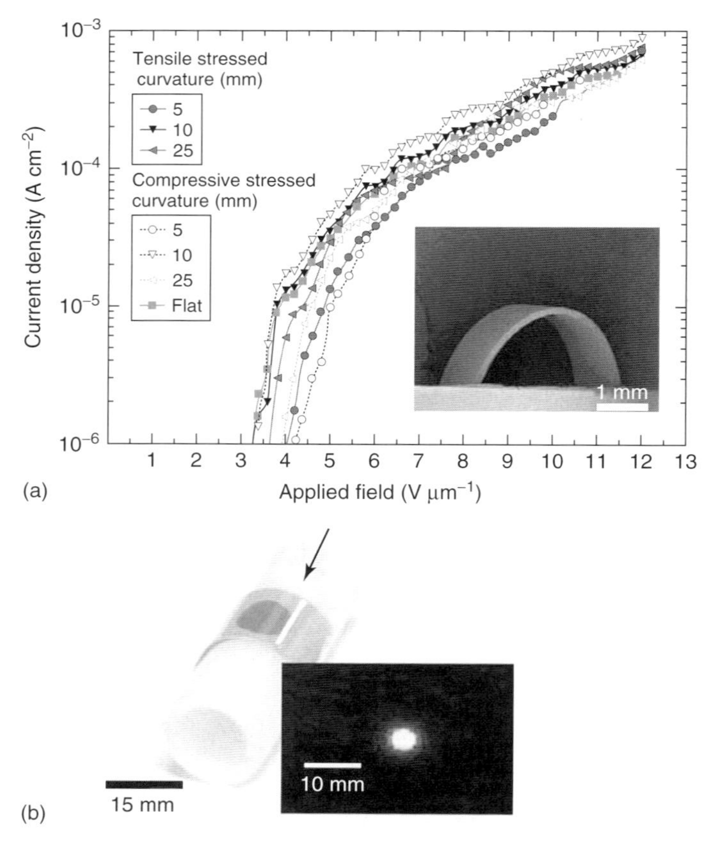

Figure 15.9 (a) FE measurement of the flexible CNF emitters at different bending curvatures, the inset showing a bent flexible CNF-FE device. (b) Photograph of the all-plastic FE device mounted on a 25-mm-diameter structure. Arrow indicates the flexible emitter. The inset shows the emission spots captured from the phosphor-coated plastic substrate at an applied field of 13 V μm^{-1}.

as long as 16 h at a field of 10 V μm^{-1} in flat and outward bending (25 mm in diameter) conditions, respectively. Thus, the emitter was quite robust regardless of planar or bent conditions.

Needless to say, an all-plastic FE display is achievable by replacing the anode copper tape with a phosphor-coated plastic, as exemplified in Figure 15.9b [37]. In this all-plastic FE display, a ZnS : Ag phosphor-coated polyester sheet (Eljen Technology) of 250 μm thickness was employed. Since the anode has to be conductive to attract electrons from the emitter, a thin Pt layer (10 nm in thickness) was deposited onto the phosphor. In the figure, the all-plastic display was mounted onto a cylindrical object 25 mm in diameter after the several times of inward and outward bending of the display. The inset of Figure 15.9b shows an emission image captured by a charge-coupled device camera with an applied field of $13\,V\,\mu m^{-1}$, showing fairly uniform emission.

Another advantage of the ion-induced CNFs is that they can be used as a template to fabricate nanoneedle and nanocone structures on any substrates [38–40]. For example, ZnO nanoneedle structures were fabricated on a ZnO-coated plastic (polyethyleneterephthalate (PET) and polyimide) at RT using the ion-induced CNFs as a template [40]. FE from the ZnO nanoneedles (100 nm in stem diameter and 700 nm in length) thus fabricated on the polyimide showed a threshold field of $4.1\,V\,\mu m^{-1}$ with a current density of $1\,\mu A\,cm^{-2}$, which is comparable with that of conventional CNTs and ion-induced CNFs, and emission-current density as high as $1\,mA\,cm^{-2}$ at $9.6\,V\,\mu m^{-1}$. Taking account of the larger work function of ZnO (5.3 eV) than that of nanocarbons (<5.0 eV), the threshold field attained is fairly low. In addition, the composition of CNFs and CNF-templated nanoneedles is controllable by a simultaneous supply of metal during their formation [16, 18, 38, 41–44]. This implies that the work function and the conductivity of ion-induced nanostructures are tunable to match practical demands. Thus, the CNFs and nanostructures produced by ion irradiation at RT are quite promising for efficient flexible field emitters.

15.4 Summary

The RT fabrication of CNFs and nanostructures by the ion-irradiation method and their applications to FE devices, especially flexible devices, were demonstrated. CNF-based flexible emission sources are promising for varieties of applications such as flexible illumination, portable X-ray computerized tomography, wearable computer displays, and flexible electric papers. Besides FE applications, ion-induced CNFs and nanostructures are applicable to probes for atomic force microscopes [46] and to ultra violet laser emission [39]. For the further expansion of application fields of FE sources, for example, to the so-called head-up displays, the optical transparency is also an important key factor. Very recently, an optically transparent and flexible FE source based on ion-induced carbon nanocones whose sizes are smaller than the wavelength of the visible light was demonstrated using a nation film as an emitter substrate [45]. Thus, the CNFs and nanostructures produced by ion irradiation at RT are quite promising for future flexible field emitters.

References

1. Iijima, S. (1991) Helical microtubules of graphitic carbon. *Nature*, **354**, 56–59.
2. Saito, Y., Hamaguchi, K., Uemura, S., Uchida, K., Tasaka, Y., Ikazaki, F., Yumura, M., Kasuya, A., and Nishina, Y. (1998) Field emission from multi-walled carbon nanotubes and its application to electron tubes. *Appl. Phys. A*, **67** (1), 95–100.
3. Choi, W.B., Chung, D.S., Kang, J.H., Kim, H.Y., Jin, Y.W., Han, I.T., Lee, Y.H., Jung, J.E., Lee, N.S., Park, G.S.,

and Kim, J.M. (1999) Fully sealed, high-brightness carbon-nanotube field-emission display. *Appl. Phys. Lett.*, **75** (20), 3129–3131.

4. Thess, A., Nikolaev, P., Dai, H.J., Petit, P., Robert, J., Xu, C.H., Lee, Y.H., Kim, S.G., Rinzler, A.G., Cocert, D.T., Scuseria, G.E., Tomanek, D.T., Fisher, J.E., and Smalley, R.E. (1996) Crystalline ropes of metallic carbon nanotubes. *Science*, **273**, 483–487.
5. Pan, Z.W., Xie, S.S., Chang, B.H., Wang, C.Y., Lu, L., Liu, W., Zhou, W.Y., and Li, W.Z. (1998) Very long carbon nanotubes. *Nature*, **394**, 631–632.
6. Ren, Z.F., Huang, Z.P., Xu, J.W., Wang, J.H., Bush, P., Siegal, M.P., and Provencio, P.N. (1998) Synthesis of large arrays of well-aligned carbon nanotubes on glass. *Science*, **282**, 1105–1107.
7. Tanemura, M., Iwata, K., Takahashi, K., Fujimoto, Y., Okuyama, F., Sugie, H., and Filip, V. (2001) Growth of aligned carbon nanotubes by plasma-enhanced chemical vapor deposition: optimization of growth parameters. *J. Appl. Phys.*, **90**, 1529–1533.
8. Boskovic, B.O., Stolojan, V., Khan, R.U.A., Haq, S., and Silva, S.R.P. (2002) Large-area synthesis of carbon nanofibres at room temperature. *Nat. Mater.*, **1**, 165–168.
9. Hofmann, S., Ducati, C., Robertson, J., and Kleinsorge, B. (2003) Low-temperature growth of carbon nanotubes by plasma-enhanced chemical vapor deposition. *Appl. Phys. Lett.*, **83** (1), 135–137.
10. Auciello, O. and Kelly, R. (eds) (1984) *Ion Bombardment Modification of Surfaces*, Elsevier, Tokyo.
11. Czanderna, A.W., Madey, T.E., and Powell, C.J. (eds) (1998) *Beam Effects, Surface Topography, and Depth Profiling in Surface Analysis*, Chapter 1, Plenum Press, New York and London.
12. Wehner, G.K. (1985) Cone formation as a result of whisker growth on ion bombarded metal surfaces. *J. Vac. Sci. Technol., A*, **3** (4), 1821–1835.
13. Tanemura, M., Yamauchi, H., Yamane, Y., Okita, T., and Tanemura, S. (2004) Controlled fabrication of Mo-seeded Si microcones by Ar+-ion bombardment. *Nucl. Instrum. Methods*, **215**, 137–142.
14. Tanemura, M., Okita, T., Yamauchi, H., Tanemura, S., and Morishima, R. (2004) Room-temperature growth of a carbon nanofiber on the tip of conical carbon protrusions. *Appl. Phys. Lett.*, **84**, 3831–3833.
15. Tanemura, M., Hatano, H., Kitazawa, M., Tanaka, J., Okita, T., Lau, S.P., Yang, H.Y., Yu, S.F., Huang, L., Miao, L., and Tanemura, S. (2006) Room-temperature growth of carbon nanofibers on plastic substrates. *Surf. Sci.*, **600**, 3663–3667.
16. Tanemura, M., Okita, T., Tanaka, J., Yamauchi, H., Miao, L., Tanemura, S., and Morishima, R. (2005) Room-temperature growth of carbon nanofibers induced by Ar+-ion bombardment. *Eur. Phys. J. D*, **34**, 283–286.
17. Tanemura, M., Tanaka, J., Itoh, K., Fujimoto, Y., Agawa, Y., Miao, L., and Tanemura, S. (2005) Field electron emission from sputter-induced carbon nanofibers grown at room temperature. *Appl. Phys. Lett.*, **86**, 113107-1–113107-3.
18. Takeuchi, D., Wang, Z.P., Yamaguchi, K., Kitazawa, M., Hayashi, Y., and Tanemura, M. (2008) Morphological and structural characterization of metal-doped carbon nanofibers synthesized at room temperature. *J. Phys. Conf. Ser.*, **100**, 012029-1–012029-4.
19. Tanemura, M., Tanaka, J., Itoh, K., Okita, T., Miao, L., Tanemura, S., Lau, S.P., Huang, L., Agawa, Y., and Kitazawa, M. (2005) Self-regenerative field emission source. *Appl. Phys. Lett.*, **87**, 193102-1–193102-3.
20. Tanemura, M., Kitazawa, M., Tanaka, J., Okita, T., Ohta, R., Miao, L., and Tanemura, S. (2006) Direct growth of a single carbon nanofiber onto a tip of scanning probe microscopy induced by ion irradiation. *Jpn. J. Appl. Phys.*, **45**, 2004–2008.
21. Zhu, W., Bower, C., Zhou, O., Kochanski, G., and Jin, S. (1999) Large current density from carbon nanotube field emitters. *Appl. Phys. Lett.*, **75** (6), 873–875.

22. Murakami, H., Hirakawa, M., Tanaka, C., and Yamakawa, H. (2000) Field emission from well-aligned, patterned, carbon nanotube emitters. *Appl. Phys. Lett.*, **76** (13), 1776–1778.
23. Chhowalla, M., Ducati, C., Rupesinghe, N.L., Teo, K.B.K., and Amaratunga, G.A.J. (2001) Field emission from short and stubby vertically aligned carbon nanotubes. *Appl. Phys. Lett.*, **79** (13), 2079–2081.
24. Tanemura, M., Filip, V., Iwata, K., Fujimoto, Y., Okuyama, F., Nicolaescu, D., and Sugie, H. (2002) Field electron emission from carbon nanotubes grown by plasma-enhanced chemical vapor deposition. *J. Vac. Sci. Technol., B*, **20**, 122–127.
25. Hofmann, S., Ducati, C., Kleinsorge, B., and Robertson, J. (2001) Direct growth of aligned carbon nanotube field emitter arrays onto plastic substrates. *Appl. Phys. Lett.*, **83**, 4661–4663.
26. Smith, R.C., Carey, J.D., Poa, C.H.P., Cox, D.C., and Silva, S.R.P. (2004) Electron field emission from room temperature grown carbon nanofibers. *J. Appl. Phys.*, **95**, 3153–3157.
27. Nicolaescu, D., Filip, V., and Okuyama, F. (1997) Proposal for a new self-focusing configuration involving porous silicon for field emission flat. *J. Vac. Sci. Technol., A*, **15** (4), 2369–2374.
28. Huang, J.Y., Kempa, K., Jo, S.H., Chen, S., and Ren, Z.F. (2005) Giant field enhancement at carbon nanotube tips induced by multistage effect. *Appl. Phys. Lett.*, **87** (5), 053110-1–053110-3.
29. Kita, S., Sakai, Y., Fukushima, T., Mizuta, Y., Ogawa, A., Senda, S., and Okuyama, F. (2004) Characterization of field-electron emission from carbon nanofibers grown on Pd wire. *Appl. Phys. Lett.*, **85** (19), 4478–4480.
30. Sim, H.S., Lau, S.P., Ang, L.K., You, G.F., Tanemura, M., Yamaguchi, K., and Yusop, M.Z.M. (2008) Field emission from a single carbon nanofiber at sub 100 nm gap. *Appl. Phys. Lett.*, **93**, 023131-1–023131-3.
31. Bonard, J.M. and Klinke, C. (2003) Degradation and failure of carbon nanotube field emitters. *Phys. Rev. B*, **67**, 115406-1–115406-10.
32. Sugie, H., Tanemura, M., Filip, V., Iwata, K., Takahashi, K., and Okuyama, F. (2001) Carbon nanotubes as electron source in an x-ray tube. *Appl. Phys. Lett.*, **78**, 2578–2580.
33. Yue, G.Z., Qiu, Q., Gao, B., Cheng, Y., Zhang, J., Shimoda, H., Chang, S., Lu, J.P., and Zhou, O. (2002) Generation of continuous and pulsed diagnostic imaging x-ray radiation using a carbon-nanotube-based field-emission cathode. *Appl. Phys. Lett.*, **81** (2), 355–357.
34. Matsumoto, T. and Mimura, H. (2003) Point x-ray source using graphite nanofibers and its application to x-ray radiography. *Appl. Phys. Lett.*, **82** (10), 1637–1639.
35. Haga, A., Senda, S., Sakai, Y., Mizuta, Y., Kita, S., and Okuyama, F. (2004) A miniature x-ray tube. *Appl. Phys. Lett.*, **84** (12), 2208–2210.
36. Tan, T.T., Sim, H.S., Lau, S.P., Yang, H.Y., Tanemura, M., and Tanaka, J. (2006) X-ray generation using carbon nanofibre based flexible field emitters. *Appl. Phys. Lett.*, **88**, 103105-1–103105-3.
37. Sim, H.S., Lau, S.P., Yang, H.Y., Ang, L.K., Tanemura, M., and Yamaguchi, K. (2007) Reliable and flexible carbon-nanofiber-based all-plastic field emission devices. *Appl. Phys. Lett.*, **90**, 143103-1–143103-3.
38. Tanemura, M., Hatano, H., Kudo, M., Ide, N., Fujimoto, Y., Miao, L., Yang, H.Y., Lau, S.P., Yu, S.F., and Kato, J. (2007) Low-temperature fabrication and random laser action of doped zinc oxide nanoneedles. *Surf. Sci.*, **601**, 4459–4464.
39. Yang, H.Y., Lau, S.P., Yu, S.F., Tanemura, M., Okita, T., Hatano, H., Teng, K.S., and Wilks, S.P. (2006) Wavelength-tunable and high-temperature lasing in ZnMgO nanoneedles. *Appl. Phys. Lett.*, **89**, 081107-1–081107-3.
40. Yang, H.Y., Lau, S.P., Yu, S.F., Huang, L., Tanemura, M., Tanaka, J., Okita, T., and Hng, H.H. (2005) Field emission from zinc oxide nanoneedles

on plastic substrates. *Nanotechnology*, **16**, 1300–1303.

41. Tanemura, M., Okita, T., Tanaka, J., Kitazawa, M., Itoh, K., Miao, L., Tanemura, S., Lau, S.P., Yang, H., and Huang, L. (2006) Room-temperature growth and applications of carbon nanofibers: a review. *IEEE Trans. Nanotechnol.*, **5**, 587–594.

42. Takeuchi, D., Wang, Z.P., Miyawaki, A., Yamaguchi, K., Suzuki, Y., Tanemura, M., Hayashi, Y., and Somani, P.R. (2008) Room-temperature synthesis and characterization of cobalt-doped carbon nanofibers. *Diamond Relat. Mater.*, **17**, 581–584.

43. Wang, Z.P., Yamaguchi, K., Takeuchi, D., Hayashi, Y., and Tanemura, M. (2009) Room-temperature synthesis and characterisation of ion-induced iron-carbon nanocomposite fibers. *Int. J. Nanotechnol.*, **6** (7-9), 753–761.

44. Sugita, Y., Kitazawa, M., Yusop, M.Z.M., Tanemura, M., Hayashi, Y., and Ohta, R. (2009) Application of ion-induced carbon nanocomposite fibers to magnetic force microscope probes. *J. Vac. Sci. Technol., B*, **27** (2), 980–983.

45. Ghosh, P., Yusop, M.Z., Satoh, S., Subramanian, M., Hayashi, A., Hayashi, Y., and Tanemura, M. (2010) Transparent and flexible field electron emitters based on the conical nanocarbon structures. *J. Am. Chem. Soc.*, **132**, 4034–4035.

46. Kitazawa, M., Ohta, R., Sugita, Y., Inaba, K., and Tanemura, M. (2009) Wafer-scale production of carbon nanofiber probes. *J. Vac. Sci. Technol. B*, **27** (2), 975–979.

16
Diamond Emitters

Shozo Kono

16.1
Field Emission from Intrinsic or p-Type Diamonds

Diamonds are semiconductors with a band gap of 5.5 eV. p-Type doping is carried out rather routinely by boron doping in chemical vapor deposition (CVD). It is known that the surfaces of hydrogen-terminated diamond have negative electron affinity (NEA). NEA of H-terminated diamonds has been regarded as best suited for field emitters. This is because electrons once excited into conduction band have essentially no potential barriers to be extracted into a vacuum. Thermal and mechanical strengths of diamond are the other factors that are suited for field emitters. The surface of a field emitter is always exposed to the bombardment of positive ions created by electron beam on its way to the anode. The strength of a diamond surface against ion sputtering is one of the best among materials [1, 2].

Field emission from H-terminated p-type diamonds has been tested, and it has found that electrons are extracted from the valence band top (E_V) as schematically shown in Figure 16.1a [3]. There was a field enhancement of the order of 10^2, but the origin of the enhancement was not clear [3]. There have been many attempts to make field emitters in which electrons are excited electronically into the conduction band of intrinsic diamond and extract electrons through NEA surface into vacuum [4, 5]. However, there was no clear evidence of electronic excitation of electrons into the conduction band. Even the mechanism of field emission in these diamonds was not clear [4, 5].

In practice, field emitters made of p-type diamonds are still attractive from the view point of durability and NEA of diamond. It is to be noted that Spindt-type field emitter arrays were made from a p-type polycrystalline diamond film, in which an emitter tip density of $\sim 4 \times 10^5$ mm^{-2} and an array area ranging from 5 to 100 μm^2 have been successfully fabricated [7]. A maximum field emission current density of 265 mA mm^{-2} was obtained. However, the operation yield of the device was rather low: it was ~82% for a 50 μm^2 emitter array and ~50% for a 100 μm^2 array [7]. The field emission mechanism of this Spindt-type emitter array was not clarified either.

Carbon Nanotube and Related Field Emitters: Fundamentals and Applications. Edited by Yahachi Saito

ISBN: 978-3-527-32734-8

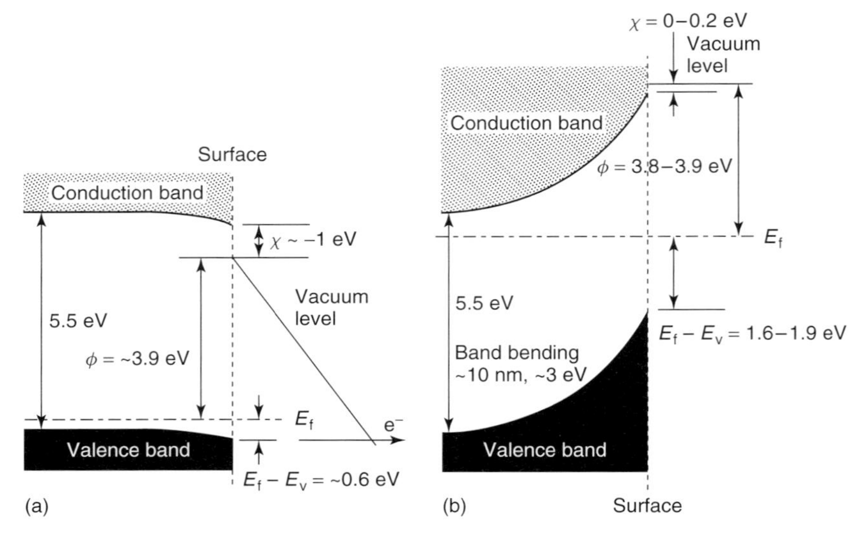

Figure 16.1 (a) Schematic band diagram of field emission from H-terminated p-type diamonds. (b) Schematic band diagram of a highly P-doped diamond (111) surface [6].

16.2 Field Emission from Nitrogen-Doped n-Type Diamonds

n-Type diamonds with an NEA surface would be the best suited for field emitters. If this type of diamonds existed in nature, there must be some potential barrier for conduction band electrons not to be lost freely into vacuum [8]. In natural diamonds, nitrogen is the source of the n-type dopant but the electronic conductivity is so low that natural diamonds are not suited for field emitters [9]. Phosphorus is another source of n-type dopant in diamond and can be doped by CVD [10–13]. A sensational demonstration of the n-type diamond field emitter was reported by Okano *et al.* [14]. The n-type diamond field emitter was made from a heavily N-doped polycrystalline diamond film deposited on an n-type Si(001) substrate. The properties of the n-type diamond emitter were later examined thoroughly. It was found that the emission sites were highly localized and that there appeared field-emitted electrons originating from valence bands and some unknown part below the valence band together with electrons from seemingly conduction band [15, 16]. These properties were not clearly understood. An N-doped polycrystalline diamond film made similarly as in [14] was very recently studied by a combination of ultraviolet photoelectron spectroscopy (UPS) and field emission spectroscopy (FES) [17]. It was found that the field-emitted electrons showed a peak at around 1.5 eV above the valence band maximum (E_V) at a certain applied field but the peak energy decreased drastically with the increase in the applied field. The energy level around 1.5 eV above E_V was inferred to be due to the donor level of N but the drastic decrease in the energy was not explained [17]. It is quite possible that the highly localized nature of field emission, as evidenced in [15] of these polycrystalline diamond films, has made the macroscopic UPS/FES measurements misleading. Namely, the FES spectrum shows an energy distribution curve (EDC) of a field emission site, whereas UPS spectrum shows an EDC of the whole area, and thus there is no direct correspondence between FES and UPS. It is worth noting that Robertson concluded that the polycrystalline diamond field emitters (as of 2003) used sp^2-like grain boundaries as internal tips for field enhancement [18].

16.3 Field Emission from Phosphorus-Doped n-Type Diamonds

Some important characteristics of P-doped n-type diamonds grown by CVD on high-pressure, high-temperature (HPHT) single-crystal (111) surfaces have been well understood [13, 19]. P-doped diamonds show very high resistivity ($\sim 10^5\ \Omega$ cm) at room temperature (RT). It was noted that the resistivity decreases as the phosphorus concentration increases although the carrier mobility decreases [20]. When highly doped with phosphorus ($>1 \times 10^{20}$ cm^{-3}), low resistive ($\sim 80\ \Omega$ cm at RT) but pseudo-n-type diamond single-crystal (111) films were realized [21]. It is, therefore, hoped that only a small electric field is needed to pull the electrons out of the surface of highly P-doped diamond (111). A field emitter tip array made of a highly P-doped diamond (111) single crystal showed an emission current density of ~ 6.5 mA mm^{-2} at a field of ~ 18 V μm^{-1} [22]. The field of ~ 18 V μm^{-1} is rather high if we consider a field enhancement factor of more than 100 at the tip. The surface energy band and electron affinity of highly P-doped diamonds (111) have been studied recently [6]. The resulting surface energy band diagrams (SEBDs) are schematically shown in Figure 16.1b. Two kinds of surface termination were studied: one was H-terminated and the other O-terminated. Neither of the two surface terminations showed predominant NEA; the H-terminated surface showed a positive value of electron affinity, $\chi = \sim 0.2$ eV, and the O-terminated surface showed $\chi = \sim 0$ eV. It was disappointing to find that the work functions ($\phi = 3.8 - 3.9$ eV) for the highly P-doped diamonds were not much different from those of H-terminated p-type diamond. This indicates that the electric field needed for a certain emission current density is about the same for both p-type and highly P-doped diamonds. For example, a typical field emission current density of 1 mA cm^{-2} can be obtained at a field of ~ 1700 V μm^{-1} for both kinds of diamonds.[1] Another surprise finding was that

1) Based on the Fowler–Nordheim equation neglecting the image potential term, $j = \frac{1.56 \times 10^5 F^2}{\phi} \exp\left[\frac{-6.83 \times 10^3 \phi^{3/2}}{F}\right]$, where j is the current density (mA cm^{-2}), F the electric field (V μm^{-1}), and ϕ the work function (eV).

the surface energy bands are bent upward toward the surface by an amount of ~3 eV with a range of ~10 nm [6]. The field needed to flatten this upward band bending would be ~300 V μm^{-1}. The upward band bending is to be contrasted with the downward band bending found for H-terminated p-type diamond of Figure 16.1a.

Nevertheless, single tips of highly P-doped CVD diamond (111) have been made and used as field emitters [23]. They were made from pillar-shaped Ib-type HTHP diamond single crystals with a $0.6 \times 0.6\,\text{mm}^2$ square base, two parallel sides of which were (111) surfaces. The tip of the pillar was cut and polished to have a triangular pyramidal shape, as schematically shown in Figure 16.2a, the upper side of which was the (111) surface. A highly P-doped layer was grown on the (111) surface by microwave CVD and the typical thickness and resistivity of the layer were ~1.5 μm and ~100 Ω cm, respectively [23]. The surface of the CVD layer was terminated with hydrogen. Field emission characteristics were studied in a high-vacuum chamber to find a field emission current of ~116 μA at a temperature of ~600 °C and at a mean field of 0.5 V μm^{-1} [23]. There was no detectable field emission current for tip temperatures below 450 °C. This indicates that the measured emission current was not due to pure field emission but thermal field emission. The energy width (full width at half maximum) and angular current density of the thermal field emission at a current of ~0.5 pA were 0.23 eV and 0.02 mA sr^{-1}, respectively [23]. Similar characteristics for LaB_6 and ZrO/W single tips were compared, and the superiority of the P-doped diamond tip was demonstrated. It is reported that the development of an electron microscope based on the P-doped single tip for electron-beam lithography is under way [23].

The mechanism of field emission from a single tip of highly P-doped diamond (111) surface has been clarified very recently [24]. A sample of a single tip emitter made similar to the highly P-doped diamond (111) sample used in [23] was installed in a commercial field emission electron microscope (FEEM)/photoemission electron microscope (PEEM) which was composed of electrostatic lens optics together with an electron energy analyzer. A microspot electron energy analysis of field/photo emitted electrons was possible at a spatial resolution of 1 μm in diameter using an iris aperture placed on the first image plane. For PEEM, a Xe discharge lamp was used whose highest photon energy was ~6.2 eV. Figure 16.2b shows a photograph of the single-tip emitter installed on the sample holder of the FEEM/PEEM apparatus. The distance between the tip end and the objective

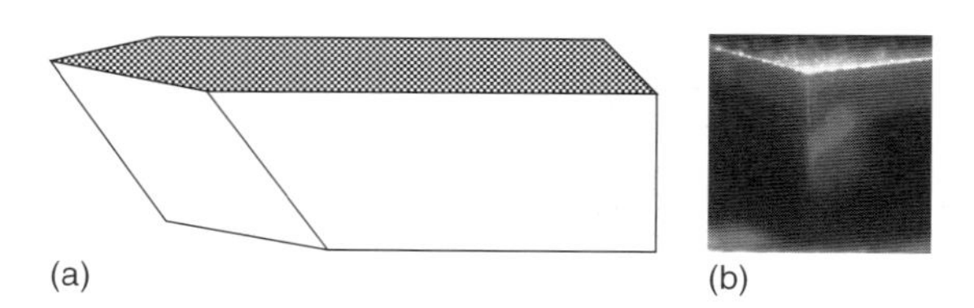

Figure 16.2 (a) Schematic illustration of a single tip of highly P-doped diamond. (b) Photograph of the tip of the single tip [24].

lens electrode was ∼1.2 mm, and 11.5 kV was applied (a nominal electric field of 9.6 V μm^{-1}) to study the image and energy distribution of field-emitted electrons at RT. It is noted that the effective field would be on the order of ∼1000 V μm^{-1} if one considers the field enhancement of ∼100 at the tip. Figure 16.3 shows a time sequence (2 s exposure and 6 s separation) of FEEM images [24]. Because of a curvature of the sample tip, the scale bar is not accurately determined but shows an order of the scale. It was clear in Figure 16.3 that the brightness of field emission spot changes with time.

Figure 16.4a shows the time dependence of EDC of field emission electrons which came through an iris aperture of parallelogram shape of ∼4 × 4 μm^2 on the FEEM of Figure 16.3. Each scan of five EDCs took about 36 s with a time separation of 90 s. One important finding in Figure 16.4a is that the field emission peaks are located 3–4 eV below the Fermi level (E_F) of substrate. The full width at half maximum of each peak is about 0.3 eV. Another important finding is that peak positions and heights fluctuate with time. The fluctuation of the peak position and height is expected to coincide with the changes in brightness of FEEM spots as in Figure 16.3. Figure 16.4b shows the time dependence of EDC of both field-emitted and photoemitted electrons which came through the iris aperture of the same size under Xe discharge illumination on the sample. Each EDC scan took 36 s with a time separation of 45 s. The appearance of the PEEM images in Figure 16.4b is not much different from that of Figure 16.3. Field emission peaks are located approximately 6 to 3 eV below the substrate E_F and photoemission peaks

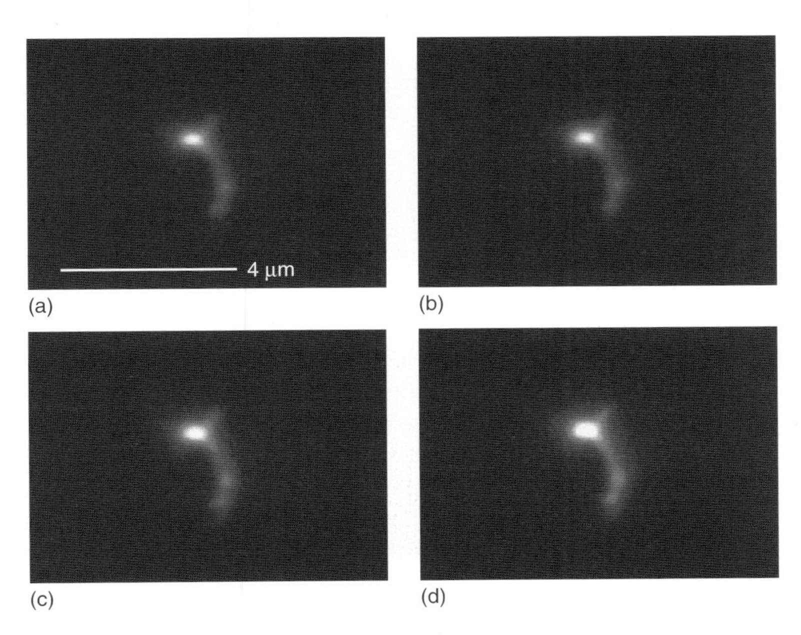

Figure 16.3 Images from a field emission electron microscope of a highly P-doped single-tip diamond emitter [24].

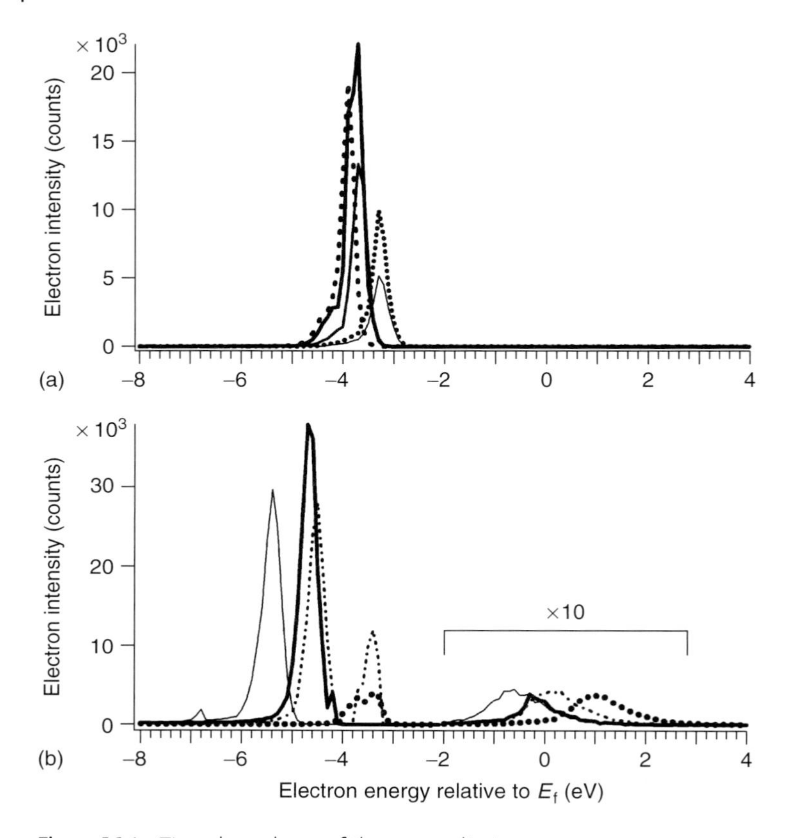

Figure 16.4 Time dependence of the energy distribution curves of (a) field-emitted electrons and of (b) photo/field-emitted electrons from the single-tip emitter of Figure 16.3 [24].

are located at −1 to +2 eV with respect to the substrate E_F. The photoemission peaks are expanded by a factor of 10 in the ordinate scale. The field emission peaks are shifted toward the lower energy side as compared to the case of the pure field emission of Figure 16.4a. The reason for this could be an effectively larger resistive potential drop due to a larger field emission current (Figure 16.5). Multiple field emission peaks are sometimes observed as in Figure 16.4b. The multiple peaks could be viewed as due to the time fluctuation of field emission during the single scan. It is to be noted that the energies of the field emission peak and the photoemission peak are correlated with each other in such a way that the separation between the two peaks is about 4–5 eV.

The mechanism of field emission of the highly P-doped diamond emitter was elucidated from the above findings together with the knowledge of work function and electron affinity of the highly P-doped diamond (111) surface [6, 24].

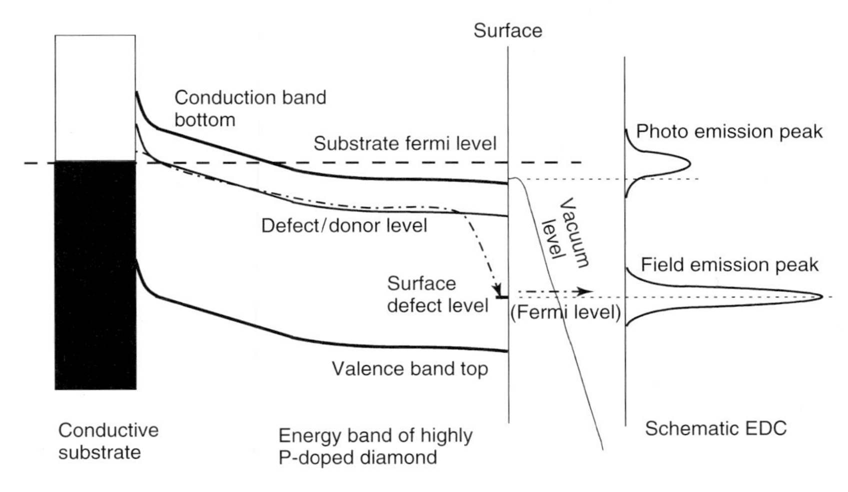

Figure 16.5 Energy diagram and schematic photo/field emission peaks of highly P-doped diamond [24].

Figure 16.5 shows the energy diagram and schematic photo/field emission peaks that explain the mechanism [24]. The diamond energy band is bent by electric field downward toward the surface, and the vacuum level at the surface is located close to the substrate E_F, since photoemission peaks are located around this energy position. The path of the field emission electrons can be conjectured as follows: electrons originating from the substrate E_F go through either the donor levels or the bulk defect levels, fall to the surface defect levels, and then tunnel-emit into vacuum as schematically shown by dash-dotted arrows in Figure 16.5. The junction property between the conductive substrate (in this case, graphite) and highly P-doped diamond is not known, but is shown as a plausible scheme. The bulk defect level due to the high concentration of P is also shown as a conjecture [24]. The origin of the fluctuation in energy and intensity of the field emission peak is not certain, but it is conjectured that irregular collisions of gas atoms/molecules or ions cause changes in surface properties at the field emission site, which in turn cause changes in field emission current and thus changes in resistive potential drop. This mechanism is not compatible with the expectation that conduction band electrons can be field-emitted from a highly P-doped n-type diamond.

16.4 Electron Emission from pn-Junction Diamond Diodes

Recently, Koizumi and coworkers reported the realization of diamond electron emitters based on pn-junction diamond diodes [25–29] (Figure 16.8). In brief, they formed diamond pn-junctions with different thickness of p-type layers on top of n-type diamond layers and examined the electron emission current in vacuum. Phosphorus-doped ($\sim 1 \times 10^{19}/cm^3$) n-type diamond films were first formed on a (111) Ib HPHT diamond surface by microwave plasma CVD with thicknesses of 5–10 μm. Boron-doped ($\sim 1 \times 10^{18}/cm^3$) p-type diamond layers were then formed in a different growth chamber on the n-type layers by CVD with a thickness variation of 0.25–1 μm. The pn-junctions were further processed by reactive ion etching (RIE) using oxygen to form mesa structures of 250 μm circular disks. A maximum emission efficiency (ratio of emission current to diode current) of ~1% was observed from a single mesa with a forward diode voltage (V_d) of several tens of volts at a diode current (I_d) of several milliamperes [28, 29].

The realization of pn-junction diamond diodes would open the door for the development of durable high-current electron sources. Kono and Koizumi [30] have examined very recently the spatial and energy distributions of emitted electrons from a pn-junction diamond emitter. The result showed that the emitted electrons indeed originated from the p-type mesa of a 250 μm circular disk and that the energy distribution was consistent with electrons emitted from a p-type NEA surface (Figure 16.8). Figure 16.6a shows an electron emission image around the p-type circular mesa under a diode condition of $V_d = 25.5$ V and $I_d = 30$ μA at the diode temperature of ~200 °C [30]. The image was observed by the FEEM/PEEM apparatus as in Figure 16.3. We can see a very bright emission part on the lower left corner, where the intensity is in fact saturated under an imaging condition for the fine structures on the terrace. We have examined the same part of the sample by an optical (Nomarski-type) microscope, the result of which is shown in Figure 16.6b. The left part is n-type diamond, the middle part is the p-type mesa, and the lower right is a metal (Au/Ti) contact. The large rectangular area enclosed by white lines roughly corresponds to the emission image of Figure 16.6a, and a defective part at the boundary of the p-type mesa can be seen in Figure 16.6b. Figure 16.6 clearly

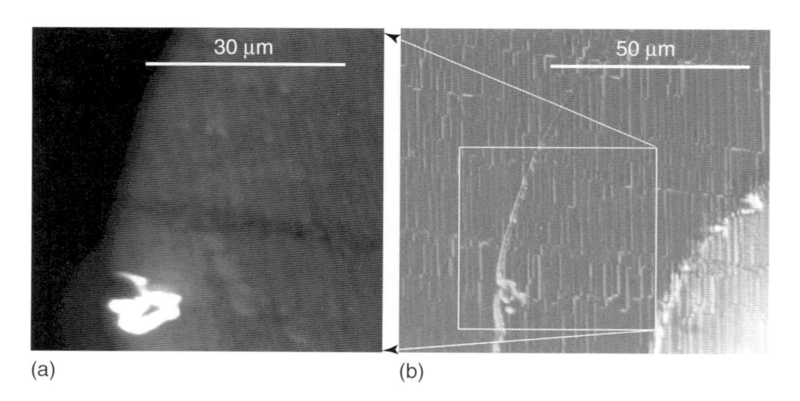

Figure 16.6 (a) An electron emission image around the p-type circular mesa of a pn-junction diamond diode [30]. (b) The corresponding optical (Nomarski-type) microscope image of an area larger than that shown in (a) [30].

shows that the electron emission in this pn-junction diode indeed occurs on the terrace of the p-type mesa and up to the boundary of the p-type mesa.

We have measured the EDCs of emitted electrons by placing the iris aperture at the first focal plane of emission images and using the electron energy analyzer of the PEEM [30]. Figure 16.7a shows the emission image. Labeled circles represent the positions where the iris aperture was placed on the image [30]. "T1" and "T2" in Figure 16.7a are the terrace sites on the p-type terrace under the same diode condition as in Figure 16.6a. "n" in Figure 16.7a is the n-type electrode site under the diode condition as in Figure 16.7a but with the Xe lamp on. "p" in Figure 16.7a represents the p-type terrace sites where photoemitted electrons, under $V_d = 0$ V with the Xe lamp on, were energy analyzed. The resulting EDCs with the maximum intensities normalized to unity are shown in Figure 16.7b [30]. The electron energy is referenced to the E_F of the electronic potential at the p-type electrode.

In Figure 16.8, we have depicted the plausible energy bands at the pn-junction and the SEBD. We have overlaid the characteristic energy positions found in Figure 16.7 on the SEBD [30]. The SEBD of a H-terminated boron-doped diamond (111) surface has been reported by Diederich *et al.* [8]. The onset energy of the

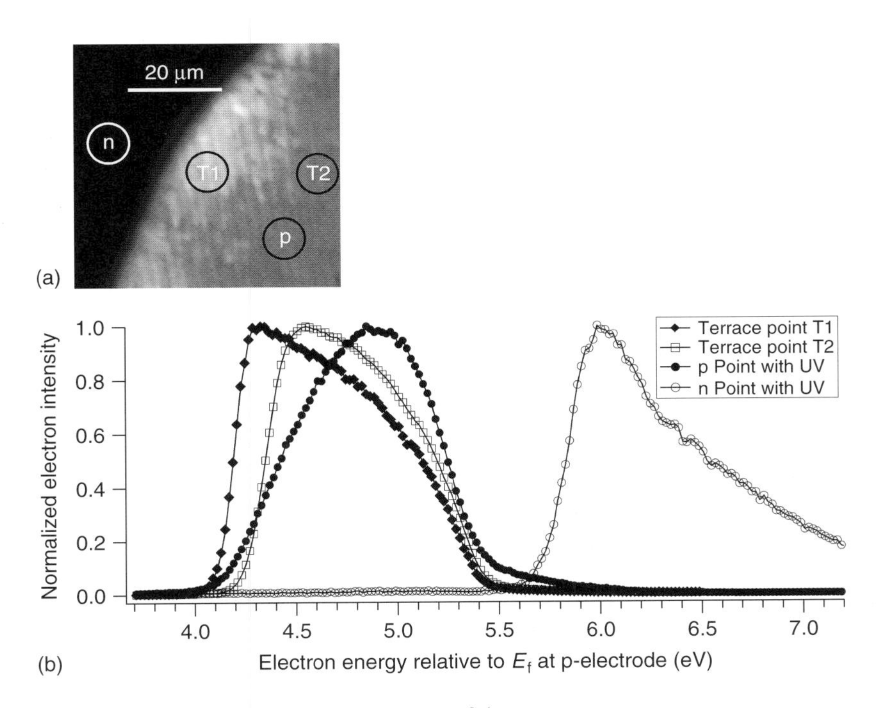

Figure 16.7 (a) An electron emission image of the pn-junction diamond diode and positions (circles) of energy analyses of electrons [30]. (b) Observed energy distribution curves (the maximum intensities are normalized to unity) of the four positions/cases of (a) [30].

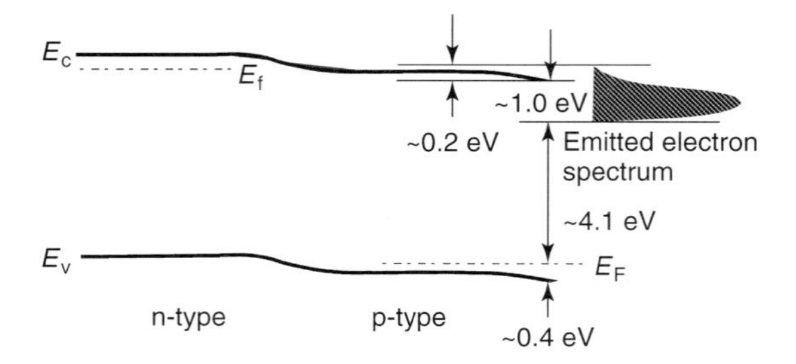

Figure 16.8 Plausible energy bands at the pn-junction and at the surface of the diamond emitter [30].

diode emitted electrons comes ~1.0 eV below the conduction band bottom E_C at the surface and the emitted electron spectrum decays off at ~0.2 eV above E_C in the bulk band. It is expected that there are unoccupied surface states up to ~1 eV below E_C for H-terminated diamond (111) surfaces [8]. Therefore, the onset of emitted electrons must correspond to either the lowest unoccupied surface states or the vacuum level. This means that most electrons transferred to the conduction band of p-type electrode are relaxed at the surface to the unoccupied surface states and come out into vacuum. Thus, a pn-junction diamond electron emitter is indeed realized. However, the EDCs of emitted electrons show a large energy spread, which indicates that there are many parameters to be improved.

16.5 Other Application of Diamond Emitter

16.5.1 Diamond Cold-Discharge Cathodes for Cold-Cathode Fluorescent Lamps

Cold-cathode fluorescent lamps (CCFLs) are now widely used as backlights for liquid crystal displays. The CCFL has a structure similar to that of hot-cathode fluorescent lamp but has no filament in the cathode. In the CCFL, electrons are emitted from the cathode by secondary electron emission generated by ion irradiation onto the cathode surface, which makes CCFL of a simple, slim, and tubular shape and with long lifetime. However, the luminous efficiency of CCFL is considerably reduced by a large voltage drop (called *cathode fall*) at the cold-cathode surface. Sakai and coworkers have developed a new cold cathode for CCFL made of Mo coated with boron-doped p-type polycrystalline diamond films [31, 32]. They have made a fully sealed discharge tube with the diamond-film-coated Mo cold cathode, a prototypical CCFL, and found ~13% reduction in the cathode fall, that is, ~13% power reduction for the discharge tube [32].

16.5.2 Low-Temperature Thermionic Emitters Based on N-Incorporated Diamond Films

Thermionic electron emission is expressed by Richardson–Dushman equation, $J = CT^2\exp(-\phi/k_BT)$, where J is the current density, C the Richardson constant, T the absolute temperature, ϕ the work function, and k_B the Boltzmann constant. Nemanich and coworkers [33] very recently made a low-temperature thermionic emitter based on a nitrogen-incorporated diamond film. The film showed substantial thermionic electron emission at temperatures as low as ~250 °C, and an analysis based on the Richardson–Dushman equation indicated a very low work function of ~1.3 eV with a Richardson constant of ~0.8 A cm^{-2} K^{-1} [33]. The diamond emitter film was made on a Mo substrate. The growth of the emitter structure was comprised of a step process initiating a nitrogen-incorporated ultra nanocrystalline diamond (UNCD) layer of a thickness of ~400 nm followed by a growth process of a nitrogen-doped diamond layer of thickness of 100–200 nm. An exposure of the final film surface to a hydrogen plasma resulted in an NEA surface characteristic [33]. The film showed considerably low resistivity of $2.5 \times 10^3\ \Omega$ cm at 100 °C which further decreased to a value of $\sim 7 \times 10^2\ \Omega$ cm at 400 °C [33]. The low work function and the low resistivity are in strong contrast with the corresponding values for P-doped single-crystal diamonds. The inclusion of the UNCD layer between the Mo substrate and the N-doped diamond film must have something to do with these low work function and resistivity. A high-performance thermionic energy conversion device may be fabricated from these diamond emitters [34].

References

1. Laegreid, N. and Wehner, G.K. (1961) *J. Appl. Phys.*, **32**, 365.
2. Blandino, J.J. *et al.* (2000) *Diamond Relat. Mater.*, **9**, 1992.
3. Bandis, C. and Pate, B.B. (1996) *Appl. Phys. Lett.*, **69**, 366.
4. Choi, W.B., Schlesser, R. *et al.* (1998) *J. Vac. Sci. Technol. B.*, **16**, 716.
5. Groning, O. *et al.* (1999) *J. Vac. Sci. Technol.*, **B17**, 1970.
6. Kono, S. *et al.* (2007) *E-J. Surf. Sci. Nanotechnol.*, **5**, 33.
7. Tatsumi, N. *et al.* (2007) *SEI Tech. Rev.*, **64**, 15.
8. Diederich, L., Küttel, O.M., Aebi, P., and Schlapbach, L. (1998) *Surf. Sci.*, **418**, 219.
9. Farrer, R.G. (1969) *Solid State Commun.*, **7**, 685.
10. Okano, K. *et al.* (1990) *Appl. Phys. A*, **51**, 344.
11. Nishimori, T. *et al.* (1997) *Appl. Phys. Lett.*, **71**, 945.
12. Koizumi, S. *et al.* (1997) *Appl. Phys. Lett.*, **71**, 1065.
13. Koizumi, S. and Suzuki, M. (2006) *Phys. Status Solidi A*, **203**, 3358.
14. Okano, K. *et al.* (1996) *Nature*, **381**, 140.
15. Matsuda, R., Okano, K., and Pate, B.B. (1998) in (eds W. Zhu *et al.*), Simultaneous Field Emission and Photoemission Characterization of N-Doped CVD Diamond. Materials Research Society, Pittsburgh, p. 59.
16. Okano, K. *et al.* (1999) *Appl. Surf. Sci.*, **146**, 274.
17. Yamaguchi, H., Yamada, T., Kudo, M., Takakuwa, Y., and Okano, K. (2006) *Appl. Phys. Lett.*, **88**, 202101.
18. Robertson, J. (2003) *IEICE Trans. Electron. E.*, **86-C**, 787.

19. Katagiri, M., Isoya, J., Koizumi, S., and Kanda, H. (2004) *Appl. Phys. Lett.*, **85**, 6365.
20. Koizumi, S. (1999) *Phys. Status Solid A*, **172**, 71.
21. Namba, A. *et al.* (2005) *Sumitomo Electr. Ind. Tech. Rev.*, **166**, 38 (in Japanese).
22. Nishibayashi, Y., Ando, Y., Furuta, H., Kobashi, K., Meguro, K., Imai, T., and Oura, K. (2003) *New Diamond Front. Carbon Technol.*, **13**, 19.
23. Namba, A., Tatsumi, N., Yamamoto, Y., Nishibayashi, Y., and Imai, T. (2006) *SEI Tech. Rev.*, **169**, 55 (in Japanese).
24. Kono, S. *et al.* (2007) *Jpn.J. Appl. Phys.*, **46**, L21.
25. Koizumi, S., Ono, T., and Sakai, T. (2006) Extended Abstracts of 20th Diamond Symposium, November 22, 2006, Tokyo, p. 262 (in Japanese).
26. Koizumi, S. (2007) Extended Abstracts of 21st Diamond Symposium, November 22, 2007, Nagaoka, p. 220 (in Japanese).
27. Koizumi, S., Suzuki, M., and Sakai, T. (2007) Extended Abstracts of Materials Research Society 2007 Fall Meeting, November 26-December 1, 2007, Boston, p. 9.5.
28. Koizumi, S. and Kono, S. (2008) Extended Abstracts of Hasselt Diamond Workshop 2008 (Surface and Bulk Defects in CVD Diamond Films XII), February 25–27, 2008, Hasselt, Belgium, p. 30.
29. Koizumi, S., Suzuki, M., and Sakai, T. (2008) Extended Abstracts of the 55th Spring Meeting of the Japan Society of Applied Physics 27a-ZN-10, March 27, 2008, Funabashi (in Japanese).
30. Kono, S. and Koizumi, S. (2009) *E-J. Surf. Sci. Nanotechnol.*, **7**, 660.
31. Ono, T., Sakai, T., Sakuma, N., Suzuki, M., Yoshida, H., and Uchikoga, S. (2006) *Diamond Relat. Mater.*, **15**, 1998.
32. Sakai, T., Ono, T., Sakuma, N., Suzuki, M., and Yoshida, H. (2007) *New Diamond Front. Carbon Technol.*, **17**, 189, and references therein.
33. Koeck, F.A.M. and Nemanich, R.J. (2009) *Diamond Relat. Mater.*, **18**, 232, and references therein.
34. Smith, J.R., Bilbro, G.L., and Nemanich, R.J. (2007) *Phys. Rev. B.*, **76**, 245327.

17
ZnO Nanowires and Si Nanowires

Baoqing Zeng and Zhi Feng Ren

17.1
Introduction

As stated in the previous chapter, one-dimensional (1D) nanostructures with high aspect ratios are considered to be ideal field emission electron sources which can emit electrons at low electric fields. Several nanostructures, including carbon nanotubes [1], ZnO nanowires (NWs) [2–4] and nanobelts [5], Si nanowires [6–8], WO nanowires [9], SiC nanowires [10], and MoO_3 nanobelts [11], are promising candidates for field emitters. Among 1D materials, carbon nanotubes have received the most attention because of their high mechanical stability and high electrical conductivity. However, some disadvantages with carbon nanotubes lie in their high-temperature synthesis method and the instability of nanotube tips at low vacuum [12].

On the other hand, ZnO, as an oxide, exhibits high mechanical strength, thermal stability, chemical stability, negative electron affinity, and resistance to further oxidation in the field emission process [13]. Therefore, ZnO-based 1D nanostructures are appropriate alternatives to carbon nanotubes for field emission devices. Meanwhile, Si nanowires exhibit a unique sp^3-bonded crystal structure and low work function (3.6 eV) [7], and Si has been the backbone of the microelectronics industry for decades. Therefore, it would be desirable to have Si field emitters to be integrated onto Si substrates along with the driving circuitry.

The organization of this chapter is as follows: in Section 17.2, we describe the growth methods for Si and ZnO nanowires or nanobelts. In Section 17.3, we describe the field emission from ZnO nanowires and Si nanowires. Finally, summary and conclusion are given in Section 17.4.

17.2
Synthesis of ZnO and Si Nanowires or Nanobelts

There are five main techniques for growth of inorganic nanowires, such as Si nanowires and ZnO nanowires, namely vapor–liquid–solid (VLS) chemical

Carbon Nanotube and Related Field Emitters: Fundamentals and Applications. Edited by Yahachi Saito

ISBN: 978-3-527-32734-8

vapor deposition (CVD) synthesis [13–18], vapor–solid (VS) synthesis [19–22], template-directed synthesis [23–25], hydrothermal reaction [26–29], and metal–organic chemical vapor deposition (MOCVD) [30]. As the VLS method is the most widely used and hydrothermal reaction can grow nanowires at low temperatures, we focus on these two methods.

17.2.1
Vapor–Liquid–Solid Nanowire Growth

The VLS process, which was proposed by Wagner in 1960s during his studies of single-crystalline silicon whisker growth [31], is one of the most popular and powerful growth methods for nanowires (Figure 17.1) such as those of Si and ZnO. According to this mechanism, there are mainly three growth stages: first, the source material from the gas phase is absorbed into a liquid droplet of a catalyst to form a liquid alloy droplet composed of the metal catalyst component (such as Au, Cu, and Sn) and the nanowire component (such as Si, III–V compound, II–VI compound, oxide) under the reaction conditions. The metal catalyst can be chosen from the phase diagram by identifying metals in which the nanowire component elements are soluble in the liquid phase but do not form solid compounds that are more stable than the desired nanowires phase. After that, upon supersaturation of the liquid alloy, nucleation generates a solid precipitate of the source material. Finally, the nanowires are longated. As a result, a nanowire obtained from the VLS process typically has a solid catalyst nanoparticle at its tip with a diameter comparable to that of the connected nanowires.

The VLS processes are usually carried out in a tube furnace, as shown in Figure 17.2. It consists of a horizontal tube (made of ceramic or quartz) furnace, a gas supply, a rotary pump system, and a control system. The right end of the ceramic (or quartz) tube is connected to a rotary pump. Both ends are sealed by rubber O-rings. The carrying gas comes in from the left end of the ceramic (or quartz) tube and is pumped out at the right end. The source material(s) is loaded on a quartz (or alumina) boat and positioned at the center of the ceramic (or quartz) tube. Several strip plates are placed downstream one by one inside the alumina (or quartz) tube, which act as substrates for collecting the growth products.

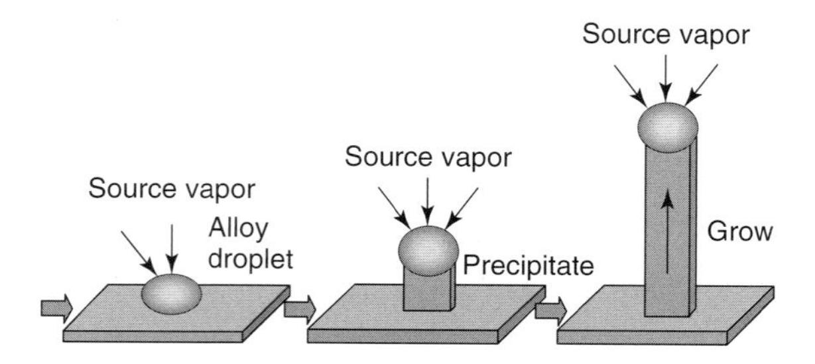

Figure 17.1 Schematic diagram showing the growth of nanowires via the VLS process.

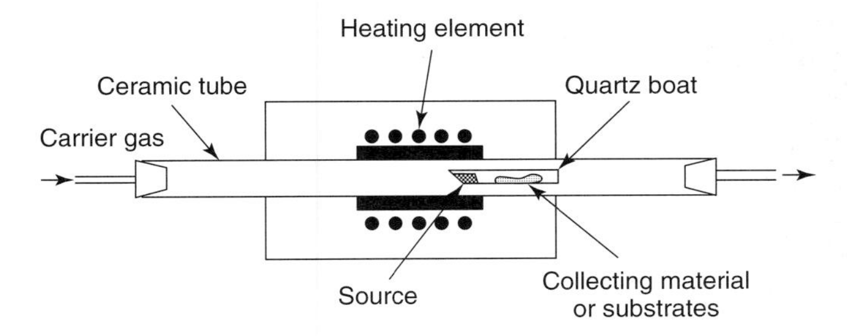

Figure 17.2 Schematic diagram of the experimental apparatus for growth of oxide nanostructures.

There are several parameters, such as the catalyst, gas species and its flow rate, substrate, temperature, pressure, and evaporation time, that should be controlled carefully.

Although the VLS mechanism has been widely used for nanowires growth, ZnO nanowire growth through the VLS mechanism could be complicated by the presence of oxygen since ZnO has a very high melting point. A modified method, involving the reduction of ZnO powder by graphite/hydrogen to form Zn and CO/H_2O vapor at a high temperature, has been suggested by Yang and coworkers [32, 33]. Mixing ZnO with carbon powder can reduce the vaporization temperature from 1300 to 900 °C,

$$ZnO(s) + C(s) \xleftrightarrow{900\,^{\circ}C} Zn(v) + CO(v)$$

The Zn vapor is transported to and reacted with the Au catalyst on substrates located downstream at a lower temperature to form alloy droplets. As the droplets become supersaturated, crystalline ZnO nanowires are formed, possibly by the reaction between Zn and CO/H_2O at a low temperature. The presence of a small amount of CO/H_2O is not expected to significantly change the Au $\pm$ Zn phase diagram; at the same time, they act as the oxygen source during the ZnO nanowire growth.

Figure 17.3a shows a typical scanning electron microscope (SEM) image of ZnO nanowires grown on a silicon substrate. The diameters of the nanowires normally range from 20 to 120 nm and the lengths are 5–20 µm. Figure 17.3b shows a typical X-ray diffraction (XRD) pattern of these ZnO nanowires. The diffraction peaks are indexed to a hexagonal structure with lattice constants of $a = 0.324$ nm and $c = 0.519$ nm.

Figure 17.4a shows a transmission electron microscope (TEM) image of a single nanowire with an alloy tip. The presence of the alloy tip is a clear indication of the VLS growth mechanism. Figure 17.4b shows a high-resolution transmission electron microscope (HRTEM) image of a single ZnO nanowire. The spacing of 0.256 ± 0.005 nm between adjacent lattice planes corresponds to the distance between two (002) crystal planes, confirming <0001> as the preferred growth direction for ZnO nanowires [33].

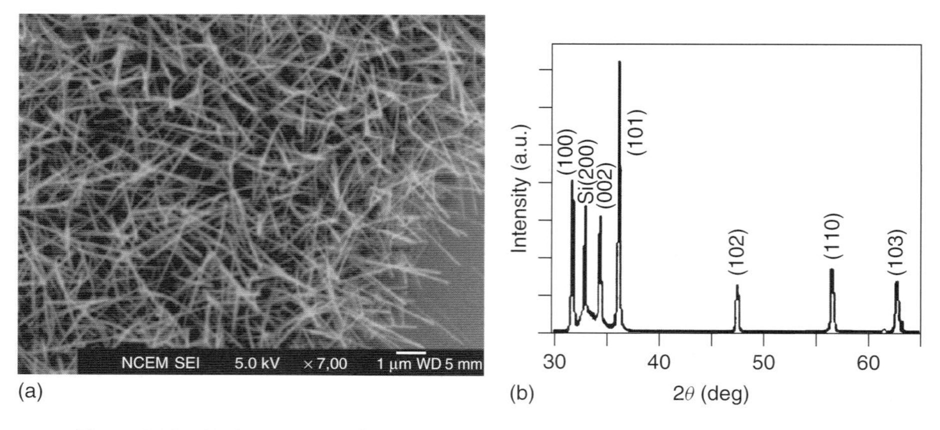

Figure 17.3 (a) SEM image of ZnO nanowires grown on a silicon substrate and (b) the XRD pattern [34].

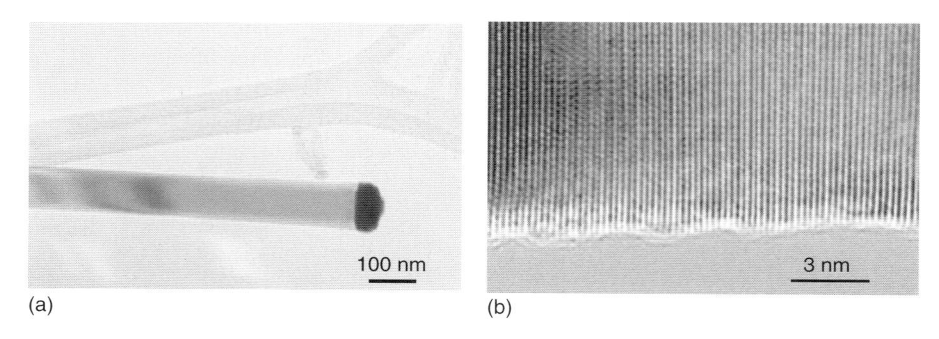

Figure 17.4 (a) TEM image of a ZnO nanowire with an alloy droplet on its tip. (b) HRTEM image of an individual ZnO nanowire showing its <0001> growth direction [33].

Even though the VLS process has been widely used, some aspects of the growth mechanism are not well understood. For example, in the conventional picture, the catalyst droplet does not change during growth, and the nanowire sidewalls are clean [34–36]. But, recently Hannon and coworkers have shown that these assumptions are not correct for silicon nanowires grown on Si(111) by observing the *in situ* growth of Si nanowires within two separate ultrahigh vacuum electron microscopes [37]. They showed that gold diffusion during growth determines the length, shape, and sidewall properties of the nanowires. Figure 17.5a and b shows that Au/Si catalyst droplets are visible at the ends of all the longer nanowires. These nanowires have a uniform length of about 750 nm. A small fraction of the nanowires (indicated by arrows in Figure 17.5a) are cone shaped, shorter, and have no catalyst droplets. High-resolution imaging revealed that the sidewalls of the nanowires were not perfectly straight. As shown in Figure 17.5b, the diameters of some wires decreased with time during growth, while those of others increased.

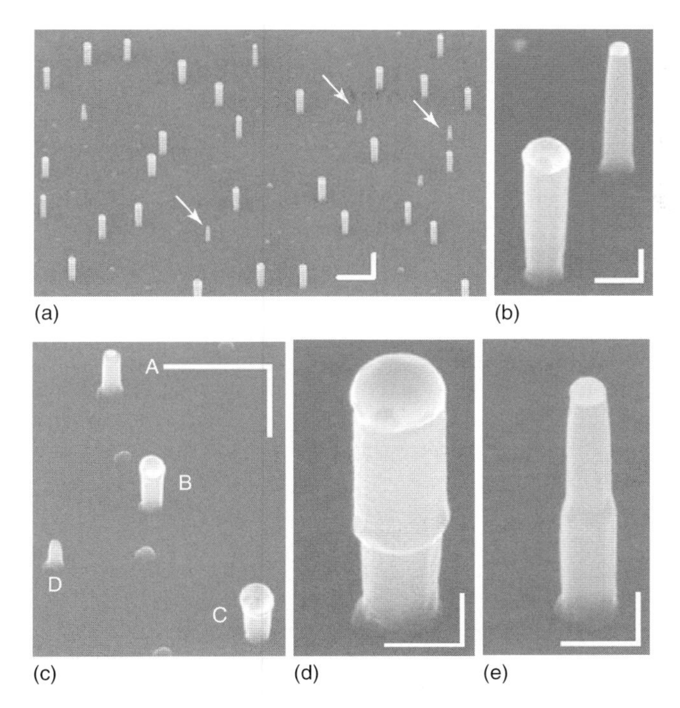

Figure 17.5 SEM images of Si nanowires grown on Si(111) recorded with a 42° angle of incidence. (a) Shorter, cone-shaped nanowires without visible catalyst droplets are indicated by arrows. (b) Coarsening of the catalyst particles leads to wires with diameters that either grow or shrink during nanowire growth. (c) Nanowires imaged after 35 min of annealing at 600 °C following nanowire growth for 35 min at 600 °C. (d and e) Nanowires imaged after 30 min of growth at 600 °C, followed by annealing at 650 °C for 20 min, and a final growth period of 25 min at 600 °C. Scale bars: (a and c) 1 μm; (b, d, and e) 200 nm [37].

These results suggest that the sizes of the catalyst droplets change during nanowire growth at 600 °C. Given that the Au/Si droplets on the surface become coarse at 600 °C before growth, it is likely that the droplets continue to become coarser during nanowire growth. That is, during nanowire growth Au preferentially migrates from the smaller catalyst droplets to the larger ones to reduce the total droplet surface area. In this process, some of the droplets coarsen away completely, leading to the shorter, cone-shaped nanowires shown in Figure 17.5a. Figure 17.5c shows a sample annealed at 600 °C for 35 min (following nanowire growth for 35 min also at 600 °C), The wires labeled A, B, and C all have the same length. They also have approximately the same diameter, indicating that they grew from catalyst droplets of similar size. However, after the postgrowth annealing, the droplet on wire C has clearly grown, but that on wire A has shrunk. The droplet on wire B is essentially unchanged. The nanowire labeled D is shorter than the others, suggesting that its droplet coarsened away during nanowire growth. The coarsening can be exploited

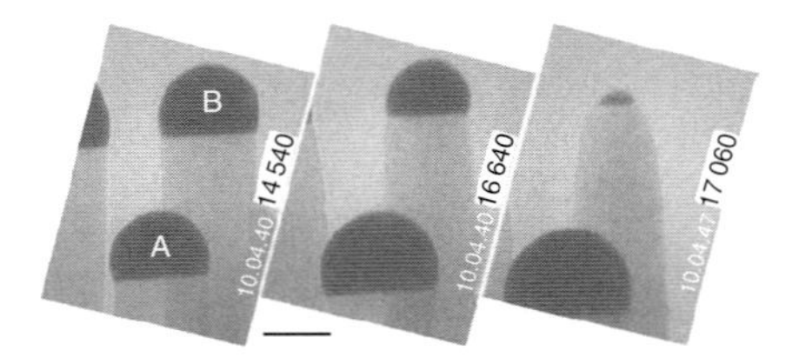

Figure 17.6 *In situ* UHVTEM images recorded during the growth of Si nanowires at 655 °C in 10^{-6} Torr disilane. Three images are labeled by the time (in seconds) after the start of growth. Scale bar: 50 nm [37].

to introduce a discontinuous change in the nanowire diameters. For example, a second stage of growth can be carried out on a postannealed sample (like that shown in Figure 17.5c). During the postgrowth annealing, the catalyst droplets coarsen and change size, so that when the growth is resumed, the nanowires grow with either a larger or smaller diameter. Examples of both types of nanowires are shown in Figure 17.5d and e.

The coarsening of the catalyst droplets during nanowire growth has been observed directly in the ultrahigh vacuum transmission electron microscope (UHVTEM). Figure 17.6 shows a sequence of images from a growth experiment at 655 °C. We find that during growth the droplet labeled A grows larger, while the one labeled B decays and eventually disappears. These results clearly show that Au can diffuse from one catalyst droplet to another during growth. That is, Au migrates over the surface as well as up and down the nanowire sidewalls. It has been shown that Au diffusion fundamentally limits the length, shape, and sidewall properties of Si nanowires.

So, gold from the catalyst droplets wets the nanowire sidewalls, eventually consuming the droplets and terminating VLS growth. These results show that the silicon nanowire growth is fundamentally limited by gold diffusion; smooth, arbitrarily long nanowires cannot be grown without eliminating gold migration.

17.2.2 Controlled Growth of Si Nanowires and ZnO Nanowires

For field emission application, it is important to control the diameter, orientation, and location of nanowires.

17.2.2.1 Diameter Control

According to the VLS mechanism, the diameter of nanowire is determined by the size of the alloy droplet, which is in turn determined by the original cluster size [34]. To control the average diameter of nanowires, substrates coated with a catalyst metal (e.g., Au) of different thicknesses could be used. Smaller catalyst droplets should favor the growth of thinner wires. A metal thin film, however, does not provide good diameter control of the resulting nanowires because of the randomness of the film breakup at the reaction temperature.

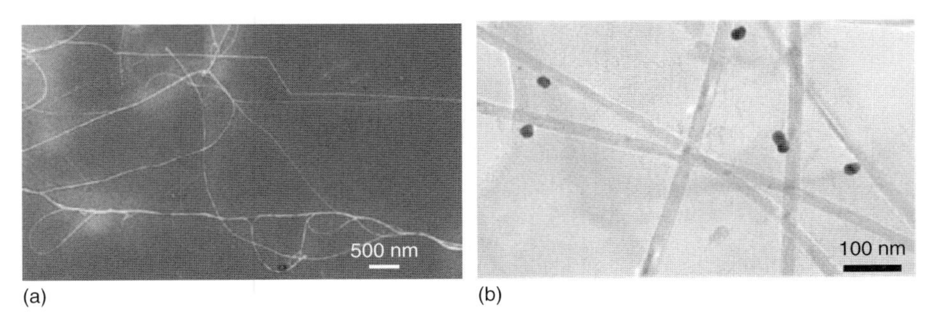

Figure 17.7 (a) FESEM image of Si nanowires grown on Au clusters embedded in a mesoporous silica thin film. (b) TEM image of the nanowires and the Au clusters dispersed on the same copper grids [38].

Yang *et al.* have suggested the use of monodispersed metal nanoclusters as the catalyst to grow uniform ZnO or Si nanowires in a CVD system. For example [38], uniform silicon nanowires with 20.6 ± 3.2, 24.6 ± 4.0, 29.3 ± 4.5, and 60.7 ± 6.2 nm diameters were grown by using Au clusters with sizes of 15.3 ± 2.4, 20.1 ± 3.1, 25.6 ± 4.1, and 52.4 ± 5.3 nm, respectively. Figure 17.7a shows a field emission scanning electron microscope (FESEM) image of uniform, long, and flexible nanowires grown by using 15-nm Au clusters as catalyst. Figure 17.7b shows a TEM picture with both Au clusters and the Si nanowires on the same TEM grid. The correlation between the size of Au clusters and the diameter of Si nanowires can been clearly observed.

17.2.2.2 Orientation Control

Growing aligned Si nanowires and ZnO nanowires is important for application in field emission. Using the conventional epitaxial crystal growth technique, it is possible to achieve precise orientation control during the Si or ZnO nanowire growth. Nanowires generally have preferred growth directions. For example, Si nanowires prefer to grow along the <111> direction, whereas ZnO nanowires prefer to grow along the <0001> direction. One strategy to grow vertically aligned nanowires is to properly select the substrate and to control the reaction conditions so that the nanowires grow epitaxially on the substrate. Usually, aligned nanowires of ZnO or Si are achieved on a lattice-matching substrate; for example, aligned ZnO nanowires have been successfully grown on sapphire, GaN, Al, GaN, and AlN substrates [39] through a VLS process, where the crystal structure of the substrate is crucial for the orientation of nanowires. The epitaxial relationship between the substrate surface and the ZnO nanowires determines whether there will be an aligned growth and how well the alignment can be. The successful alignment of ZnO nanowires on sapphire and nitride substrates is attributed to the very small lattice mismatches between the substrates and ZnO. For example, ZnO nanowires have wurzite structure with lattice constants $a = 0.324$ nm and $c = 0.519$ nm and

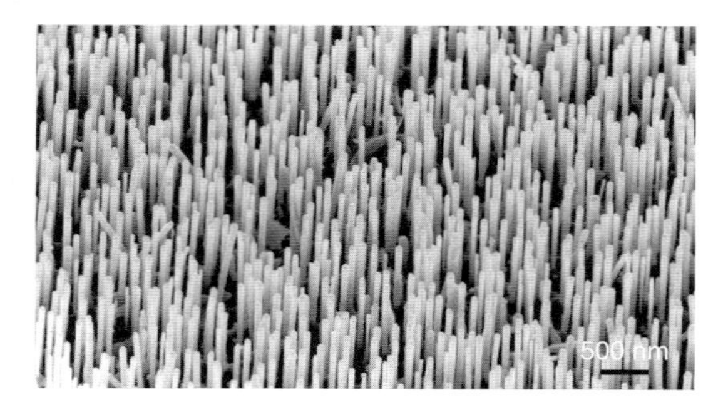

Figure 17.8 SEM image of aligned ZnO nanowires grown on a sapphire substrate using a thin layer of gold as catalyst [42].

prefer to grow along the <0001> direction. A ZnO nanowire can grow epitaxially on the (110) plane of sapphire, because the ZnO *a* axis and sapphire *c* axis are related by a factor of 4 (mismatch less than 0.08% at room temperature). Figure 17.8 shows vertical ZnO nanowire arrays grown on the sapphire substrate [40]. Another example is the Si nanowires grown on a <111> Si wafer; Si nanowires will grow epitaxially and vertically on the substrate and form a nanowire array [41].

17.2.2.3 **Positional Control**

The positions of nanowires can be controlled by the initial positions of catalyst droplet, such as Au clusters or thin films. Various lithographic techniques including, for example, e-beam, photolithography, and nanosphere lithography (NSL) can be used to create patterns of the Au thin film for the subsequent Si or ZnO nanowire growth. Among them, NSL has been proven to be a very simple and cost-effective technique for large-scale fabrication of particle arrays with long-range periodicity [43, 44]. Ren's and Wang's groups have demonstrated the large-scale preparation of ZnO nanorod arrays templated by NSL [42, 45]. Recently, Xie and coworkers have developed a strategy for fabricating hexagonally patterned Au particle arrays to grow individually patterned and vertically aligned ZnO nanorod arrays with modified NSL [46]. The main fabrication process is as follows: (i) Au films with thickness of 1.5–10 nm were sputtered onto the clean substrates. (ii) The Au-coated substrates were immersed in an ethanol solution of octadecanethiol (ODT, 10 mM) at 50 °C for 1 h and at room temperature for another 24 h to form a densely packed ODT self-assembled monolayer (SAM) on the gold surface. (iii) The polystyrene (PS) microsphere monolayer was prepared by using the self-assembly technique on a water surface [47]. For example, 120 ml of monodispersed PS microsphere suspension (diameter 1.39 μm, 10 wt% aqueous dispersion) was diluted by mixing with an equal volume of ethanol and slowly applied to the surface of deionized water in a Petri dish (ϕ15 cm). Before distributing the PS microspheres on the surface of water, the ODT-modified Au-coated substrates were placed at the bottom of the Petri dish. The PS microsphere SAM could be promoted by slow and careful vessel tilting. The PS microsphere SAM was deposited on the substrates by slow water drainage. After the drying process, the substrates covered with PS microsphere SAM were heated on a 110 °C hot plate for 2 min to melt the PS microspheres slightly and make them stick to the substrates (it provides a good protection for the ODT SAM under the PS microspheres in the subsequent oxygen plasma etching; glass transition temperature for PS, 95–105 °C); (iv) Oxygen plasma etching was used to reduce the size of the PS microspheres and to burn off the ODT molecules not covered by the PS microspheres; (v) Upon removal of the PS microspheres by ultrasonication in ethanol, the gold particle pattern was fabricated by wet etching with an aqueous Fe^{3+}/thiourea (20 mM/30 mM) solution because the ODT SAM on Au provides enough etch resistance to this etch bath and prevents the etchants from dissolving the Au below [48]; (vi) Finally, a periodic ZnO nanorod or nanowire array was grown on the patterned substrate by a VSL process.

Figure 17.9a shows a typical SEM image of a PS microsphere SAM on the substrate. After etching with oxygen plasma, the morphology of the PS microsphere SAM turned into a non-close-packed pattern from a close-packed one (Figure 17.9b).

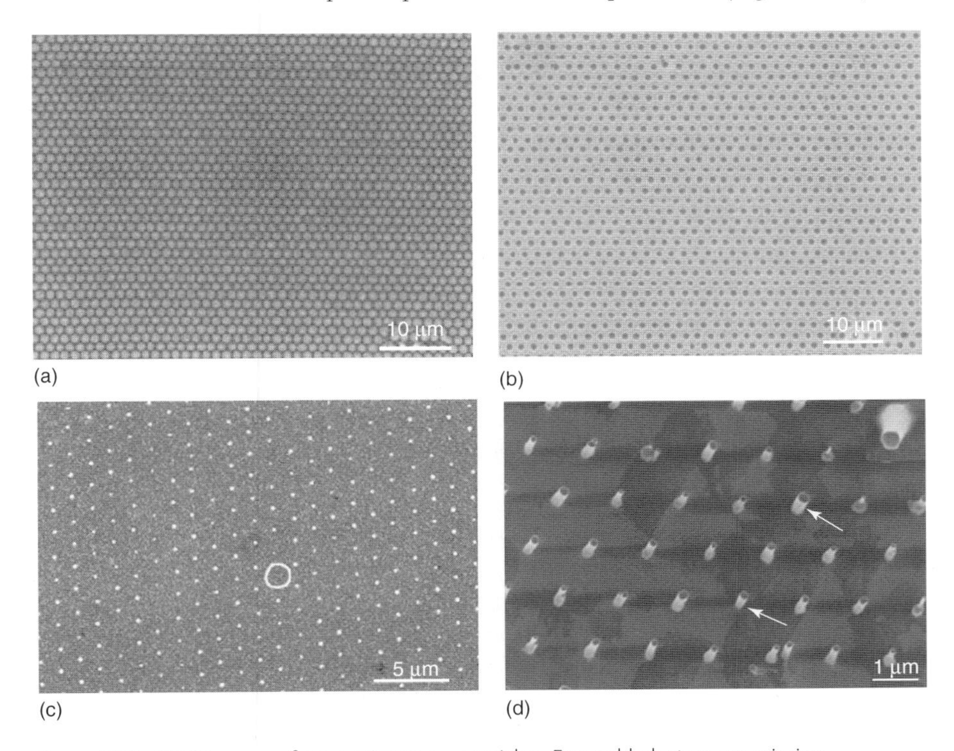

Figure 17.9 SEM images of (a) a PS microsphere SAM deposited on the ODT-modified gold-coated sapphire (0001) substrate, (b) PS microsphere SAM after oxygen plasma etching, (c) the obtained gold particle pattern, and (d) a ZnO nanorod array grown from the patterned gold particles. Few gold clusters are missing (see circle), and most of the rods have a bigger base, but a few have a uniform size along the axial direction (marked by arrows). The inset in (d) is an enlarged view, which clearly shows the hexagonal flat end and the big base of the nanorod [46].

Figure 17.9c is an SEM image of the obtained Au particle patterns. The average size of the Au particles is about 250 nm. (The size can be controlled by either the time of heating on the hot plate or the time of oxygen plasma etching, or both.) Some of the Au particles are missing (see circle). Three factors account for this phenomenon: (i) point defects in the original PS microsphere SAM; (ii) some PS microspheres that are not tightly adhered to the substrate after heating and cannot prevent oxygen plasma from burning off the ODT SAM underneath; and (iii) a few undesired PS microspheres above the PS microsphere SAM drawn away the below microsphere from its original site during the oxygen plasma etching. Figure 17.9d presents a tilted SEM image of the obtained ZnO nanorod arrays. It is obvious that the nanorods possess a hexagonal lattice, similar to the initial PS microspheres. The length of the rod can be regulated easily by the growth time. To increase the aspect ratio for field emission, we should increase the length and keep the diameter constant. But, it was found that it was not easy to control the diameter of the rod by the size of Au particles or their thickness, so it is a challenge.

17.2.3
Hydrothermal-Based Chemical Approach

Hydrothermal synthesis is another approach for the synthesis of ZnO nanowires with the use of ZnO seeds in the forms of thin films or nanoparticles [49, 50].

Wang and coworkers have reported a chemical approach for fabricating density-controlled, aligned ZnO nanowires arrays without using ZnO seeds [51]. By adjusting the precursor concentration, the density of ZnO NW arrays could be controlled. This synthesis technique does not require ZnO seeds or an external electrical field, and it can be carried out at low temperatures and on a large scale on any substrates irrespective of whether they are crystalline or amorphous.

For simplicity of description, a Si substrate is used to describe the experimental procedure. The main steps are as follows: (i) A 50-nm-thick layer of Au was deposited on a cleaned Si wafer by sputtering. Between the Si wafer and Au layer, 20 nm Ti was deposited as an adhesion layer to buffer the large lattice mismatch between the Si(100) surface with native oxide and Au(111) surface and to improve the interface bonding. (ii) The substrate was annealed at 300 °C for 1 h. (iii) The next step is to prepare the nutrient solution. The nutrient solution was composed of a 1 : 1 ratio of zinc nitrate and hexamethylenetetramine (HMTA). (iv) The substrate was placed face down on top of the nutrient solution surface. Due to surface tension, the substrate would float on top of the solution surface. Growth of ZnO nanowires was carried out in a convection oven.

The annealing process helps the as-deposited Au layer to form a uniform crystalline thin layer on the surface of the Si substrate, which is critical in the oriented growth of aligned ZnO nanowires. The chemistry of the growth is well documented [52, 53]. Zinc nitrate salt provides Zn^{2+} ions required for building up ZnO nanowires. Water molecules in the solution provide O^{2-} ions. Even though the exact function of HMTA during the ZnO NW growth is still unclear, it is believed to act as a weak base, which would slowly hydrolyze in the water solution

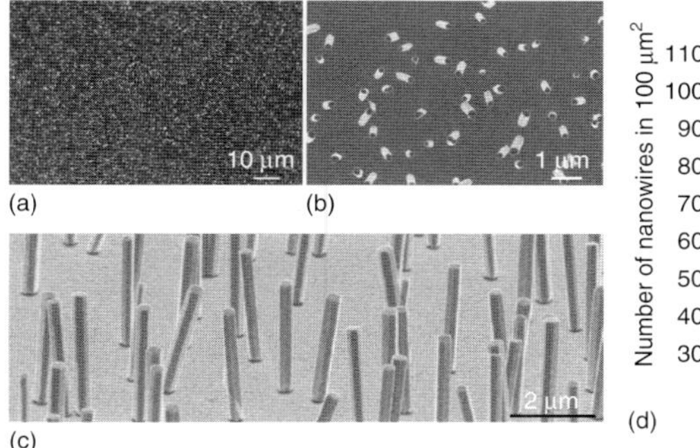

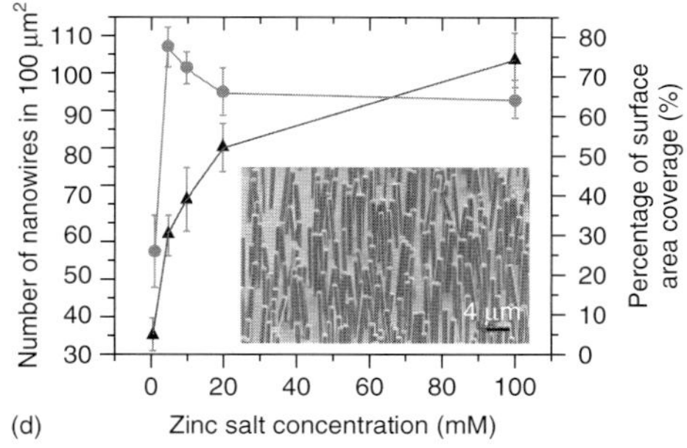

Figure 17.10 A general view of the as-grown ZnO nanowire arrays at 5 mM, grown for 24 h at 70 °C: (a) top view; (b) enlarged top view; (c) with a 60° tilt. (d) Density variation with concentration: plot of ZnO nanowire density in a 100 µm^2 area (line with filled circle data points) and plot of area percentage covered by ZnO nanowires (line with filled triangle data points). Each data point was obtained from four different areas. Inset is a typical image of ZnO nanowires grown at 5 mM [51].

and gradually produce OH^-. This is critical in the synthesis process, because, if the HMTA hydrolyzes very fast and produces a large quantity of OH^- in a short period, the Zn^{2+} ions in the solution would precipitate out very quickly because of the high pH environment, which would have little contribution to the ZnO NW oriented growth, and eventually result in fast consumption of the nutrient and prevents further growth of ZnO nanowires.

The growth process of ZnO nanowires can be controlled through the five chemical reactions listed in [51]. All five reactions are actually in equilibrium and can be controlled by adjusting the reaction parameters, such as precursor concentration, growth temperature, and growth time, in order to push the reaction equilibrium forward or backward. In general, precursor concentration determines the NW density (Figure 17.10a–d). Growth time and temperature control the ZnO NW morphology and aspect ratio, and the density of ZnO arrays is closely related to the precursor concentration [51].

17.3
Field Emission of Si and ZnO Nanowires

Most reports on field emission describe the fabrication of ZnO or Si nanostructure films and present typical I–V curves. Basically, all studies show that field emission is excellent for nearly all types of Si or ZnO nanostructure, such as ZnO nanowires, ZnO nanobelts, Si nanowires, Si nanotubes, and Si nanotips. An emission current density of 1 mA cm^{-2}, which can be used in field emission displays, has been observed at macroscopic fields of several to tens of volts per micrometer, and the turn-on field may be as low as 1 V µm^{-1}. Tables 17.1 and 17.2 summarize the results reported in the literature. Several parameters such as the diameter of the NWs, crystallization, and surface treatment have an impact on the emission

Table 17.1 Representative results on ZnO nanostructures with different shapes and their corresponding field emission performances.

ZnO emitter	Synthesis method	Turn-on field ($V\,\mu m^{-1}$)	Threshold field ($V\,\mu m^{-1}$)	Field enhancement factor(β)	Stability: testing time and fluctuation	Reference
Nanowires	Vapor phase growth	15.5 and 9.5	–	1 188 and 1 334	–	[54]
	Carbon thermal reduction vapor transport	4–4.5	–	1 180	24 h, <10%	[55]
	Cu-catalyzed vapor–liquid–solid process	0.8	–	7 180	–	[56]
Nanowires grown on C cloth	Carbon thermal vapor transport	–	0.7	41 100	–	[57]
Nanotubes	Hydrothermal reaction	11.2	17.8	910	24 h, <10%	[58]
	Template-based chemical vapor deposition	9	11.5	570	–	[59]
Nanobelts or nanoribbons	Molten-salt-assisted thermal evaporation	1.3	3	14 000	–	[5]
Nanoneedles	Metal organic chemical vapor deposition (MOCVD)	1.5	5	8 328	–	[60]
	Vapor phase growth	3	6.5	1 464	–	[61]

Nanonails and nanopencils	Thermal evaporation process	7.9 and 7.2	–	–	–	[62]
	Thermal evaporation without catalysts	3.7	4.5	2 300	–	[63]
Nanorods	Thermal evaporation process	3.4	7.1	3 990	–	[64]
	Hydrothermal growth	3.83–6.33	5.65–8.58	1 103–2 612	–	[65]
	Noncatalytic thermal evaporation process	7.2	–	2 018	–	[66]
Nanoscrews	Vapor phase growth	3.6	11	–	30 min, $<10\%$	[67]
Tetrapod-like nanostructure	Thermal evaporation process	2.3	4.5	6 285	–	[68]
Nanopins	Vapor transport on copper-coated silicon wafer	2.3	5.9	657	–	[69]
Nanocavity	Vapor phase growth	5	11.6	1 035	–	[61]
Nanowalls	Electrodeposit	4–7	–	4 700	–	[70]

The turn-on field and threshold field are defined at a field producing an emission current density of 10 $\mu A\,cm^{-2}$ and 1 $mA\,cm^{-2}$, respectively.

Table 17.2 Representative results on Si nanostructures with different shapes and their corresponding field emission performances.

Si emitter	Synthesis method	Turn-on field (V μm^{-1})	Threshold field (V μm^{-1})	Field enhancement factor (β)	Stability: testing time and fluctuation	Reference
Nanowires	Laser ablation	–	–	500	–	[7]
Nanowires	CVD	2	3.4	–	–	[6]
Nanowires	CVD	26.6	49.3	–	1000 s, <±1.1%	[80]
Nanowires	High-temperature annealing	6.3–7.3	9–10	1000		[8]
Nanowires	CVD template method	14	–	–	24 h, ~5%	[81]
Nanowires	VLS reaction	5.2	5.37	500	–	[82]
NW arrays	Chemical vapor deposition	0.8	5	450	–	[83]
Nanowires grow on carbon cloth	Vapor–liquid–solid reaction	0.3	0.7	61 000	–	[84]
Diamond-coated nanowires	Chemical vapor deposition	4.4	8.94	367	–	[85]
Cone arrays	Ion-beam sputtering	13	–	–	–	[86]
Carbon-coated Si	Plasma etching	2.7	–	6350	–	[87]
Nanotip arrays	Si-based PAAM as a mask	8.5		1100	–	[88]
Nanotubes	Multistep template replication route	5.1	7.3	740	–	[89]

The turn-on field and threshold field are defined at a field producing an emission current density of 10 μA cm^{-2} and 1 mA cm^{-2}, respectively.

as in the case of carbon nanotubes; on the other hand, as a semiconductor, there are other parameters impacting the emission.

17.3.1
ZnO Nanowires

Field emission properties of ZnO nanorod arrays should have a strong dependency on the aspect ratio and their density, as with carbon nanotubes. Qian and coworkers

Table 17.3 The morphological characteristic and field emission property of ZnO nanorod arrays (r, average radius; L, length of nanorod; β, field enhance factor; s, spacing of nanorods) [65].

Sample	r (nm)	L (μm)	L/r	s (nm)	Turn-on field (V μm^{-1})	Threshold field (V μm^{-1})	β
A	90 ± 2	2.5 ± 0.05	28	195 ± 10	6.33	8.58	1 103
B	125 ± 5	4.2 ± 0.05	34	183 ± 10	5.16	7.43	1 772
C	136 ± 5	5.3 ± 0.05	39	167 ± 10	3.83	5.65	2 612
D	150 ± 5	6.3 ± 0.05	42	143 ± 10	4.16	6.57	2 382
E	168 ± 5	7.4 ± 0.05	44	126 ± 10	4.65	6.97	1 760

synthesized five nanorod arrays with different aspect ratios [65] and densities through a multistep hydrothermal process [71]. The structural parameters such as diameter, length, and spacing of the five ZnO nanorods arrays were measured by an SEM, which are shown in Table 17.3. The field emission property of the ZnO nanorod arrays was measured by using a two-parallel-plate configuration in a vacuum chamber at a base pressure of about 1.0×10^{-6} Pa at room temperature. Figure 17.11 shows the current density versus electric field (J–E) curves, and the corresponding Fowler–Nordheim (FN) plots as well as the turn-on and threshold fields can be found from this figure. Assume the work function $\phi = 5.3$ for ZnO nanorods [72]. The enhancement factor β can be calculated by fitting the slope of the FN curve, which is shown in Figure 17.11b. The field enhancement factors of samples A, B, C, D, and E are 1103, 1772, 2612, 2382, and 1760, respectively. The relationship of β and L/r is shown in Figure 17.12. The interesting result is that β is not increasing linearly with the aspect ratio. The apparent reason for this is the screening effect [73–77]. An empirical model can be used to explain this phenomenon [65]:

$$\beta_0 = b(L/r + h)^{0.9} \tag{17.1}$$

$$\beta = \beta_0 \left[1 - \exp\left(-a\frac{s}{L}\right)\right] \tag{17.2}$$

L and r are the length and radius of ZnO nanorods, respectively, h is an alterable parameter which can be adjusted to fit the experimental data. β_0 is the intrinsic field enhancement factor for a single emitter, which is determined by the aspect ratio [74]. β is the field enhancement factor of the emitter array, which can be determined by the aspect ratio and the interspacing of the nanorods (density effect). When the aspect ratio increases gradually, β_0 will keep up with it. Then β will also increase gradually. When the interspacing s decreases, there is a negative effect on the increment of β. When $s \ll L$, Eq. (17.2) has the following approximate

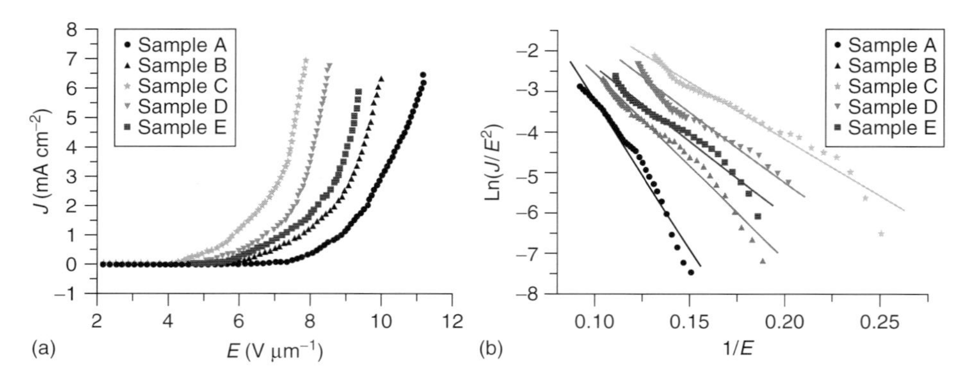

Figure 17.11 The field emission properties of ZnO nanorod arrays: (a) J–E plots. (b) The corresponding FN plots [65].

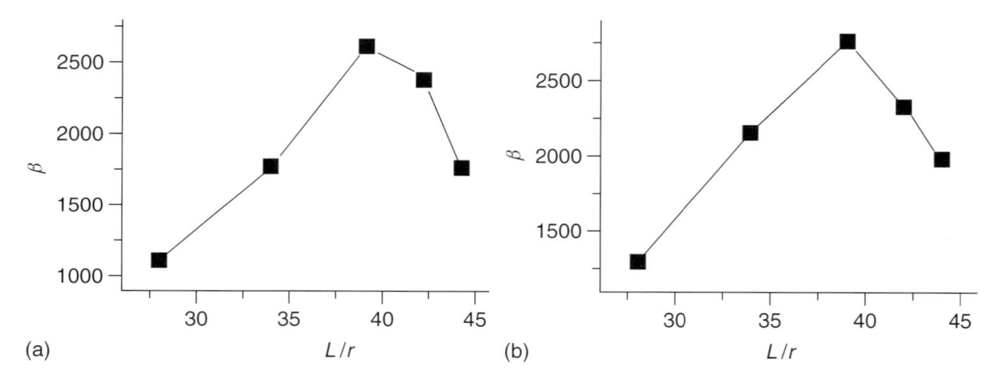

Figure 17.12 The relationship between β and L/r: (a) experimental results and (b) The simulation results [65].

expression:

$$\beta = \beta_0 \times a \times \frac{s}{L} = a \times b \times \left[\frac{L}{r} + h\right]^{0.9} \times \frac{s}{L} = B \times \left[\frac{L}{r} + h\right]^{0.9} \times \frac{s}{L} \quad (17.3)$$

where B is an adjustable parameter. The adjustable parameter h is chosen as −26 to fit the data of samples A and B. Then, using the formula (17.3), the simulated relationship of β and L/r is as shown in Figure 17.11b, which shows that sample C ($\beta = 2885$) has the highest field enhancement factor. This simulation value is only slightly higher than the actual value in Table 17.3. In this case, β is not only related to the aspect ratio but is also dependent on s. For a low density, the interspacing is large and the screening effect is weak, resulting in the field enhancement factor β increasing with the increase of aspect ratio. For a small interspacing, the screening effect is strong, leading to a decrease in the field enhancement factor. These two opposite effects take place simultaneously; the enhancement factor shows a maximum value at the balance point (aspect ratio: 38.9, interspacing: 178 nm), which is consistent with the sample C (aspect ratio: 39, interspacing: 167 ± 10 nm), for producing an optimization of field emission properties.

Recently, Wang *et al.* reported aligned ultralong ZnO nanobelts synthesized by employing molten-salt-assisted thermal evaporation [5]. In this approach, traditionally called *molten flux synthesis*, Zn metal powder is evaporated into a liquid environment of molten sodium chloride (NaCl) salt. Figure 17.13a shows a photograph of the as-grown ZnO nanobelts on a Au substrate, demonstrating that the technique can be effectively used to grow nanobelts whose length can be up to several millimeters. The SEM image (Figure 17.13b) indicates that the product exhibits belt-like structures with a width of up to 6 μm. Figure 17.13c shows a typical TEM image of a single belt. Its transparency to the electron beam (one can see a copper TEM grid beneath the nanostructure) clearly reflects the thinness of the belts as compared to their widths. The ripple-like contrast variation in the image suggests the presence of strain in these structures. The corresponding diffraction spots in the rectangular selected area electron diffraction (SAED) pattern, shown in the inset, can be indexed to a hexagonal wurtzite cell (lattice parameters: $a = 0.325$ nm and $c = 0.521$ nm, JCPDS No. 36-1451). The HRTEM image of this belt is shown in Figure 17.13d. The clear lattice fringes imply perfect crystallinity and defect-free nature of the nanobelts. The field emission performance of the aligned, ultralong ZnO nanobelts is shown in Figure 17.13e. The turn-on electric field is measured as 1.3 V, and the as-grown ZnO nanobelts have a very high field enhancement factor of $1.4\ 6 \times 10^4$. It is believed that the enhanced field emission of the aligned, ultralong ZnO nanobelts is the result of the extremely high aspect ratio of the obtained emitter geometry.

As we know, electrons are more easily emitted from ZnO nanostructures with sharp tips [9, 10, 16] or surface perturbations [13] than from nanowires

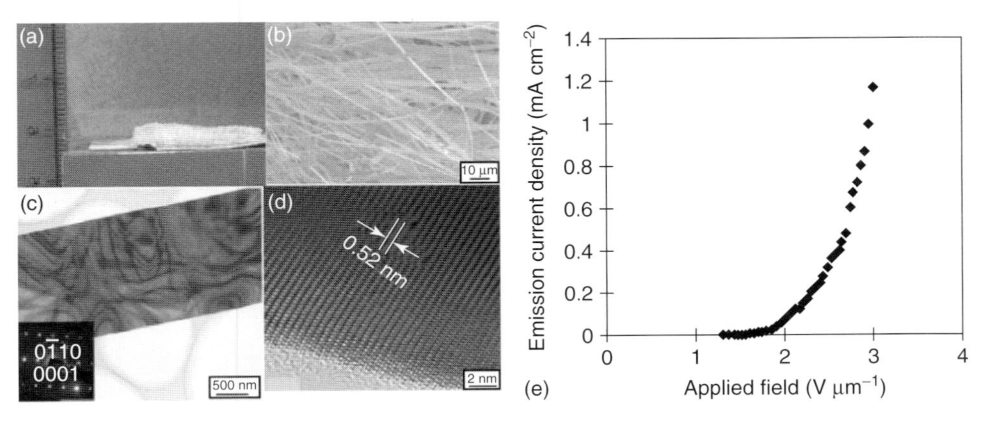

Figure 17.13 (a) A side view of the optical photograph of as-grown ZnO nanobelts on Au, showing the material alignment over a length of 3.3 mm, (b) typical SEM image showing the belt-like structures; the nanobelts have a width of up to 6 μm, (c) TEM image and SAED pattern (inset) of a single nanobelt, (d) HRTEM image of a nanobelt showing its high crystallinity, and (e) field emission characteristics of the aligned ZnO nanobelts [5].

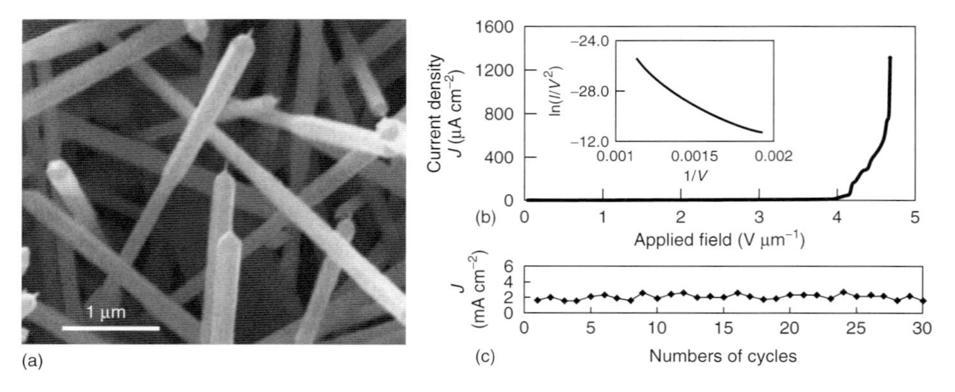

Figure 17.14 (a) SEM image of the nanopencils, showing pen tips on the nanopencils, (b) field emission property of the ZnO nanopencils: current density versus electric field characteristics with the inset for the corresponding FN plot, and (c) current density reproducibility recorded at an applied field of 5 V μm^{-1} [63].

with a uniform diameter [8, 11]. Recently, a simple two-step, pressure-controlled, catalyst-free vapor phase deposition method for producing ZnO nanopencils at a low temperature of 550 °C has been demonstrated [63].

Figure 17.14a shows the SEM image of ZnO nanopencils. Figure 17.14b indicates that the ZnO nanopencils have a turn-on field of 3.7 V μm^{-1}, which is lower than for other normal ZnO nanostructures [4, 8, 9, 14–16]. The current density is 1.3 mA at 4.6 V μm^{-1} and shows no saturation. The corresponding FN plot is shown in the inset of Figure 17.14b. From the average slope of the FN plot, the field enhancement factor β of the ZnO nanopencils was estimated to be about 2300 by assuming the work function of $\phi = 5.3$ eV [72], which is good enough for various applications of field emission. The average β of the ZnO nanopencils is related to the geometry, structure, and density of the nanopencils grown on the substrate. The grown nanopencils exhibit high crystalline quality and sharp tips with nanosized surface perturbations, resulting in a high field enhancement factor. Furthermore, the nanopencils could be subjected to the field emission test at least 30 times under the sweeping electric field from 0 to 5 V μm^{-1} without obvious changes in the results. Figure 17.14c shows the current density reproducibility at an electric field of 5 V μm^{-1}. The average current density is 2.1 mA cm^{-2}.

17.3.2
Si Nanowires

Si has been the backbone of the microelectronics industry for decades and it is desirable to have Si field emitters integrated onto Si substrates along with the driving circuitry. Recently, Si nanowires have been synthesized by several methods, and field emission from various Si nanostructures has been reported; the results obtained from the literature are summarized in Table 17.3.

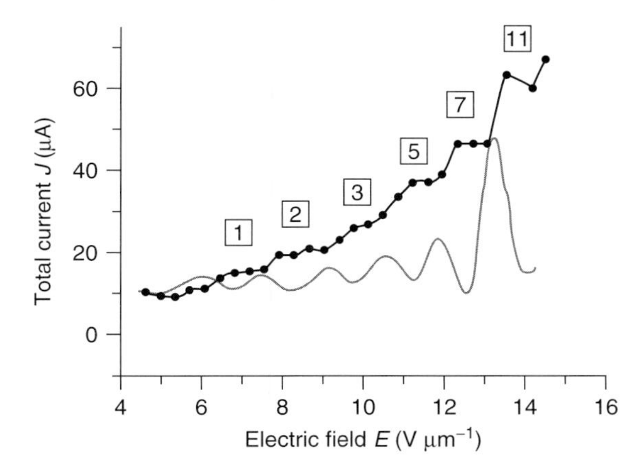

Figure 17.15 Electric field versus current characteristic (with data points) and its derivative (data points omitted) [78].

It is well known that there are some quantum effects in semiconductor nanowires, and quantized field electron emission has been found by Klimovskaya and coworkers in self-assembled arrays of Si nanowires [78, 79]. Figure 17.15 shows the typical current–voltage characteristic measured on undoped silicon nanowires at 300 K. The most interesting feature of the characteristics is the stepwise increase of the current with increasing voltage that does not disappear with repeated measurements. The magnitudes of the current on each step, normalized by the current on the first step, are indicated in boxes in Figure 17.15. The first derivative of the current is shown in the same figure.

Both time fluctuations (noise) of the current at a fixed voltage and dependence of the magnitude of these fluctuations on the voltage have been observed, but on each plateau the magnitude of the fluctuations is small. Figure 17.16 shows the current fluctuations versus the voltage (Figure 17.16a) and, for comparison, the derivative of the current (Figure 17.16b). Positions of the maxima of the fluctuations correspond to the positions of the derivative maxima, while the magnitude of the fluctuations is vanishingly small on the plateaus. Furthermore, the larger the voltage, the larger the magnitude of the fluctuations; however, its value normalized by the corresponding value of the current decreases with increasing voltage. The current noise observed in their experiment is typical of quantum systems with a small number of charge carriers. It is usually observed in quantum nanoelectromechanical systems with a stepwise current–voltage characteristics, where the increase in voltage gives rise to a stepwise increase in the number of shuttling electrons [90]. The transition from one step to another leads to enhancement of the noise, while in the regions of plateaus the noise is vanishingly small because of Coulomb blockade. Experimental observation of the specific behavior of the current and current noise suggests a quantum origin of the electron emission. It is possible that the mechanism of quantization is based on a collective behavior of nanowires, while to elucidate a

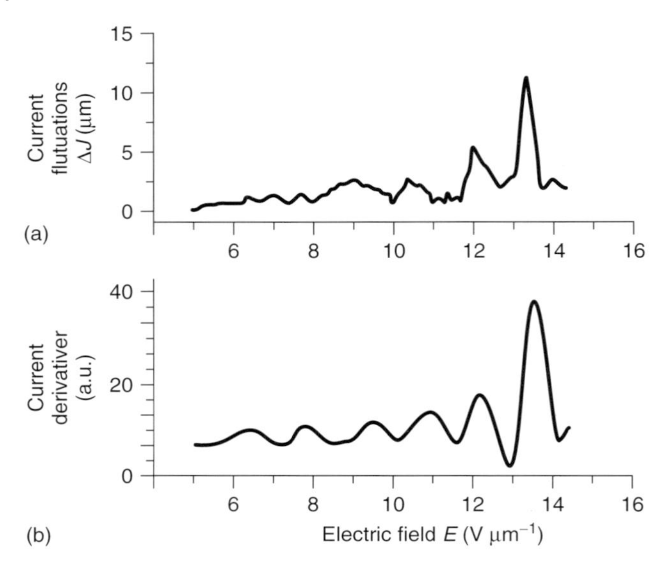

Figure 17.16 Current fluctuations (a) and the current derivative (b) versus the electric field [78].

detailed mechanism of the current quantization and specific behavior of the current noise further theoretical and experimental investigations are needed.

In order to achieve a lower electric field, we grew Si nanowires on carbon cloth as a field emitter [84]. An emission current density of 1 $mA\,cm^{-2}$ was achieved at an electric field as low as 0.7 $V\,\mu m^{-1}$, as shown in Figure 17.17. This significantly improved field emission property can be explained by the special geometric arrangement of Si nanowires grown on the carbon cloth. The *field enhancement factor*, which is defined as the ratio of the local electric field at the tip of a nanowire to the macroscopic electric field, is about 6.1×10^4 calculated from the slope of the FN plot. Carbon cloth, which is a textile-like material consisting of long carbon fibers (each about 10 μm in diameter oriented in at least two directions) significantly contributes to the high field enhancement factor [91]).

Xu and coworkers [83] have reported an approach to fabricate Si-nanowires-based field emission microtriode arrays based on a self-assembled nanomask, together with hydrogen ion etching for synthesis of vertically aligned Si nanowires.

The synthesis of the aligned Si nanowires was performed in a microwave plasma-enhanced CVD system. Take Si(100) wafers as the substrates. First, a CH_4 (20 sccm)/H_2(80 sccm) plasma was used to treat the substrate for 1 h to generate nanomasks. This is followed by a pure H_2 (80 sccm) plasma etching. It was found that the length of the formed nanowires increases with increasing H_2 plasma etching time. In both steps, the substrate temperature and bias were 500 °C and −200 V, respectively. Figure 17.18a shows a typical SEM image of the fabricated nanowires. The diameter and the length of the nanowires are 10–40 nm and 1–1.2 μm, respectively. All the nanowires are aligned vertically to the substrate

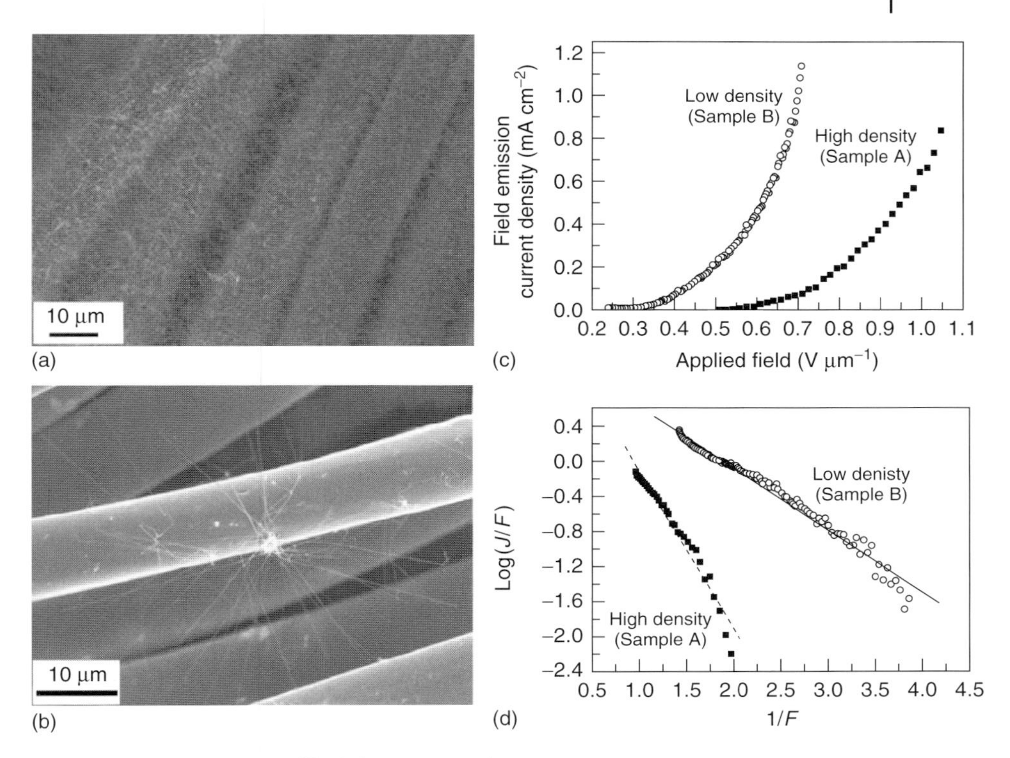

Figure 17.17 SEM images of high-density (a) and low-density (b) Si nanowires grown on carbon cloth by chemical vapor deposition, (c) relationship of the emission current versus electric field, and (d) FN plot, indicating the true nature of field emission [84].

surface. Their typical density is 10^9 cm^{-2}. The typical J–E curve for the as-fabricated Si nanowires is shown in Figure 17.18b, giving a turn-on field as low as 0.8 V μm^{-1} and the threshold field of 5.0 V μm^{-1}. Based on this Si nanowire synthesis, gated Si nanowire field emission microtriodes have been fabricated. Figure 17.19 shows the procedure of fabrication of the gated Si nanowire device. First, SiO_2 (0.4 μm) and Cr (0.2 μm) thin films were deposited on the Si substrate to form a Si/SiO_2/Cr sandwich structure (Figure 17.19a). By using photolithography, an array of holes (5 μm in diameter) was defined. After removing the Cr and SiO_2, respectively, gated structures were obtained (Figure 17.19b). By performing the Si nanowire preparation processes described above, Si nanowires were fabricated on the gated structures (Figure 17.19c). Field emission tests on arrays of 40 × 40 gated devices were performed with the configuration illustrated in Figure 17.19d. Figure 17.19g shows the typical field emission J–V characteristics of one array. J_A and J_C are the current densities of anode and cathode, respectively. Here J_A and J_C are obtained by dividing the measured current by the gate hole area. Both J_A and J_C increase exponentially with increasing applied voltage. From the inset of Figure 17.19g,

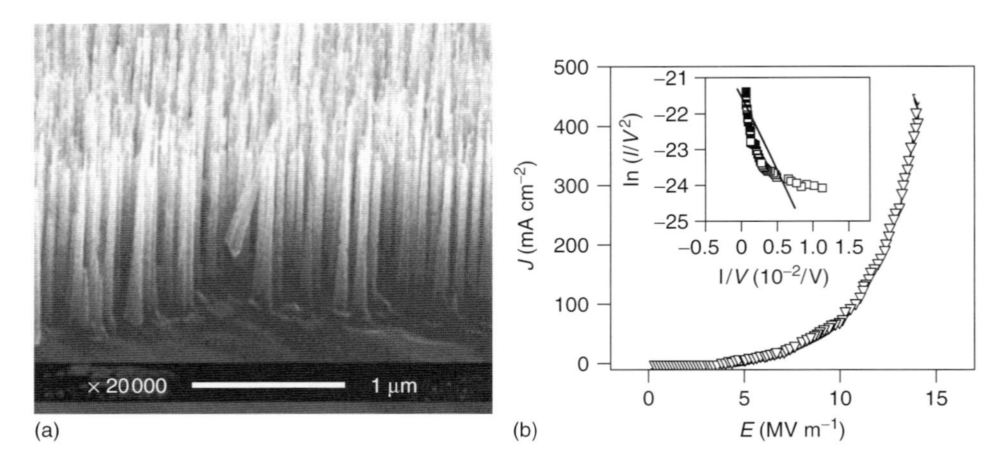

Figure 17.18 (a) SEM image with 75° view angle of Si nanowires arrays, (b) field emission J–E curve of the Si nanowires; the low turn-on and threshold fields are 0.8 and 5.0 V μm^{-1}, respectively. The maximum current density is 442 mA cm^{-2} [83].

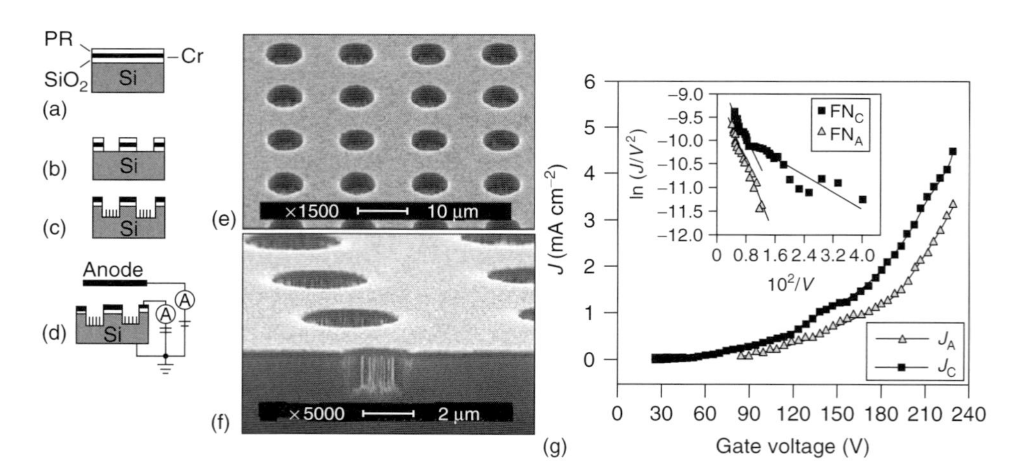

Figure 17.19 (a–c) The fabrication procedure of the gated Si nanowire device; (d) the arrangement for field emission test of the gated Si nanowire device; (e and f) typical SEM images of the gated Si nanowires devices, and (g) typical field emission characteristics of a gated Si nanowires emitter array (40 × 40) [83].

we can see that the plot of $\ln(J_A/V^2) \sim 1/V(\mathrm{FN_A})$ appears to be a straight line, confirming that the collected current is from field emission. In addition, the plot of $\ln(J_C/V^2) \sim 1/V(\mathrm{FN_C})$ shows a nonlinear behavior; at the high field region, leakage current between the gate electrode and the cathode must have contribution to J_C.

17.4 Summary

In this chapter, we have presented a review on recent advances in studies of ZnO and Si nanowires and their field emission properties. Tables 17.1 and 17.3 give an up-to-date summary of the important reports on field emission of ZnO and Si nanostructures. The results show that ZnO and Si nanostructures have become a popular subject of research owing to the richness of their physical and chemical properties and the wide range of their possible applications. Significant challenges still exist in enhancing the field emission property and for applying ZnO and Si nanostructures in vacuum microelectronic devices. There are varieties of nanostructure morphologies, such as nanorods, nanowires, nanobelts or nanoribbons, nanotubes, nanocables, nanocones, nanoneedles, nanotips, nanonails, nanopenciles, nanoscrews, nanoarrays, nanorings, nanohelices, nanosheets, nanowalls, and nanopyramids. They can provide sharp emitters, which are very important for developing next-generation field emission devices. But the growth kinetics and thermodynamics involved in the synthesis of ZnO and Si nanostructures are extremely complex and assume different mechanisms under different growth conditions. So more work should be done on their syntheses, which includes, but is not limited to, reliable control in the diameter, length, orientation, density, crystallization, and hierarchical assembly.

There is still plenty of opportunity for the development of ZnO and Si nanostructures and their field emission applications. We believe that future work in this direction should continue to focus on the synthesis of ZnO or Si nanostructures in a more controlled, predictable, and simple way, and on enhancing their field emission properties up to the level desirable for many real industrial applications.

Acknowledgment

The work performed by BQZ is partially supported by NSFC (Grant No. 60071043 and 600532010), the National Laboratory for Vacuum Electronics. The work performed at Boston College is supported by DOE DE-FG02-00ER45805.

References

1. De Heer, W.A., Chatelain, A., and Ugarte, D. (1995) A carbon nanotube field-emission electron source. *Science*, **270** (5239), 1179–1180.

2. Lee, C.J., Lee, T.J., Lyu, S.C., Zhang, Y., Ruh, H., and Lee, H.J. (2002) Field emission from well-aligned zinc oxide nanowires grown at low temperature. *Appl. Phys. Lett.*, **81** (19), 3648–3650.
3. Jo, S.H., Lao, J.Y., Ren, Z.F., Farrer, R.A., Baldacchini, T., and Fourkas, J.T. (2003) Field-emission studies on thin films of zinc oxide nanowires. *Appl. Phys. Lett.*, **83** (23), 4821–4823.
4. Dong, L.F., Jiao, J., Tuggle, D.W., Petty, J.M., Elliff, S.A., and Coulter, M. (2003) ZnO nanowires formed on tungsten substrates and their electron field emission properties. *Appl. Phys. Lett.*, **82** (7), 1096–1098.
5. Wang, W.Z., Zeng, B.Q., Yang, J., Poudel, B., Huang, J.Y., and Ren, Z.F. (2006) Aligned ultralong ZnO nanobelts and their enhanced field emission. *Adv. Mater.*, **18**, 3275–3278.
6. Zeng, B.Q., Xiong, G.Y., Chen, S., Jo, S.H., Wang, W.Z., Wang, D.Z., and Ren, Z.F. (2006) Field emission of silicon nanowires. *Appl. Phys. Lett.*, **88**, 213108-1–213108-3.
7. Frederick, C.K.Au., Wong, K.W., Tang, Y.H., Zhang, Y.F., Bello, I., and Lee, S.T. (1999) Electron field emission from silicon nanowires. *Appl. Phys. Lett.*, **75**, 1700–1703.
8. Chueh, Y.L., Chou, L.J., Cheng, S.L., He, J.H., Wu, W.W., and Chen, L.J. (2005) Synthesis of taper-like Si nanowires with strong field emission. *Appl. Phys. Lett.*, **86** (13), 133112-1–133112-3.
9. Zhou, J., Ding, Y., Deng, S.Z., Gong, L., Xu, N.S., and Wang, Z.L. (2005) Three-dimensional tungsten oxide nanowire networks. *Adv. Mater.*, **17**, 2107–2110.
10. Wong, K.W., Zhou, X.T., Au, F.C.K., Lai, H.L., Lee, C.S., and Lee, S.T. (1999) Field-emission characteristics of SiC nanowires prepared by chemical-vapor deposition. *Appl. Phys. Lett.*, **75** (8), 2918–2920.
11. Li, Y.B., Bando, Y., Golberg, D., and Kurashima, K. (2002) Field emission from MoO_3 nanobelts. *Appl. Phys. Lett.*, **81** (26), 5048–5050.
12. Ham, H., Shen, G.Z., Cho, J.H., Lee, T.J., Seo, S.H., and Lee, C.J. (2005) Vertically aligned ZnO nanowires produced by a catalyst-free thermal evaporation method and their field emission properties. *Chem. Phys. Lett.*, **404** (1-3), 69–73.
13. Wang, Z.L. (2008) Oxide nanobelts and nanowires: growth, properties and applications. *J. Nanosci. Nanotechnol.*, **8** (1), 27–55.
14. Xu, S., Wei, Y.G., Kirkham, M., Liu, J., Mai, W.J., Davidovic, D., Snyder, R.L., and Wang, Z.L. (2008) Patterned growth of vertically aligned ZnO nanowire arrays on inorganic substrates at low temperature without catalyst. *J. Am. Chem. Soc.*, **130** (45), 14958–14959.
15. Banerjee, D., Rybczynski, J., Huang, J.Y., Wang, D.Z., Kempa, K., and Ren, Z.F. (2005) Large hexagonal arrays of aligned ZnO nanorods. *Appl. Phys. A: Mater. Sci. Process.*, **80** (4), 749–752.
16. Huang, M.H., Wu, Y.Y., Feick, H., Tran, N., Weber, E., and Yang, P.D. (2001) Catalytic growth of zinc oxide nanowires by vapor transport. *Adv. Mater.*, **13** (2), 113–116.
17. He, J.R., Hsu, J.H., Wang, C.W., Lih, H.N., Chen, L.J., and Wang, Z.L. (2006) Pattern and feature designed growth of ZnO nanowire arrays for vertical devices. *J. Phys. Chem. B*, **110** (1), 50–53.
18. Kolasinski Kurt, W. (2006) Catalytic growth of nanowires: Vapor– liquid– solid, vapor– solid– solid, solution– liquid– solid and solid– liquid– solid growth. *Curr. Opin. Solid State Mater. Sci.*, **10**, 182–191.
19. Wang, Z.L. (2009) ZnO nanowire and nanobelt platform for nanotechnology. *Mater. Sci. Eng. R: Rep.*, **64** (3-4), 33–71.
20. Gao, P.X., Ding, Y., Mai, W.J., Hughes, W.L., Lao, C.S., and Wang, Z.L. (2005) Materials science: conversion of zinc oxide nanobelts into superlattice-structured nanohelices. *Science*, **309** (5741), 1700–1704.
21. Umar, A., Kim, S.H., Lee, Y.S., Nahm, K.S., and Hahn, Y.B. (2005) Catalyst-free large-quantity synthesis of ZnO nanorods by a vapor-solid growth mechanism: structural and optical properties. *J. Cryst. Growth*, **282** (1-2), 131–136.

22. Campos, L.C., Tonezzer, M., Ferlauto, A.S., Grillo, V., Magalhães, P.R., Oliveira, S., Ladeira, L.O., and Lacerda, R.G. (2008) Vapor-solid-solid growth mechanism driven by epitaxial match between solid AuZn alloy catalyst particles and ZnO nanowires at low temperatures. *Adv. Mater.*, **20** (8), 1499–1504.
23. Jung, S., Cho, W., Lee, H.J., and Oh, M. (2009) Self-template-directed formation of coordination-polymer hexagonal tubes and rings, and their calcination to ZnO rings. *Angew. Chem. Int. Ed.*, **48** (8), 1459–1462.
24. Fan, H.J., Lee, W., Hauschild, R., Alexe, M., Rhun, G.L., Scholz, R., Dadgar, A., Nielsch, K., Kalt, H., Krost, A., Zacharias, M., and Gösele, U. (2006) Template-assisted large-scale ordered arrays of ZnO pillars for optical and piezoelectric applications. *Small*, **2** (4), 561–568.
25. Müller, T., Heinig, K.-H., and Schmidt, B. (2002) Template-directed self-assembly of buried nanowires and the pearling instability. *Mater. Sci. Eng. C*, **19** (1-2), 209–213.
26. Zheng, W.W., Guo, F., and Qian, Y. (2005) Growth of bulk ZnO single crystals via a novel hydrothermal oxidative pressure-relief route. *Adv. Funct. Mater.*, **15** (2), 331–335.
27. Ku, C.H. and Wu, J.J. (2006) Aqueous solution route to high-aspect-ratio zinc oxide nanostructures on indium tin oxide substrates. *J. Phys. Chem. B*, **110** (26), 12981–12985.
28. Ho, G.W. and Wong, A.S.W. (2007) One step solution synthesis towards ultra-thin and uniform single-crystalline ZnO nanowires. *Appl. Phys. A: Mater. Sci. Process.*, **86** (4), 457–462.
29. Bin Liu, B. and Zeng, H.-C. (2003) Hydrothermal Synthesis of ZnO Nanorods in the Diameter Regime of 50 nm. *J. Am. Chem. Soc.*, **125** (15), 4430–4431.
30. Kar, J.P., Jeong, M.C., Lee, W.K., and Myoung, J.M. (2008) Fabrication and characterization of vertically aligned Zn-MgO/ZnO nanowire arrays. *Mater. Sci. Eng. B: Solid-State Mater. Adv. Technol.*, **147** (1), 74–78.
31. Wagner, R.S. and Ellis, W.C. (1964) Vapor-liquid-solid mechanism of single crystal growth. *Appl. Phys. Lett.*, **4** (5), 89–89.
32. Huang, M.H., Wu, Y., Feick, H., Tran, N., Weber, E., and Yang, P. (2006) Catalytic growth of zinc oxide nanowires by vapor transport. *Adv. Mater.*, **13** (2), 3275–3278.
33. Yang, P., Yan, H., Mao, S., Russo, R., Johnson, J., Saykally, R., Morris, N., Pham, J., He, R., and Choi, H.J. (2002) Controlled growth of ZnO nanowires and their optical properties. *Adv. Funct. Mater.*, **12** (5), 323–331.
34. Wagner, R.S. (1970) in *Whisker Technology* (ed. A.P. Levitt), Wiley-Interscience, New York, pp. 47–119.
35. Givargizov, E.I. (1975) Fundamental aspects of VLS growth. *J. Cryst. Growth*, **31**, 20–230.
36. Law, M., Goldberger, J., and Yang, P. (2004) Semiconductor nanowires and nanotubes. *Annu. Rev. Mater. Res.*, **34**, 83–122.
37. Hannon, J.B., Kodambaka, S., Ross, F.M., and Tromp, R.M. (2006) The influence of the surface migration of gold on the growth of silicon nanowires. *Nature*, **440** (7080), 69–71.
38. Wu, Y. and Yang, P. (2001) Direct observation of vapor-liquid-solid nanowire growth. *J. Am. Chem. Soc.*, **123** (13), 3165–3166.
39. Wang, X.D., Song, J.H., Li, P., Ryou, J.H., Dupuis, R.D., Summers, C.J., and Wang, Z.L. (2005) Growth of uniformly aligned ZnO nanowire heterojunction arrays on GaN, AlN, and Al0.5Ga0.5N substrates. *J. Am. Chem. Soc.*, **127** (21), 7920–7923.
40. Huang, M., Mao, S., Feick, H., Yan, H., Wu, Y., Kind, H., Weber, E., Russo, R., and Yang, P. (2001) Room-temperature ultraviolet nanowire nanolasers. *Science*, **292** (5523), 1897–1899.
41. Wu, Y., Yan, H., Huang, M., Messer, B., Song, J.H., and Yang, P. (2002) Inorganic semiconductor nanowires: rational growth, assembly, and novel properties. *Chem.-A Eur. J.*, **8** (6), 1261–1268.
42. Wang, X.D., Summers, C.J., and Wang, Z.L. (2004) Large-scale hexagonal-patterned growth of aligned

ZnO nanorods for nano-optoelectronics and nanosensor arrays. *Nano Lett.*, **4** (3), 423–426.

43. Rybczynski, J., Ebels, U., and Giersig, M. (2003) Large-scale, 2D arrays of magnetic nanoparticles. *Colloids Surf. A: Physicochem. Eng.*, **219** (1-3), 1–6.

44. Kempa, K., Kimball, B., Rybczynski, J., Huang, Z.P., Wu, P.F., Steeves, D., Sennett, M., Giersig, M., Rao, D.V.G.L.N., Carnahan, D.L., Wang, D.Z., Lao, J.Y., Li, W.Z., and Ren, Z.F. (2003) Photonic crystals based on periodic arrays of aligned carbon nanotubes. *Nano Lett.*, **3** (1), 13–18.

45. Rybczynski, J., Banerjee, D., Kosiorek, A., Giersig, M., and Ren, Z.F. (2004) Formation of super arrays of periodic nanoparticles and aligned ZnO nanorods – simulation and experiments. *Nano Lett.*, **4** (10), 2037–2040.

46. Liu, D.F., Xiang, Y.J., Wu, X.C., Zhang, Z.X., Liu, L.F., Song, L., Zhao, X.W., Luo, S.D., Ma, W.J., Shen, J., Zhou, W.Y., Wang, G., Wang, C.Y., and Xie, S.S. (2006) Periodic ZnO nanorod arrays defined by polystyrene microsphere self-assembled monolayers. *Nano Lett.*, **6** (10), 2375–2378.

47. Kosiorek, A., Kandulski, W., Chudzinski, P., Kempa, K., and Giersig, M. (2004) Shadow nanosphere lithography: simulation and experiment. *Nano Lett.*, **4** (7), 1359–1363.

48. McLellan, J.M., Geissler, M., and Xia, Y.N. (2004) Edge spreading lithography and its application to the fabrication of mesoscopic gold and silver rings. *J. Am. Chem. Soc.*, **126** (35), 10830–10831.

49. Hsu, H.C., Tseng, Y.K., Cheng, H.M., Kuo, J.H., and Hsieh, W.F. (2004) Selective growth of ZnO nanorods on pre-coated ZnO buffer layer. *J. Cryst. Growth*, **261** (4), 520–525.

50. Ma, T., Guo, M., Zhang, M., Zhang, Y.J., and Wang, X.D. (2007) Density-controlled hydrothermal growth of well-aligned ZnO nanorod arrays. *Nanotechnology*, **18** (3), 035605–035607.

51. Xu, S., Lao, C., Weintraub, B., and Wang, Z.L. (2008) Density-controlled growth of aligned ZnO nanowire arrays by seedles chemical approach on smooth surfaces. *J. Mater. Res.*, **23** (8), 2072–2077.

52. Zhang, J., Sun, L.D., Jiang, X.C., Liao, C.S., and Yan, C.H. (2004) Shape evolution of one-dimensional single-crystalline ZnO nanostructures in a microemulsion system. *Cryst. Growth Des.*, **4** (2), 309–313.

53. Vayssieres, L. (2003) Growth of arrayed nanorods and nanowires of ZnO from aqueous solutions. *Adv. Mater.*, **15** (5), 464–465.

54. Cao, B.Q., Teng, X.M., Heo, S.H., Li, Y., Cho, S.O., Li, G.H., and Cai, W.P. (2007) Different ZnO nanostructures fabricated by a seed-layer assisted electrochemical route and their photoluminescence and field emission properties. *J. Phys. Chem. C*, **111**, 2470–2476.

55. Jang, H.S., Kang, S.Q., Nahm, S.H., Kim, D.H., Lee, H.Y., and Kim, Y.I. (2007) Enhanced field emission from the ZnO nanowires by hydrogen gas exposure. *Mater. Lett.*, **61** (8-9), 1679–1682.

56. Li, Y.S., Lin, P., Lee, C.Y., and Tseng, T.Y. (2004) Field emission and photofluorescent characteristics of zinc oxide nanowires synthesized by a metal catalyzed vapor-liquid-solid process. *J. Appl. Phys.*, **95** (7), 3711.

57. Banerjee, D., Jo, S.H., and Ren, Z.F. (2004) Enhanced field emission of ZnO nanowires. *Adv. Mater.*, **16** (22), 2028–2032.

58. Wei, A., Sun, X.W., Xu, C.X., Dong, Z.L., Yu, M.B., and Huang, W. (2006) Stable field emission from hydrothermally grown ZnO nanotubes. *Appl. Phys. Lett.*, **88** (21), 213102-1–213102-3.

59. Shen, X.P., Yuan, A.H., Hu, Y.M., Jiang, Y., Xu, Z., and Hu, Z. (2005) Fabrication, characterization and field emission properties of large-scale uniform ZnO nanotube arrays. *Nanotechnology*, **16** (10), 2039–2043.

60. Park, C.J., Choi, D.K., Yoo, J., Yi, G.C., and Lee, C.J. (2007) Enhanced field emission properties from well-aligned zinc oxide nanoneedles grown on the Au/Ti/n-Si substrate. *Appl. Phys. Lett.*, **90** (8), 083107-1–083107-3.

61. Zhao, Q., Zhang, H.Z., Zhu, Y.W., Feng, S.Q., Sun, X.C., Xu, J., and

Yu, D.P. (2005) Morphological effects on the field emission of ZnO nanorod arrays. *Appl. Phys. Lett.*, **86** (20), 203115-1–203115-3.

62. Shen, G.Z., Bando, Y., Liu, B.D., Golberg, D., and Lee, C.J. (2006) Characterization and field-emission properties of vertically aligned ZnO nanonails and nanopencils fabricated by a modified thermal-evaporation process. *Adv. Funct. Mater.*, **16** (2), 410–416.

63. Wang, R.C., Liu, C.P., Huang, J.L., Chen, S.J., Tseng, Y.K., and Kung, S.C. (2005) ZnO nanopencils: efficient field emitters. *Appl. Phys. Lett.*, **87** (1), 013110-1–013110-3.

64. Lin, C.C., Lin, W.H., Hsiao, C.Y., Lin, K.M., and Li, Y.Y. (2008) Synthesis of one-dimensional ZnO nanostructures and their field emission properties. *J. Phys. D: Appl. Phys.*, **41** (4), 045301–045306.

65. Qian, X.M., Liu, H.B., Guo, Y.B., Song, Y.L., and Li, Y.L. (2008) Effect of aspect ratio on field emission properties of ZnO nanorod arrays. *Nanoscale Res. Lett.*, **3** (8), 303–307.

66. Umar, A., Kim1, S.H., Lee, H., Lee, N., and Hahn, Y.B. (2008) Optical and field emission properties of single-crystalline aligned ZnO nanorods grown on aluminium substrate. *J. Phys. D: Appl. Phys.*, **41** (6), 065412-1–065412-6.

67. Liao, L., Li, J.C., Liu, D., Liu, H.C., Wang, D.F., Song, W.Z., and Fu, Q. (2005) Self-assembly of aligned ZnO nanoscrews: growth, configuration and field emission. *Appl. Phys. Lett.*, **86** (8), 083106-1–083106-3.

68. Wan, Q., Yu, K., Wang, T.H., and Lin, C.L. (2003) Low-field electron emission from tetrapod-like ZnO nanostructures synthesized by rapid evaporation. *Appl. Phys. Lett.*, **83** (11), 2253–2255.

69. Xu, C.X. and Sun, X.W. (2003) Field emission from zinc oxide nanopins. *Appl. Phys. Lett.*, **83** (18), 3806–3808.

70. Pradhan, D., Kumar, M., Ando, Y., and Leung, K.T. (2008) Efficient field emission from vertically grown planar ZnO nanowalls on an ITO–glass substrate. *Nanotechnology*, **19** (3), 035603-1–035603-6.

71. Greene, L.E., Law, M., Goldberger, J., Kim, F.K., Johnson, J.C., Zhang, Y.F., Saykally, R.J., and Yang, P. (2003) Low-temperature wafer-scale production of ZnO nanowire arrays. *Angew. Chem. Int. Ed.*, **42** (26), 3031–3034.

72. Wang, X.D., Zhou, J., Lao, C.S., Song, J.H., Xu, N.S., and Wang, Z.L. (2007) In situ field emission of density-controlled ZnO nanowire arrays. *Adv. Mater.*, **19** (12), 1627–1631.

73. Jo, S.H., Tu, Y., Huang, Z.P., Carnahan, D.L., Wang, D.Z., and Ren, Z.F. (2003) Effect of length and spacing of vertically aligned carbon nanotubes on field emission properties. *Appl. Phys. Lett.*, **82** (20), 3520–3522.

74. Patra, S.K. and Rao, G. Mohan. (2006) Field emission current saturation of aligned carbon nanotube-effect of density and aspect ratio. *J. Appl. Phys.*, **100** (2), 024319.

75. Nilsson, L., Groening, O., Emmenegger, C., Kuettel, O., Schaller, E., Schlapbach, L., Kind, H., Bonard, J.-M., and Kern, K. (2000) Scanning field emission from patterned carbon nanotube films. *Appl. Phys. Lett.*, **76** (15), 2071–2073.

76. Bonard, J.M., Weiss, N., Kind, H., Stöckli, T., Forró, L., Kern, K., and Châtelain, A. (2001) Tuning the field emission properties of patterned carbon nanotube films. *Adv. Mater.*, **13** (3), 184–188.

77. Chhowalla, M., Ducati, C., Rupesinghe, N.L., Teo, K.B.K., and Amaratunga, G.A.J. (2001) Field emission from short and stubby vertically aligned carbon nanotubes. *Appl. Phys. Lett.*, **79** (13), 2079–2081.

78. Klimovskaya, A.I., Litvin, Y.M., Moklyak, Y.Y., Dadykin, A.A., Kamins, T., and Sharma, S. (2006) Field-electron emission at 300 K in self-assembled arrays of silicon nanowires, *Appl. Phys. Lett.*, **89** (9), 093122-1–093122-3.

79. Klimovskaya, A.I., Raichev, O.E., Dadykin, A.A., Litvin, Yu.M., Lytvyn, P.M., Prokopenko, I.V., Kamins, T.I., Sharma, S., and Moklyak, Yu. (2007) Quantized field-electron emission at 300 K in self-assembled arrays of silicon nanowires. *Phys. E: Low-Dimens. Syst. Nanostruct.*, **37** (1-2), 212–217.

80. Riccitelli, R., Di Carlo, A., Fiori, A., Orlanducci, S., Terranova, M.L., Santoni, A., Fantoni, R., Rufoloni, A., and Villacorta, F.J. (2007) Field emission from silicon nanowires: conditioning and stability. *J. Appl. Phys.*, **102** (5), 054906-1–054906-5.

81. Lu, M., Li, M.K., Kong, L.B., Guo, X.Y., and Li, H.L. (2004) Synthesis and characterization of well-aligned quantum silicon nanowires arrays. *Compos. Part B: Eng.*, **35**, 179–184.

82. Kulkarni, N.N., Bae, J., Shih, C.K., Stanley, S.K., Coffee, S.S., and Ekerdt, J.G. (2005) Low-threshold field emission from cesiated silicon nanowires. *Appl. Phys. Lett.*, **87**, 213115-1–213115-3.

83. She, J.C., Deng, S.Z., Xu, N.S., Yao, R.H., and Chen, J. (2006) Fabrication of vertically aligned Si nanowires and their application in a gated field emission device. *Appl. Phys. Lett.*, **88**, 013112-1–013112-3.

84. Zeng, B.Q., Xiong, G.Y., Chen, S., Jo, S.H., Wang, W.Z., and Ren, Z.F. (2007) Field emission of silicon nanowires. *Appl. Phys. Lett.*, **90** (3), 033112-1–033112-3.

85. Tzeng, Y.F., Lee, Y.C., Lee, C.Y., Lin, I.N., and Chiu, H.T. (2007) On the enhancement of field emission performance of ultrananocrystalline diamond coated nanoemitters. *Appl. Phys. Lett.*, **91**, 063117-1–063117-3.

86. Shang, N.G., Meng, F.Y., Au, F.C.K., Li, Q., Lee, C.S., Bello, I., and Lee, S.T. (2002) Fabrication and field emission of high-density silicon cone arrays. *Adv. Mater.*, **14** (18), 1308–1311.

87. Wang, Q., Li, J.J., Ma, Y.J., Bai, X.D., Wang, Z.L., Xu, P., Shi, C.Y., Quan, B.G., Yue, S.L., and Gu, C.Z. (2005) Field emission properties of carbon coated Si nanocone arrays on porous silicon. *Nanotechnology*, **16**, 2919–2922.

88. Huang, G.S., Wu1, X.L., Cheng, Y.C., Li1, X.F., Luo, S.H., Feng, T., and Chu, P.K. (2006) Fabrication and field emission property of a Si nanotip array. *Nanotechnology*, **17**, 5573–5576.

89. Mu, C., Yu, Y.X., Liao, W., Zhao, X.S., and Xu, D.S. (2005) Controlling growth and field emission properties of silicon nanotube arrays by multistep template replication and chemical vapor deposition. *Appl. Phys. Lett.*, **87**, 113104-1–113104-3.

90. Raichev, O.E. (2006) Coulomb blockade of field emission from nanoscale conductors. *Phys. Rev. B Condens. Matter Mater. Phys.*, **73**, 195328–195329.

91. Jo, S.H., Wang, D.Z., Huang, J.Z., Li, W.Z., Kempa, K., and Ren, Z.F. (2004) Field emission of carbon nanotubes grown on carbon cloth. *Appl. Phys. Lett.*, **85**, 810–812.

Part IV
Applications of Carbon Nanotubes

Carbon Nanotube and Related Field Emitters: Fundamentals and Applications. Edited by Yahachi Saito

ISBN: 978-3-527-32734-8

18
Lamp Devices and Character Displays

Sashiro Uemura

18.1
Introduction

Many prototypes of field emission displays (FEDs) have been realized by various methods, but the principal issues are the limited display area and manufacturing cost. We have tried to overcome these issues by the new material, that is carbon nanotubes (CNTs), which were discovered in 1991 [1]. The CNT emitters have the following properties [2]:

1) a sharp tip allowing low driving voltages;
2) a high aspect ratio and high mechanical strength ensuring a long life;
3) high chemical stability providing emission stability in ambient gases;
4) high melting point ensuring high emission density.

The above-listed features suggest that the CNT emitter should be suitable for large-area FEDs. The excellent performance of CNTs as field emitters was demonstrated using high-voltage FED elements at society for information displays (SID) in 1998, where a light source tube for outdoor large-size display and diode-type flat-panel display were presented with a screen-printed nanotube cathode [3]. Figure 18.1 shows the photographs of the first experimental light source tube and diode-type flat panel.

18.2
Lamp Devices for Light Sources

18.2.1
Structure of the Lighting Element

Figure 18.2a shows the schematic structure of the lighting element presented in 1998. The CNT cathode was prepared by screen-printing technology. The cathode is covered with a grid electrode which controls emission of electrons from nanotubes. Both the cathode and the grid are fixed on a ceramic board. The distance between the top surface of nanotube cathode and the grid is about 0.1–0.2 mm. The

Carbon Nanotube and Related Field Emitters: Fundamentals and Applications. Edited by Yahachi Saito

ISBN: 978-3-527-32734-8

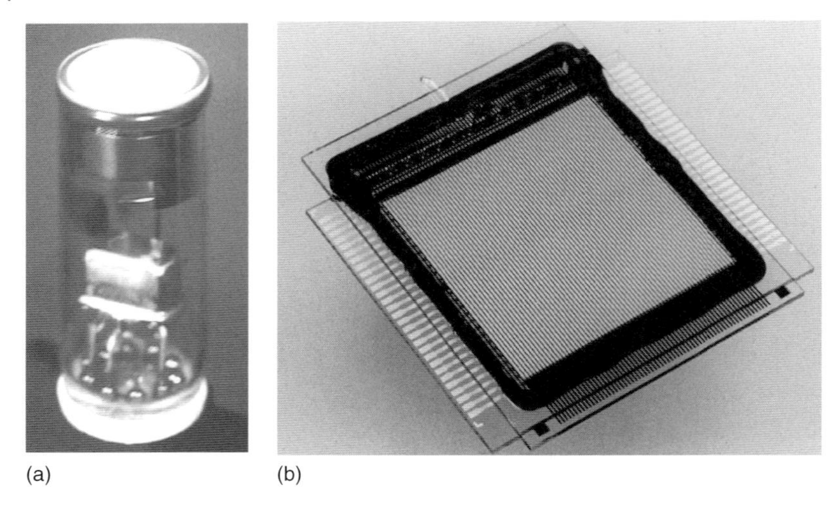

Figure 18.1 Photographs of the first CNT experimental devices: (a) Lighting element. (b) Flat-panel display. (Reprinted with permission from S. Uemura *et al.*, *J. Soc. Inf. Disp.* **11** (2003) 145. Copyright 2003, Society for Information Display.)

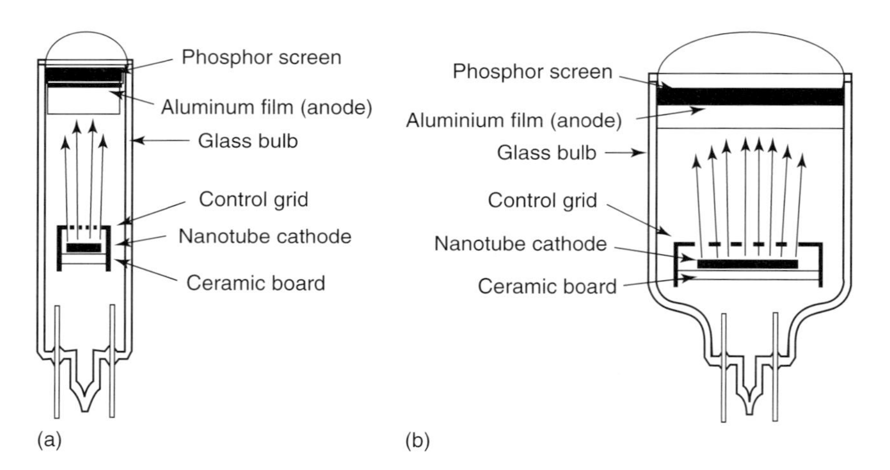

Figure 18.2 Schematic structure of a lighting element. (a) Conventional FED lighting element, CNT; about 3 mm diameter and (b) new high-luminous-flux FED lighting element, CNT; about 10 mm diameter.

phosphor screen, being the anode, is backed by a thin aluminum film (about 100 nm in thickness) to give electrical conductivity and also to reflect fluorescence in the forward direction. After sealing the vacuum tube, a getter material was flashed to attain a high vacuum of 10^{-5}–10^{-6} Pa. The cathode is grounded (0 V) and the control grid is biased to a positive voltage. The current density on the

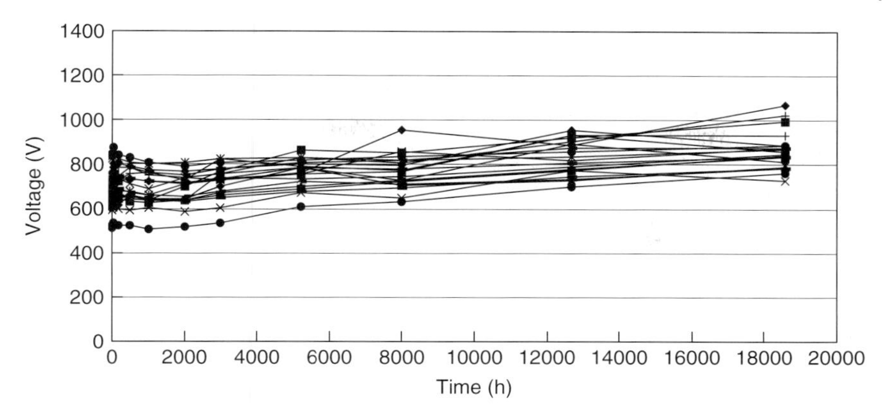

Figure 18.3 Life characteristics of the field emission from CNT emitters under constant current at DC driving condition. The emission current was 410 μA. (Reprinted with permission from H. Kurachi *et al.*, *IEEJ Trans. SM*, **127** (2007) 170. Copyright 2007, The Institute of Electrical Engineers of Japan.)

cathode surface was 100 mA^{-1} A cm^{-2} on average. The threshold electric field of emission is about 0.8 V μm^{-1} and the driving voltage is about 300 V. A high voltage is applied to the anode so as to accelerate the electrons, which excite the phosphor screen. The device has been undergoing lifetime tests under DC driving conditions (anode voltage 10 kV, emission current 410 μA). The electron emission is still continuing, with almost no diminution after about 20 000 h under constant current driving (Figure 18.3). For high-luminous-flux lighting element, we investigated a CNT emitter of large emission area to increase emission current with low driving voltage. Figure 18.2b shows the structure of the new FED lighting element, which emits higher luminous flux. The fabrication processes of the new FED lighting element are almost the same as that of the conventional FED lighting element. The outer diameter of the new device was enlarged up to 30 mm. The phosphor screen inner diameter was 27 mm. However, the lead-out structure should be the same as the conventional one. Because glass parts are used for the stem, plug, and socket of both devices, the outer diameter must be shrunk to 20 mm at the lead-out area. We can use the same conventional lead-outs for both devices. The device height is almost the same for both devices, that is, 80 mm [4].

18.2.2
Carbon Nanotube Emitter for the Lighting Element

We developed a uniform CNT emitter by chemical vapor deposition (CVD) to increase the emission current [5]. Also, we succeeded in making a large emission area with less edge emission from the periphery. Figure 18.4a shows a portion of deposited CNT on an electrode. Figure 18.4b shows the emission

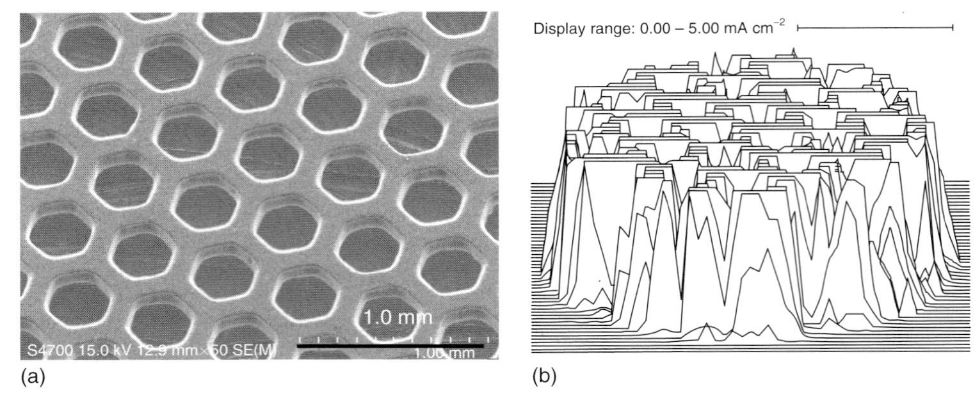

Figure 18.4 (a) A portion of deposited CNT cathode on an electrode and (b) emission distribution profiles (0–5 mA cm^{-2}). (Reprinted with permission from S. Uemura *et al.*, *J. Soc. Inf. Disp.* **11** (2003) 145. Copyright 2003, Society for Information Display.)

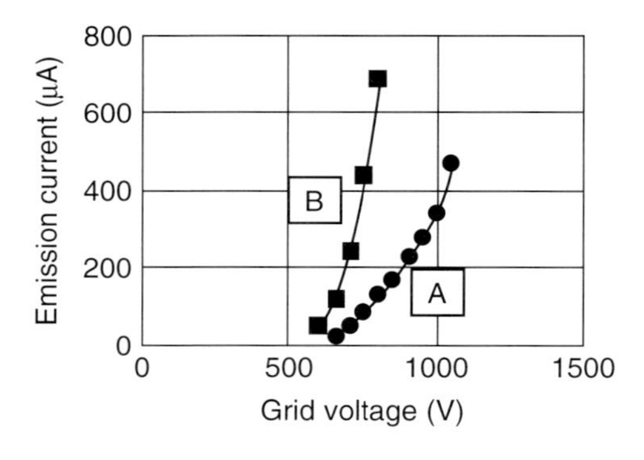

Figure 18.5 Emission current versus grid voltage characteristics under the same *j/V*. A: conventional and B: large-area cathode.

distribution profile from the deposited CNT layer on a patterned electrode, which shows a periodic pattern due to uniform emission from the electrode. The figure shows a cross section with a current density of 5 mA cm^{-2}. The CNT electrode has round edges and smooth surfaces to prevent concentration of the electrical field. The cathode electrode was formed in a special pattern to have uniform electron distribution. Another special design was introduced to prevent edge emission.

We tried to make a large and uniform cathode by the CVD method. Two kinds of cathode samples with different sizes were experimentally fabricated. The emission current from each cathode in a vacuum tube was measured. Figure 18.5 shows the characteristics of the emission current versus the applied grid voltage. Sufficient emission current was observed by the larger cathode with a low driving voltage [6].

18.2.3
Performance of the Lighting Elements

Figure 18.6 shows the high-luminous-flux lighting element (Figure 18.6b) in comparison with the first CNT FED element (Figure 18.6a). The cathode is grounded (0 V), and the control grid is biased to a positive voltage under the phosphor electrode which is DC-biased to a high positive voltage. A 10 kV voltage is applied to the anode so as to accelerate electrons which excite the phosphor screen. Figure 18.7 shows the driving characteristics, that is, emission current versus the applied grid voltage. The anode current is approximately 60% of the total emission. The displayed colors depend on phosphor materials printed on the anode. The luminance of the phosphor screens is about 3.5×10^4 cd m^{-2} for green, about 2.2×104 cd m^{-2} for red, about 1.0×10^4 cd m^{-2} for blue, about 3.5×10^4 cd m^{-2} for white, and about 3.3×10^4 cd m^{-2} for orange. The power consumption of the device is 3–3.5 W.

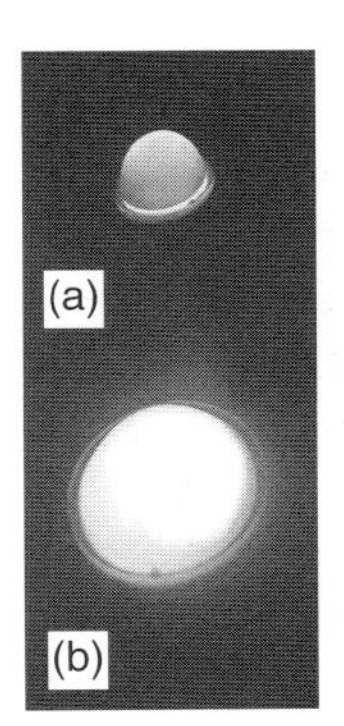

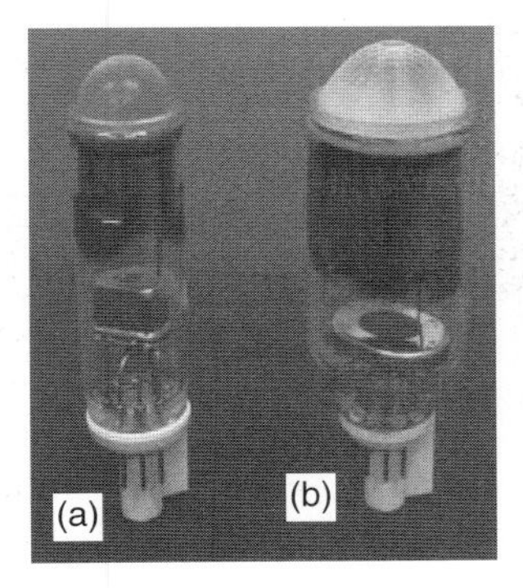

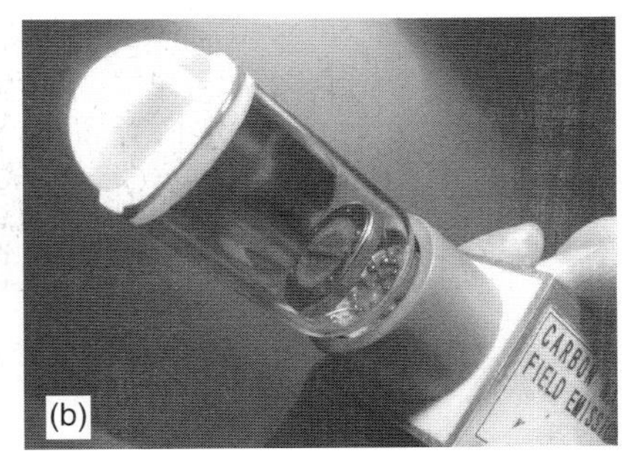

Figure 18.6 Photograph of two kinds of CNT FED lighting element. (a) conventional type, (b) large high-luminous-flux type.

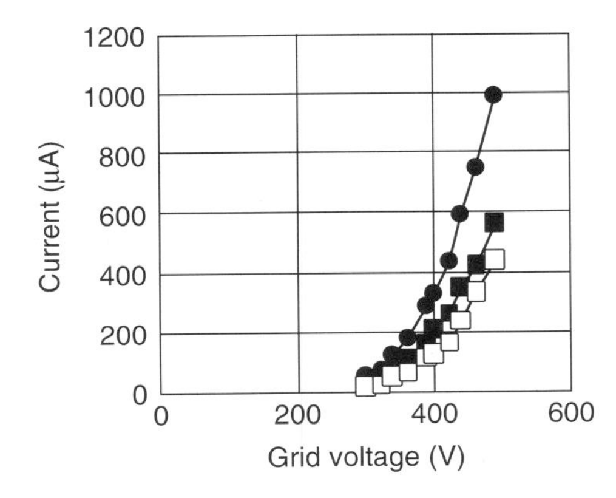

Figure 18.7 Current versus grid voltage characteristics of a large lighting element. Total emission current (●), anode current I_p(■), and grid current I_g(□).

18.3
Super-High-Luminance Light Source Device

For the super-high-luminance light source applications, special multiwall carbon nanotubes (MWNTs) were selected for the field emitters. MWNTs produced by different processes have their own characteristics as field emitters. The special MWNTs we used in the present study were named nanografibers (NGF)s. Figure 18.8 shows a TEM image of the NGF. NGFs have an extremely narrow channel at their center; the diameter of the innermost layer is only 0.5–0.7 nm. The NGF possesses the following properties favorable for field emitters: they have a smaller outer diameter and are more graphitized than other MWNTs [7].

The NGF was prepared by DC arc discharge between pure graphite electrodes in hydrogen gas. The size of purified NGF disk is about 2.0–2.5 mm in diameter and about 0.2 mm in thickness. Figure 18.9 shows a SEM image of the NGF disk surface. The NGF is almost straight and intertwined with each other. The network configuration of the NGF is well developed all over the surface of the disk. The junctions between the NGFs are connected with some nanoparticles, which may help make the surface conductivity uniform.

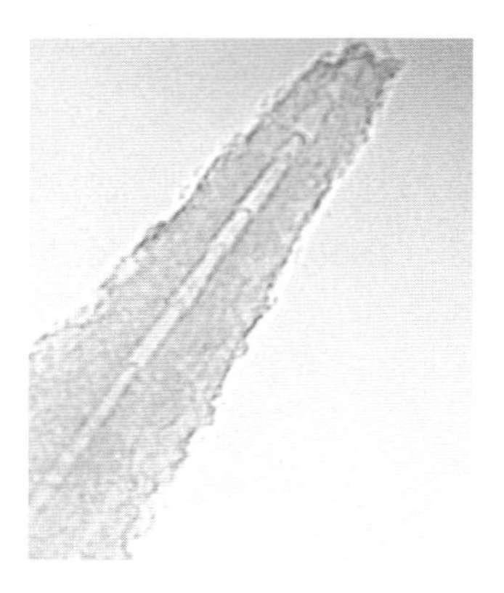

Figure 18.8 TEM image of a nanografiber.

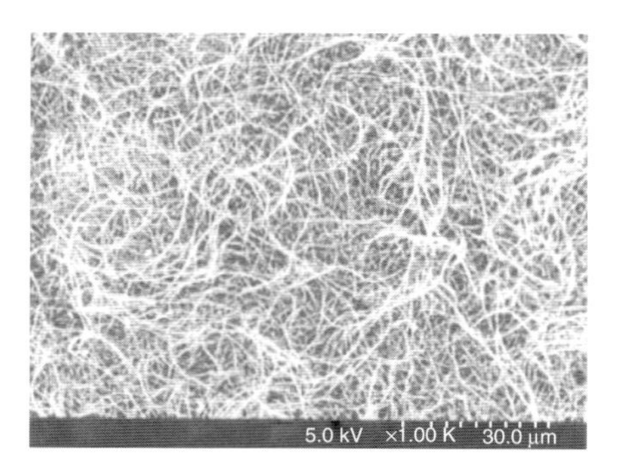

Figure 18.9 SEM image of nanografiber cathode. (Reprinted with permission from J. Yotani *et al.*, *J. Vac. Soc. Jpn*, **44** (2001) 956. Copyright 2001, Vacuum Society of Japan.)

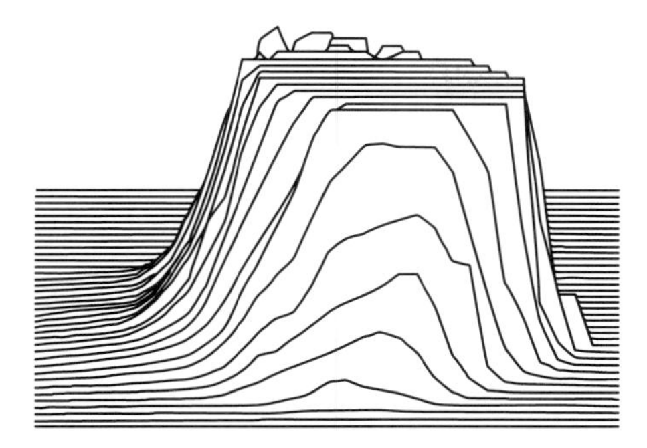

Figure 18.10 Emission distribution profiles. Display range: 0.00–100.00 mA cm^{-2}. (Reprinted with permission from J. Yotani *et al.*, *J. Vac. Soc. Jpn*, **44** (2001) 956. Copyright 2001, Vacuum Society of Japan.)

The electron emission distribution from the NGF was measured by a profiling instrument with a flat electrode. Figure 18.10 shows the emission distribution profiles. Emission profiles of this sample are leveled at 100 mA cm^{-2}. This result shows that the NGF cathode emits sufficient electrons with a high current density [8].

18.3.1 Device Structure of the Super-High-Luminance Light Source Device

Figure 18.11 shows the structure of a device with X-ray shield and cooling system. The fabrication processes of a light source are almost the same as for the vacuum fluorescent displays (VFDs).

The NGF disk as an emitter is covered with a grid electrode. The grid controls the emission of electrons from the NGF cathode. Around this control grid, one more electrode was fabricated as a shield electrode. This shield electrode protects the NGF emitter against unexpected high-voltage discharge in the production process. Moreover, the cylindrical electrode controls the electron beam to irradiate uniformly the surface of phosphor plane.

The phosphor screen, being the anode, was formed by screen printing. And the luminous screen size is about 24 mm in diameter. Colors of the emitted light depend on phosphors: that is, ZnS : Cu,Al, Y_2SiO_5 : Tb and $Y_3(Al,Ga)_5O_{12}$: Tb for green; Y_2O_3 : Eu for red; ZnS : Ag,Al for blue. Considering the luminance of white balance, luminance of about 3×10^5 cd m^{-2} is required for red and $1–2 \times 10^5$ cd m^{-2} is required for blue in comparison with a green luminance of about 1×10^6 cd m^{-2} [9].

Figure 18.12 shows the photograph of a primary color light source device. The length of the device is 100–150 mm, and the outer diameter is 36.5 mm. For X-ray shielding, the light source tube is covered with a sheath of lead glass. A cooling liquid fills the space between the vacuum tube and the sheath. The structure is plugged by an insulator bung.

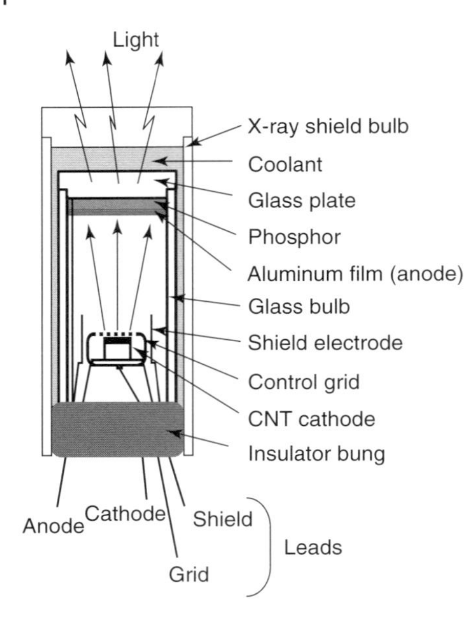

Figure 18.11 Structure of a light source device. (Reprinted with permission from J. Yotani *et al.*, *J. Vac. Soc. Jpn*, **44** (2001) 956. Copyright 2001, Vacuum Society of Japan.)

Figure 18.12 Photographs of primary color light source devices. (Reprinted with permission from J. Yotani *et al.*, *J. Vacuum Soc. Jpn*, **44** (2001) 956. Copyright 2001, Vacuum Society of Japan.)

18.3.2
Performance of the Super-High-Luminance Light Source Device

Figure 18.13 shows the total emission current from the NGF cathode, which was measured as a function of the voltage applied to the control grid. The current is divided between the anode and the grid; about 65% of total current irradiates the phosphor anode. It indicates that electron emission sufficient to realize super-high luminance is available. The grid voltage can be lowered by reducing the spacing between the grid and the cathode.

The shield electrode is grounded (0 V), and the control grid is biased to a negative voltage to spread the electron beam to the phosphor electrode, which is DC-biased to a high positive voltage (25–30 kV). The NGF cathode is driven to a higher negative voltage than the grid potential by a DC or pulse driving condition.

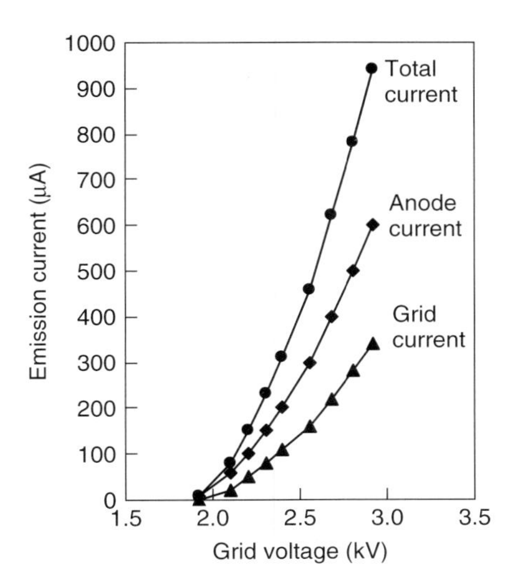

Figure 18.13 Emission current as a function of grid voltage. Grid to cathode distance: about 1.5 mm. (Reprinted with permission from J. Yotani *et al.*, *J. Vac. Soc. Jpn*, **44** (2001) 956. Copyright 2001, Vacuum Society of Japan.)

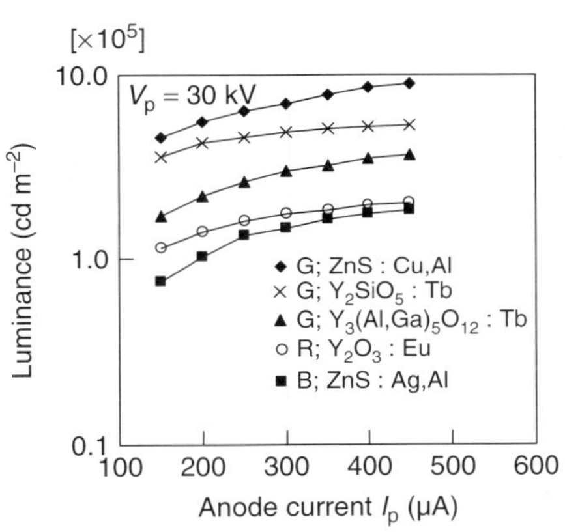

Figure 18.14 Luminance as a function of the anode current. Anode voltage V_p= 30 kV. (Reprinted with permission from J. Yotani *et al.*, *J. Vac. Soc. Jpn*, **44** (2001) 956. Copyright 2001, Vacuum Society of Japan.)

Figure 18.14 shows the luminance as a function of the anode current under an applied anode voltage of 30 kV, and Figure 18.15 shows the luminance as a function of the anode voltage under an adjusted anode current of 400 μA. The practical luminous spot size was 15–20 mm in diameter. Uniformity of the electron irradiation on the phosphor screen was not sufficiently good, though stable driving was performed under a high anode voltage. These results indicate that a target luminance of about 1×10^6 cd m^{-2} for the ZnS : Cu,Al green phosphor will be achieved at the anode voltage of about 30 kV and anode current of 400 μA. For the ZnS : Ag,Al blue phosphor and the Y_2O_3 : Eu red one, the luminance of about 1.5×10^5 and 2×10^5 cd m^{-2}, respectively, was achieved at an anode voltage of about 30 kV and anode current of 400 μA.

Figure 18.16a shows the lifetime test for the $Y_3(Al,Ga)_5O_{12}$: Tb phosphor under a constant DC of 300 μA with a driving voltage of 25 kV. Figure 18.16b shows the shift of the grid voltage during the lifetime test. Although the $Y_3(Al,Ga)_5O_{12}$: Tb

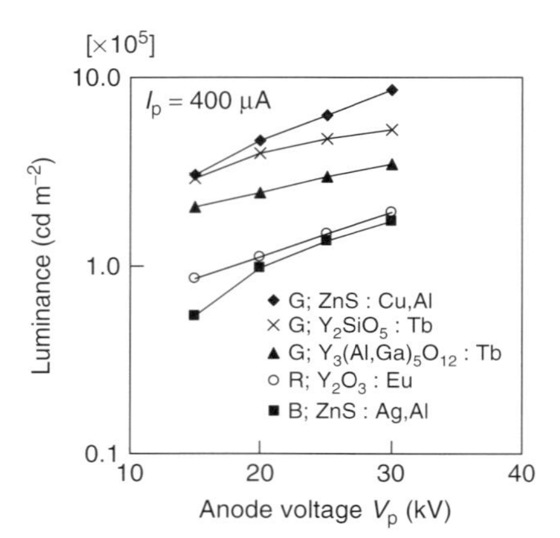

Figure 18.15 Luminance as a function of the anode voltage. Anode current I_p = 400 μA. (Reprinted with permission from J. Yotani *et al.*, *J. Vac. Soc. Jpn*, **44** (2001) 956. Copyright 2001, Vacuum Society of Japan.)

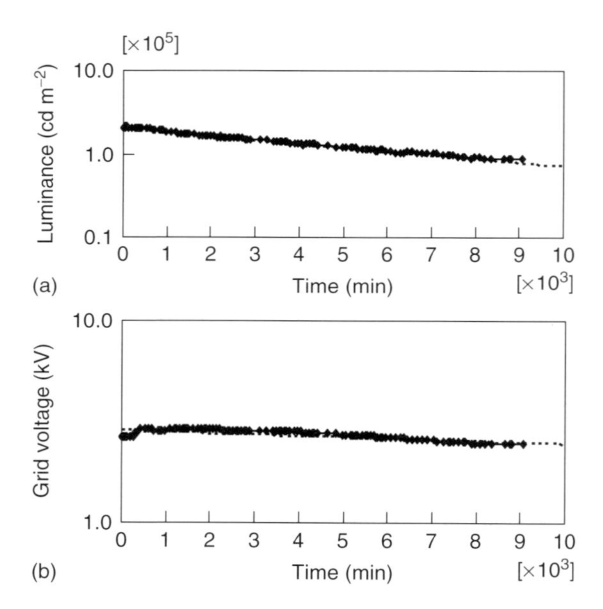

Figure 18.16 (a) Lifetime test of phosphor. (b) Lifetime test of a nanografiber emitter.

phosphor does not show high luminous efficiency, the lifetime test suggests that this phosphor is the toughest green phosphor against electron irradiation. These results indicate that the NGF cathode provides sufficient emission current during the test. On the other hand, the phosphor was degraded by heavy electron irradiation. The phosphor screen was damaged by partial concentration of the irradiating electron beam. In the case of pulse driving, it was confirmed that degradation of the phosphor screen by electron irradiation decreased compared to DC driving [10].

18.4 Summary of Lamp Devices

These results revealed that the nanoscale material (CNT) was useful as a field emitter for display devices. A high aspect ratio and small radius of curvature of their tips, together with their high chemical stability and mechanical strength, are the advantages for their use as field emitters.

The high flux was realized by enlarging a lighting area with high luminance. The outer diameter of the device was enlarged, but the cost of a device should be reduced by using the same lead-out as the conventional one. For this new device, a special bulb with different diameters was developed. Figure 18.17 shows one of the applications of large-size character displays. A uniform CNT emitter was realized by CVD technology to increase the emission current. The NGF cathodes emit sufficient electrons with higher current densities than the usual MWNT cathodes.

Long-term luminous stability limited by the phosphor material is essential for more long-time applications. Super-high luminance with long life will be achieved by the improvement of following techniques: (i) reduction of glass browning, (ii) more uniform electron irradiation to the phosphor screen, (iii) high-voltage and low-electron-density driving, (iv) higher density phosphor screen preparation, and (v) tough phosphor material.

The CNT FED has merits of high luminance and low power consumption for large-screen-size flat-panel light sources, flat-panel displays, and various types of light source applications.

Figure 18.17 One of the applications of large-size character displays.

18.5
Carbon Nanotube Field Emission Displays for Low-Power Character Displays

The first experimental FEDs were manufactured with CNT emitters in 1998 [2, 3], and the CNT emitters showed excellent field emission properties because of the high chemical stability and high mechanical strength of CNTs [11]. The greatest benefit of a display using CNT technology is high luminance performance with low power consumption. So, we intend to develop a CNT FED for half-meter-sized character displays, which will be used for message displays with color. One of the applications of CNT FED character displays is message display in vending machines. In Japan, vending machines are set up everywhere, indoors and outdoors. As they can be connected by ubiquitous communication technology, the information board in a vending machine is expected to display important messages to allow evacuation of people from a disaster area in an emergency. Usually, in an emergency the electric power lines would be shut down to prevent fire. So, the display should be battery driven. Figures 18.18 and 18.19 show the luminance characteristics and power consumption of the CNT FED compared with those of other character displays, respectively. In this work, we investigated a middle-size color character display with high luminance with low power consumption. The size was required to fit the vending machine, which could not be realized by light emitting diodes (LEDs).

This feature will also significantly contribute to energy conservation when applied to ubiquitous displays. In addition, the high-definition capability of a CNT display can provide good visibility when installed in outdoor locations, enabling public use. The development and practical application of this new technology is expected to benefit an advanced information society.

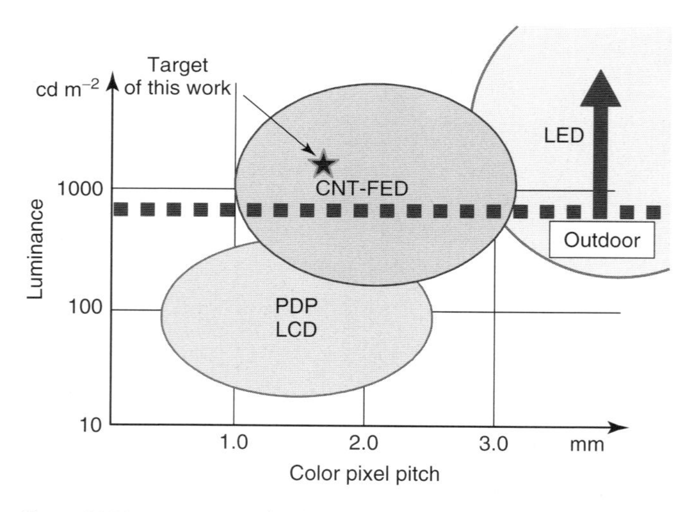

Figure 18.18 Luminance characteristics of CNT FED compared with other character displays. (Reprinted with permission from S. Uemura *et al.*, *J. Soc. Inf. Disp.* **17** (2009), 361. Copyright 2009, Society for Information Display.)

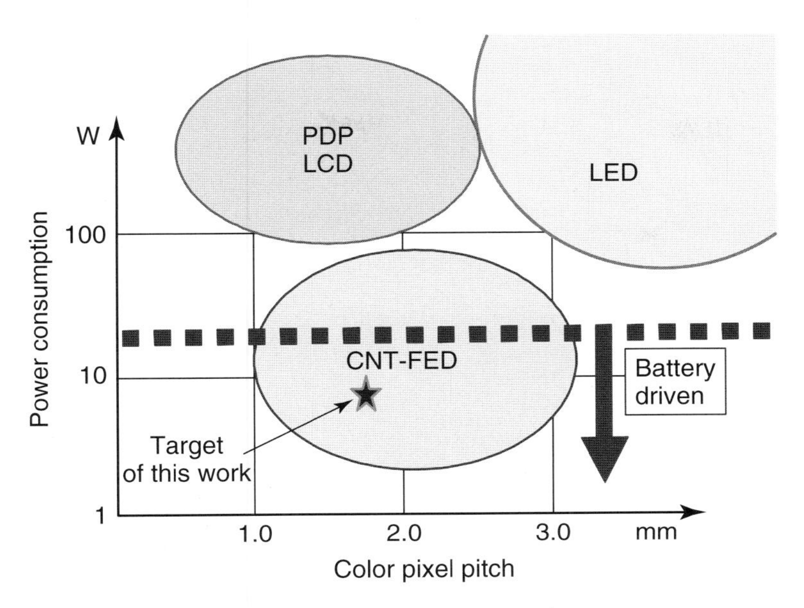

Figure 18.19 Power consumption of CNT FED compared with other character displays. (Reprinted with permission from S. Uemura *et al.*, *J. Soc. Inf. Disp.* **17** (2009), 361. Copyright 2009, Society for Information Display.)

18.5.1
Panel Structure and Rib Design

Figure 18.20 shows the schematic structure of the CNT FED. In order to realize high luminance with long-life characteristics, a high anode voltage is required. We intended to increase the anode voltage up to 7 kV, which required 2-mm-tall spacers to keep the distance between the anode and gate electrodes. In 2007, we presented a new structure with a practical process technology to form 2-mm-tall ribs [12]. The structure was designed to apply the new low-cost process technology [13].

The 2-mm-tall ribs as spacers were formed on a sheet of metal that acted as the electric-potential-control electrode. Insulating ribs were also formed on the reverse side of the metal sheet for supporting the metal gate electrodes. The anode substrate needed only very shallow ribs that touched the tops of the tall ribs on the gate-substrate ribs. The panel structure can be constructed by low-cost processes.

The anode electrode, which is composed of 256×3 lines of red, green, and blue (RGB) color phosphors backed with a thin aluminum film, is operated at a high positive potential.

The phosphor screen, 10–15 μm in thickness, was formed by the screen-printing method. The displayed color can be selected by the phosphors, and P22 phosphors used for the color cathode ray tubes (CRTs) were used in this panel.

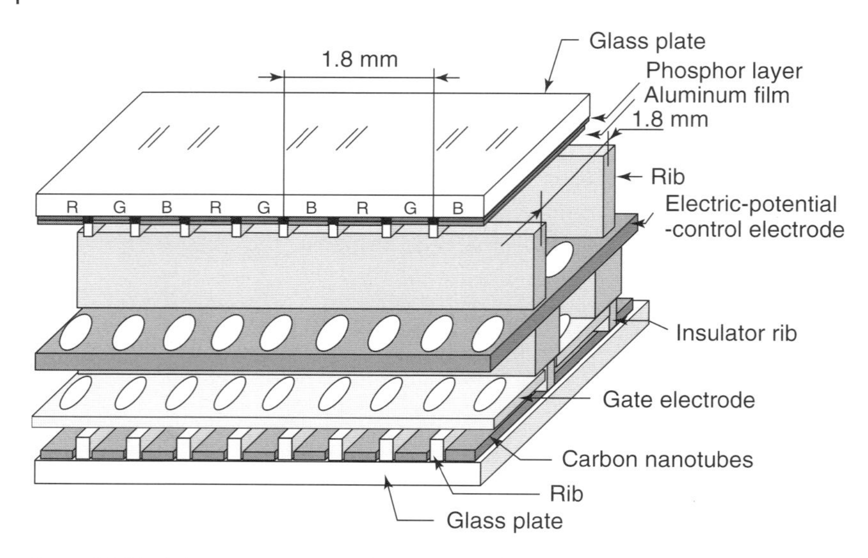

Figure 18.20 Schematic structure of a 1.8-mm-pixel-pitch CNT FED using line rib spacers. (Reprinted with permission from S. Uemura *et al.*, *J. Soc. Inf. Disp.* **17** (2009), 361. Copyright 2009, Society for Information Display.)

18.5.2 Pixel Design

Since 2003, we have been developing one half-meter-size color character displays. The phosphor line was of 1 mm pitch, and the panel had 48 × 480 dots. For the color display, the color pixel size was 3 mm × 3 mm. And the panel had 16 × 160 color pixels. In 2007, we presented a prototype device, shown in Figure 18.21, which displayed 10 Japanese characters [12].

Furthermore, we developed a high-luminance 1.8-mm-pitch CNT FED for a color character display, which had 32 × 256 color pixels. The subpixel size was 0.6 mm × 1.8 mm, which displayed 32 small or 8 large Japanese characters [14].

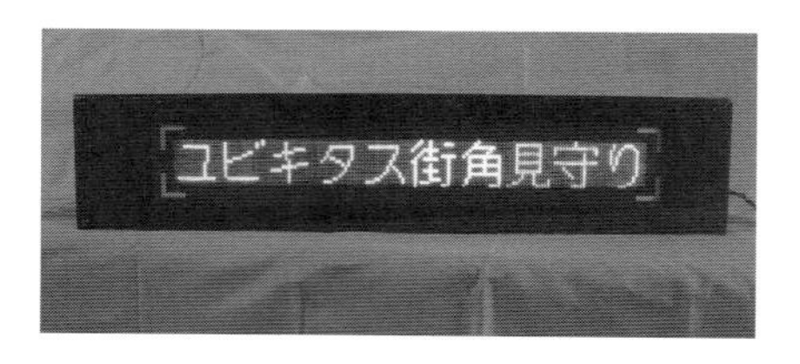

Figure 18.21 Photograph of a CNT FED character display. The color pixel size was 3 mm × 3 mm and the display area was 48 mm × 480 mm. (Reprinted with permission from S. Uemura *et al.*, *J. Soc. Inf. Disp.* **17** (2009), 361. Copyright 2009, Society for Information Display.)

18.5.3
CNT Electrode for the Display Panel

In 1998, we manufactured the first experimental devices with screen-printed CNT emitters [3]. The surface of screen-printed films, after being dried and sintered, had to be abraded to expose the CNTs buried in the films. One of the effective methods of removing surface binder materials was laser ablation, by which numerous CNT tips were exposed from the sintered layer. Since then, several surface treatment techniques over the screen-printed CNT layer have been reported. The screen-printing method was suitable for fabricating a large-area cathode, but we considered that it was not suitable for fabricating a uniform emitter. For image displays, a more uniform emission at a low applied voltage is required.

As a first trial of fabricating a uniform CNT cathode by the CVD technique, we prepared an aligned CNT layer on a metal substrate by plasma-enhanced CVD. Even though the CNTs thus prepared were vertically aligned, electron emission density and uniformity were unsatisfactory. This might have been caused by the nonuniform concentration of the electric field over individual CNTs. A more uniform electric field over the cathode surface will be formed when the surface of the CNT layer is smooth and round around the electrode edges and the surface resistivity is uniform. Figure 18.22 shows the schematic drawings illustrating the two types of surface morphologies [11].

18.5.4
Uniform Emission from the CNT Electrode

We succeeded in obtaining a surface morphology favorable for the CNT field emitter by the newly developed thermal CVD [15]. The CNTs grown on metal electrode frames by the new method formed a web-like networked structure, and the tips of the CNTs were found throughout the small-mesh network-like structure. The spatial distribution of the electron emission from the new CNT emitter was moderately uniform, indicating that a uniform electric field is formed over the CNT electrode. The technology was also applicable to a large area up to 40 in. size by arranging many metal frames on a glass substrate [11, 15].

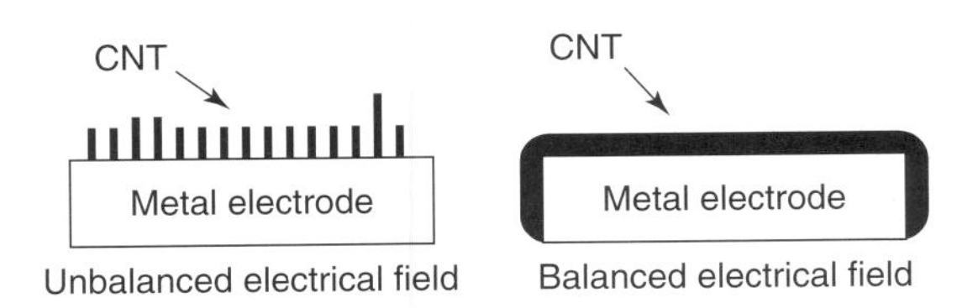

Figure 18.22 Illustration of the surface morphology of a CNT layer (two types). (Reprinted with permission from S. Uemura *et al.*, *J. Soc. Inf. Disp.* **11** (2003) 145. Copyright 2003, Society for Information Display.)

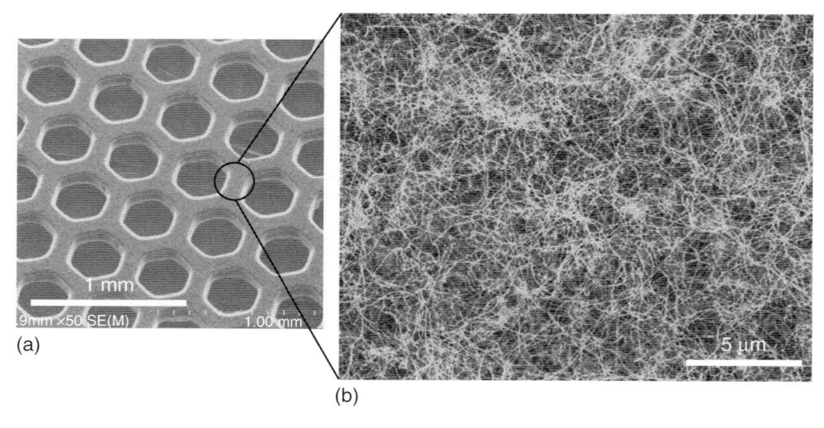

Figure 18.23 SEM images of a CNT electrode: (a) Portion of the CNT emitter and (b) surface morphology of the CNT layer. (Reprinted with permission from H. Kurachi *et al.*, *Jpn.J. Appl. Phys.* **45**, (2006) 5307. Copyright 1996, The Institute of Pure and Applied Physics.)

The CNT layer was prepared using thermal CVD equipment on a patterned plate of 426 alloy (Fe–42wt%Ni–6wt%Cr) [5]. Metal catalysts necessary for CNT growth are included in the alloy. The alloy has a special thermal property: that is, its thermal expansion coefficient coincides with that of glass which is used as the material of vacuum vessels. The tested electrode made of this alloy was 11.8 mm in diameter and 0.1 mm in thickness. An electrode of honeycomb pattern, as shown in Figure 18.23a, was realized by chemical etching. The side of the hexagonal hole was 200 μm and the width of the metal strip that remained after the etching was 100 μm.

The metal electrode was placed in the CVD chamber evacuated below 1 Pa, and 30% CO in hydrogen gas was introduced into the chamber up to a working pressure of 1 atm. The total flow rate of CO and hydrogen was 1500 sccm. The metal electrode was then heated to approximately 700 °C by halogen lamps and kept at this temperature for 30 min.

Figures 18.23a and 18.6b show the SEM images of the electrode covered with a CNT layer, revealing that the cathode has round edges and smooth surfaces. The high-magnification image (Figure 18.23b) shows that the CNTs form a networked structure. The thickness of the CNT layer was approximately 15 μm. The outer diameter of the CNTs was about 40 nm.

Field emission current versus voltage ($I-V$) characteristics and the electron emission distribution over the CNT cathode were measured using a cathode emission profiler (Tokyo Cathode Laboratory) [16]. The profiling instrument, based on an anode-hole scanning method, works under a pressure of less than 1×10^{-7} Pa. Emission characteristics were measured using a fixed anode that collected the emission current from the entire surface of the cathode. Emission profiles (two-dimensional distributions of electron emission) were obtained by

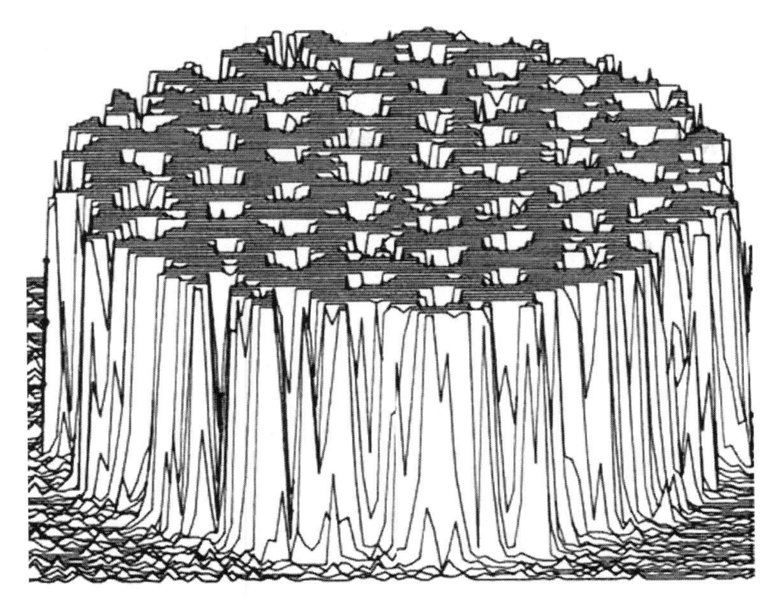

Figure 18.24 Field emission distribution (0–1 mA cm^{-2}) from honeycomb-patterned CNT electrodes. Emission measurement was performed at an anode–emitter spacing of 300 μm through an anode hole of 20 μm over a scanning area of 4 mm^2. (Reprinted with permission from S. Uemura *et al.*, *J. Soc. Inf. Disp.* **13** (2005), 727. Copyright 2005, Society for Information Display.)

scanning the anode hole (20 μm in diameter) over the emitter surface. The specimen was attached to the cathode stage, which was covered with a thin metal foil with a 4-mm-diameter hole in order to measure the central area of the metal electrode. The thickness of the metal cover was 100 μm and the spacing between the cover and the anode was 200 μm; therefore, the spacing between the anode and the surface of the CNT emitter was 300 μm.

Figure 18.24 shows the emission distribution image from the CNT layer deposited over the patterned electrode, which reveals a uniform periodic emission pattern of 1 mA cm^{-2}. The results described above indicate that the finely meshed network structure of CNTs provides a uniform electric field to the CNT edges. Figure 18.25 shows the I–V characteristic with a turn-on electric field of about 1.5 V μm^{-1}, and the spacing between electrodes was 300 μm.

18.5.5 Preparation of CNT Selectively Deposited Lead Frame

For practical applications, the lifetime of the panel is vital. The lifetime data of our CNT emitter has been published [17]. The data shows that the increase of gate voltage was only 20% after about 20 000 h under DC driving condition at about 1.5 mA cm^{-2} emission density, and so that the 1/16 duty driving gives a lifetime long enough for practical applications.

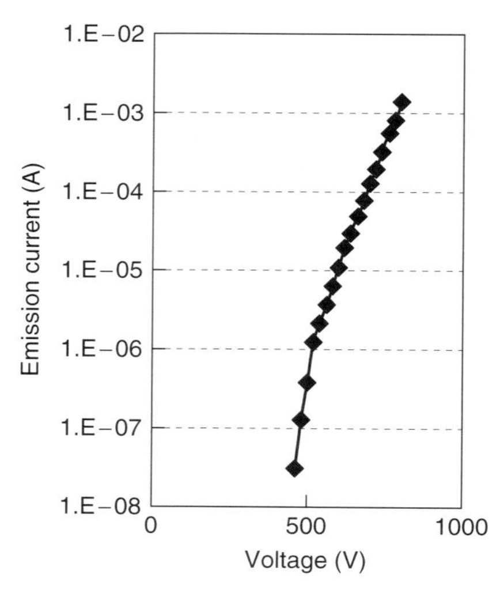

Figure 18.25 *I–V* characteristics of the cathode shown in Figure 18.25. The spacing between electrodes is 300 μm. (Reprinted with permission from S. Uemura *et al.*, *J. Soc. Inf. Disp.* **16** (2008) 273. Copyright 2008, Society for Information Display.)

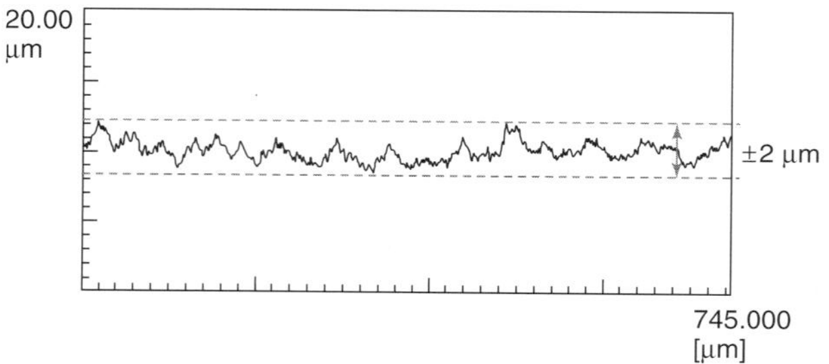

Figure 18.26 Surface flatness of the CNT electrode. (Reprinted with permission from S. Uemura *et al.*, *J. Soc. Inf. Disp.* **16** (2008) 273. Copyright 2008, Society for Information Display.)

The CNT layer is required to have the desired surface morphology, that is, round edges and a smooth surface, for obtaining uniform emission. It is also important to produce CNT emitters with good uniformity over a large area and high yield for manufacturing FEDs. So, we have developed the direct growth of CNT on metal electrodes of the 426 alloy (Fe −42wt%Ni −6wt%Cr) as in the case of a frame by thermal CVD. The technique has some advantages for deposition temperature and manufacturing yield by arranging many metal frames on a glass substrate. Figure 18.26 shows the surface flatness of the CNT electrode. It was within ±2 μm by the improved CVD process [18].

And also, the CNT layer could be selectively deposited on the desired area of the metal frame by applying a barrier film to prevent undesired CNT growth. Figure 18.27a shows a completed CNT-deposited metal lead frame, and Figure 18.27b and c shows partial SEM images of the CNT-deposited metal electrode. The lead wire pitch of the lead frame was 0.6 mm, which was equal to

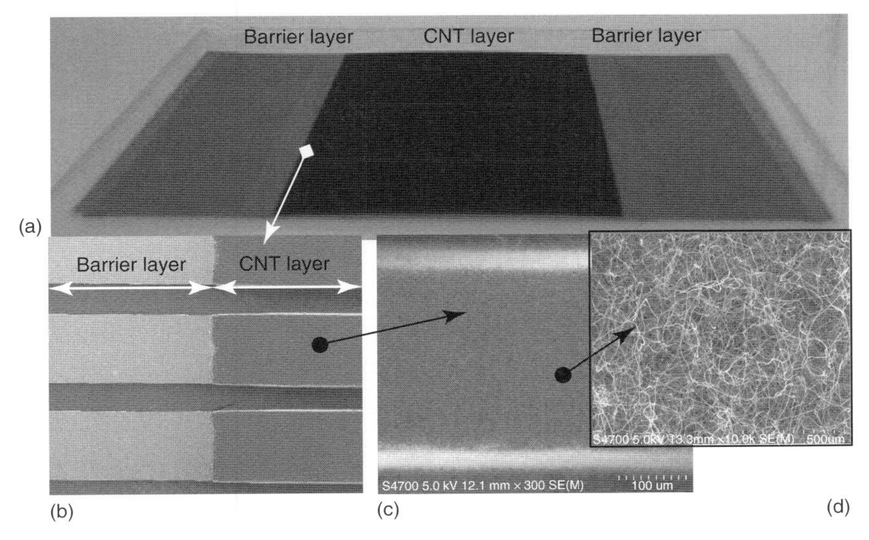

Figure 18.27 SEM images of CNT-deposited metal lead frame: (a) Photograph of the CNT-deposited metal frame, (b) SEM image – boundary of barrier area, (c) SEM image – surface of the CNT layer, and (d) SEM image – surface morphology of CNT. (Reprinted with permission from S. Uemura *et al.*, *J. Soc. Inf. Disp.* **17** (2009), 361. Copyright 2009, Society for Information Display.)

phosphor-line pitch. Each line electrode was arranged between ribs on the glass substrate.

18.5.6
Fabrication Process for the Display Panel

The fabrication process of the CNT FED is shown in Figure 18.28. The assembly for sealing process is only stacking three substrates with an accurate alignment tool.

The 2-mm-tall ribs were formed by a new process. It was carried out by only pushing the rib paste out of a multislit nozzle. After squeezing out the paste, a part of the rib lines was cured by irradiating with ultraviolet light [12].

The cathode electrodes consist of metal electrodes with CNT layers. The electrical isolation was performed by many shallow ribs on the cathode glass. Cathode lines were arranged parallel to the cathode ribs.

The CNT layer was deposited on several cathode lead frames by the newly developed CVD equipment. The cathode lead frames should be processed automatically, so we developed new equipment for mass production. Figure 18.29a and b shows a schematic and a photograph of the automatic CVD equipment. Trays were installed in a cassette; a tray was automatically set in the chamber, a CNT layer was automatically deposited on metal lead frames, and the tray was stocked in another cassette. Such CNT deposition is continuously repeated up to fill another cassette.

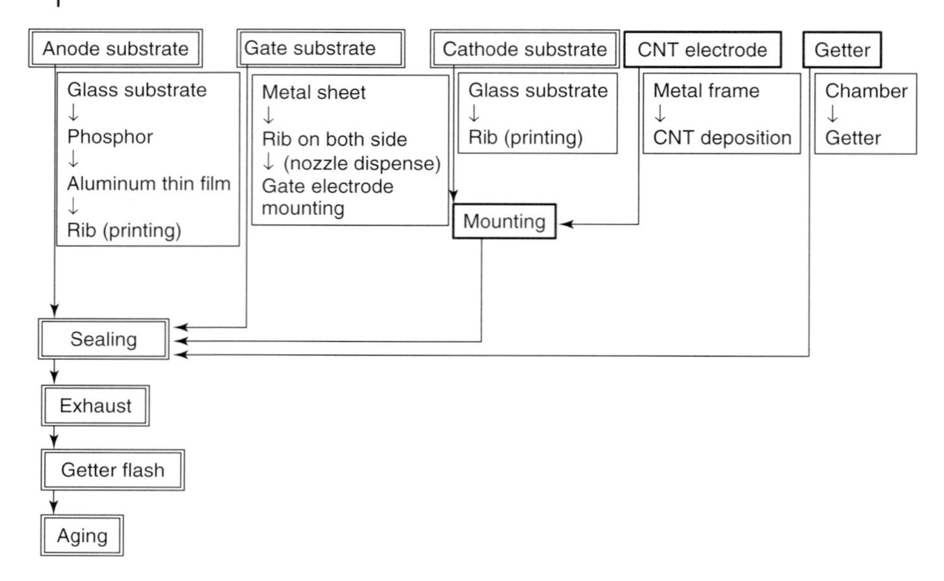

Figure 18.28 CNT FED fabrication processes. (Reprinted with permission from S. Uemura *et al.*, *J. Soc. Inf. Disp.* **17** (2009), 361. Copyright 2009, Society for Information Display.)

18.5.7
Performance of the Display Panel

Since 2003, we have been developing one half-meter-size color character displays. The phosphor line was of 1 mm pitch, and the panel had 48 × 480 dots. For color display, the color pixel size was 3 mm × 3 mm, and the panel had 16 × 160 color pixels. Figure 18.30a shows a photograph of the panels comparing the industrial model and the prototype. The outer size of industrial model was 508 mm × 76 mm and the display area was 480 mm × 48 mm, and the model had the only 14 mm frame width including the hermetic-shield area. Figure 18.30b shows a photograph of the module, which is displaying scrolled Japanese characters by color.

Besides that, we developed a high-luminance 1.8-mm-pitch CNT FED for a color character display which had 32 × 256 color pixels. The subpixel size was 0.6 mm × 1.8 mm, which displayed 32 small or 8 large Japanese characters [14]. Figure 18.31a shows the photograph of the 1.8 mm pixel pitch panel. The outer size was 85.6 mm × 488.8 mm and the display area was 57.6 mm × 460.8 mm. The device has also the compact configuration of 14 mm frame width. Figure 18.31b shows the experimental display module.

Figures 18.32 and 18.33 show the luminance characteristics of each subpixel. The luminance of green dot was about 10 000 cd m^{-2} under 1/16 duty cycle driving at 6.0 kV anode voltage. And the green luminance of another panel was about 4000 cd m^{-2} under 1/32 duty cycle driving at 6.0 kV anode voltage under an adjusted anode current density of about 600 μA cm^{-2} at peak current. The

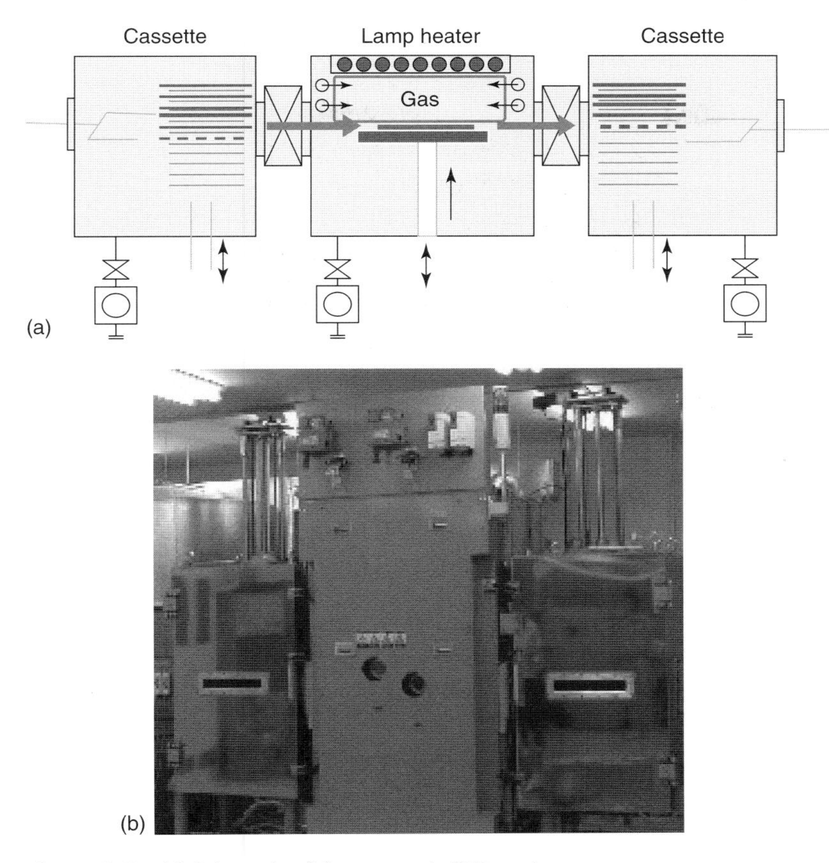

Figure 18.29 (a) Schematic of the automatic CVD equipment. (b) Photograph of the developed automatic CVD equipment. (Reprinted with permission from S. Uemura *et al.*, *J. Soc. Inf. Disp.* **17** (2009), 361. Copyright 2009, Society for Information Display.)

total power consumptions were about 4.5 and 5.5 W under 6.0 kV anode voltage, respectively. The experimental panel could be driven by a small battery.

The devices and the display characteristics were tested in various conditions from −40 °C to +85 °C. Vibration, mechanical and thermal shock, heat cycle tests were conducted. Additionally, 1000 h aging test was conducted under 85 °C, 85% RH, and −40 °C, respectively. The results showed that the device was suitable for ubiquitous display in any environment.

Table 18.1 shows the characteristics of a 1.8-mm-pixel-pitch CNT FED compared with LEDs. The features are fine pixel pitch, large viewing angle, and low power consumption. The power consumption is one order less than in the case of LEDs. Also, the operating temperature range is wide enough to use as a ubiquitous display. Lifetime estimation is under the field testing.

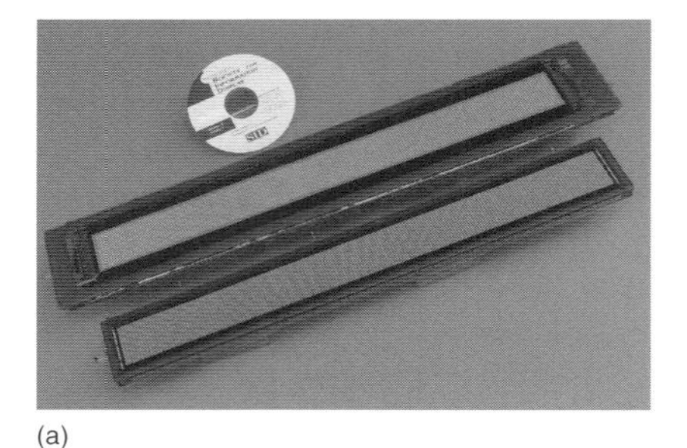

(a)

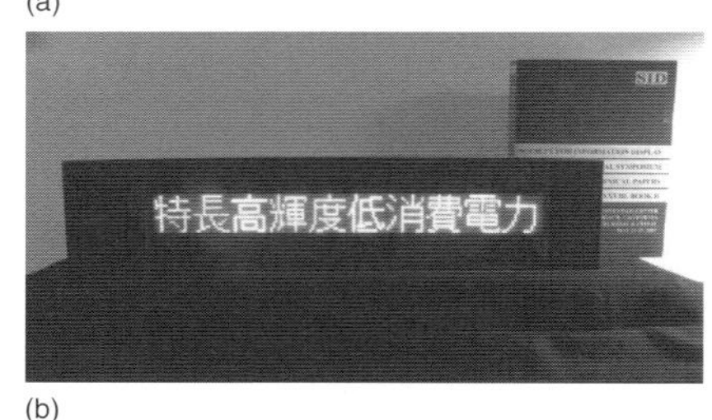

(b)

Figure 18.30 (a) Photograph of the panels. The display area is 480 mm × 48 mm. The upper device is a prototype device. The outer size was 580 mm × 112 mm. Lower device is an industrial model. The outer size was 508 mm × 76 mm. (b) Photograph of a battery driven demo display.

(a)

(b)

Figure 18.31 (a) A photograph of the panel. The outer size is 85.6 mm × 488.8 mm and the display area is 57.6 mm × 460.8 mm. (b) Photograph of a displayed color character pattern. The color pixel size was 1.8 mm × 1.8 mm. (Reprinted with permission from S. Uemura *et al.*, *J. Soc. Inf. Disp.* **17** (2009), 361. Copyright 2009, Society for Information Display.)

18.6
Summary of the Display Panel

We developed the high-luminance 3.0- and 1.8-mm-pixel-pitch CNT FEDs for color character displays, which had 16 × 160 and 32 × 256 color pixels, respectively. The developed panel structure, CNT lead frame emitter, and manufacturing processes had the advantages of size flexibility and high production yield. The greatest

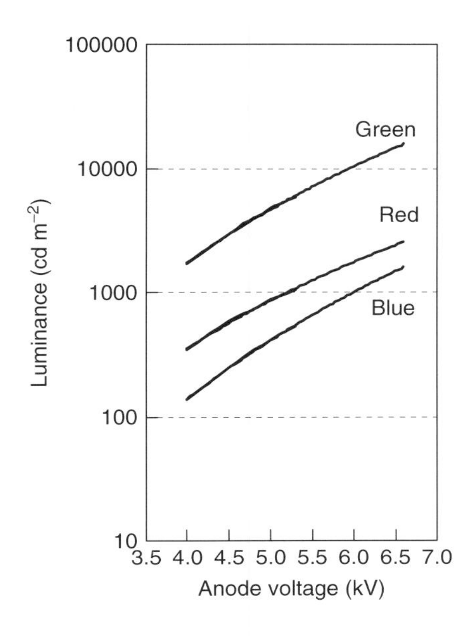

Figure 18.32 Luminance for green, red, blue dots as a function of anode voltage (duty factor 1/16). (Reprinted with permission from S. Uemura *et al.*, *J. Soc. Inf. Disp.* **16** (2008) 273. Copyright 2008, Society for Information Display.)

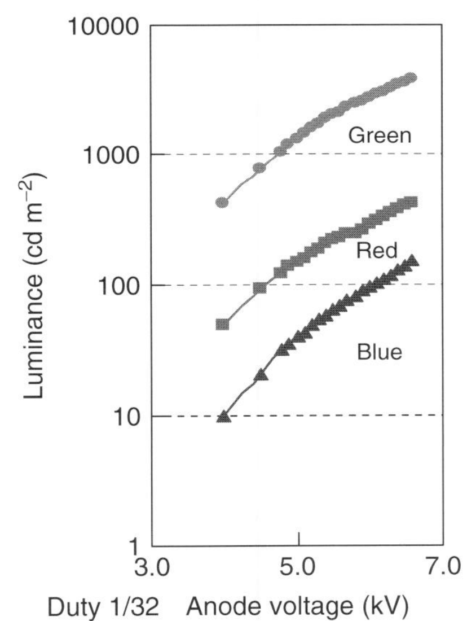

Figure 18.33 Luminance for green, red, blue dots as a function of anode voltage (duty factor 1/32). (Reprinted with permission from S. Uemura *et al.*, *J. Soc. Inf. Disp.* **17** (2009), 361. Copyright 2009, Society for Information Display.)

benefit of a display using the CNT technology is high luminance performance with low power consumption. The high luminance of a display panel can provide good visibility when installed even in outdoor locations. One of the promising applications of the CNT FED technology presented here is for public displays showing important messages for inhabitants to be evacuated from disaster areas, even under emergency no-power conditions [19].

Table 18.1 Characteristics of 1.8-mm-pixel-pitch CNT FED compared with LEDs.

Character display	CNT FED	RGB LED	RG LED
Pixel pitch (mm)	1.8	6.0	2.5
Multicolor	○	○	▲
Luminance (cd m^{-2})	About 1000	About 1500	About 1000
Viewing angle (°)	170	120	90
Power consumption per character (white) (W)	About 0.3 (about 0.6)	About 6 (about 15.0)	About 3.2 (about 8.0)
Operating temperature (°C)	−40 to +85	−20 to +60	−10 to +60
Weight per character (g)	About 35	About 100	About 100
Lifetime (h)	About 50 000 (expected)	50 000	50 000

Reprinted with permission from S. Uemura *et al.*, *J. Soc. Inf. Disp.* **17** (2009), 361. Copyright 2009, Society for Information Display.

Acknowledgments

The large-area CNT emitter by CVD technology was a result of collaboration with Dr. Hirohoko Murakami of ULVAC. The NGF emitter was the result of collaboration with Prof. Yoshinori Ando and Dr Xinlou Zhao of Meijo University. We thank the New Energy and Industrial Technology Development Organization of Japan for financial support. The research work for low-power character displays was supported by New Energy and Industrial Technology Development Organization of Japan (NEDO. We also thank to Nanotechnology and Material Technology Department of the NEDO.

References

1. Iijima, S. (1991) *Nature*, **354**, 56.
2. Saito, Y., Uemura, S., and Hamaguchi, K. (1998) *Jpn. J. Appl. Phys.*, **37**, L346.
3. Uemura, S., Nagasako, T., Yotani, J., Shimojo, T., and Saito, Y. (1998) Society for Information Display International Symposium. Digest of Technical Papers, p. 1052.
4. Ezaki, T., Uemura, S., Yotani, J., Nagasako, T., Kurachi, H., Yamada, H., Maesoba, T., Tatsuda, K., Seko, Y., and Saito, Y. (2001) Asia Display/IDW'01 (The 20th International Display Research Conference in Conjunction with The 8th International Display Workshops), p. 1245.
5. Uemura, S., Yotani, J., Nagasako, T., Kurachi, H., Yamada, H., Murakami, H., Hirakawa, M., and Saito, Y. (2000) Conference Record of the 20th International Display Research Conference, p. 398.
6. Uemura, S. (2001) Society for Information Display 2001 International Symposium. Digest of Technical Papers, p. 142.
7. Ando, Y., Zhao, X., Kataura, H., Achiba, Y., Kaneto, K., Uemura, S., and Iijima, S. (2000) *Trans. of Mater. Res. Soc. Jpn.*, **25**, 817.
8. Yotani, J., Uemura, S., Nagasako, T., Kurachi, H., Yamada, H., Ezaki, T., Saito, Y., Ando, Y., Zhao, X., and Yumura, M. (2001), Society for Information Display 2001 International

Symposium. Digest of Technical Papers, p. 312.

9. Yotani, J., Uemura, S., Nagasako, T., Kurachi, H., Yamada, H., Ezaki, T., Saito, Y., Ando, Y., Zhao, X., and Yumura, M. (2001) *J. Vac. Soc. Jpn.*, **44**, 956 (in Japanese).
10. Uemura, S., Yotani, J., Nagasako, T., Kurachi, H., Yamada, H., Ezaki, T., Maesoba, T., Nakao, T., Ito, M., Saito, Y., Ando, Y., Zhou, X., and Yumura, M. (2004) LIGHT SOURCE 2004, Proceedings of 10th International Symposium on the Science and Technology of Light Sources, pp. 125–134.
11. Uemura, S., Yotani, J., Nagasako, T., Kurachi, H., Yamada, H., Ezaki, T., Maesoba, T., Nakao, T., Saito, Y., and Yumura, M. (2003) *J. Soc. Inf. Disp.*, **11**, 145.
12. Yotani, J., Uemura, S., Nagasako, T., Kurachi, H., Ezaki, T., Maesoba, T., Nakao, T., Ito, M., Sakurai, A., Shimoda, H., Yamada, H., and Saito, Y. (2007) Society for Information Display International Symposium. Digital Technical Papers 38, p. 1301.
13. Yabe, M., Iwashima, M., and Omoto, K. (2007) Society for Information Display International Symposium. Digital Technical Papers, p. 1201.
14. Yotani, J., Uemura, S., Nagasako, T., Kurachi, H., Nakao, T., Ito, M., Sakurai, A., Shimoda, H., Ezaki, T., Fukuda, K., and Saito, Y. (2008) Society for Information Display International Symposium. Digital Technical Papers 39, p. 151.
15. Uemura, S., Yotani, J., Nagasako, T., Kurachi, H., Yamada, H., Ezaki, T., Maesoba, T., Nakao, T., Saito, Y., and Yumura, M. (2002) Society for Information Display International Symposium. Digital Technical Papers, p. 1132.
16. Kai, J., Kanai, M., Tama, M., Ijima, K., and Tawa, Y. (2001) *Jpn. J. Appl. Phys.*, **40**, 4696.
17. Kurachi, H., Uemura, S., Yotani, J., Nagasako, T., Yamada, H., Ezaki, T., Maesoba, T., Nakao, T., Ito, M., Sakurai, A., Shimoda, H., Saito, Y., and Shinohara, H. (2007) *IEEJ Trans. SM*, **127**, 170.
18. Yotani, J., Uemura, S., Nagasako, T., Kurachi, H., Ezaki, T., Maesoba, T., Nakao, T., Ito, M., Sakurai, A., Shimoda, H., Yamada, H., and Saito, Y. (2008) *J. Soc. Inf. Disp.*, **16**, 273.
19. Yotani, J., Uemura, S., Nagasako, T., Kurachi, H., Nakao, T., Ito, M., Sakurai, A., Shimoda, H., Ezaki, T., Fukuda, K., and Saito, Y. (2009) *J. Soc. Inf. Disp.*, **17**, 361–367.

19
Screen-Printed Carbon Nanotube Field Emitters for Display Applications

Yong Churl Kim, In Taek Han, and Jong Min Kim

19.1
Introduction

Preparation tools for carbon nanotube (CNT) electron emitters are central to every field of field emission (FE) devices such as field emission displays (FEDs) and lamps including backlight units (BLUs) for liquid crystal displays (LCDs). From the initial development for FEDs, two main approaches have been developed: One is the selective growth of CNT arrays on a lithographically patterned catalyst film, that is, chemical vapor deposition (CVD). The other is screen printing of a CNT paste. The CVD method typically yields fine gate hole structures of dimension 1–10 μm, leading to a dense integration of gate holes. Various advantages that emerge from the CVD approach, however, will be valuable only when they are implemented with techniques that can be scaled for large substrate sizes over 40 in. diagonal at low cost. The alternative approach using a CNT paste can offer low-cost operation and high throughputs, suitable for mass production, especially in large substrate sizes.

During the past 30 years, carbon-based electron emitter structures have been exploited intensively. Synthetic CVD diamond [1–3], amorphous carbon (a-C), or diamond-like carbon (DLC) films [4–6] were the initial candidates because of their possible (effective) negative electron affinities. Most recently, CNTs were found to be more powerful in electron FE because of their very large geometrical aspect ratio of over 1000. The first recognition of FE from CNTs was reported in the mid-1990s [7, 8], even before the superior FE properties of CNTs were generally recognized. During this time, an interest in FEDs based on arrays of the Spindt-tip was driven by its incredibly thin feature and the possibility of future displays that could be an alternative to the 100-year-old cathode ray tube (CRT). The drawbacks of the CRT are that they are bulky and fragile and the cathode must be heated up to about 1000 °C to operate the device. Tubes for vacuum electronics have, however, some advantages. They are immune to ambient fluctuations of temperature (−269 to 450 °C) and ionizing radiation, and are not destroyed by even electromagnetic pulses from nuclear explosions and external intense microwaves as originally proposed by Ken Shoulders [9].

Carbon Nanotube and Related Field Emitters: Fundamentals and Applications. Edited by Yahachi Saito

ISBN: 978-3-527-32734-8

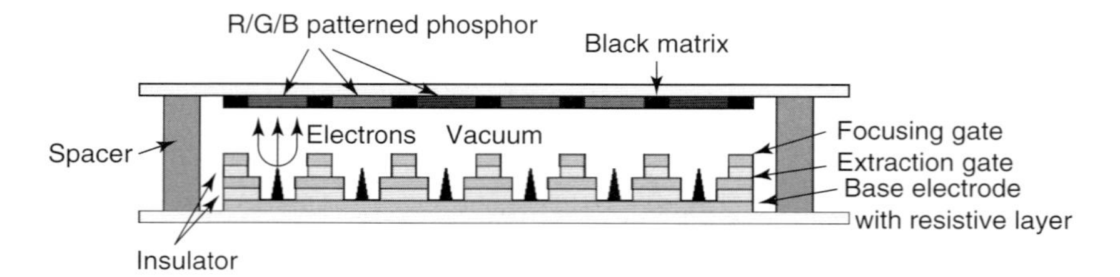

Figure 19.1 Schematic cross-sectional view of a triode field emission display based on the Spindt cathode, where the tip array and focusing gate are oversimplified for illustration. In a real FED, an ensemble of thousands of tips makes an array to correspond to a subcolor pixel, and the focusing electrode is located around the circumference of the ensemble.

Their structure is very similar to that of traditional triode tubes: The field-emitting cathodes generate electrons, whose current is controlled by a gate electrode, focused by a built-in focusing electrode, and collected at an anode as illustrated in Figure 19.1. For display applications, a few modifications are required, such as coating a phosphor layer on the anode and providing a ballast (resistive layer) in each cathode group (a color pixel). Ken Shoulders and Dudley Buck at MIT in the 1950s proposed [10] an innovative idea to fabricate integrated vacuum field effect transistors that employ thin film deposition and micromachining techniques. This device used very sharp tips, of $\sim 20-50$ nm radius of curvature, to enhance the local electric field and thus obtain field electron emission. In 1968, as a part of Shoulder's research program, the first working cathode developed by Capp Spindt had a tip of molybdenum, which produced detectable electron emission with only 20 V applied between the gate and tips, and extracted a few microamperes of current with 100 V [11, 12]. An important vacuum microelectronic parameter is its transconductance, g_{m}, defined as $(\partial I_{\mathrm{a}}/\partial V_{\mathrm{g}})$ at constant V_{a}, where I_{a}, V_{g}, and V_{a} are the anode current, gate voltage, and anode voltage, respectively. In the Spindt cathode, $g_{\mathrm{m}} = 50\ \mathrm{S\ cm^{-2}}$ is easily obtained, which is roughly $\sim$10 000 times larger than that obtained from classical thermionic triodes. Many shapes and configurations of tip arrays including wedges [13] and volcanoes [14] have been proposed. Initial effort to expand this technology into a display device was started in 1985 by a French group at LETI, led by Robert Meyer. They announced preliminary results to produce an FED based on the Spindt cathode [15]. LETI also generated the idea of a built-in resistive layer, which usually adds a resistance in series with the emitter [16, 17]. This layer converts a steep, exponentially rising current into a not-so-steep one. The series resistance can substantially enhance the inter-pixel uniformity resulting from negative feedback through voltage drops along the emitters, being directly proportional to the emission currents. On the contrary, adding a resistance can raise the required driving voltage. Alternatively, the use of a series capacitance [18] was proposed for enhanced uniformity, but this raises the driver voltage, too. Unfortunately, the higher driver voltages will increase the cost of the driver integrated circuit, which is a substantial part of the total cost of the FED module. All these technological breakthroughs for the Spindt FED took place by borrowing a technique from integrated circuit construction to fabricate small cathodes using FE to be an array electron source. Since Chris Holland of SRI reported the first full-color FED panel in 1987 [19], and following the international alliance in 1992 among the companies Pixtech, Motorola, Texas Instrument, Raytheon, and Futaba to produce a commercial FED panel, the microtip-based FED is now available on the market for small-size niche products by Futaba. An example of the process flow for the fabrication of the Spindt cathode is shown in Figure 19.2, and a scanning electron microscope (SEM) image of Spindt-tip arrays for FED manufactured by Futaba is presented in Figure 19.3a [20]. Also, a captured image of DVD player equipped with the Spindt FED which was demonstrated at the International Vacuum Microelectronics Conference (IVMC)'99 (Darmstadt, Germany) by Candescent technologies, CA, USA, is shown in Figure 19.3b.

The essential processes in the building of Spindt-tip arrays are based on electron beam lithography to form fine, submicrometer apertures and electron beam evaporation, which, directed at an angle, deposits a thin release layer over each hole. Metal (Mo or W) tips are automatically centered in the gate holes by this process. Finally, the release layer is etched away, leaving the completed cathodes and gates. In spite of its potential advantages for vacuum electronic applications, the microtip process has two critical limitations to be a successful technology for the application to a large-size display panel: it needs eight or more photomasks and an unconventionally huge evaporation facility for processing over 20 in diagonal

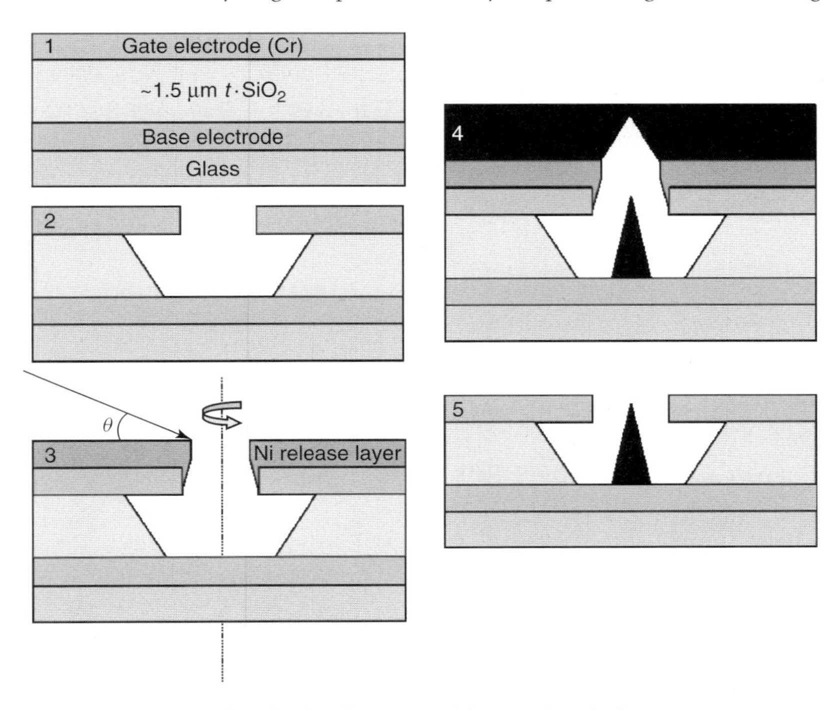

Figure 19.2 Process flow for the fabrication of the Spindt cathode.

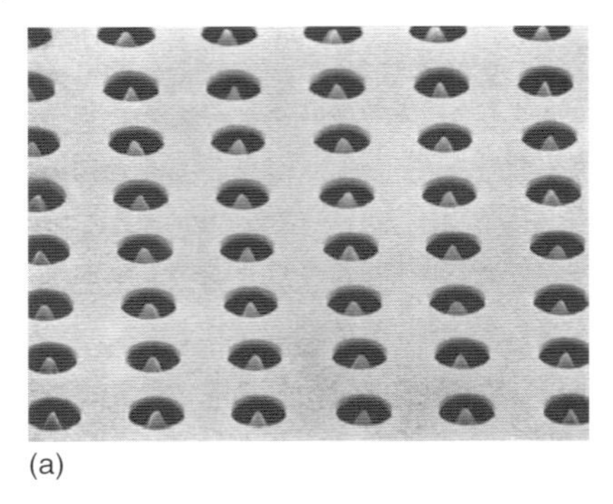

(a)

(b)

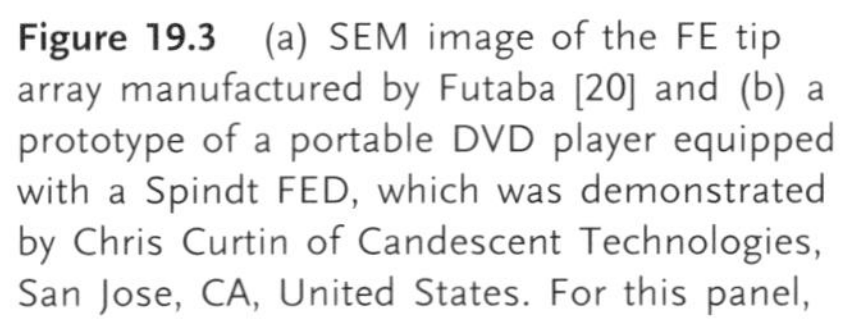

Figure 19.3 (a) SEM image of the FE tip array manufactured by Futaba [20] and (b) a prototype of a portable DVD player equipped with a Spindt FED, which was demonstrated by Chris Curtin of Candescent Technologies, San Jose, CA, United States. For this panel, 5000–10 000 Spindt tips are formed for a subcolor pixel, but only about 5% of them are working to emit electrons, which is sufficient to meet the required display specifications such as global uniformity and lifetime.

panel. To make uniform cone angles over a large panel, the distance between the metal source and the substrate must be large enough. The evaporator used at Motorola for panels as large as $370 \times 470\ \text{mm}^2$ (20 in. diagonal, Gen. II) is shown in Figure 19.4, which has a throw of about 170 cm. Cho *et al.* at Samsung Advanced Institute of Technology (SAIT) reported [21] that, for the 72 cm throw, only 0.8° angular variation in incidence of evaporant resulted in a change in the tip radius of curvature of +2 nm along the 6 cm distance in the substrate which resulted in a decrease in emission current down to ~75% for tips at the edge compared to those in the center. Such limitations, as well as the growing demand for larger displays in the market, made Motorola turn their interest to a cost-competitive CNT field emitter as an alternative to the metallic tip array.

Figure 19.4 Electron beam evaporator used to manufacture field emitter arrays available for 20 in. diagonal Spindt-type FEDs (Motorola, Tempe, AZ, United States). [39].

The basic Fower–Nordheim (FN) equation for a metallic field emitter [22] can be written as follows:

$$I = aV^2 \exp\left(\frac{-b}{V}\right)$$

$$a \sim 1.5 \times 10^{-6} \left(\frac{A}{\phi}\right) \exp\left(\frac{10.4}{\phi^{1/2}}\right) \beta^2$$

$$b \sim 6.44 \times 10^7 \frac{\phi^{3/2}}{\beta}$$

where A is the effective emission area (cm^2), ϕ is the work function of emitter surface (eV), and β is the geometric field enhancement factor in (cm^{-1}) that determines the virtual electric field E at the cathode surface such that $E = \beta V$. The major parameters that govern the FN equation are ϕ and β. In the CNT cathode, the β value plays a major role in determining the FE characteristics. To extract physically meaningful parameters from the FN fit – ln (I/V^2) versus $1/V$, especially for data obtained from randomly distributed multiple emitting tips like the surface of CNT arrays, it should be noted that the best that can be expected is an average of the tip distribution. Normally, the FN plot generates only two available parameters, namely, offset a and slope b, but three interesting physical parameters are left to be interpreted, namely, ϕ, β, and effective emission area A. Thus, it is difficult to make any meaningful estimation of the emitter system solely from the I–V data because the parameters are usually coupled with each other. Nevertheless, it is typically accepted that the electric field required for a certain current density level, for example 1 μA cm^{-2}, can be used as a figure of merit to rank the variously prepared CNT emitters. To better understand the basic FN theory, detailed accounts can be found in a review article by Gadzuk [23] and in books by Gomer [24] and by Modinos [25].

Both theoretical predictions and experimental observations have reported that CNTs possess unique physicochemical properties [26–31]. In particular, the extraordinarily high aspect ratio in geometry, coupled with their high chemical stability, makes CNTs natural electron field emitters. Despite their large work function (~5 eV) [32] associated with that of graphite, extreme concentration of electric field can be induced on the sharp CNT tips, which allows electrons to penetrate surface barrier at low macroscopic applied fields (approximately a few volts per micrometer). Although the tip ends of CNTs can be terminated with a variety of structures (open, closed, and so on) with different radii (1–50 nm), unseen large β factors can generally be obtained. Even with CNTs, it is more effective to use thinner ones for maximizing β and emission site density in the application to FED. With aspect ratios in the range of 10^2–10^3, CNT emitters present the same high order of β values. The field enhancement at the tip of a CNT is well established both in theory and experiment [33–37]. Generally, β of CNT can be approximated by $l\ r^{-1}$, where r is the radius (~nm) and l is length (~μm) of the CNT emitter. In an ensemble of CNT arrays, β depends on the spacing between neighboring emitters (d) because of the field screening effect. Nilsson *et al.* [38] showed by first-approximation calculations that β can be maximized at $d \geq 2\,l$. Under this criterion, assuming $r = 1$ nm and $l = 1$ μm, the practically attainable emitter density with minimum loss in β by field screening can be estimated as 2.5×10^7 cm^{-2}. This is obviously an overestimated value in a real situation owing to the lack of precise control in height and spacing in a CNT ensemble. Because the FE of electrons from a tip apex is governed exponentially by its β value, the overall display uniformity is obviously determined by distribution of the individual emitters formed over the whole cathode area. Theoretically, only 10% difference in the tip height leads to almost 90% difference in emission currents according to FN picture. Although the arrays of printed CNTs have been manipulated through various posttreatment methods, which remarkably increase the population of emission sites, they still pocess huge room for improvement to meet the uniformity standard for high-definition television (HDTV) application even with the assistance of a built-in integrated resistive layer (approximately tens of megohms per subcolor pixel). Various issues on FED cathodes, including MIM (metal–insulator–metal), metal–semiconductor–insulator, nanofissure using PbO film, and printable carbon emitters, are intensively reviewed in [39].

This chapter reviews FE of printed CNTs, beginning with the formulation of CNT paste focused on a photoimageable (UV-curable) paste. The surface of the printed layer is usually treated with a special technique, the so-called posttreatment, to protrude CNT tips out from the material being patterned. The choice of the posttreatment method depends on the cathode structure. The latter part of the chapter covers practical cathode architecture for FED based on printed CNT emitters.

19.2
Formulation of Photoimageable CNT Paste

CNT raw materials can be produced using various techniques, including arc discharges between two carbon electrodes, pulsed laser ablation on a carbon target containing Co/Ni catalysts, and chemical reaction from carbonaceous gas species [40–42]. CVD and paste are usually used to produce CNT-based FEDs. Along with a benefit in manufacturing cost for large-size panels, paste mixing of CNTs has key advantages over other technologies in forming field emitters such that *ex situ* prepared CNTs of high quality are available without process limitations. By contrast, the CVD method is limited by its process temperature, practically under 500 °C, for the use of a glass substrate. This section covers materials and formulation methods for the photoimageable CNT pastes. Screen-printing technique is well established in plasma panel industry to form the associated electrodes, barrier ribs, and phosphor layers. Thus, initial development skills of the CNT paste were borrowed from plasma panel technologies.

Historically, the first preparation of CNT field emitters and measurement on their FE properties were challenged in the middle of the 1990s. Rinzler *et al.* [7] observed a 100-fold increase in emission current by opening the CNT cap by either laser irradiation or annealing in an oxygen ambient. De Heer and coworkers made CNT films from nanotube suspensions from which they obtained an emission current density of over 100 mA cm^{-2}. The first CNT–organic mixtures were prepared by Wang and coworkers [43] and Collins and Zettl [44, 45] who mixed CNTs with epoxy resins. Though it was primitive, they also tried the first form of "posttreatment" on the surface of CNT–epoxy mixtures to activate the tips, which is described in the next section in detail.

The CNT paste system is usually based on acrylate binders which are applied to substrates by screen printing. The UV-curing system requires a photoinitiator (PI) in the paste formulation that starts the radical polymerization. In UV-curing, two types of PI are normally employed, which form radicals by hydrogen abstraction from a hydrogen donor (RH), for example benzophenone, or by homolytic breakage of C–P bonds as in the case of TPO (2,4,6-trimethylbenzoyldiphenyl phosphine oxide). The PIs are usually added at 1–5% concentration for the paste formulation. The photoreaction steps in these two typical PIs are listed in Figure 19.5.

The sequential reaction steps of radical chain polymerization initiated by PI are briefly summarized in Figure 19.6. The most commonly used reactive functional group is based on acrylate, which is attached to oligomer and monomer components. Trimethylolpropane triacrylate (TMPTA), pentaerythritol triacrylate (PETIA), and/or hexandiol diacrylate (HDDA) are typically used functional monomers. The monomer concentration in the UV system is normally a small percentage. Epoxy/polyester/urethane acrylates are available for the oligomeric moieties in the photoimageable CNT paste. The working viscosity, coating properties, and residual ash proportion after firing are mainly dependent on the crosslinking densities, which are controlled by functionalities of the chosen monomer and oligomer. For the alkali-development process, acid-terminated polymer binders are also added to the paste system. Such a prepared organic binder system including PI, monomer, oligomer, and binder polymers forms the photoimageable CNT paste. Meanwhile, a well-dispersed CNT solution is prepared in specific solvents such as isopropyl alcohol (IPA), terpineol, texanol, and/or butyl carbitol acetate (BCA) with a small amount of surfactant. The proportion of CNTs with respect to the organic binder system depends on the purity of the CNT powder, usually

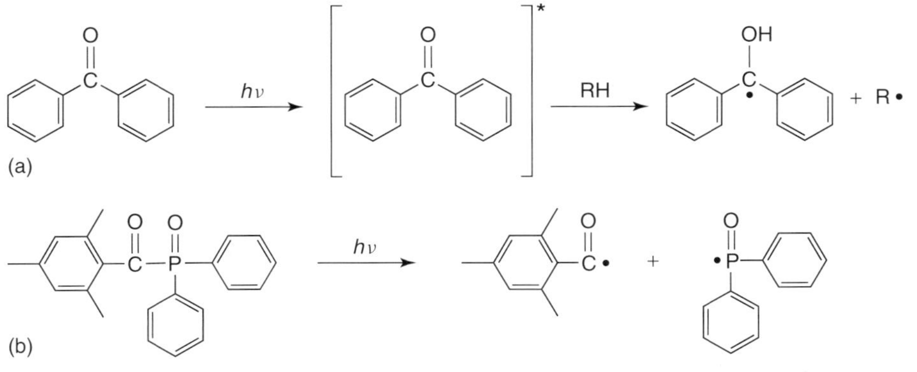

Figure 19.5 Photoreaction steps in PIs: (a) benzophenone and (b) TPO.

Initiation	Photoinitiation	PI	$\xrightarrow{h\nu}$	R•
	Chain-start	R• + M	$\longrightarrow$	R–M•
Propogation		R–M• + M	$\longrightarrow$	$R{-}M_n\bullet$
Transfer		$R{-}M_n\bullet$ + TH	$\longrightarrow$	$R{-}M_n{-}H$ + T•
Termination	Recombination	$R{-}M_n\bullet$ + T/R•	$\longrightarrow$	$R{-}M_n{-}T/R$
	Quenching	$R{-}M_n\bullet$ + Q–H	$\longrightarrow$	$R{-}M_n{-}H$ + Q
	Disproportionation	$2R{-}CH_2{-}CHX\bullet$	$\longrightarrow$	$R{-}CH_2{-}CH_2X$ + R–CH=CHX

Figure 19.6 Reaction steps of radical chain polymerization initiated by PI. Here, M denotes monomer and T/R/Q and RX represent termination/recombination/quenching agent and alkyl halide, respectively.

ranging between 1 and 10 wt%. Due to their high specific surface area, addition of too much CNT powder exceeding 10 wt% can raise some technical difficulties during paste formulation such as nonhomogeneous agglomeration among the CNTs themselves and uncontrollably high viscosity that makes the printing process troublesome.

Also, nanosized conductive fillers such as SnO_2 and In_2O_3 with smaller than 1 μm diameter are frequently added to the paste, which make up the body of the printed dot even after firing the organic components, act as spacers to distribute CNTs uniformly in the matrix and supply auxiliary conductivity in the CNT dots. The fillers in CNT paste should be carefully chosen so as not to accelerate oxidation of CNTs during the firing process. Silver and frit glasses (PbO) are unfavorable because they possibly lower the oxidation energy barrier of the CNTs, leading to fatal damage of CNTs [46, 47]. An example of the SEM images showing the effect of fillers on the survival of CNTs after firing in air ambient is given in Figure 19.7.

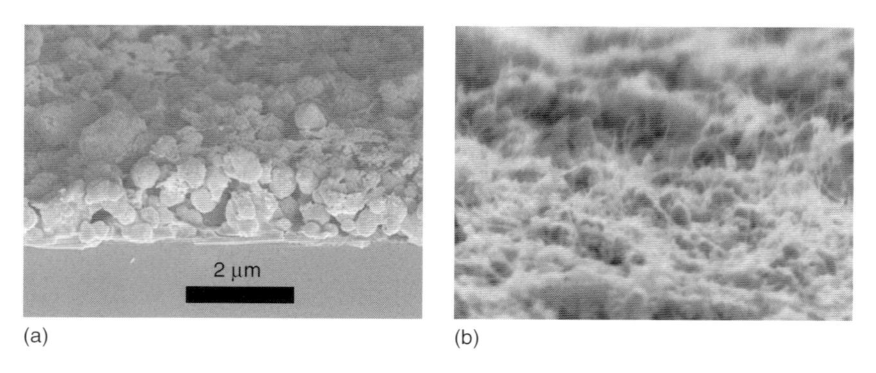

Figure 19.7 SEM images showing the effect of fillers on the CNTs after firing in air ambient: (a) Ag filler degrading the CNTs during heat process at 350 °C. (b) Many CNTs surviving in case of using SnO_2 as a filler, even after heat processing at 400 °C in air ambient.

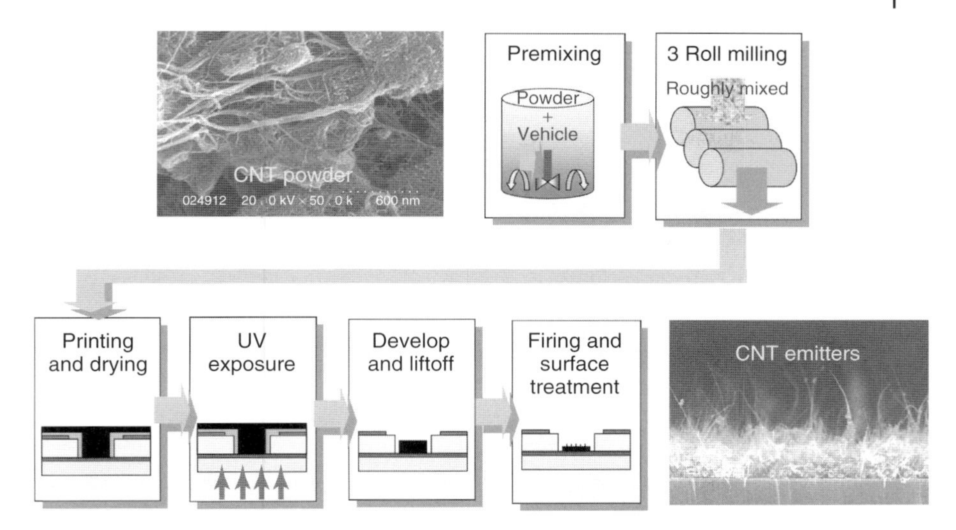

Figure 19.8 Formulation process of CNT paste and its application to the FE cathode. The device fabrication process is described in Section 19.4.

The mixture of CNTs and fillers are dispersed in the solvent, assisted by surfactants and high-speed mechanical agitation using a homogenizer or an ultrasound machine. The prepared CNT solution is then mixed with the previously prepared UV-curable organic binders, resulting in the CNT paste. An intensive three-roller milling then follows for these mixed moieties by which the volatile solvents are dried, and the extensive shear force during roll milling leads to a paste with well-dispersed CNTs. In Figure 19.8, the formulation process of CNT paste is illustrated schematically.

The viscoelastic properties (rheology) of the mixtures involving such a material with high aspect ratio like CNT has not been well established, but a couple of valuable articles [48, 49] are now available on this subject. The pioneering companies in this material, E. I. DuPont of United States and the Japanese-based Toray, are now producing their own commercial CNT pastes with single-walled, double-walled, and thin multiwalled CNTs. Alternative deposition methods for CNT field emitters can be found in various articles on the subjects of widely used CVD method [50], electrophoretic deposition [51], sol–gel process [52], and spray coating [53].

19.3 Posttreatment

After firing the CNT paste, a few CNT tips are extruded out of the printed surface, so a specific treatment of the surface is indispensable to activate the latent tips that are buried under the printed layer to be sprouted up. In the middle of the 1990s, pioneering research on FE from CNT–organic matrix was carried out by Collins and Zettl at the University of California, Berkeley [44, 45], in which multiwalled CNTs were mixed with epoxy resins. Here, the matrix surface was treated with mechanical sanding or burning out the organic materials by high voltage arcing to produce CNT emitters from the matrix, which is the first "posttreatment" technology appeared in open literature. At the end of the 1990s, SAIT in Korea reported the first phosphor image on the prototype FED based on CNT paste, in which the printed surface was scribed mechanically to expose the CNT tips from the surface [54]. So far, activation techniques using adhesive tape [55], thermal/UV-curable liquid elastomer [56], or ablation by laser [57–59] have also been developed to produce large-area FEDs. The adhesive tape detaches the top layer of the printed surface, so the internal CNTs are exposed to the surface, as shown in Figure 19.9, leading to far improved FE characteristics (Figure 19.10). This method is very simple and cost effective and

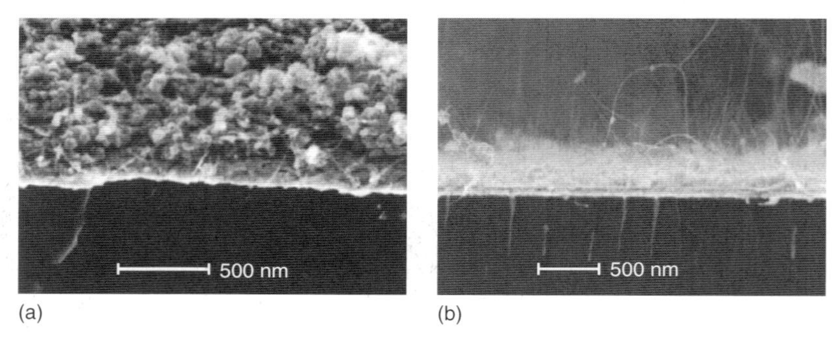

Figure 19.9 Cross-sectional SEM images of screen-printed cathode layers of SWNTs and binder material onto Si substrates: (a) as-deposited and (b) after removing the top part of the layer by an adhesive tape [55].

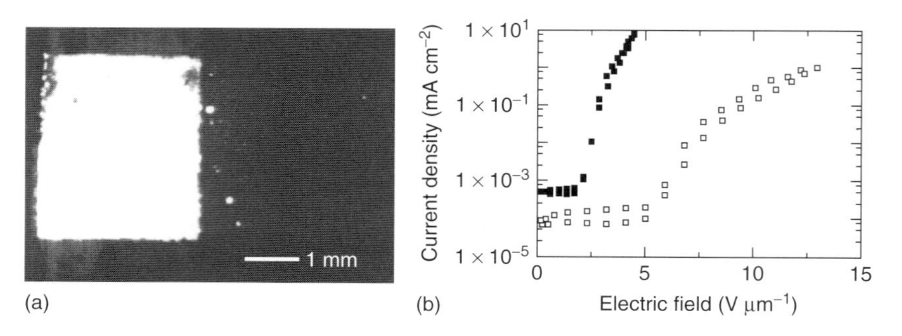

Figure 19.10 (a) Emission image of a screen-printed SWNT. The left half of the sample (4 × 3 mm^2), showing the white image, was activated by an adhesive tape, but the right half, showing the black image, was not activated. (b) *I–V* curves from the activated (solid points) and the nonactivated sample (open points) [55].

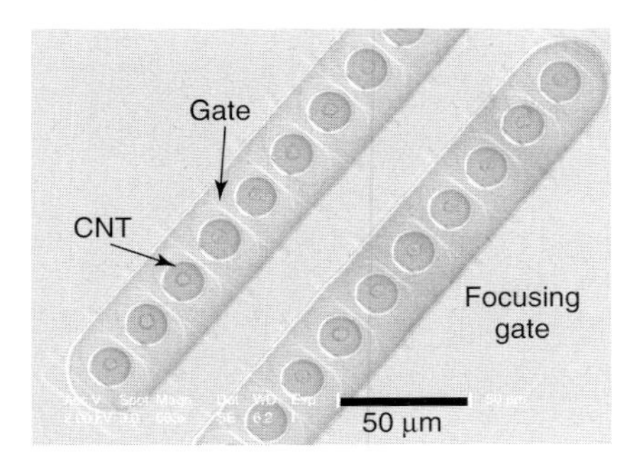

Figure 19.11 Top view of a subcolor pixel of FED cathode based on screen-printed CNT emitters, where a total of 20 CNT dots of 6 μm diameter make a pixel [60].

obviously increases the number of emission sites. The thickness of detached layer can be controlled by the thickness of adhesive glue on the tape and the applied pressure during the lamination process. One of the advantages of tape activation is that it can be applied to a cathode with a top-gate structure, normally with ~15 μm gate width and ~10 μm depth, as shown in Figure 19.11, by controlling the glue thickness and the laminating pressure. Smearing the adhesive glue onto the printed surface by pressure is the key process in this method. The exposed CNTs by taping tend to get damaged during this detaching process. It should also be noted that, in the tape process, because the adhesion strength between the CNT paste and substrate is sustained by residual organic materials after firing, sometimes the printed layer gets completely peeled off in case insufficient residues are left in the CNT layer. Therefore, the CNT paste should be carefully controlled to successfully use the tape technique.

The laser treatment is also a type of destructive method in which an intense laser beam exposes CNT tips, which is accomplished by cutting through the surface of CNTs by a UV laser scanning the printed surface. In Figure 19.12, an example of laser-treated CNT paste and its effect on the FE characteristics are presented. Here, the choice of laser wavelength, control of the laser power, and beam focusing are key parameters.

At the same time, another approach in posttreatment of CNT paste was developed which uses elastic force [46]. A thick roller covered by soft rubber is rolled under pressure over the printed surface, and a small portion of the surface layer is removed by the elastic elongation and shrinkage process of the rolling rubber. The roller should be made of very soft and elastic rubber so that the roller face can uniformly cover all over the CNT pixels. Additionally, the residual organic binders after firing should be controlled to be minimal because the elastic force involved in this process is much weaker than that of tape method. This method not only produces very little damage on the CNTs but also is more cost effective compared to the technique using the adhesive tape. This technique is powerful for FE devices with a flat surface, such as the so-called under-gate [61] or lateral-gate cathode [62]. However, it is very hard to apply to the top-gate cathode with a small aperture size like the structure shown in Figure 19.11. Also by this technique, it was shown that

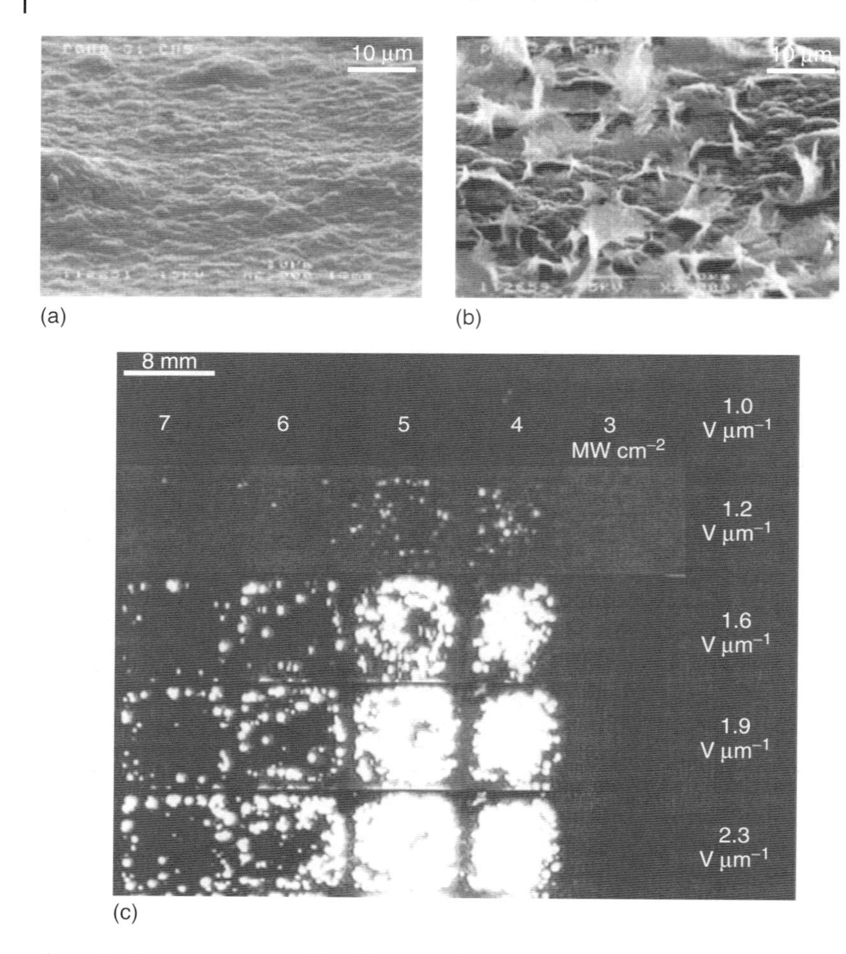

Figure 19.12 SEM images of the CNT cathode surfaces (a) before and (b) after the laser treatment [58]. (c) Emission of CNT cathodes after KrF square beam irradiation at various power densities: from right to left the spot power was 3, 4, 5, 6, and 7 MW cm^{-2} [59].

the printed CNTs can be aligned in the vertical direction with a repetition of FE [46] as shown in Figure 19.13.

Although the arrays of printed CNTs have been manipulated through various activation methods, which remarkably increase the population of the emission sites, they still have large room for improvement to meet the uniformity standard for HDTV application even with the assistance of a built-in ballast layer. This lack of uniformity (height and spacing) in the prepared CNT tips limits further improvement in CNT-based FEDs. When CNT field emitters are operated, a few

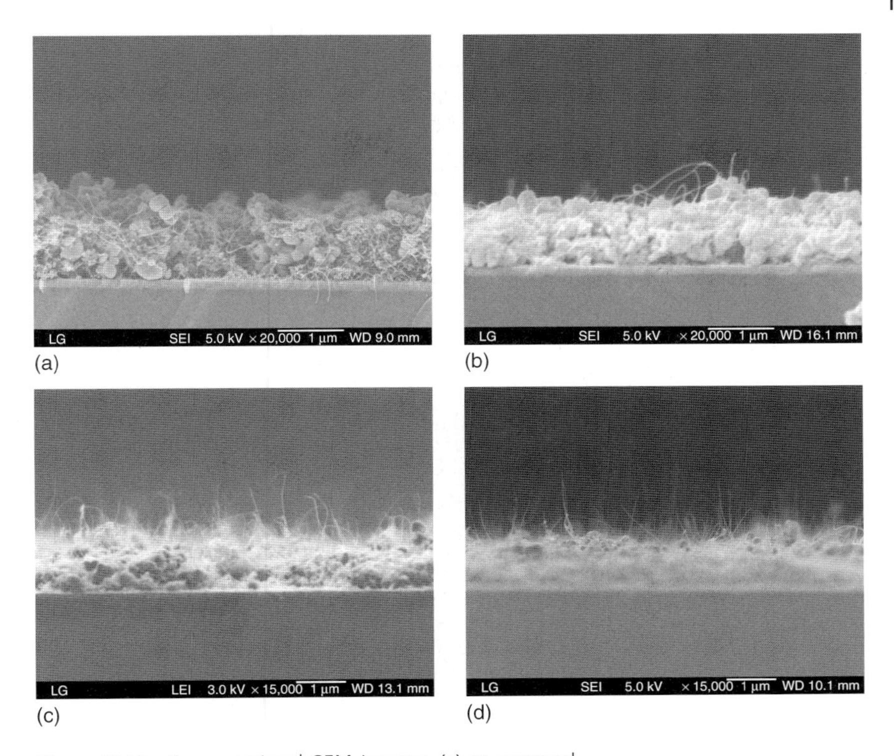

Figure 19.13 Cross-sectional SEM images: (a) as-prepared, (b) after activated by a rubber roller, (c) after multiple FE cycles, and (d) more additional FE cycles.

highly protruding tips generate most of the emission current, which causes spatial nonuniformity and fast decay in lifetime behavior.

Oxygen exposure with relatively high partial pressure ($\sim 8 \times 10^{-4}$ Torr) to the field-emitting CNTs can lead to enhanced uniformity and lifetime behavior [60]. Generally, O_2 adsorption on the CNT emitters degrades the FE properties [63–67], such as decreasing the emission currents, increasing the turn-on voltage, and eventually etching away the graphene layers from the top of the CNT tips. Based on these observations, it is expected that the oxidative etching by O_2 is accelerated when O_2 is supplied with a high partial pressure during the FE operation. In addition, this process is also predicted to occur selectively on the highly emitting CNTs with higher β among others by the assistance of heat generation (Joule heating) [68] along the length of those nanotubes. In spite of such negative effects caused by O_2, a positive role of O_2 is to be used as "scissors" to trim or shorten the highly emitting CNTs, leading to improved uniformity and emission site density. This process shows effective improvement in uniformity and emission stability as shown in Figures 19.14 and 19.15. These results are interpreted as due to the increased number of effective emission sites resulting from selective trimming/shortening

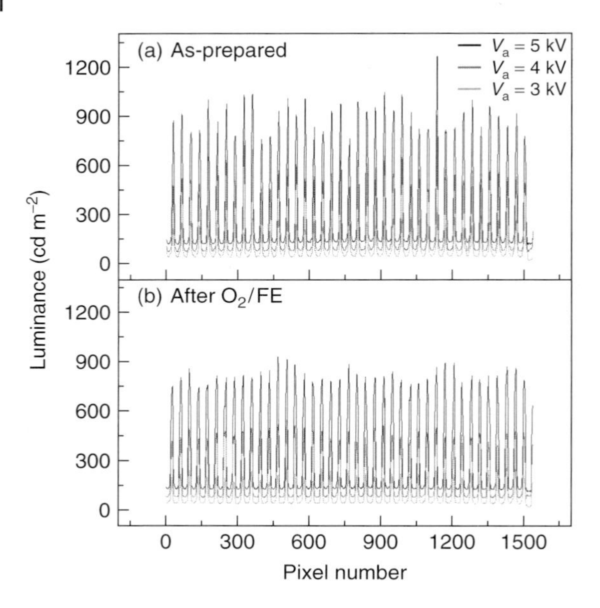

Figure 19.14 Luminance data along a scan line of a 2 in. triode FED (the cathode structure is shown in Figure 19.11). (a) As-prepared sample with the applied gate voltage of 45 V. (b) After O_2 treatment on a field-emitting cathode for 10 min. A gate voltage 50 V with 1/1000 duty for the same 65 μA of emission current was applied. Data are obtained at anode voltages of 3, 4, and 5 kV, respectively [60].

of the highly protruded CNT tips assisted by oxygen. As a result, this uniform array delocalizes the current load over an increased number of effective emitters (not being concentrated on a few elite tips) leading to long-term stability in lifetime behavior.

19.4 Field Emission Display Based on Printed CNTs

19.4.1 Cathode

The FED cathode using CNTs grown by hot-filament-assisted CVD was mainly developed by Motorola and LETI (Figure 19.16) [69, 70]. Motorola is pursuing a top-gate cathode for FEDs, wherein CVD-grown multiwalled CNTs emit electrons at low applied voltages. The rest of the process uses conventional photolithography to define gate holes, focusing gates, and the resistive layer.

On the other side, SAIT and Samsung SDI have demonstrated their prototype of FEDs based on printed CNTs. During the initial stage, prototype FEDs with diode structure using electrode edge [54, 61] and mesh gated triode structure [71] were demonstrated by SAIT. In 1999, they developed a 9-in. diagonal diode-operated FED

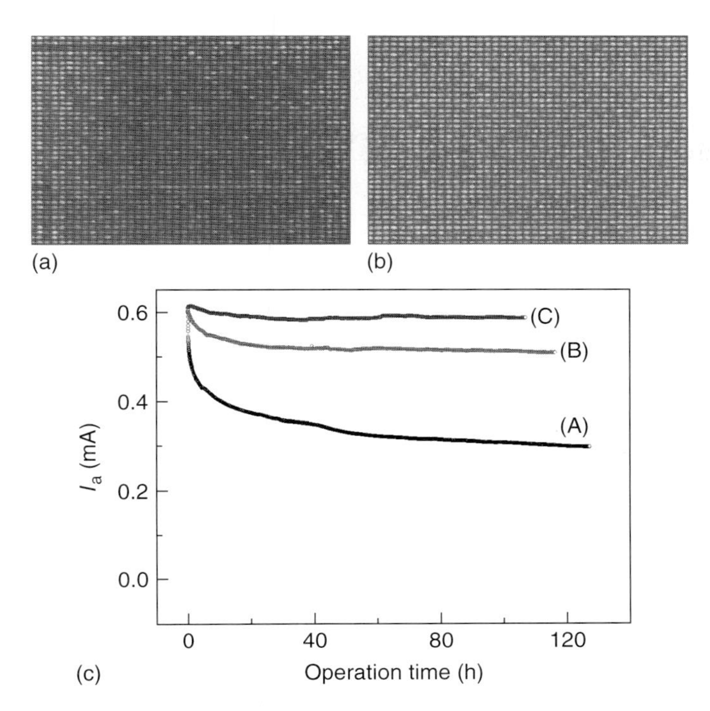

Figure 19.15 Improved emission uniformity and lifetime behavior by O_2 treatment during the FE process observed on a 2 in. triode FED: (a) the as-prepared image and (b) image after O_2 treatment during the FE process for 20 min. (c) Plots show the changes in lifetime behavior measured in a fully sealed 5 in. FED panel with degree of O_2 treatment; (A) as-prepared, (B) 10 min treated, and (C) 20 min treated cathode. For the same starting emission current, I_a (0.6 mA), the three triode cathodes are operated by applying gate voltages, $V_g = 61$, 64, and 66 V, respectively [60].

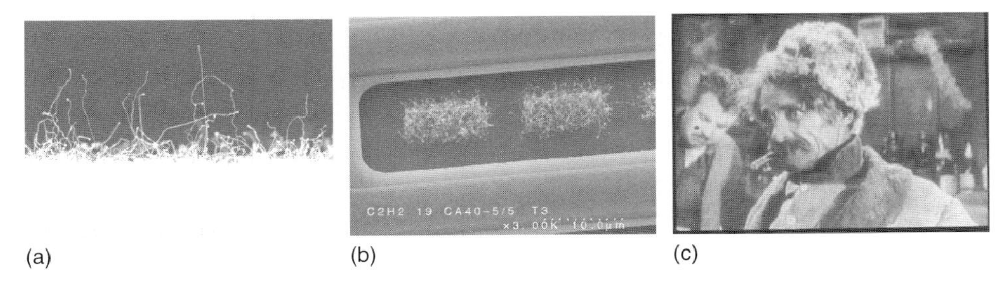

Figure 19.16 (a) SEM image of CVD-grown CNT emitters in LETI. (b) An image showing CNT arrays for FED cathode with gate structure. (c) A captured image from a 6 in. monochrome QVGA CNT FED of LETI, driven at an anode voltage of 2.5 kV, in which 85 V was applied to the gate, and the switching voltage was 40 V [70].

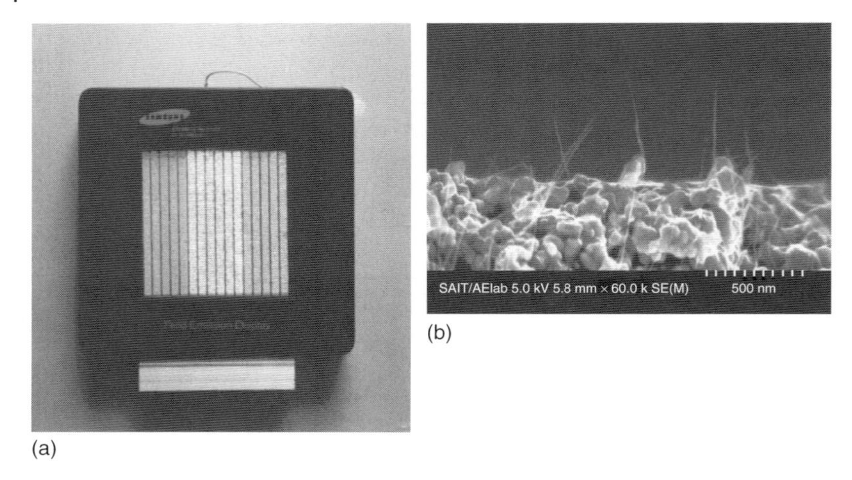

(a)

(b)

Figure 19.17 (a) Emission image of fully sealed SWNT FED in the color mode with red, green, and blue phosphor columns. (b) An SEM image of the single-walled CNT emitters used in the left device [54].

and demonstrated color images with fairly good uniformity (Figure 19.17). After this period of feasibility test on printed CNTs, they focused on triode prototypes operated with an under-gate cathode [61, 72] for the next generation (Figure 19.18). At the same time, Japan-based ISE electronics and Mie University reported a prototype FE lamp using printed multiwalled CNTs [73, 74] which is operated with a mesh gate (Figure 19.19).

The under-gate cathode does not need the conventional gate well structure. The electric fields are concentrated on the edges of CNT dots which are aligned along the cathode electrode as shown in Figure 19.18c [61]. The under-gate cathode is easily extendable to large panels with low cost and may offer a significant reduction in manufacturing cost and capital investment over Spindt-based displays, provided the display specifications can be met. After an intensive effort to overcome a few difficulties to meet the TV standard in this structure up to 38 in. diagonal panel (Figure 19.18d), Samsung turned their interest to top-gate structures based on thin/thick film process, mainly because of the uniformity issue. The top-gate cathode structure based on printed CNTs is shown in Figure 19.11, and images from a prototype top-gate FED produced by Samsung is presented in Figure 19.20. They obtained over 95% average pixel uniformity (a-PU) and 88% peak pixel uniformity (p-PU), which are high enough levels for a commercial display. To evaluate uniformity, a statistical estimation is usually conducted for the luminance data observed by a spectrophotometer, which allows analysis of individual luminance from every single pixel (Figure 19.21). The 6×6 adjacent pixels are selected as one group, and then five groups are randomly chosen to make a set. Two definitions of uniformity are used as follows: $(\text{a-PU}, \%) = [1 - (\sigma/\mu)] \times 100$ and

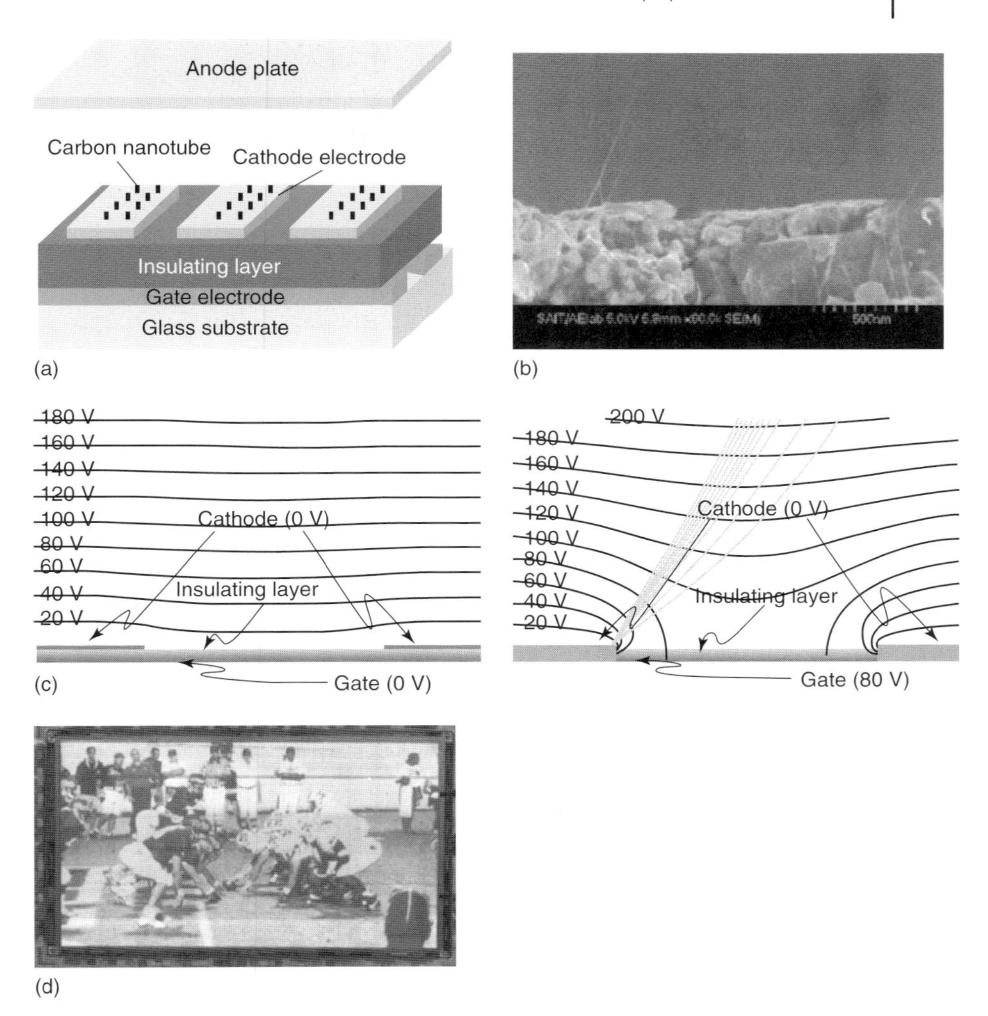

Figure 19.18 (a) Schematic diagram of an under-gate structure for FEDs with carbon nanotube emitters. (b) A cross-sectional SEM image of printed CNTs. (c) Electric potential distributions and trajectories of the emitted electrons in the under-gate structure for gate voltages of (left) 0 and (right) 80 V. Here, the gate electrodes are located under the cathode electrodes, separated by a thick insulator film [61]. (d) A demonstrated image of a 38 in. lateral-gate FED manufactured by Samsung (2003).

(p-PU, %) = $(L_{\min}/L_{\max}) \times 100$, where σ is standard deviation, μ is numerical average, and $L_{\min}/L_{\max}$ is the minimum/maximum luminance, respectively.

On the other hand, in Samsung, effort on CVD deposition of CNTs for FEDs with full HD resolution was also made to overcome the limitations of screen-printed emitters. Using the CNT paste, integration of gate holes of over 20 per color pixel is limited by the process margin in 40 in. panel having full HD resolution. Plasma-enhanced thermal CVD method was used for low-temperature (420 °C)

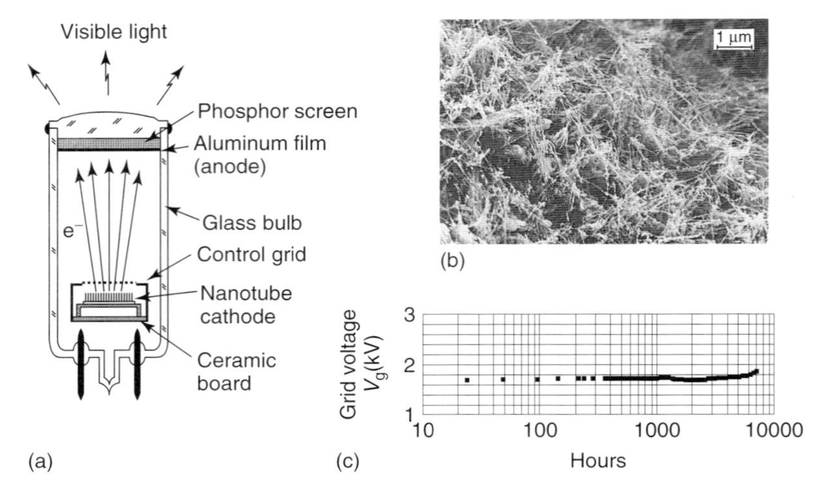

Figure 19.19 (a) CRT fluorescent display created by ISE electronics with a field emission cathode made of CNTs. (b) SEM image of printed CNTs in this device. (c) Lifetime behavior showing ∼10 000 hours, measured in this device [73].

Figure 19.20 (a) Fifteen inch diagonal FED panel based on top-gated CNT emitters produced by Samsung. (b) An enlarged image of color pixels in the top-gate FED showing excellent uniformity. The color gamut is 93% by the NTSC standard, 10 kV of anode bias, and gate and data switching voltages are 80 and 26 V, respectively.

CNT growth [75], and the cathode structure and emission image are shown in Figure 19.22, where they succeeded in achieving well-centered CNT dots in the gate holes of 4 μm diameter. A total of 108 holes were formed for a color pixel, which can be compared to the number 20 of the screen-printed CNT FED.

For a decade, noteworthy advancements in CNT-based FEDs have been achieved. Many challenging issues including lifetime and uniformity also have been successfully resolved. Nevertheless, the high capital investment and manufacturing cost for FED manufacturing create significant risk for this device entering major display

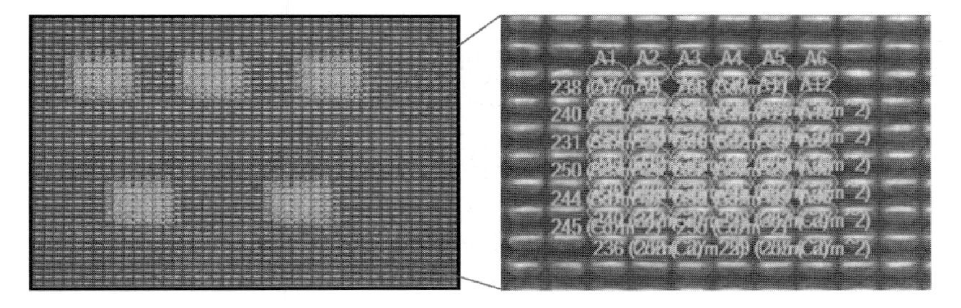

Figure 19.21 Example of uniformity measurement using spectrophotometer, where five randomly chosen groups of 6 × 6 adjacent pixels yield a measurement set.

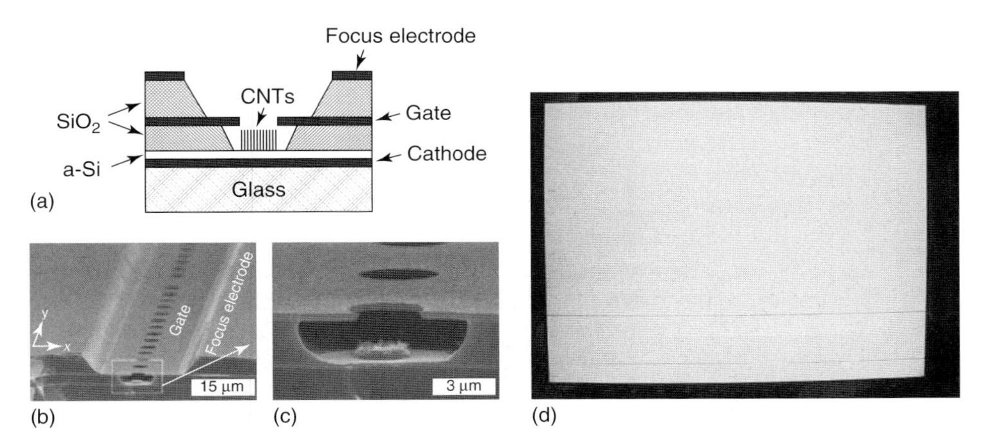

Figure 19.22 (a) Top-gate CNT FED with focusing electrode developed by Samsung. (b) SEM image with low magnification for this device. (c) A magnified SEM image showing gate diameter of 4 μm with well-centered, selectively grown CNTs. Here, a total of 108 gate holes make a color pixel. (d) An emission image on phosphor from 4.75 in. top-gate FED, where the emission current was adjusted to 0.2 mA by applying gate voltage 68.5 V, anode bias 6 kV, and focusing bias 0 V [75].

markets. Moreover, recent rapid reduction in cost and significant improvement in display quality of plasma panel and LCD TV make companies hesitate to invest in FEDs. In the future, even better display performance with much lower cost of production of this device will make the decision in the display market of FED inclusive of CNT cathode.

19.4.2 Anode: Phosphors and Phosphor Plate

Since the early stages of FED development, standard CRT phosphors have been employed with minor modifications. The conventional CRT phosphors show their optimum performance in the operation anode voltage range of 20–30 kV.

However, due to the smaller thickness of about 2–3 mm of the FED, much lower anode voltages of about 10 kV are desired to prevent flashovers and arcing events that can occur at such high voltages. The stable operation over 30 000 h of the FED produced by SED, operated at 10 kV with 2 mm gap between cathode and anode plates, has been announced in consumer electronics exhibitions since 2004. They used the conventional P22 CRT phosphors. Prior to this technological achievement by Canon, Candescent's 8-kV operation of the Spindt FED was the highest voltage ever.

The efficacy of cathodoluminescent phosphors, η, can be defined as [76], $\eta = \pi LA/P$ (lm W^{-1}), where L is the measured luminance (cd m^{-2}), A is the electron bombarding area (m^2), and P is the operation power in watts by multiplying the anode voltage by the emission current. The constant π appears from the solid escape angle (corresponding to $\sim 120^{\circ}$ in radian), under the assumption of spherical arrays of phosphor particles. The difference in dimensions of the both sides in the equation (lm vs. cd) comes by the definition that a point light source with 1 cd generates a luminous flux 4π lm all over the solid angle ($4\pi \sim 360^{\circ}$).

The screen luminous efficacy in FED is typically lower than that of the intrinsic phosphor efficacy obtained from the phosphor powders because of the residual organic binders that come from the preparation slurry. The residues can not only absorb parts of the electron dose but also degrade phosphors by chemical reaction. The screen efficacy in FED is normally measured in the transmission mode [76] by measuring the luminance on the opposite side of the electron source.

When the phosphor plate is operated at lower anode voltages, higher current densities than those at higher voltage are needed to obtain comparable brightness. Unfortunately, under an electron dose with high density, phosphors undergo luminance saturation, which is caused by the shallow penetration depth, and thus a thinner excited phosphor volume. Additionally, phosphors decrease in efficacy under high current densities, known as *Coulomb aging* [77]. Many reports on phosphors for FEDs are available in the literature, including low-voltage phosphors [77–81].

19.5 Conclusion

We have reviewed the status of CNT field emitters, especially focused on the printing technique. Additionally, the formulation of photoimageable CNT paste was briefly summarized for the fabrication of FEDs. This novel technology potentially offers an advantage of low-cost manufacturing in large-size panels. The FE characteristics of CNT emitters and the applications based on these materials are still being actively explored. However, critical issues associated with stringent uniformity and reliability remain to be overcome for CNT-based FEDs to be implemented in the major display market.

References

1. Wang, C., Garcia, A., Ingram, D.C., Lake, M., and KorKordesch, M.E. (1991) *Electron Lett.*, **27**, 1459.
2. Geis, M.W., Efremow, N., Woodhouse, J., Mcalese, M., Marchywka, M., Socker, D., and Hochedez, J. (1991) *IEEE Elect. Dev. Lett.*, **12**, 456.
3. Zhu, W., Kochanski, G.P., and Jin, S. (1998) *Science*, **282**, 1471.
4. Becker, R.S., Higashi, G.S., Chabal, Y., and Becker, A.J. (1990) *Phys. Rev. Lett.*, **65**, 197.
5. Djubua, B.C. and Chubun, N.N. (1991) *IEEE Trans. Elect. Dev.*, **38**, 2314.
6. Missert, N., Friedmann, T.A., Sullivan, J.P., and Copeland, R.G. (1997) *Appl. Phys. Lett.*, **70**, 1995.
7. Rinzler, A.G., Hafner, J.H., Nicolaev, P., Lou, L., Kim, S.G., Tomanek, D., Nordlander, P., Colbert, D.T., and Smalley, R.E. (1995) *Science*, **269**, 1550.
8. de Heer, W.A., Chatelain, A., and Ugarte, D. (1995) *Science*, **270**, 1179.
9. Shoulders, K.R. (1961) *Adv. Comput.*, **2**, 135.
10. Buck, D.A. and Shoulders, K.R. (1958) *Proceedings of the Eastern Joint Computer Conference*, American Institute of Electrical Engineers, New York, p. 55.
11. Spindt, C.A. and Shoulders, K.R. (1966) Eighth IEEE Conference on Tube Techniques, p. 143.
12. Spindt, C.A. (1968) *J. Appl. Phys.*, **39**, 3504.
13. Spindt, C.A., Holland, C.E., Rosengreen, A., and Brodie, I. (1991) *IEEE Trans. Elect. Dev.*, **38**, 2355.
14. Spindt, C.A. (1992) *Surf. Sci.*, **266**, 145.
15. Meyer, R., Ghis, A., Rambaud, P., and Muller, F. (1985) *Proc. Japan Display*, 513.
16. Ghis, A., Meyer, R., Rambaud, P., Levy, F., and Leroux, T. (1991) *IEEE Trans. Elect. Dev.*, **38**, 2320.
17. Levine, J.D., Meyer, R., Baptist, R., Felter, T.E., and Talin, A.A. (1995) *J. Vac. Sci. Technol. B*, **13**, 474.
18. Kochanski, G.P. (1994) Flat panel field emission display apparatus. US Patent 5,283,500.
19. Holland, C.E., Spindt, C.A., Brodie, I., Mooney, J., and E.R. Westerberg, (1987) International Display Conference, London.
20. Itoh, S., Tanaka, M., and Tonegawa, T. (2004) *J. Vac. Sci. Technol. B*, **22**, 1362.
21. Cho, J.H., Zoulkarneev, A.R., Kim, J.W., Hong, J.P., and Kim, J.M. (1997) Technical Digest of the 10th International Vacuum Microelectronics Conference, Kyongju, Korea, p. 401.
22. Fowler, R.H. and Nordheim, L. (1928) *Proc. R. Soc. London, Ser. A*, **119**, 173.
23. Gadzuk, J.W. and Plummer, E.W. (1973) *Rev. Mod. Phys.*, **45**, 487.
24. Gomer, R. (1993) *Field Emission and Field Ionization*, American Institute of Physics, New York.
25. Moldinos, A. (1984) *Field, Thermionic, and Secondary Emission Spectroscopy*, Plenum, New York.
26. Dresselhaus, M.S., Dresselhaus, G., and Eklund, P.C. (1996) *Science of Fullerenes and Carbon Nanotubes*, Academic, New York.
27. Yakobson, B.I. and Smalley, R.E. (1997) *Am. Sci.*, **85**, 324.
28. Frank, S., Poncharal, P., Wang, Z.L., and de Heer, W.A. (1998) *Science*, **280**, 1744.
29. Odum, T.W., Huang, J.L., Kim, P., and Lieber, C.M. (1998) *Nature*, **391**, 62.
30. Wildoer, J.W.G., Venema, L.C., Rinzler, A.G., Smalley, R.E., and Dekker, C. (1998) *Nature*, **391**, 59.
31. Dekker, C. (1999) Physics Today, May.
32. Suzuki, S., Bower, C., Watanabe, Y., and Zhou, O. (2000) *Appl. Phys. Lett.*, **76**, 4007.
33. Filip, V., Nicolaescu, D., Tanemura, M., and Okuyama, F. (2001) *Ultramicroscopy*, **89**, 39.
34. Bonard, J.M., Dean, K.A., Coll, B.F., and Klinke, C. (2002) *Phys. Rev. Lett.*, **89**, 197602.
35. Forbes, R.G., Edgcombe, C.J., and Valdrè, U. (2003) *Ultramicroscopy*, **95**, 57.
36. Wang, X.Q., Wang, M., He, P.M., Xu, Y.B., and Li, Z.H. (2004) *J. Appl. Phys.*, **96**, 6752.

37. Smith, R.C., Forest, R.D., Carey, J.D., Hsu, W.K., and Silva, S.R.P. (2005) *Appl. Phys. Lett.*, **87**, 013111.
38. Nilsson, L., Groening, O., Emmenegger, C., Kuettel, O., Schaller, E., Schlapbach, L., Kind, H., Bonard, J.-M., and Kern, K. (2000) *Appl. Phys. Lett.*, **76**, 2071.
39. Talin, A.A., Dean, K.A., and Jaskie, J.E. (2001) *Solid-State Electron.*, **45**, 963.
40. Iijima, S. (1991) *Nature*, **354**, 56.
41. Ebbesen, T.W. (ed.) (1997) *Carbon Nanotubes: Preparation and Properties*, CRC Press, Cleveland.
42. Saito, R., Dresselhaus, G., and Dresselhaus, M.S. (1998) *Physical Properties of Carbon Nanotubes*, Imperial College Press, London.
43. Wang, Q.H., Corrigan, T.D., Dai, J.Y., Chang, R.P.H., and Krauss, A.R. (1997) *Appl. Phys. Lett.*, **70**, 3308.
44. Collins, P.G. and Zettl, A. (1996) *Appl. Phys. Lett.*, **69**, 1969.
45. Collins, P.G. and Zettl, A. (1997) *Phys. Rev. B*, **55**, 9391.
46. Kim, Y.C., Sohn, K.H., Cho, Y.M., and Yoo, E.H. (2004) *Appl. Phys. Lett.*, **84**, 5350.
47. Lee, J.H., Shin, J.H., Kim, Y.H., Park, S.M., Alegaonkar, P.S., and Yoo, J.B. (2009) *Adv. Mater.*, **21**, 1257.
48. Kharchenko, S.B., Douglas, J.F., Obrzut1, J., Grulke, E.A., and Migler, K.B. (2004) *Nature Mater.*, **3**, 564.
49. Xu, D.H., Wang, Z.G., and Douglas, J.F. (2008) *Macromolecules*, **41**, 815.
50. Li, Y.J., Sun, Z., Lau, S.P., Chen, G.Y., and Tay, B.K. (2001) *Appl. Phys. Lett.*, **79**, 1670.
51. Choi, W.B., Jin, Y.W., Kim, H.Y., Lee, S.J., Yun, M.J., Kang, J.H., Choi, Y.S., Park, N.S., Lee, N.S., and Kim, J.M. (2001) *Appl. Phys. Lett.*, **78**, 1547.
52. Ren, Z.F., Huang, Z.P., Xu, J.W., Wang, J.H., Bush, P., Siegal, M.P., and Provencio, P.N. (1998) *Science*, **282**, 1105.
53. Bower, C., Zhou, O., Zhu, W., Ramirez, A.G., Kochanski, G.P., and Jin, S. (2000) *Mater. Res. Soc. Symp. Proc.*, **593**, 215.
54. Choi, W.B., Chung, D.S., Kang, J.H., Kim, H.Y., Jin, Y.W., Han, I.T., Lee, Y.H., Jung, J.E., Lee, N.S., Park, G.S., and Kim, J.M. (1999) *Appl. Phys. Lett.*, **75**, 3129.
55. Vink, T.J., Gillies, M., Kriege, J.C., and van de Laar, H.W.J.J. (2003) *Appl. Phys. Lett.*, **83**, 3552.
56. Lee, H.J., Lee, Y.D., Cho, W.S., Ju, B.K., Lee, Y.H., Han, J.H., and Kim, J.K. (2006) *Appl. Phys. Lett.*, **88**, 093115.
57. Sawada, A., Iriguchi, M., Zhao, W.J., Ochiai, C., and Takai, M. (2003) *J. Vac. Sci. Technol. B*, **21**, 362.
58. Hosono, A., Shiroishi, T., Nishimura, K., Abe, F., Shen, Z., Nakata, S., and Okuda, S. (2006) *J. Vac. Sci. Technol. B*, **24**, 1423.
59. Rochanachirapar, W., Murakami, K., Yamasaki, N., Abo, S., Wakaya, F., Takai, M., Hosono, A., and Okuda, S. (2005) *J. Vac. Sci. Technol. B*, **23**, 765.
60. Kim, Y.C., Nam, J.W., Hwang, M.I., Kim, I.H., Lee, C.S., Choi, Y.C., Park, J.H., Kim, H.S., and Kim, J.M. (2008) *Appl. Phys. Lett.*, **92**, 263112.
61. Choi, Y.S., Kang, J.H., Park, Y.J., Choi, W.B., Lee, C.J., Jo, S.H., Lee, C.G., You, J.H., Jung, J.E., Lee, N.S., and Kim, J.M. (2001) *Diamond Relat. Mater.*, **10**, 1705.
62. Kim, Y.C. and Yoo, E.H. (2005) *Jpn. J. Appl. Phys.*, **44**, L454–L456.
63. Zhu, X.Y., Lee, S.M., Lee, Y.H., and Frauenheim, T. (2000) *Phys. Rev. Lett.*, **85**, 2757.
64. Kim, C., Choi, Y.S., Lee, S.M., Park, J.T., Kim, B., and Lee, Y.H. (2002) *J. Am. Chem. Soc.*, **124**, 9906.
65. Akdim, B., Duan, X., and Pachter, R. (2003) *Nano Lett.*, **3**, 1209.
66. Wadhawan, A., Stallcup, R.E.II, Stephens, K.F., Perez, J.M., and Akwani, I.A. (2001) *Appl. Phys. Lett.*, **79**, 1867.
67. Dean, K.A. and Chalamala, B.R. (1999) *Appl. Phys. Lett.*, **75**, 3017.
68. Purcell, S.T., Vincent, P., Journet, C., and Binh, V.T. (2002) *Phys. Rev. Lett.*, **88**, 105502.
69. Dijon, J., Fournier, A., De Monsabert, T.G., Levis, M., Meyer, R., Bridoux, C., Montmayeul, B., and Sarrasin, D. (2006) SID Symposium Digest of Technical Papers 37, p. 1744.
70. Dijon, J. (2005) NT′ 05, Gothenburg, Sweden.
71. Lee, N.S., Chung, D.S., Han, I.T., Kang, J.H., Choi, Y.S., Kim, H.Y., Park, S.H.,

Jin, Y.W., Yi, W.K., Yun, M.J., Jung, J.E., Lee, C.J., You, J.H., Jo, S.H., Lee, C.G., and Kim, J.M. (2001) *Diamond Relat. Mater.*, **10**, 265.

72. Choi, Y.S., Kang, J.H., Kim, H.Y., Lee, B.G., Lee, C.G., Kang, S.K., Jin, Y.W., Kim, J.W., Jung, J.E., and Kim, J.M. (2004) *Appl. Surf. Sci.*, **221**, 370.
73. Saito, Y., Uemura, S., and Hamaguchi, K. (1998) *Jpn. J. Appl. Phys.*, **37**, L346.
74. Uemura, S., Yotani, J., Nagasako, T., Saito, Y., and Yumura, M. (1999) Euro Display '99, Late News, p. 93.
75. Choi, Y.C., Jeong, K.S., Han, I.T., Kim, H.J., Jin, Y.W., Kim, J.M., Lee, B.G., Park, J.H., and Choe, D.H. (2006) *Appl. Phys. Lett.*, **88**, 263504.
76. Shea, L.E. (1998) *Electrochem. Soc. Interface*, **7**, 24.
77. Holloway, P.H., Sebastian, J., Trottier, T., Swart, H., and Peterson, R.O. (1995) *Solid State Technol.*, **38**, 47.
78. Jacobson, S. (1996) *J. SID*, **4**, 331.
79. Stoffers, C., Yang, S., Jacobsen, S.M., and Summers, C.J. (1996) *J. SID*, **4**, 337.
80. Bukesov, S.A., Nikishin, N.V., Dmitrienko, A.O., Shmakov, S.L., and Kim, J.M. (1997) Proceedings of IVMC, p. 676.
81. Summers, C.J. and Wagner, B.K. (1998) Workshop Digest of Asia Display, p. 261.

20
Nanotube Field Emission Displays: Nanotube Integration by Direct Growth Techniques

Kenneth A. Dean

20.1
Introduction

Nanotube field emission displays (FEDs) hold the promise of delivering cathode ray tube (CRT) picture quality with the thin form factor of a liquid crystal display or plasma display. In order to enter the display marketplace, FEDs must be superior to existing flat screen displays in both performance and large-volume cost. A key element of the cost and performance is the method in which nanotubes are deposited into the display structure. Display developers have pursued two main approaches to integrate nanotubes. The first approach focuses on the low cost of screen-printing technology to print the nanotubes into the structure, which is covered in another chapter. The advantages are a low-cost process compatible with plasma display factories. However, the nanotubes do not emit after printing because they are embedded in a matrix. The matrix must be ripped up to expose or "activate" the nanotubes using processes such as adhesive taping, laser ablation, or abrasion. The primary disadvantages are the following: (i) screen printing cannot achieve feature sizes small enough to realize low-voltage structures, thereby necessitating costly high-voltage electronics; (ii) lack of control of the nanotube positions leads to excess leakage current; (iii) the postprocessing "activation" step generates high defectivity (poor yield) and cannot be used with many traditional device designs; and (iv) residual organics in the paste mixture could lead to long-term reliability challenges after the display is sealed. The second approach trades increased deposition cost with tight control of the nanotube placement and lower defectivity. A catalyst material is deposited and patterned onto the surface, generally using conventional thin film lithography techniques, and then nanotubes are grown into place in the device structure with chemical vapor deposition (CVD) direct growth techniques. The processing cost of the second approach is higher than in the first approach, but fabrication is compatible with liquid crystal display manufacturing lines and the automated tooling in their factories. The lithography steps produce device features on the order of 5 μm, thereby delivering a low-voltage device structure requiring less expensive driver electronics. The cleanliness of CVD is another advantage. Nanotubes are grown where intended, so defectivity is low

Carbon Nanotube and Related Field Emitters: Fundamentals and Applications. Edited by Yahachi Saito

ISBN: 978-3-527-32734-8

and yield is high. The catalyst material has no residual chemicals to contaminate the display after it is sealed. Finally, the reducing atmosphere of CVD growth removes oxides and contaminants from the entire device, promoting long-term display reliability after the final seal. In this chapter, we describe in detail displays fabricated by the chemical vapor growth of nanotubes.

While the potential advantages of using chemical vapor techniques to grow nanotubes into displays are clear, there are key technological challenges that must be overcome. Economically viable large-area displays must be fabricated on commercial glass substrates such as those used for liquid crystal displays or plasma display panels. These glasses have strain point temperatures between $\sim$510 °C (sodalime silicate) and $\sim$666 °C (aluminoborosilicate), but deform measurably at lower temperatures given sufficient time. In general, the physics of CVD and catalytic growth produces small-diameter, dense nanotubes at high temperature, and sparser, large-diameter, highly defective nanotubes at temperatures below 700 °C. Unfortunately, a low-voltage, low-cost display requires a small diameter, but somewhat sparse nanotubes, deposited below 650 °C. Simply growing the right nanotubes is not enough. Additional challenges include the following: (i) selecting a device structure that tolerates variation of nanotube heights, (ii) creating homogenous nanotube properties between pixels such that the pixel-to-pixel emission variation is less than 4% (for high-definition television), (iii) achieving sufficient nanotube current density for adequate brightness (>1 000 000 emitting nanotubes per square centimeter of screen), (iv) controlling the electron beam spread to achieve color purity between subpixels, (v) minimizing pixel defectivity, and (vi) achieving display longevity. Several organizations have generally overcome these challenges, and several complete system solutions have been demonstrated.

20.2
Field Emission Display Design and Drive Voltage

An FED sends electrons into a phosphor, using the same physics as CRTs to produce similarly bright and colorful video images. Obtaining a thin form factor display requires using hundreds of electron sources per pixel as well as spacers supporting the vacuum envelope between the plates of glass. An example structure is shown in Figure 20.1. Device designs optimize display brightness, color, uniformity, lifetime, and drive voltage while minimizing the fabrication cost. The cost of the display electronics is an important part of the overall display cost. The drive electronics costs are minimized by keeping the gate voltage low. The cost of the display drivers is tiered, with high voltage (>100 V) costing significantly more than mid-range voltage (<50 V), which in turn is significantly more than the low voltage (<20 V) used by liquid crystal displays (LCDs) in great volumes. The drive voltage depends on the local electric field produced by the device geometry and the field-enhancing geometry of the nanotube. To a first-order approximation, the local applied field at the nanotube tip must be on the order of 5000 V μm^{-1} to extract substantial field emission current from a material with a graphite-like work function [2] The local

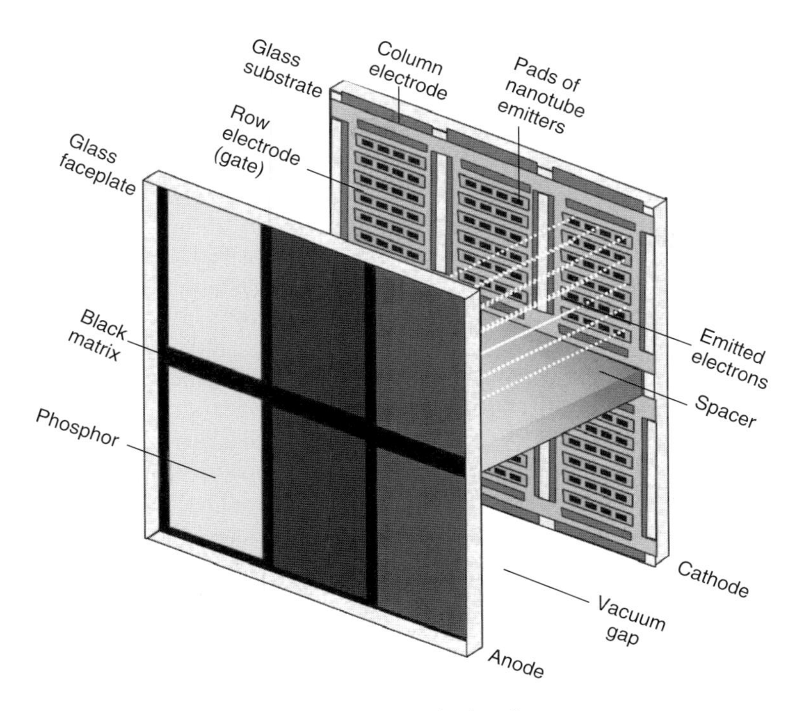

Figure 20.1 Schematic of a field emission display showing phosphor subpixels and pads of nanotube emitters behind them. The size of the nanotube pads has been exaggerated for illustration. The display device typically contains hundreds of pads per subpixel [1]. (Reprinted with permission from the Society for Information Display.)

electric field at the nanotube surface is approximately given by [1, 3]

$$\left[\alpha\frac{V_a}{d}+\delta\frac{(V_g-V_c)}{s}\right]\cdot\frac{0.7\,h}{r}$$

where V_g is the gate (scan) electrode voltage, V_c is the cathode (data) electrode voltage (usually zero in the "on" state), s is the nanotube to gate electrode spacing, V_a is the anode voltage, d is the anode to cathode spacing, h is the nanotube height, r is the nanotube radius, and α and δ are electrode geometry constants which are near unity. As an example, "full-on" field emission conditions ($>5000\ \mathrm{V}\ \mu\mathrm{m}^{-1}$) can nominally be obtained with a device structure comprising an anode at 5000 V and a 2 mm spacing, a scan electrode at 70 V, nanotubes having an h/r ratio of 500, a gate spacing s of 5 μm , and $Vc = 0$. An "off" condition is generally obtained at half the scan voltage ($Vc = Vg/2$), if the nanotube distribution is tight. Thus, for the above example, the swing voltage on the data lines is approximately 35 V. For a typical design that accommodates nanotubes from 0.5 to 2 μm in length, the radius must be nominally 1–4 nm. This order of magnitude estimate is sufficient to show that single-walled nanotubes and small multiwalled nanotubes are needed for these device designs. Multiwalled nanotubes in the 20–60 nm diameter range (typical of those grown at low temperatures) simply will not work at reasonable voltages unless the gate spacing is smaller or the nanotubes have sharper tips.

It should be noted that it is very easy to grow a few nanotubes with aspect ratios exceeding 1000 or even 3000. These nanotubes are readily observed in diode measurements of nanotube films. However, achieving sufficient current density in a display geometry generally requires tens of thousands of these nanotubes emitting electrons per square centimeter of anode.[1] In a display application, the nanotube-containing cathode plate also contains row and column address lines, gaps for spacers, lateral ballast resistors, and gaps between emitter regions to prevent color cross talk. Consequently, nanotubes can occupy only a few percent of the total cathode area, so the actual film area density for these nanotubes with an $h/r \sim 500$ needs to be closer to 1 000 000 per square centimeter of nanotube film. Achieving such a high density of high-aspect-ratio nanotubes is more challenging. Moreover, the nanotube density cannot be too high. When nanotubes are nominally closer together than they are tall, they screen the electric field, reducing the field enhancement because of their geometries.

While the gate-to-nanotube spacing needs to be nominally 5 μm or less to keep the drive voltage down, reducing this spacing below 5 μm drives up fabrication costs and impacts device defectivity. Already, fabrication of 5 μm features is well beyond the reach of low-cost screen-printing fabrication techniques used for plasma displays. It is also beyond the reach of photodefinable screen-printed materials. Photolithography of thin film materials is required for these geometries. However, if the critical geometry can be kept above 1–3 μm, the critical features can be fabricated with low-cost photolithography equipment, rather than the high-resolution steppers used by today's semiconductor industry. Consequently, keeping the spacing near 5 μm requires the same techniques and tool sets that are used to fabricate LCDs. Figure 20.2 shows examples of several device structures that were fabricated with critical feature sizes in the 3–5 μm range.

Smaller gate-to-nanotube spacing makes the devices susceptible to several kinds of device defects. In low-temperature CVD systems, nanotubes are routinely grown to lengths of 1 μm. However, some nanotubes grow as long as 5 μm. As the length of some nanotubes approaches the gate-to-nanotube spacing, nanotubes can bridge the spacing to the gate, causing pixel shorts that degrade device performance with leakage current [6]. With hundreds to thousands of nanotubes per pixel, only one exceptionally long nanotube can short-out the entire pixel. Motorola found that these long-nanotube defects could be eliminated by burning the shorted nanotubes in air. They heated the cathode up to approximately 200 °C to make the ballast resistors conductive (nonprotecting) and, in air, applied reverse bias voltage to the cathode. Nanotubes burned up in air when currents exceeded several microamperes in each nanotube.

1) With an average nanotube current of 100 nA, peak anode current densities between 3 and 10 mA cm^{-2} are realized with 30 000–100 000 emitting nanotubes per square centimeter of anode. Individual nanotubes risk destruction at currents on the order of 1 μA, so 100 nA per nanotube is a realistic design.

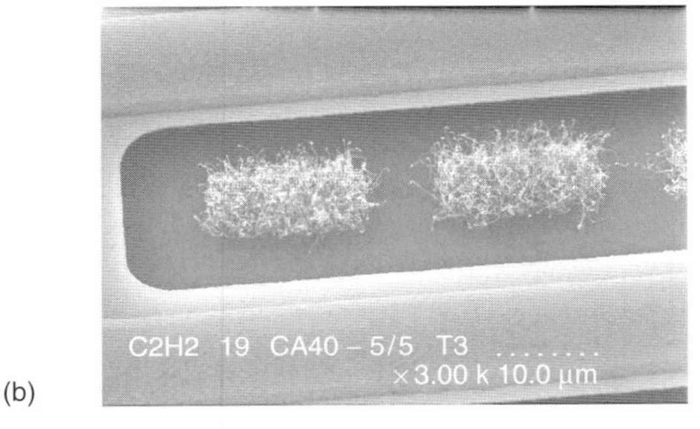

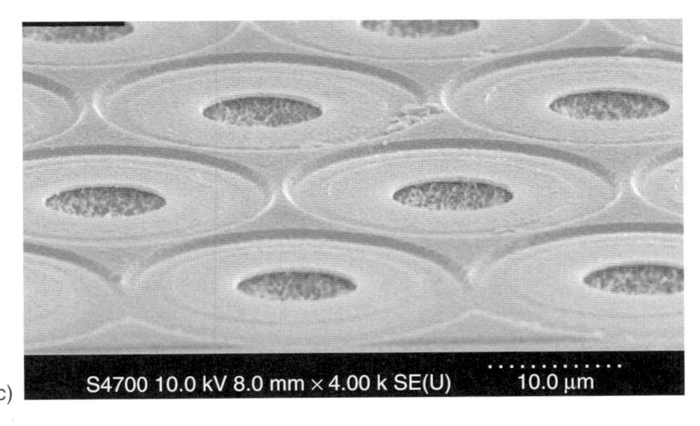

Figure 20.2 Various cathode designs and nanotube densities from (a) Motorola (cross section) [4], (b) LETI (top view) [3], and (c) cDreams (low angle view) [5]. (Reprinted with permission from the Society for Information Display.)

Another important device design constraint comes from the mechanical properties of nanotubes themselves. Nanotubes with an h/r ratio of 500 bend under electric field (that aspect ratio is similar to that of human hair) [7]. Long nanotubes can be pulled in the direction of the gate; thereby the emitting electrons travel either straight into the gate electrode, or at high angles and into phosphors of the wrong color. A solution for this problem is to design the gate electrode to produce a very symmetrical electric field at the nanotubes such that there is negligible lateral electric field. This keeps the nanotubes pointing toward the anode and the emitting electrics in a field with minimal lateral components.

20.3 Fabricating the Display

20.3.1 Building the Structure

FEDs comprise two plates: a cathode plate containing the emitters and an anode plate containing the phosphors. The cathode glass substrate must contain scan (row) and data (column) addressing electrodes, a dielectric between them, and carbon nanotubes. The electrodes are typically made of a refractory metal such as molybdenum. Most commonly, a ballast resistor is put in electrical series between the column electrodes and the nanotubes to limit current, improve uniformity, and provide stability. For devices built using the direct growth of carbon nanotubes, a catalyst material is deposited and patterned on the surface.

As for anodes, the structures of low-voltage and high-voltage anodes differ. The low-voltage anode fabrication sequence begins with depositing a transparent conductor such as indium tin oxide. A black matrix is then deposited and patterned. Colored phosphors in a binder are then screen-printed onto the anode sequentially and fired. For the high-voltage anodes, a black matrix is deposited onto the glass substrate and patterned. Colored phosphors in a binder are then screen-printed onto the anode sequentially and fired. Finally, a lacquer layer is applied over the phosphor layer, and a thin (500 nm) aluminum layer is deposited onto the lacquer. Upon firing, the lacquer burns out, leaving the Al layer as a parabolic reflector. The efficiency gain of the reflective aluminum offsets the loss in electron energy through the aluminum when the anode voltage is higher than about 4000 V.

20.3.2 Growth of Carbon Nanotubes on Glass

Growing nanotubes in place via CVD, rather than using screen-printing or electrodeposition techniques, provides a number of key advantages including control of nanotube location, spacing, length, orientation, and vertical alignment. Achieving good display performance requires growing small-diameter nanotubes on glass. Yet, growth at temperatures below 600 °C or even below 650 °C often produces no growth [8], or growth of large-diameter defective nanotubes [9, 10] (or graphitic nanofibers, bamboo, etc.). There are many reasons for this, and growing high-quality nanotube field emitters requires a strong understanding of nanotube growth parameters. Important growth parameters include the substrate temperature, the catalyst material, the substrate surface conditions, and the reactant gas mixture. In a typical CVD growth system, nanotubes are grown from a catalyst film on a heated substrate. The carbon is provided from a carbon-containing feedstock gas (i.e., methane, ethylene, acetylene). Often, a reducing gas (H_2 or NH_3) is also introduced to increase the catalytic gas reaction, to minimize the catalyst "poisoning", and to etch any deposited amorphous carbon or graphitic byproducts. In order to grow tubes, the carbon source gas must be dissociated into highly reactive species [8]. Thermal dissociation of methane, for example, occurs above 900 °C, and is not practical for glass substrates.

Instead, plasmas and hot filaments are typically employed to break the source gases and enhance their reactivity [8,11–13]. Under many conditions, scientists believe that CH_3 is the primary growth species, supplying the carbon atoms that build up to form the CNT lattice. Atomic hydrogen also plays a major role in the growth chemistry, creating gas-phase radicals, breaking surface bonds, stabilizing carbon bonds at the growth front, and preferentially etching amorphous and graphitic phases. Plasmas (microwave, RF, and DC) are traditional means for generating active species for low-temperature growth. The electric field induced by a plasma may align the nanotubes, but care must be taken not to damage nanotubes by bombardment with highly energetic species. Hot filament CVD also generates reactive species and atomic hydrogen, but contains minimal electric field (except in the case of "plasma-enhanced" hot filament CVD). In all cases, the top layer of the substrate may be at a substantially different effective temperature than the typically reported substrate temperature, because of the arrival of energetic species. The glass substrate is a poor thermal conductor, and it can support a substantial temperature difference between its front and back surfaces.

The other critical factor is the use of transition metal catalysts such as Ni, Fe, and Co deposited on the substrate. These metals will crack hydrocarbons, but very little catalytic activity is observed below 550 °C. Catalyst materials also have a strong effect on the nanotube diameter, growth rate, wall thickness, morphology, and microstructure. Many have reported that the nanotube growth rate depends on the diffusion of carbon across or through the catalyst particle, as has been established for vapor-grown carbon fibers [14]. This rate is very slow at low temperature (below 700 °C), and a sharp drop-off in the rate of nanotube growth has been reported at low temperatures [15]. Slow rates at low temperature can lead to the buildup of an amorphous carbon coating on the catalyst that shuts down all catalytic activity.

The simple presence of transition metal catalysts is not sufficient to grow a good field emissive film at low temperate. These transition metal catalysts must be presented during growth as tiny nanoparticles. As a rough generalization, the diameter of a catalyst particle sets the diameter of the nanotube [13–15]. Small catalyst particles are also more reactive than a bulk catalyst, having activation energies and diffusion rates that support growth at lower platen temperatures. The size of the catalyst particles depends on many deposition variables, as does their spatial density. One common catalyst strategy is to deposit a thin (<20 nm) film of metal. During growth, the metal film is subjected to a higher temperature anneal in hydrogen, ammonia, or other reducing agents, causing the metal film to break into small particles [15–18]. The initial film thickness determines the size of the particles, their spatial density, and therefore the size and density of

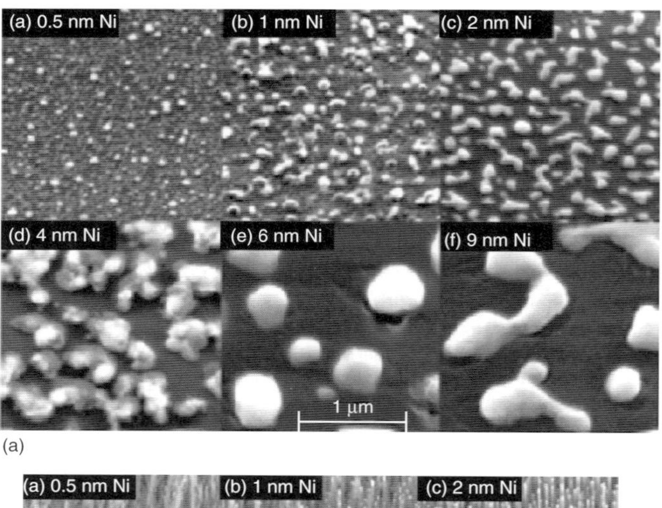

Figure 20.3 (A) Fragmentation of thin (50 nm) Ni catalyst films at 750 °C in hydrogen and (B) the resulting nanotubes grown by PECVD. (Reprinted with permission from Chhowalla *et al.* [15], copyright 2001, American Institute of Physics.)

the resultant nanotubes (Figure 20.3) [15]. The extreme sensitivity of the nanotube properties to 1 nm of film thickness makes this process very difficult to control in a manufacturing setting.

Moreover, the catalyst particle size and the spatial density are tied together by conservation of volume of the catalyst material. This process produces thin nanotubes at high spatial density, or large-diameter nanotubes at lower density. Neither is optimal for low-voltage, high-current field emission. In particular, nanotubes that are spaced more closely than their height screen the applied electric field, resulting in reducing field enhancement (lower effective h/r ratio) [19, 20]. In the case of Figure 20.3, the smallest diameter nanotubes are subject to intense field-screening.

A more robust catalyst approach includes a catalyst support with the transition metals [11, 21, 22]. Supports such as silica or alumina stabilize the catalyst particle sizes. This allows the diameter of the nanotube to be determined independent of the film thickness or the spatial density. In addition, the size and density of catalyst particles can be controlled by setting the initial catalyst size and spatial density in the catalyst deposition step. It should be noted that catalyst support oxides are often unintentionally present in CVD growth of nanotubes, often as a diffusion barrier layer.

Several investigators have reported using the right combinations of catalysts and CVD techniques to grow nanotubes on glass which emit in triode display structures [23–25]. (It should be noted that in the literature, the definition of a nanotube has often been stretched to include highly defective fibers with graphitic walls.) Samsung reported fabricating a module that they paired with a display anode to demonstrate display switching functionality. They reported using an Invar alloy catalyst and CVD to fabricate an array of pixels on glass, and they demonstrated switching color pixels [26]. These devices demonstrate basic feasibility.

Several teams have reported developing full-motion video displays with directly grown carbon nanotube cathodes. cDream Corporation reported depositing 5 and 10 nm films of Ni catalyst, annealing in hydrogen to form small catalyst islands, and then growing the nanotubes under inductively coupled plasma [5]. In Figure 20.4a, they demonstrated full-color video operation with a low gate driving voltage of 46 V (using a gate-to-nanotube spacing of less than 1 μm) [27]. Motorola reported using hot filament CVD to crack the source material [11]. The hot filaments, located ~2 cm above the substrate, provided a local cracking surface with a temperature exceeding 1600 °C. Motorola also reported employing a thick (100–400 nm) supported catalyst, a mixture of catalyst nanoparticles less than 3 nm in diameter embedded in a supporting oxide such as aluminum oxide. The catalyst support prevented the aggregation of small catalyst particles into larger particles, thereby allowing growth of nanotubes <3 nm in diameter below 600 °C (and as low as 480 °C). Motorola demonstrated color video operation with a data voltage of 35–45 V and gate (scan) voltage of 75–90 V (Figure 20.4b). In addition, LETI reported using thin (10 nm) Ni catalyst films which were converted to islands and used to grow a dense array of nanotubes in the 30 nm diameter range [3]. They ultimately reported full-motion video displays driven with an 80 V scan voltage and a 40 V swing (data) voltage (Figure 20.4c).

It is clearly worth discussing an FED technology developed by Canon in partnership with Toshiba. They dubbed the technology as *surface emission display* or SED. To date, very few papers have been published by Canon or Toshiba on the technology, but they have demonstrated 36″ and 55″ diagonal high-definition television displays at major trade shows with excellent image quality [29]. An early paper describes the technology as a laterally disposed gate electrode and cathode electrode structure with a nanogap between them and PdO emitting material [30]. However, the most recent paper describes an additional step of conditioning each pixel in a hydrocarbon atmosphere by driving electrons between the electrodes [31].

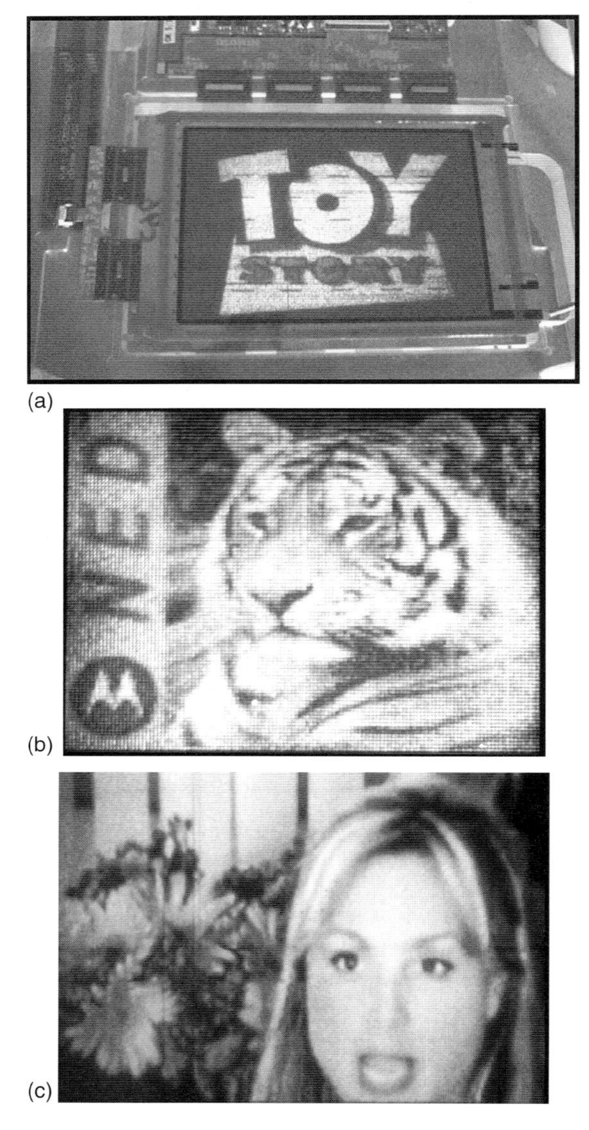

Figure 20.4 Full-color display prototypes (5–6 in.) from (a) cDreams [27], (b) Motorola, and (c) LETI [28]. (Reprinted with permission from the Society for Information Display.)

The electrons crack the hydrocarbon gas, depositing carbon on the electrodes. After conditioning, the emitter is a carbon material. It is unclear if the carbon is nanostructured or is, in fact, a nanotube, but it is deposited by a form of CVD. This technology is clearly the most mature of the FED technologies that employ *in situ* direct growth of carbon emitters.

20.4 Luminance Uniformity and Control and Nanotube Distributions

LCDs light up pixels with a broad area light source that is made highly uniform with diffusers. As a result, the luminance uniformity between adjacent pixels is typically better than 98%. CRTs illuminate pixels by rastering an electron source across them. The same electron source illuminates every pixel so the resulting luminance uniformity between adjacent pixels is generally better than that of LCDs. In contrast, FEDs and plasma displays use millions of individual electron sources to produce illuminated pixels. Slight differences in uniformity between neighboring electron sources can lead to a short-range variation in luminance between pixels and their nearest (and second-nearest) neighbors. To display viewers, significant short-range luminance variations make a display looks dirty. For a 42″ 1280 × 720 HDTV viewed at the proper viewing distance, the eye can detect short-range luminance variations of 4% and larger. This means that a subpixel short-range uniformity above 96% is required. FEDs can reduce short-range variations below the eye's detection threshold, just as plasma displays do.

The source of luminance variation in nanotube FEDs is the nanotubes themselves. Since both the electron emission current and the resulting luminance are exponentially dependent on the height to radius (h/r) ratio, the field emission process amplifies the effects of the population distribution. For CVD-grown nanotubes, the population's distributions in radius and length depend on the catalyst and the CVD growth conditions. It should be noted that screen-printed nanotube FEDs share this short-range uniformity problem. While the nanotube starting material can be fabricated and purified to a tighter initial distribution, the acts of printing and "activation" create a population with variations in length and embedded angle. In both cases, good luminance uniformity may be obtained by using a tight h/r distribution, statistical averaging in the device, and appropriate resistive ballasting.

A ballast resistor network is commonly employed in FEDs to improve uniformity [32]. In this approach, emitters are electrically separated from each other by a resistive path. This resistor network has the effect of limiting the current from the high h/r nanotubes, in effect truncating the "high h/r" end of the distribution. In practice, small areas of nanotubes, often called *pads* are networked through the ballast resistor film. Note that the primary role of the ballast resistor is to improve stability of the emitters by preventing them from evaporating or arcing (completely necessary with Spindt-tip emitters, less so with nanotubes that have intrinsic, noncatastrophic current-limiting properties.) With pads of nanotubes, larger ballast resistance improves uniformity, but it also produces a large decrease in field emission current. There is a trade-off between current and uniformity using ballast resistors.

Dean *et al.* report that within a display, the approximate distribution in h/r ratio can be measured using devices with only one emitter pad per subpixel and pad sizes which produce typically one emitting nanotube per pad [33]. They then measure the distribution of voltage to reach a given current for each pad. Using the Fowler–Nordheim equation, this result is converted to an h/r distribution. This technique interrogates only the highest h/r ratio nanotubes which field-emit, so they measure only the high h/r tail of a presumably Gaussian distribution. With a measurement of the nanotube population distribution, Dean *et al.* report the ability to modify the nanotube catalyst and growth conditions to improve uniformity.

The device structure and ballast network design strongly influence the short-range uniformity. Dean *et al.* reported inputting the measured h/r distribution into a computer model which assigned nanotubes to pads, and then calculated the pad current, pixel current, display current, and short-range uniformity variations within [33]. By varying the h/r distribution, the device geometry, the number of ballasted emitter pads, and the ballast resistance, they examined the most important factors for improving uniformity. They concluded that the primary factor in obtaining short-range uniformity for a given h/r distribution was an increase in the number of nanotubes in the system, whether it was with increased pad number or increasing nanotubes per pad. Adding multiple nanotubes per pad had the effect of producing an effective "pad" h/r distribution with a truncated "low h/r" side. (They did not incorporate field-screening into the computation, so no cases with more than 15 high-aspect-ratio nanotubes per pad were considered). They concluded that 5000 or more nanotubes per subpixel, having the population distribution of their direct growth nanotubes, are required to achieve >96% short-range uniformity.

Dijon *et al.* developed a model to describe and optimize short-range uniformity as a function of device design [6, 34]. First, they employed an expression for the density distribution of the exponentially distributed field-enhancement factor for nanotubes (which is proportional to h/r). Next, they assumed that only the most efficient nanotubes were emitting per pad. Combining the field-enhancement factor probability with device geometry-dependent expressions for applied electric field, and an expression for the current from each ballasted pad (including the ballast resistor drop), they arrived at an expression for the standard deviation of luminance between pixels (i.e., uniformity) for their device:

$$\overline{U} = \frac{\sigma_{\text{pixel}}}{\overline{I_{\text{pixel}}}} = \frac{1}{\sqrt{n_{\text{pad}}}}\sqrt{\frac{1-p}{p} + \frac{\sigma_{\text{pad}}^2(V_g)}{(\overline{I_{\text{pad}}})^2}}$$

where σ_{pad} is the standard deviation of the pads current, n_{pad} is the number of pads per pixel, p is the number of working pads (accounting for some yield loss due to pad shorting or poor electrical connection), and I is the current. This equation predicts the short-range uniformity variation between pixels decreases with increasing pad number and increasing current. Dijon *et al.* demonstrate this functionality with measurements of FEDs with varying numbers of pads, showing a good match to the model and measured short-range uniformity above 98% for test designs. They report a uniformity metric which is semiempirical fit to their data with the same format as the above equation:

$$\overline{U} = \sqrt{\frac{1-p}{pn_{\text{pad}}} + AJ^{-\lambda} + \sigma_{\text{anode}}^2}$$

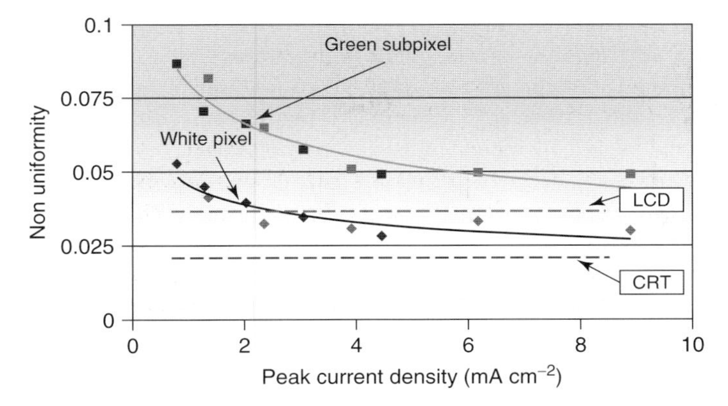

Figure 20.5 Nonuniformity of the display versus the peak current density for a 0.6 mm × 0.6 mm pixel comprised of 0.6 mm × 0.2 mm subpixels [28]. (Reprinted with permission from the Society for Information Display.)

where J is the current density, n_{pad} is the number of pads per pixel, p is the number of working pads (accounting for some yield loss due to pad shorting or poor electrical connection), σ_{anode} is luminance nonuniformity due to the anode, and A and λ are constants. For HDTV pixel designs, they report green subpixel uniformity above 95% and white pixel uniformity above 97% (Figure 20.5) [28].

Several investigators have shown that sufficient field emission short-range uniformity can be achieved given sufficient numbers of nanotubes and ballasted nanotube pads, a tight distribution of directly grown nanotubes, and sufficient extraction voltage and subpixel current. Short-range uniformity can be further improved using a look-up table in the electronics. In this scheme, the luminance from each subpixel at the operating gate voltage is recorded. Since the display is pulse-width modulated, a calibration factor is applied to the pulse width of each subpixel to remove the short-range variation. Dropping costs of integrated circuits and memory chips have made this simple and inexpensive to do, and, in fact, all types of displays employ substantial digital manipulation to improve picture quality.

20.5 Display Performance

20.5.1 Luminance

The required luminance of a display depends on the ambient light in the surrounding environment. As a rough guideline, office computer monitor luminance typically ranges from 80 to 120 cd m^{-2}. Portable devices with limited battery power are often dimmer, with notebook computers in the 80–100 cd m^{-2} range.

Displays for indirect sunlight require greater than 150 cd m^{-2}, but recent trends in cell phones and smartphones have resulted in brightness increases into the 200–300 cd m^{-2} range. Sunlight-viewable applications require 600 cd m^{-2} and higher depending on the antireflective properties of the screen. Consumer televisions are primarily viewed in indoor office-like environments, but the home environment is periodically brighter during the day. Televisions today readily provide 300–500 cd m^{-2}, which appears to be sufficient for the home environment. To achieve adequate contrast in a high ambient light room, a contrast enhancement filter is often employed. This cuts the light by as much as half, requiring the display panel to provide twice as much light as the finished television luminance presented above. Note that television product literature often quotes panel luminance (i.e.,1000 cd m^{-2}) that the televisions cannot actually achieve as shipped, because of contrast-enhancing filters and small power supplies.

FEDs produce light when electrons bombard a phosphor, the same mechanism used for CRTs. A CRT raster scans one high-intensity, pixel-sized beam across line 1, and then line 2, and so on, until the entire frame is painted. Due to the extremely high brightness of the beam and the phosphor persistence, the eye integrates this scanning to perceive a continuous image. In contrast, an FED scans an entire line at a time. This allows the electron beam to address each pixel for a longer time, generating a similar brightness, but with either a smaller current density or lower anode voltage than a CRT when using CRT phosphors.

FED luminance is determined primarily by the anode voltage, anode electron current, the duty cycle, and the phosphor. An expression from Oguchi *et al.* [29] with slight modification by Dean *et al.* [33] describes the pixel luminance:

$$L(\text{cd m}^2) = \frac{1}{\pi \cdot S_{\text{pixel}}(\mu m^2)} \cdot \eta(\text{lm W}^{-1}) \cdot I_e(\mu\text{A}) \cdot (V_a - V_{\text{Al}})(\text{V}) \cdot \text{PW}(\mu\text{s}) \cdot F(\text{Hz})$$

where S_{pixel} is the subpixel area, η is the emission efficiency, I_e is the peak subpixel current, V_a is the anode voltage, V_{Al} is the electron energy loss in the Al layer, PW is the pulse width, and F is the driving frequency (60 Hz). P22 CRT phosphors can show an efficiency decrease at high current densities and low anode voltages. The term η, therefore, is a function of current density. It also includes the phosphor fill factor (~50% for most investigators). The Al layer adds efficiency by reflecting all the light generated by the phosphor to the viewer, rather than half the light without it. However, it produces a typical electron energy loss of about 2500 V. While this loss is small for CRT voltages (~25 kV), it is large for lower voltage FEDs (3000–10 000 V), and it improves efficiency only for sufficiently high anode voltages.

cDreams, Motorola, and LETI each demonstrated full-color, full-motion video displays using directly grown of carbon nanotubes. cDreams reported a white screen luminance of 136 cd m^{-2}, but did not report anode voltage, current density, or duty cycle [27]. Motorola reported demonstrating a white screen luminance of 600 cd m^{-2} at 6 kV with an aluminized anode, 3 mA cm^{-2} of peak current density on the anode, P22 phosphors, a 1.7 mm cell gap, and a 1/360 duty cycle for the

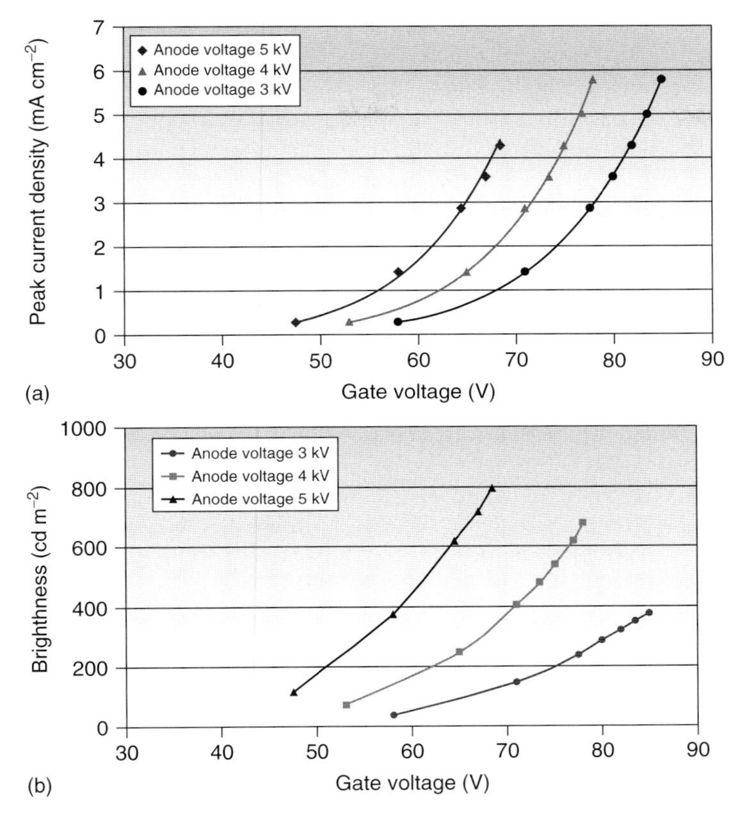

Figure 20.6 Performance characteristics of a LETI display prototype: (a) Peak current density versus gate voltage and anode voltage and (b) luminance versus gate voltage and anode voltage [28]. (Reprinted with permission from the Society for Information Display.)

HDTV 720p format [33]. LETI reported a white screen luminance of 600 cd m^{-2} at 4 kV with a nonaluminized, 4.5 mA cm^{-2} of peak current density on the anode, a 1 mm anode–cathode spacing, and a 1/312.5 duty cycle (Figure 20.6) [28]. All of these values are reported without neutral density filters in place. In general, directly grown nanotube FEDs can deliver the required luminance for television applications.

20.5.2 Color Purity

In FEDs, colors shades are produced from combinations of three primary color phosphors whose color coordinates define the maximum saturated color obtainable from the display. Many displays today are designed to deliver the SMPTE-C color standard. The first goal in color purity is achieving this video standard. Another color purity metric for measuring the success of the device design is the percentage of the phosphor color gamut triangle that can be achieved by the display.

The color of the FEDs is primarily limited by the size of the electron beams from the cathode. A beam that grows significantly larger than its target phosphor excites the neighboring phosphor color, creating an impure or erroneous color. A field emission source imparts an initial lateral velocity component to the electrons. The longer it takes for the electrons to reach the anode, the larger the electrons spread out. Also, the greater the distance between anode and cathode, the farther the electrons spread out. There is an engineering trade-off between luminance driven by high anode voltage (which is facilitated by a large anode to cathode spacing) and color purity.

A black matrix is employed on the anode to separate the phosphors and to provide contrast. When the electrons spread beyond the intended phosphor and hit the black surround region, electrons are wasted and the power efficiency suffers. If these electrons spread beyond the black surround to the neighboring phosphor, a color error occurs. For this reason, FEDs often incorporate a separate focusing electrode over the cathode to ensure that electrons hit their intended targets [35–38]. This adds complexity and cost.

LETI demonstrated a full-motion video display with good color using pixels sized for high-definition television [28]. At a gap of 1 mm and an anode voltage of 4000 V, they reported that color purity was close to target specifications without the need for a focusing grid.

Motorola reported designing their cathode structure for maximum color purity by minimizing the electric field component in the direction of the color subpixels (*X* direction) [1]. The gate fields were applied predominantly in the *Y* direction, where the pixel separations were larger and the wide beams would not cause a color error. With that structure and pixels designed for high-definition television, Dean *et al.* measured the resulting beam spread versus anode spacing, anode voltage, and gate voltage in both the *X* direction (color subpixel) and in the *Y* direction. They also created a model for computing the electron beam spreading as a function of these variables. The model required a novel adaptive step size approach to accurately compute the local electric fields and their effects at drastically different lengths scales: at the nanotube surface (nm), near the gate electrode (μm), and near the anode (mm). They reported good agreement between the model and the experiments, showing little change in the *X* direction beam size with gate voltage, decreasing beam size with anode voltage, and increasing beam size with anode to cathode spacing. Motorola reported achieving 88% of the area of our phosphor gamut in frit-sealed displays with a 5000 V anode voltage and a 1.7 mm gap between the anode and the cathode (Figure 20.7) [1]. In order to obtain 100% of the SMPTE-C color triangle, Motorola needed to change to a different P22 green phosphor with better color coordinates better matching the SMPTE-C gamut, and make either a small change to the anode layout, or an increase in anode voltage to 6 kV.

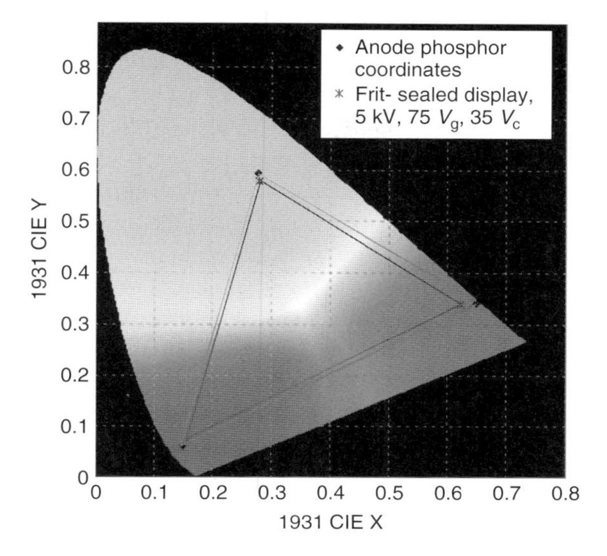

Figure 20.7 1931 CIE color triangle with the anode phosphor coordinates and color performance of the display design [1]. (Reprinted with permission from the Society for Information Display.)

20.6 Sealing

FEDs are formed from their constituent parts by sealing together the field-emitting cathode plate and a phosphor-coated anode plate. The plates are sealed with a low-melting-temperature glass frit. The space between the plates is evacuated into the high vacuum range. To keep the plates from collapsing under atmospheric pressure, spacers are placed between the plates prior to sealing. These spacers must fit within the black matrix region of the anode so that they are not visible to the viewer. The spacers must also insulate the anode plate from the cathode plate while maintaining performance under electron and ion bombardment.

As an example, Motorola reported using a multistep process to seal FEDs fabricated on borosilicate glass (NEG OA-10). First, both the phosphor-coated anode plate and the nanotube-containing cathode plate were fabricated by typical means. Next, a hole was drilled into the completed cathode using a dental drill. Frit was dispensed and preglazed onto a glass pinch-off tube/getter cavity assembly. A barium evaporable getter was placed into the getter cavity. The pinch-off tube/getter cavity was aligned with the hole and the cathode and the parts were heated in a vacuum oven to 480 °C to bond the parts together. In parallel, a glass frame was prepared comprising a frit/glass frame/frit stack with a lower temperature frit material. Additionally, thin strips of spacer material were then aligned and affixed to the anode (Figure 20.1). For the final seal, the anode, frame stack, and cathode were placed into a fixture and aligned. These parts were then fritted together in a vacuum oven. In the final step, the display was placed on a pump station via the pinch-off tube. It was pumped down, baked out for several hours, and sealed with a glass tube pinch-off process. The getter was then flashed by RF induction.

The nanotubes were particularly sensitive to several steps in this process. Subjecting nanotubes to even small partial pressures of oxygen above 450 °C caused degradation of the field emission behavior. The degradation was consistent with oxidation of the nanotubes, reducing both the overall current and the uniformity. While the cathode and anode were relatively clean parts, the glass frit contained dissolved gases which were liberated in a burst when the frit melted. The frit must be outgassed prior to the final seal. The oven is also a potential source of oxygen, and the walls and fixtures must be kept clean. With these procedures, Motorola observed that the degradation of emission during sealing was eliminated.

Obtaining a good vacuum environment in a sealed display is very important for ensuring a long device lifetime. This means that both gaseous species and oxygen-containing surfaces need to be carefully removed. The metal parts that go into vacuum devices such as CRTs and vacuum tubes are often baked out in a hydrogen reducing atmosphere prior to assembly. This step improves the device lifetime of vacuum devices by reducing the metal, outgassing the metal (especially oxygen-containing species), and loading the metal with dissolved hydrogen. One of the strengths of FED fabrication by direct growth of nanotubes is that the nanotube growth process is a reducing process. LETI and Motorola performed the nanotube growth process as the last step before sealing. Consequently, immediately prior to sealing, the cathode and all of its metal layers were subjected to an energetic reducing atmosphere. However, a direct link to lifetime was not reported, in part due to the inability to create a control sample with nanotubes but no reducing atmosphere.

20.7 Operating Lifetime

The operating lifetime of FEDs has historically been a challenging issue. *Emissive display lifetime* is generally defined as the time required for the luminance to fall to 50%. Products have different lifetime requirements. For example, portable consumer electronics displays and automotive displays often require lifetimes of 2000–3000 h. CRT televisions were generally specified to last for 10 000 h, but often did much better. FED programs have generally targeted 10 000 h of lifetime for television applications.

FEDs have two main degradation mechanisms: phosphor aging and emitter aging. Aging of CRT-type phosphors occurs as the efficiency of a phosphor decreases over time under electron bombardment, and is a function of the overall coulomb load. Phosphor aging is minimized by generating light with a high bias voltage and low electron current, as opposed to operating at a low voltage with a high electron current.

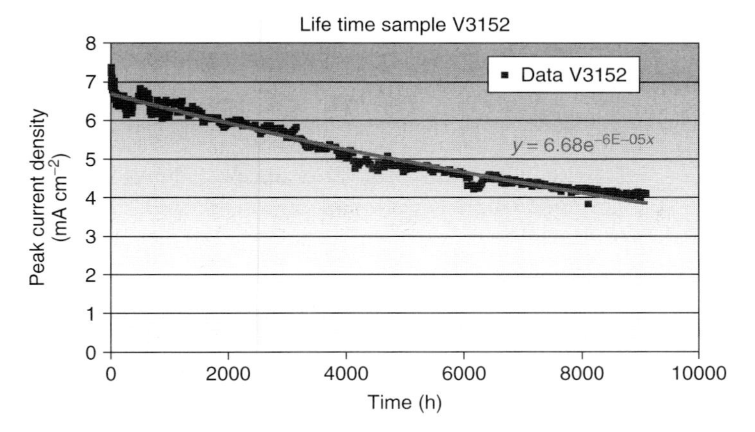

Figure 20.8 Lifetime measurements of a 1-in. frit-sealed display with a CNT grown by CVD. After 9100 h of continuous operations, the current decreases by 45%, leading to an extrapolated 50% lifetime of more than 10 000 h. The phosphor is Y2SiO5 : Tb and the getters are ST171 from SAES [3]. (Reprinted with permission from the Society for Information Display.)

The aging of metal field emitters has been attributed to nanoprotrusion destruction via ion bombardment and work function change through oxidation [39, 40]. Unlike metals, nanotubes are covalently bonded, and do not form stable surface oxides. Consequently, the above metallic degradation mechanisms are absent, and nanotube researchers had high optimism that nanotubes would show the desired lifetime behavior.

Dijon *et al.* demonstrated the lifetime performance of a nanotube FED with directly grown nanotubes. They reported a lifetime exceeding 10 000 h in a vacuum-sealed device with Y_2SiO_5: Tb phosphor at 3000 V (Figure 20.8) [3]. The initial peak current density was aggressive at 7 mA cm^{-2} of anode, and the test was performed at full brightness with 100% duty cycle (as opposed to a typical video signal with a 20% duty cycle).

Dijon *et al.* also investigated the degradation mechanism through current–voltage curve changes [3]. They observed that the slope of the Fowler–Nordheim curve remained unchanged but the intercept decreased over time. They concluded that aging does not occur through work function changes, but rather through a reduction in the number of emitting nanotubes at a given voltage. It remains unclear whether nanotubes are breaking [7, 41], shortening [7, 42], evaporating [43], opening caps [3], or subject to another mechanism, although each of these have been described in various studies.

20.8 Conclusions

Direct growth of nanotubes into FEDs was proposed as a means for better controlling the electrical performance characteristics of the displays. Several teams demonstrated solutions to the key challenge of growing high-quality field-emitting films on glass substrates at low temperature. The resulting nanotubes possessed small radii, adequate height, sufficient density, and tight distributions, thereby enabling devices to meet targets of drive voltage, current density, and uniformity. Investigators also demonstrated complete display systems capable of meeting display performance requirements for brightness, color purity, and longevity. The performance of FEDs fabricated by direct growth of carbon nanotubes has been validated. The next milestone will be the commercialization of this technology.

References

1. Dean, K.A., Coll, B.F., Howard, E., Johnson, M.R., Li, H., Marshbanks, L., and Dworsky, L. (2007) *J. Soc. Inf. Disp.*, **15**, 1047.
2. Ago, H., Kugler, T., Cacialli, F., Salaneck, W.R., Shaffer, M.S.P., Windle, A.H., and Friend, R.H. (1999) *J. Phys. Chem. B*, **103**, 8116.
3. Dijon, J., Bridoux, C., Fournier, A., Geffraye, F., De Monsabert, T.G., Montmayeul, B., Levis, M., Sarrasin, D., Meyer, R., Dean, K.A., Coll, B.F., Johnson, S.V., Hagen, C., and Jaskie, J.E. (2004) *J. Soc. Inf. Disp.*, **12**, 373.
4. Coll, B.F., Dean, K.A., Howard, E., Johnson, S.V., Johnson, M.R., Li, H., Jordan, D.C., Tisinger, L.H., Hupp, M., Smith, S.M., Young, S.R., Baker, J., Weston, D., Dauksher, W.J., Wei, Y., and Jaskie, J.E. (2005) Proceedings of EuroDisplay, 2005, pp. 144–147.
5. Kang, S., Brae, C., Son, W., Kim, M.H., Yi, J., Lee, S.T., Chang, A., Kim, J.J., Lee, C.R., Moon, J.H., Lim, S.H., Kim, H.S., and Jang, J. (2003) *Soc. Inf. Disp. Symp. Digest*, **34**, 802.
6. Dijon, J., Fournier, A., Goislard De Monsabert, T., Levis, M., Meyer, R., Bridoux, C., Montmayeul, B., and Sarrasin, D. (2006) *Soc. Inf. Disp. Symp. Digest*, **37**, 1744–1747.
7. Wei, Y., Xie, C.G., Dean, K.A., and Coll, B.F. (2001) *Appl. Phys. Lett.*, **79**, 4527.
8. Meyyappan, M., Delzeit, L., Cassell, A., and Hash, D. (2003) *Plasma Sources Sci. Technol.*, **12**, 205–216.
9. Ren, Z.F., Huang, Z.P., Wang, D.Z., Wen, J.G., Xu, J.W., Wang, J.H., Calvet, L.E., Chen, J., Klemic, J.F., and Reed, M.A. (1999) *Appl. Phys. Lett.*, **75**, 1086.
10. Li, Y.J., Sun, Z., Lau, S.P., Chen, G.Y., and Tay, B.K. (2001) *Appl. Phys. Lett.*, **79**, 1670.
11. Coll, B.F., Dean, K.A., Howard, E., Johnson, S.V., Johnson, M.R., and Jaskie, J.E. (2006) *J. Soc. Inf. Disp.*, **14**, 477.
12. Park, K.H., Lee, K.M., Choi, S., Lee, S., and Koh, K.H. (2001) *J. Vac. Sci. Technol., B V*, **19**, 946–949.
13. Ren, Z.F., Huang, Z.P., Xu, J.W., Wang, J.H., Bush, P., Siegal, M.P., and Provencio, P.N. (1998) *Science*, **282**, 1105.
14. Rodriguez, N.M. (1993) *J. Mater. Res.*, **8**, 3233–3250.
15. Chhowalla, M., Teo, K.B.K., Ducati, C., Rupesinghe, N.L., Amaratunga, G.A.J., Ferrari, A.C., Roy, D., Robertson, J., and Milne, W.I. (2001) *J. Appl. Phys.*, **90**, 5308.
16. Cui, H., Zhou, O., and Stoner, B. (2000) *J. Appl. Phys.*, **88**, 6072.
17. Chhowalla, M., Ducati, C., Rupesinghe, N.L., Teo, K.B.K., and Amaratunga, G.A.J. (2001) *Appl. Phys. Lett.*, **79**, 2079.
18. Yudasaka, M., Kikuchi, R., Matsui, T., Ohki, Y., and Yoshimura, S. (1995) *Appl. Phys. Lett.*, **67**, 2477.
19. Suh, J.S., Jeong, K.S., Lee, J.S., and Han, I. (2002) *Appl. Phys. Lett.*, **80**, 2392.
20. Nilsson, L., Groening, O., Emmenegger, C., Kuettel, O., Schaller, E., Schlapbach, L., Kind, H., Bonard, J.-M., and Kern, K. (2000) *Appl. Phys. Lett.*, **76**, 2071.
21. Kind, H., Bonard, J.-M., Emmenegger, C., Nilsson, L.-O., Hernadi, K., Maillard-Schaller, E., Schlapbach, L.,

Forro, L., and Kern, K. (1999) *Adv. Mater.*, **11**, 1285.

22. Cassell, A.M., Verma, S., Delzeit, L., Meyyappan, M., and Han, J. (2001) *Langmuir*, **17**, 260.
23. Uh, H.S., Lee, S.M., Jeon, P.G., Kwak, B.H., Park, S.S., Kwon, S.J., Cho, E.S., Ko, S.W., Lee, J.D., and Lee, C.G. (2004) *Thin Solid Films*, **462-463**, 19.
24. Shiratori, Y., Hiraoka, H., Takeuchi, Y., Itoh, S., and Yamamoto, M. (2003) *Appl. Phys. Lett.*, **82**, 2485.
25. Han, I.T., Kim, H.J., Park, Y.J., Lee, N., Jang, J.E., Kim, J.W., Jung, J.E., and Kim, J.M. (2002) *Appl. Phys. Lett.*, **81**, 2070.
26. Choi, Y.C., Jeong, K.S., Han, I.T., Kim, H.J., Jin, Y.W., Kim, J.M., Lee, B.G., Park, J.H., and Choe, D.H. (2006) *Appl. Phys. Lett.*, **88**, 263504.
27. Kang, S., Bae, C., Son, W., and Kim, J.J. (2005) *Soc. Inf. Disp. Symp. Digest*, **36**, 1940.
28. Dijon, J., Fournier, A., Levis, M., Meyer, R., Bridoux, C., Montmayeul, B., Muller, F., Nicolas, P., Sarrasin, D., Adasmski, J.R., Bellanger, J.L., Bellissens, D., Lefort, M., and Ricaud, J.L. (2007) *Soc. Inf. Disp. Symp. Digest Tech. Papers*, **38**, 1313–1316.
29. Ouguchi, T., Yamaguchi, E., Sasaki, K., Suzuki, K., Uzawa, S., and Hatanaka, K. (2005) *Soc. Inf. Disp. Symp. Digest Tech Papers*, **36**, 1929.
30. Yamaguchi, E., Sakai, K., Nomura, I., Ono, T., Yamanobe, M., Abe, N., Hara, T., Hatanaka, K., Osada, Y., Yamamoto, H., and Nakagiri, T. (1997) *J. Soc. Inf. Display*, **5**, 345.
31. Yamamoto, K., Nomura, I., Yamazaki, K., Uzawa, S., and Hatana, K. (2005) *Soc. Inf. Disp. Symp. Digest*, **36**, 1933.
32. Ghis, A., Meyer, R., Rambaud, P., Levy, F., and Leroux, T. (1991) *IEEE Trans. Electron Devices*, **38**, 2320.
33. Dean, K.A., Coll, B.F., Dworsky, L., Howard, E., Li, H., Johnson, S.V., Johnson, M.R., Johnson, S.V., and Jaskie, J.E. (2007) Performance of nanotube field emission displays. Proceedings of the International Display Manufacturing Conference, 2007, p. 108.
34. Dijon, C., Bridoux, C., Fournier, A., Goislard De Monsabert, T., Montmayeul, B., Levis, M., Sarrasin, D., and Meyer, R. (2005) Proceedings of International Display Workshops (IDW)/Asia Display, 2005, pp. 1635–1638.
35. Tanaka, M., Obara, Y., Naito, Y., Kobayashi, H., Toriumi, M., Niiyama, T., Sato, Y., Itoh, S., and Kawasaki, H. (2004) *Soc. Inf. Disp. Symp. Digest Tech. Papers*, **35**, 832.
36. Choi, Y.C., Jeong, K.W., Han, I.T., Kim, H.J., Kim, Y.W., Kim, J.M., Lee, B.G., Park, J.H., and Choe, D.H. (2006) *Appl. Phys. Lett.*, **88**, 263504.
37. Xie, C.G., Wei, Y., and Smith, B.G. (2002) *IEEE Trans. Electron Devices*, **49**, 324.
38. Curtin, C.J. and Iguchi, Y. (2000) *Soc. Inf. Disp. Symp. Digest Tech. Papers*, **31**, 1263.
39. Chalamala, B.R., Wallace, R.M., and Gnade, B.E. (1998) *J. Vac. Sci. Technol. B*, **16**, 2859–2865.
40. Wei, Y., Chalamala, B.R., Smith, B.G., and Penn, C.W. (1999) *J. Vac. Sci. Technol. B*, **17**, 233–236.
41. Bonard, J.-M., Klinke, C., Dean, K.A., and Coll, B.F. (2003) *Phys. Rev. B*, **67**, 115406.
42. Saito, Y., Seko, K., and Kinoshita, J. (2005) *Diamond Relat. Mater.*, **14**, 1843.
43. Dean, K.A., Burgin, T.P., and Chalamala, B.R. (2001) *Appl. Phys. Lett.*, **79**, 1873.

21
Transparent-Like CNT-FED

Takeshi Tonegawa, Masateru Taniguchi, and Shigeo Itoh

21.1
Diode-Type CNT-FED

Triode-type CNT-FEDs are well discussed in the previous chapters. A triode-type CNT-FED has CNT cathode electrodes, gate electrodes, and phosphor-coated anode electrodes in a vacuum panel. The electric potential of each electrode can be independently controlled. On the other hand, there are problems to be solved in the uniformity and emission efficiency, which depend on the accurate locations of the CNT tips and the effective gate edge.

In a diode-type CNT-FED, only CNT cathodes and phosphor-coated anodes are located in parallel in a vacuum panel. As the anode in the diode panel plays both as an extraction electrode for field emission and as an exciting electrode for the phosphor, the lower the extraction voltage, the lower the excitation voltage of the phosphor. Therefore, if a low-voltage-excitation phosphor that has high efficiency is available, the diode-type CNT-FED can be realized. Besides, the diode-type CNT-FED has the merit of simple structure, and the extraction field plane is formed parallel to the cathode surface, which makes the extraction field more effective than that for an edge gate type. The results and problems revealed through the investigation of diode-type CNT-FEDs, which have a simple construction and several merits, are reported in this chapter.

21.2
Structure of Diode-Type CNT-FED

The structure of a diode-type CNT-FED is shown in Figure 21.1. A phosphor layer of 15 μm thickness is coated on the anode electrode and a CNT emitter layer with a few micrometers thickness is printed on the cathode electrode. The gap between the anode and the cathode substrate is 35 μm obtained using spacers. Therefore, taking thickness of the phosphor and CNT layer into consideration, the real gap between the anode and the emission sites is estimated at about 15 μm. In the diode-type CNT-FED, the distance between the emission sites of the CNTs and the

Carbon Nanotube and Related Field Emitters: Fundamentals and Applications. Edited by Yahachi Saito

ISBN: 978-3-527-32734-8

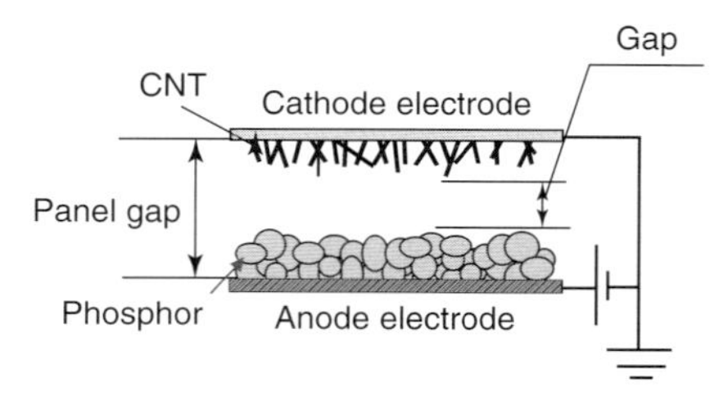

Figure 21.1 Diode-type CNT-FED.

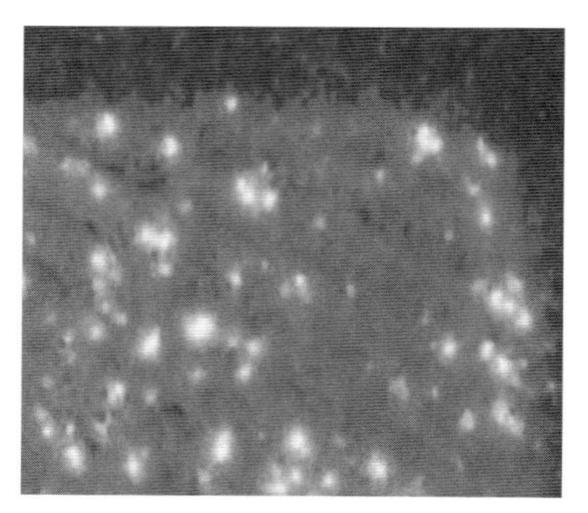

Figure 21.2 Picture showing the luminescent appearance (about 1 mm^2).

Figure 21.3 SEM image showing the fluff state of a CNT emitter layer.

surface of the phosphor layer is important. How the uniformity and reproducibility of the gap is maintained is a fundamental technical subject.

An example of luminescent appearance is shown in Figure 21.2, exhibiting the gathered cathodoluminescent points form the luminescent area. The scanning electron microscopy (SEM) image in Figure 21.3 shows the fluffy state of the CNT emitter layer. The CNT layer is also formed by the screen-printing method, and surface treatment is applied to make its surface fluffy to increase the efficiency of field emission.

The emitter ink paste was made of multiwalled CNTs which were produced by the arc-discharge method. The SEM picture also shows that there is a rather large variation in the length of each CNT. The observation of the luminescent appearance suggests a tendency of the distribution of luminescent points is in accordance with protruded portions of the phosphor layer. That is, the luminescent

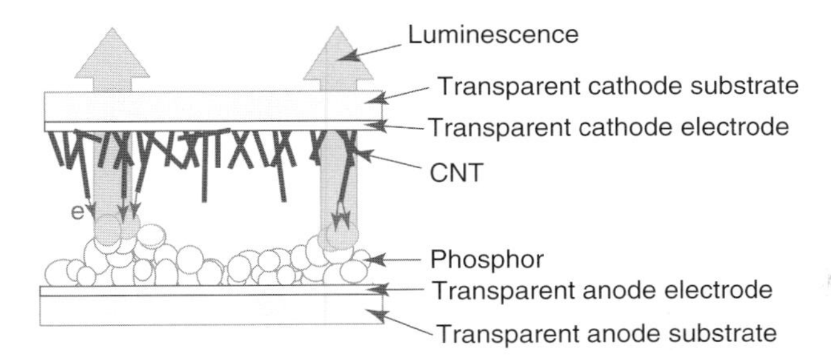

Figure 21.4 Schematic of transparent-like CNT-FED.

points are composed of the spots where emission sites of CNTs and the top surface of phosphor grains are close to each other accidentally, as illustrated in Figure 21.4. As the CNT emitter layer can supply electrons even though the thickness is only a few micrometers, a highly transparent cathode layer can be obtained. So, a transparent-like CNT-FED panel is expected to result by using a transparent cathode electrode, anode electrode, and substrates. The concept of the transparent-like CNT-FED is shown in Figure 21.4.

21.3 Characteristics of CNT-FED

The luminescent state, emission, and luminance characteristics of the CNT-FEDs with two different panel gaps (A, 35 μm; B, 40 μm) are shown in Figures 21.5–21.7, respectively.

These data show that the panel with the smaller gap shows a higher density of luminescent points, steeper rise in $J_a - V_a$ characteristics, and higher brightness under the same condition than the panel with a larger gap. The current density shown in Figures 21.6 and 21.7 were measured under a pulsed voltage drive with 1/16 duty ratio.

The luminescent efficiencies for the two panels with the different gaps are plotted as a function of electric power, as shown in Figure 21.8, where the power represents that when the pulse is on (i.e., the power when the duty ratio is assumed to be unity).

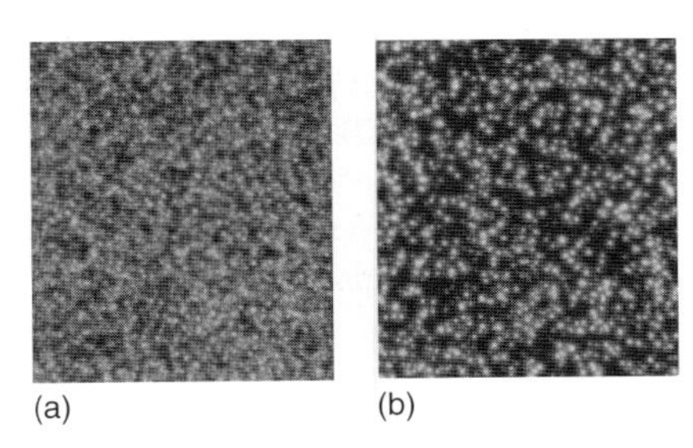

Figure 21.5 Microscopic observation of luminescent state (120 V). Short side of the pictures is 3 mm long.

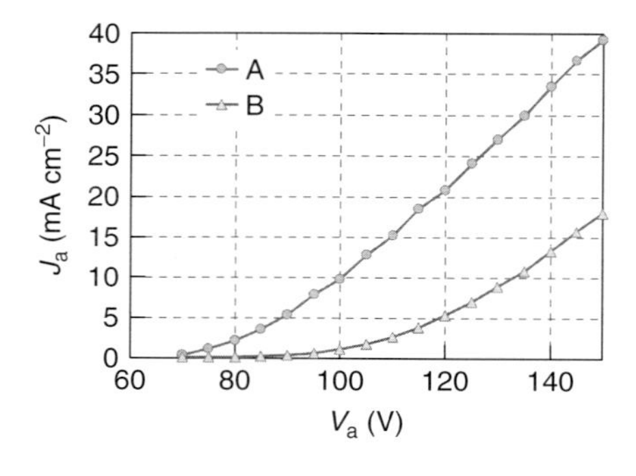

Figure 21.6 Emission current density versus applied voltage.

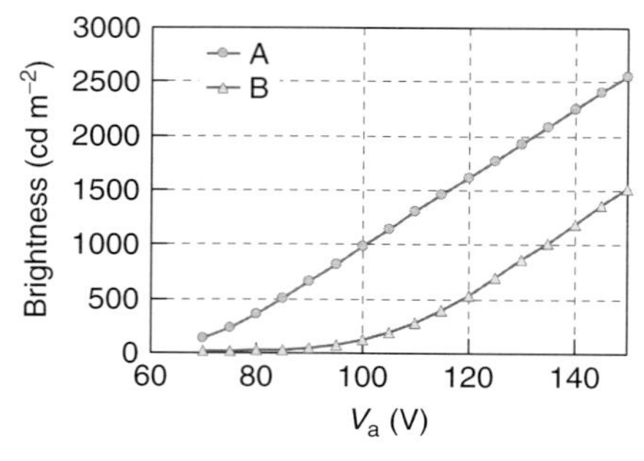

Figure 21.7 Brightness versus applied voltage.

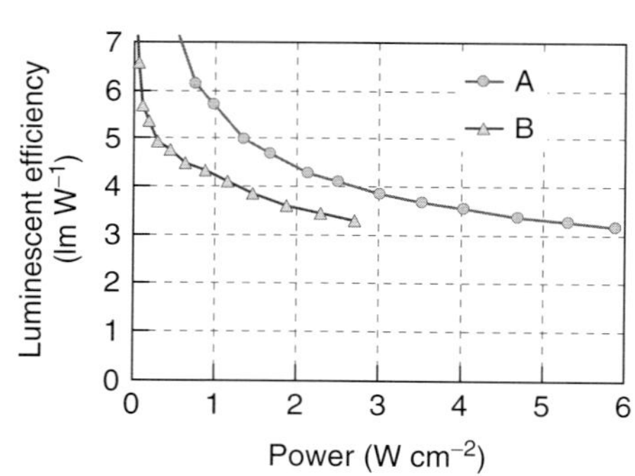

Figure 21.8 Luminescent efficiency versus power.

It is found that as the electric power increases, the luminescent efficiency decreases. This phenomenon is considered to originate from the physical property of the phosphor material.

As the gap increases, both the luminance point density and luminescent efficiency decrease. The reason for this is that the luminescent area on the phosphor layer becomes small. We tried to quantify the luminescent area with a picture analysis. The luminescent point pictures were recorded and analyzed by a binarization processing method. The results obtained from the pictures in Figure 21.5 are shown in Figure 21.9. The resultant luminescent area is plotted as a function of the electric power, as shown in Figures 21.10 and 21.11.

It is clearly revealed that, as the gap becomes smaller, the luminance starts from a lower voltage and the luminescent area grows larger under the same electric power.

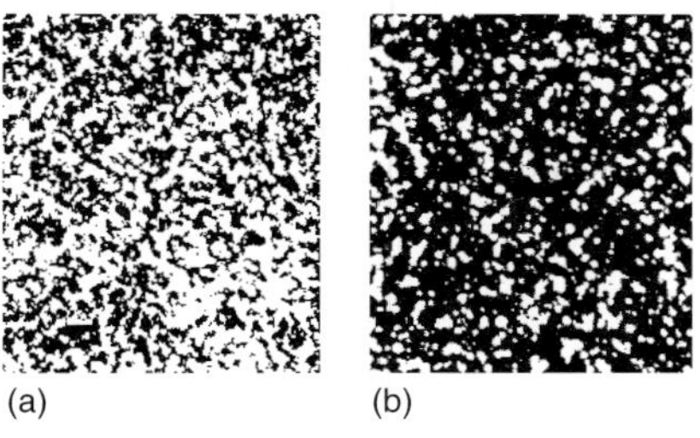

Figure 21.9 Binarization analysis of Figure 21.5. (a) 35 μm gap and (b) 40 μm gap.

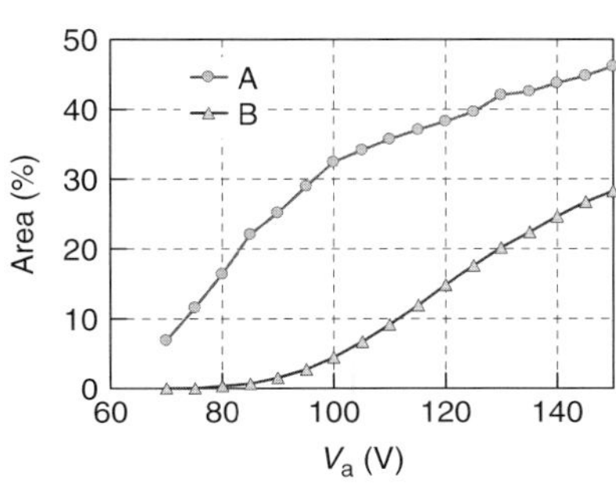

Figure 21.10 Luminescent area versus applied voltage.

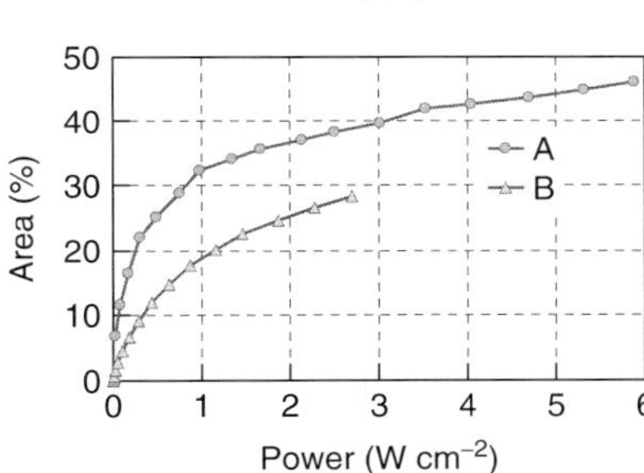

Figure 21.11 Luminescent area versus power.

The formula of luminescent efficiency is as follows:

$$\eta = \frac{\pi \times L}{V \times \left(\frac{I}{S}\right)} \tag{21.1}$$

where η is luminescent efficiency (lm W^{-1}), L is the brightness (cd m^{-2}), V is the anode voltage (V), I is the anode current (A), and S is the area (m^2).

When the luminescent area increases, the current density (I/S) becomes small and thus the luminescent efficiency improves. By increasing the luminescent area and improving the uniformity, the luminescent efficiency can be raised close to the intrinsic efficiency of the phosphor material.

21.4
Relation between Gap and Emission

The dispersion of current density versus gap for the test panels is shown in Figure 21.12. Values of the gap used in the figure are the calculated distances

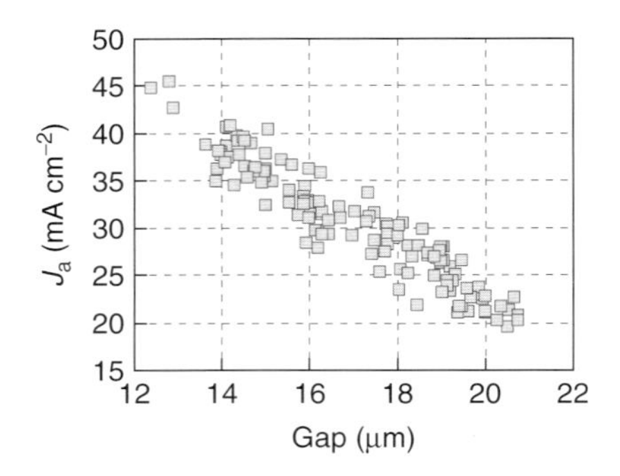

Figure 21.12 Current density versus gap.

from the panel gap minus the thickness of a phosphor layer. A strong correlation between the gap and the emission current is found. When the gap varies in a range of 6 μm, the corresponding emission current changes by a factor of about 2.

The dispersion of emission was about 15% even at a given gap. This is considered to be due to the dispersion of the height of CNTs which are projecting out of the cathode surface. This mean that the characteristic of the diode-type CNT-FED depends on the accuracy and uniformity of spacer thickness and the height of CNTs layer.

21.5 Property of CNT-FED

Figure 21.13 shows the diode-type CNT-FED panel unit, and Table 21.1 shows the properties of the unit panel.

This panel has the advantage of low power consumption, which is due to the high luminance of the low-voltage phosphor ZnO : Zn, wide viewing angle, small thickness, and light weight. As the luminescent area is about 40%, there is still room for improving the luminescent property and lowering the power consumption by increasing the luminescent area. All the driving lines are laid out on one side of the panel, as shown in Figure 21.13. This design enables the panels to be combined side by side in order to make a large screen.

21.6 Nonevaporable Getter

The Ba getter which is often used in vacuum tube or vacuum fluorescent device is shown in Figure 21.14a. In the case of using an evaporable Ba getter of 0.5 mm thickness, it is necessary to keep sufficient space around it in order to make a sufficiently wide getter mirror film by evaporation.

To accomplish a thin CNT-FED with as much thickness as the two glass plates, it is indispensable to develop a nonevaporable getter which can be screen-printed in the gap of the anode and cathode substrates. The photograph of nonevaporable

Figure 21.13 The panel unit.

Table 21.1 Properties of the unit panel.

Size (mm)	96 × 96
Pixel pitch (mm)	6
The number of pixels	16 × 16
Thickness (mm)	2.2
Anode voltage (V)	120
Cut-off voltage (V)	70
Emission current (mA)	32 (all pixels lit)
Power consumption (W)	3.8 (all pixels lit)
Brightness (cd m^{-2})	1800
Viewing angle (°)	170°
Phosphor	ZnO : Zn

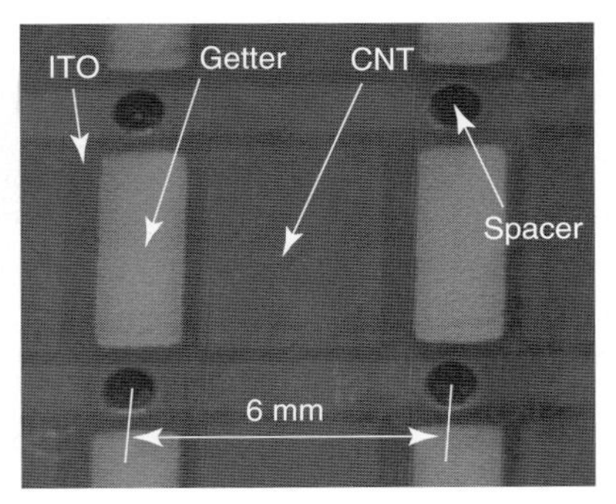

Figure 21.14 (a) Ba getter and (b) nonevaporable getter.

getters that were fabricated on the cathode substrate is shown in Figure 21.14b. The fabricated CNT-FEDs are shown in Figure 21.15.

The CNT-FED in Figure 21.15a has a getter space in which Ba getters are contained, while those in Figure 21.5b,c, the thickness of which are 2.2 and 0.8 mm, respectively, have nonevaporable getters inside the gap. For CNT-FEDs with nonevaporable getters, the CNT emitter layer, spacers, and the getters are fabricated with the screen-printing method.

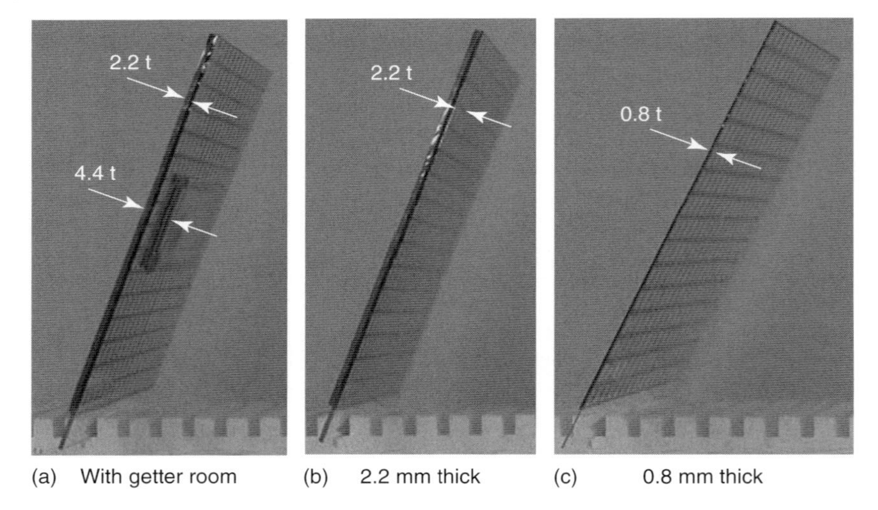

Figure 21.15 (a) Fabricated CNT-FEDs with Ba getter space and (b, c) with nonevaporable getter.

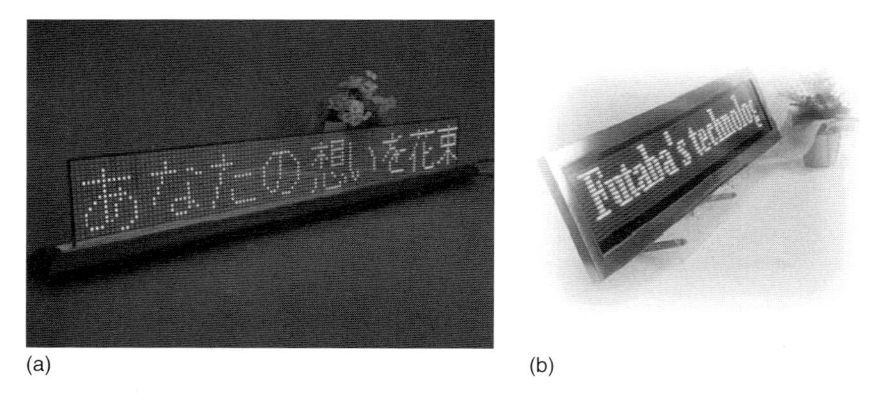

Figure 21.16 (a) Transparent-like CNT-FED panel and (b) a normal diode-type CNT-FED panel.

By employing nonevaporable getters and substrates which have more transparent areas, transparent-like CNT-FEDs can be fabricated, as shown in Figure 21.16.

21.7 Summary

Diode-type CNT-FEDs have been developed. This panel has several excellent features, such as low power consumption, wide viewing angle, small thickness, and light weight, in spite of its simple structure. It is found that the characteristics of the diode-type CNT-FEDs depend on the accuracy and uniformity of the spacer thickness and the height of CNT layer. By improving the characteristics and the manufacturing technology, the luminescent efficiency of the CNT-FEDs can be improved further, which would enable these novel display panels to be used as energy-saving devices.

Moreover, the transparent-like CNT-FEDs have been developed. Continuing the improvement of CNT-FEDs will advance the development of new display devices which can be applied to new-generation media so as to convey information to the public in real time.

References

1. Itoh, S., Tanaka, M., and Tonegawa, T. (2004) Development of field emission displays. *J. Vac. Sci. Technol.*, **B 22** (3), 1362–1366.
2. Itoh, S., Tanaka, M., Tonegawa, T., Obara, Y., Naito, Y., Niiyama, T., Kobayashi, H., Sato, Y., Toriumi, M., Takeya, Y., Taniguchi, M., Namikawa, M., Yamaura, T., and Kawasaki, H. (2004) Development of field emission display. Proceedings of the IDW'04, Niigata, 2004, pp. 1189–1192.
3. Tonegawa, T., Taniguchi, M., Itoh, S., Nawamaki, K., Marushima, Y., Kubo, Y., Fujimura, Y., and Yamaura, T. (2005) Proceedings of the IDW'05, Takamatu, 2005, pp. 1659–1662.

22
CNT-Based FEL for BLU in LCD

Yoon-Ho Song, Jin-Woo Jeong, and Dae-Jun Kim

22.1
Introduction

The carbon nanotube (CNT) has been intensively studied as a field emitter cathode because of its high aspect ratio with a nanometer-scale diameter and high chemical inertness [1–5]. The excellent electron emission from CNT field emitters allowed them to be used in field emission displays (FEDs) for a relatively long time [6–10]. However, the CNT-based FED (abbreviated to CNT-FED) has still a serious technical problem in uniformity: especially in short-range pixel-to-pixel uniformity compared with the long-range one. The short-range uniformity required for a commercial graphic display is over 98%, which is rarely achieved with CNT field emitters owing to their irregularity in height and diameter. Of course, a high anode voltage for electron acceleration along with an electron beam focusing imposes additional difficulties on the CNT-FED.

Recently, a CNT-based field emission lamp (FEL) has attracted much attention, rather than FEDs [11–25]. The technical huddles in CNT-FED can lessened in the FEL technology; however, other performance factors such as luminance, efficiency, and lifetime are crucial for the FEL application. The CNT-FEL for backlight unit (BLU) (abbreviated to CNT-BLU) in liquid crystal display (LCD) has been developing to improve the image quality of LCD. Figure 22.1 shows the evolution of BLU in LCD TV in the future. The conventional BLU, at present, adopts a cold cathode fluorescent lamp (CCFL) or an external electrode fluorescent lamp (EEFL) based on gas discharge and provides a constant light source to the LCD panel, resulting in low contrast ratio, motion blur, and high power consumption along with an intrinsic liquid crystal property. There is always a small amount of transmitted light even when an LCD presents a black image. Since the CCFL-BLU illuminates the whole display area including the black image area, it is not possible to represent a black image with a very low gray scale. Another critical limit of the conventional LCD with a CCFL-BLU is the motion blur phenomenon. It is caused by its sample-and-hold nature, that is, the liquid crystal remains in the same state after addressing during a whole frame. When displayed objects move, as in the case of TV images, it produces a blurred

Carbon Nanotube and Related Field Emitters: Fundamentals and Applications. Edited by Yahachi Saito

ISBN: 978-3-527-32734-8

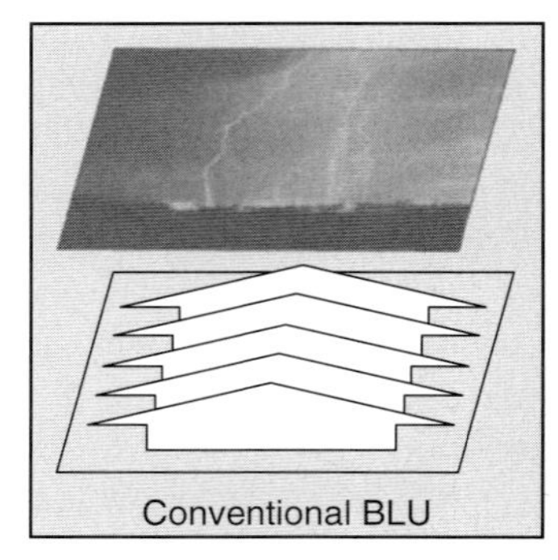

Figure 22.1 Evolution of BLU in TFT-LCD. The conventional BLUs provide a constant light source to the LCD panel while the local dimming BLU controls the luminance locally according to LCD images.

image of the objects on the retina of a viewer [26]. The local dimming BLU controls the luminance locally according to LCD images, enhancing the contrast ratio, and also another function of an impulsive scanning of BLU can overcome the motion blur, showing a cathode ray tube (CRT)-like image with the LCD panel.

A light emitting diode (LED)-BLU with local dimming has been reported for improving the contrast ratio of the LCD [27–32]. However, the size of the dimming block in the LED-BLU (a few tens of square centimeters) is not small enough to represent fine luminance control, causing light leakage in some black areas close to a bright image. In addition to the dimming, a local brightening is also required to increase the contrast ratio further. However, it is very difficult for the LED-BLU to increase the luminance locally. In contrast with LED-BLU, the CNT-BLU can be made with a very fine pitch to have a very small local block (finally down to the pixel pitch of LCD panel) and have the functions of both locally controllable luminance (dimming and brightening) and impulse-type scanning. Therefore, the LCD image with a CNT-BLU can exhibit ideal image characteristics like CRT, as shown in Figure 22.2. Also, the CNT-BLU can greatly reduce the number of optical films, finally to a single diffusion plate or sheet, while the CCFL- and LED-BLUs require many optical components such as diffusion plate, diffusion sheet, and double brightness enhancing film (DBEF) and brightness enhancing film (BEF). Table 22.1 summarizes and compares various BLUs in thin-film transistor (TFT)-LCD including the CNT-FEL as a next-generation one.

In this chapter, we will review the CNT-BLU technology including the basic structure and fabrication process for CNT cathode, anode, and vacuum packaging. Also, the driving scheme will be depicted for stable operation of the CNT-BLU.

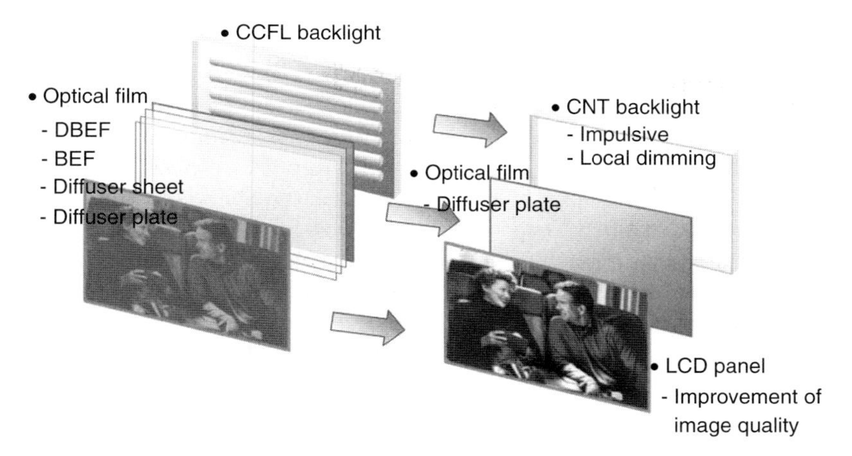

Figure 22.2 Comparison of CCFL- and CNT-based BLUs. The CCFL-BLU requires many optical components such as a diffusion plate, diffusion sheet, and (double) brightness enhancing film (DBEF, BEF), while the CNT-BLU can greatly reduce the optical films, finally to a single diffusion plate or sheet.

Table 22.1 Comparison of various BLUs in TFT-LCD.

Light source	Main luminescent mechanism	Efficiency (lm W^{-1})	Advantages	Disadvantages
CCFL	Glow discharge	60–70	High efficiency Low cost	Linear lamp Mercury
EEFL	Wall charge	60–70	Parallel driving Low cost	Linear lamp Mercury
LED	Charge recombination	30–40	High color gamut Mercury-free Local dimming, impulsive driving	Point source (many optical components) High cost Heat/complicate driving
CNT-FEL	Field emission/ cathodo-luminescence	30–40	Surface emitting lamp Mercury-free Very fine local dimming/brightening, impulsive driving Simple optical sheet	Under development (efficiency, lifetime, large area)

22.2
CNT-FEL Structure

Figures 22.3 and 22.4 shows perspective views of a CNT-FEL with normal and common gate structures, respectively [16–20, 24]. The CNT-FEL consists of cathode, gate, and anode parts presenting a typical vertical triode structure. The gate part including the gate insulator and electrode may be integrated onto the cathode substrate or formed separately from the cathode and then combined with the cathode during a packaging process. A soda lime glass is normally used as cathode and anode substrates for easy vacuum packaging like in a plasma display panel (PDP).

The CNT emitters reside in the gate holes with an insulator of several micrometers in thickness on the cathode electrodes. In the normal gate structure, the cathode and gate electrodes with a stripe shape are arranged in a matrix format like FED and address the dimming signals to each local block. Meanwhile, the common gate structure shown in Figure 22.4 has only one gate sheet like a metal mesh with many openings for gate holes, in which the dimming signals are addressed through only the cathode electrodes connected to each local block separately. The gate insulator in the common gate structure supports the spacing between the cathode electrodes and the common gate sheet and so may be formed in a discrete insulation pillar made of glass or ceramic instead of conformal (entire) deposition for the case of the normal gate.

The anode is mainly composed of cathodoluminescent (CL) phosphors with a back metal of aluminum (Al). Occasionally, a transparent electrode such as of indium tin oxide (ITO) may be interposed between the anode substrate and the phosphor layer. The CL phosphors for FEL applications take from the CRT

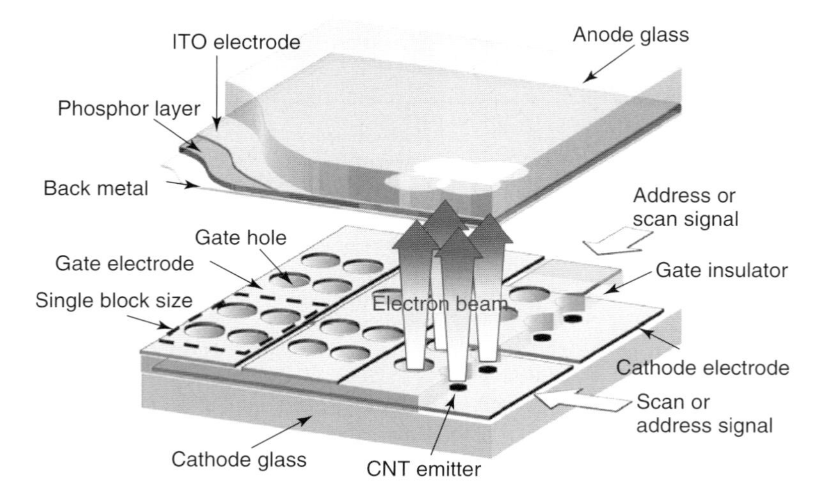

Figure 22.3 Perspective view of CNT-FEL with a normal gate structure. In the normal gate structure, the cathode and gate electrodes with a stripe shape are arranged in a matrix format and address the dimming signals to each local block.

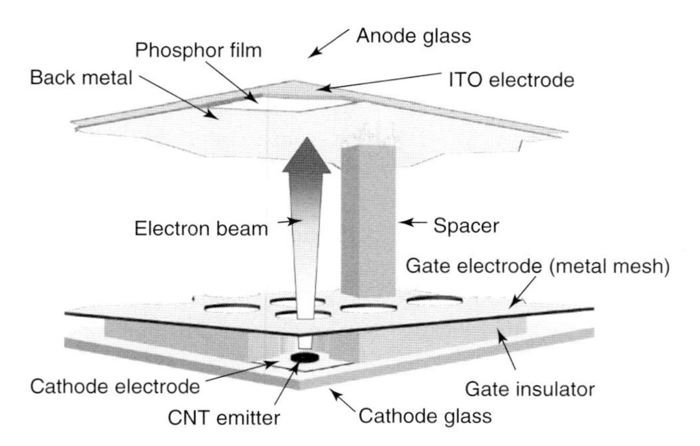

Figure 22.4 Perspective view of CNT-FEL with a common gate structure. The common gate structure has only one gate sheet like a metal mesh, in which the dimming signals are addressed through only the cathode electrodes connected to each local block separately.

technology although several researchers have reported on a low-voltage phosphor [20, 33, 34]. However, the efficiency and lifetime of the low-voltage phosphor proposed are not high enough so far to apply it to FEL and so a high-voltage accelerated electron beam over 10 kV is needed to achieve the luminance, efficiency, and lifetime from the CL phosphors for commercial use. Furthermore, the CL phosphor has an electrical insulating property so that the Al back metal is required to discharge the charges on the phosphor and also to reflect the backward emitted CL to the forward direction for enhancing the luminance and efficiency. The high-voltage architecture with the Al back metal should be developed unless low-voltage phosphors are successfully developed. The phosphor layer, in general, is formed from a mixture of red (R), green (G), and blue (B) emitting CL particles, resulting in a white light source for a BLU or lighting application. In special applications like a color-tunable lighting or a field sequential LCD, the primary color of R, G, and B phosphors are separately formed and operated with a different duty cycle or addressed sequentially.

The spacer, which is made of ceramic or glass in a pillar or rib type, is a prerequisite to the FEL supporting the spacing between the cathode (including the gate) and anode parts from atmospheric pressure. In the application of FED, the spacer has been a major technical problem, limiting the panel resolution and the luminance by a charging phenomenon [35]. The charging might be caused by the electrons from the CNT emitter bombarding the spacer and the successive emission of secondary electrons from the spacer. The charged (mainly positively charged) spacer deflects the electron trajectory and limits the electron acceleration voltage applied to the anode, deteriorating the panel resolution and the luminance. However, this spacer problem can be easily avoided in the FEL application by designing the spacer-loaded area to be wide enough for the electrons not to bombard to the spacer. The darkening at the spacer can be resolved with the help of an additional diffusion sheet in the BLU, giving a uniform luminance with hardly sacrificing the thickness of BLU.

Other FEL structures besides the vertical cathode-to-gate architectures shown in Figures 22.3 and 22.4 have been proposed. In the planar structure [13, 14], the cathode and gate are formed in the same plane and so can be easily fabricated. However, the gate in the planar structure cannot shield the CNT emitters from the anode electric field, thereby giving rise to an anode-induced field emission and limiting the applied anode voltage. As described above, the limitation in the anode voltage restricts the use of efficient high-voltage CL phosphors in FEL applications. Also, the diode configuration in a flat panel or tube-type FEL [15, 21] hardly can control the field emission current independently of the acceleration voltage, which, in turn, strongly limits the anode voltage for acceleration along with difficult control of the field emission current.

22.3 CNT Cathode

Although a variety of carbon molecular structures such as carbon nanofiber, nanowall, and graphene sheets have been reported for their good field emission [36–41], CNT is still a good candidate as a field emitter. There have been two methods to fabricate CNT or CNT-like field emitters for FEL applications. One is the direct growth technique on the cathode substrate by using chemical vapor deposition (CVD) [42, 43]. Up to now, the growth temperature is not low enough to grow CNT emitters on sodalime glass. So, several researchers have grown CNT emitters on metal stripes first and then attached the metal stripes onto the cathode substrate to obtain an excellent CNT cathode [44, 45]. The other is a paste-printing technique which is very useful for large-area, low-cost fabrication of the CNT cathode [18–20]. The paste formulation and printing process with a screen mask are commonly used in PDP fabrication [46, 47]. We will introduce the paste-printing technique for CNT emitters in detail.

Figure 22.5 shows an example of process flow for CNT emitters by the paste-printing technology. A single-walled or multiwalled CNT powder synthesized by an arc discharge or thermal CVD is first dispersed in a solvent such as isopropyl alcohol (IPA). The dispersed CNTs are then mixed with a nanosize metal as an inorganic filler, an organic binder based on acrylate, monomers, and negative photoinitiators using a high-speed homogenizer. The acrylate binder with the monomers and negative photoinitiators are chosen for lithographic patterning after screen printing with the CNT paste. As the final step of CNT paste, the CNT mixture is sufficiently milled by using a three-roller mill. The produced CNT paste is screen-printed on the cathode substrate with a screen mask and then exposed to UV light from the backside of the cathode. The development process that follows using a solvent forms a fine and even CNT pattern. The patterned CNT paste on the cathode is fired at a high temperature of 250–350 °C to burn out the organic

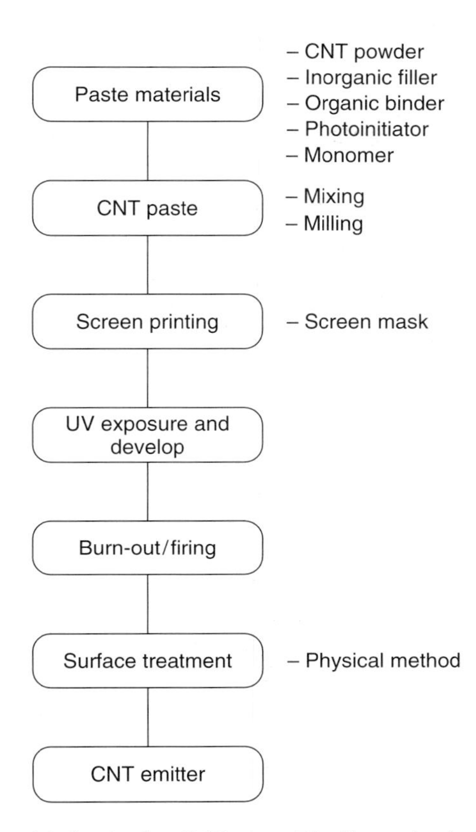

Figure 22.5 Process flow of CNT emitters by paste-printing technology.

binder in the CNT paste. Finally, a physical surface treatment is performed using a weak adhesive roller, laser, plasma, or liquid elastomer to form CNT emitters with vertical protrusion [48–58]. The dispersion of CNTs in the paste formulation is still not satisfactory because of its physical (a high aspect ratio with a diameter of several nanometers) and chemical properties (aggregation even with the use of a surfactant). The uniformity of field emission current, however, can be improved through post-processing such as trimming of tall CNT emitters under various environments [59–61].

The field emission properties of the CNT emitters formed by the paste-printing technique strongly depend on the paste formulation along with the emitter process. In order to investigate the compositional effect on the field emission from the CNT paste, we prepared four types of CNT pastes with various compositional ratios as shown in Figure 22.6. The CNT pastes were controlled by two relative content ratios. One is the relative content between the CNT and inorganic filler, and the other is the relative monomer quantity to the photoinitiator. Here, in order to enhance the adhesion of CNT emitters to the cathode electrode, a nanosized metal particle with a low melting temperature was used as the inorganic filler. In general, the nanosized material has many physical and chemical properties different from those of the bulk due to its large specific surface. When the metal particle gets smaller, it has a

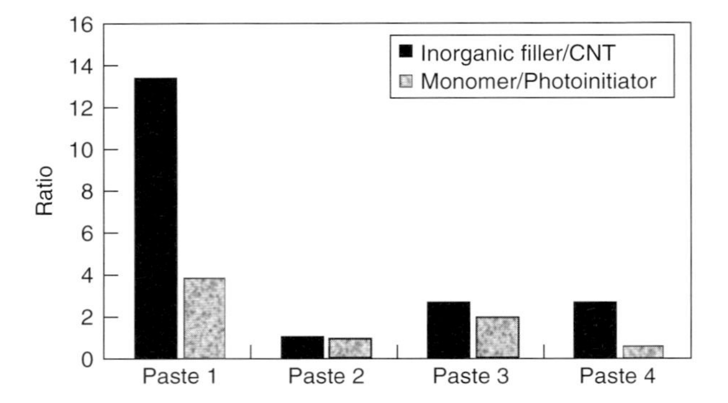

Figure 22.6 Four types of CNT pastes with two relative content ratios of CNT to inorganic filler and monomer to photoinitiator. The optimized composition corresponds to paste 4.

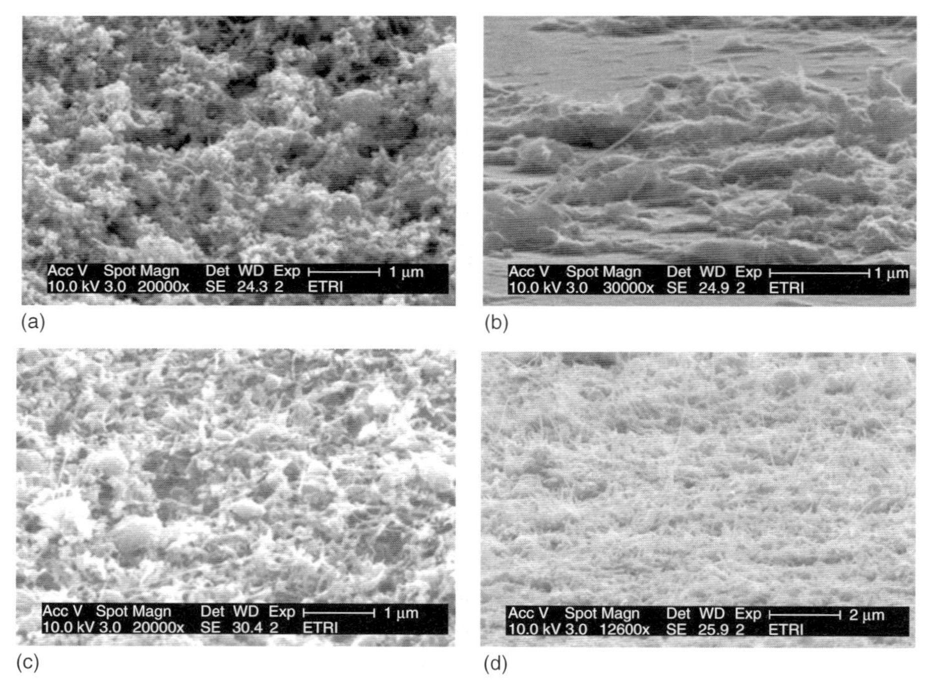

Figure 22.7 SEM images of CNT emitters after surface treatment for (a) paste 1, (b) paste 2, (c) paste 3, and (d) paste 4 with the optimized composition. The surface treatment was performed by using a rolling adhesive.

lower melting temperature. Therefore a strong adhesion between CNT emitters and cathode electrode can be expected by using a proper nanosized metal particle as the inorganic filler. The optimized composition in Figure 22.6 corresponds to the paste 4, and the comparative results of the other compositions (pastes 1–3) will be described. For each CNT paste, the screen printing was performed with a screen mask having 4800 dot patterns of 300 μm by 250 μm each on a sodalime glass coated with ITO. The burn-out/firing process was executed at 260 °C in an air ambient using a furnace. In this firing step, the organic binder was burned out and the nano metal was melted. The physical surface treatment using a rolling adhesive was applied to project and align CNTs vertically. Finally, in order to carbonize the remaining binder after the surface treatment, a second firing was done at around 400 °C in a vacuum chamber of below 10^{-5} Torr.

The surface morphology of the CNT emitter samples is presented in Figure 22.7, in which paste 4 has the optimum composition ratio. Pastes 1–3 showed that the polymer binders wrapped around the CNT surface excessively, and the attachment of the CNT emitters on the cathode electrode was poor. In the case of paste 4 with the optimized composition, a pure CNT surface could be observed along with better and smooth morphology, attachment, and uniformity. Also, paste 4 had a relatively enhanced vertical alignment of CNT emitters as shown in Figure 22.7d. Figure 22.8 shows the field emission images from the four CNT emitter samples with the anode having a green phosphor. The field emission was measured in a vacuum

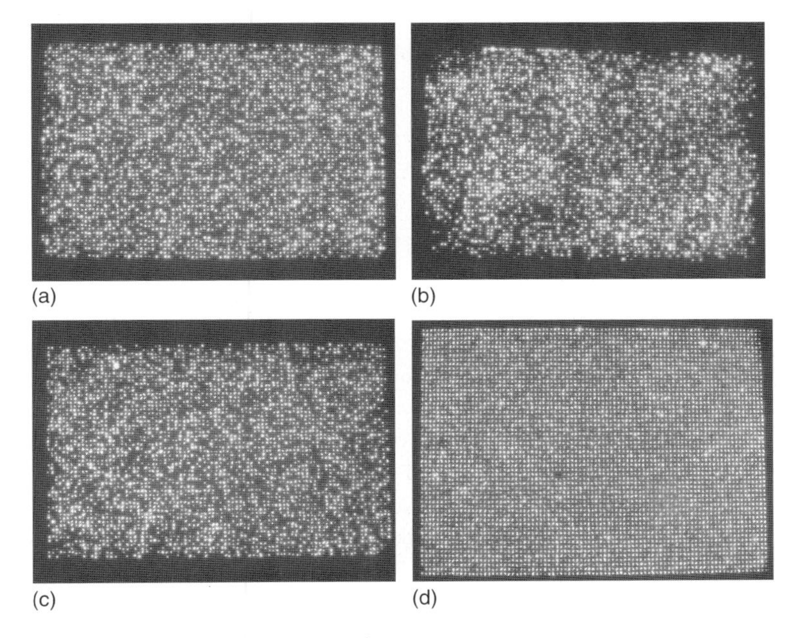

Figure 22.8 Field emission images by (a) paste 1, (b) paste 2, (c) paste 3, and (d) optimized paste 4. The field emission was measured in the diode configuration with an anode-to-cathode spacing of 300 μm.

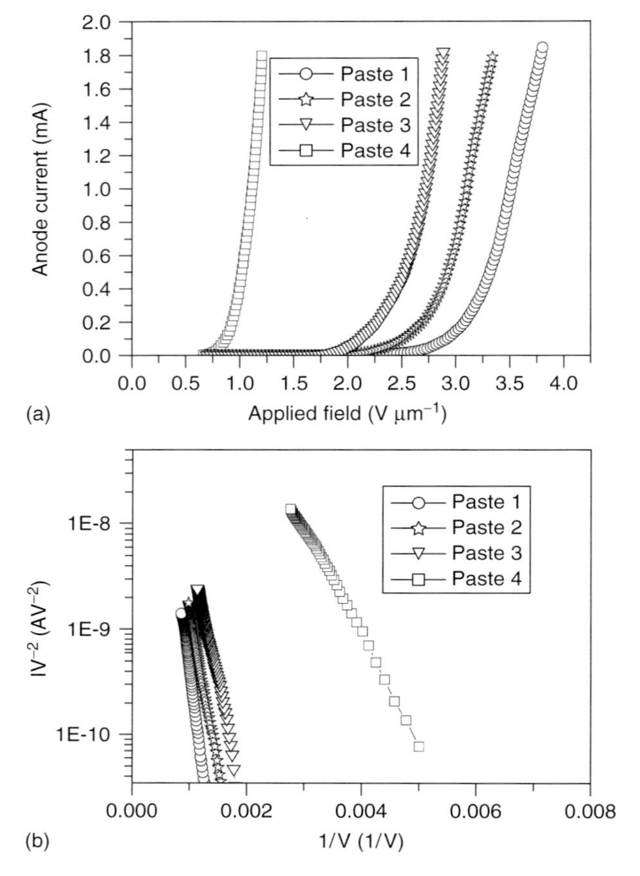

Figure 22.9 (a) Field emission characteristics of CNT emitters from paste 1 to optimized paste 4 measured in the diode configuration with an anode-to-cathode spacing of 300 μm and (b) the corresponding Fowler–Nordheim plots.

chamber of about 5×10^{-5} Torr in the diode configuration with an anode-to-cathode spacing of 300 μm. In order to protect the device and measuring equipments from instantaneous arcing in a high-voltage regime, a serial resistor of 200 kΩ was connected to the anode. The optimized CNT paste has a relatively uniform field emission, while the others have randomly, spatially distributed field emission sites with a very large variation in emission properties. The optimized CNT paste showed a relatively uniform field emission property. The improvement of emission properties in paste 4 is clearly seen from the field emission current versus voltage characteristics shown in Figure 22.9. Paste 4 was observed to have sufficient emission current under an apparent electric field of 2.5 V μm^{-1} along with a good uniformity, as shown in Figure 22.8d. The corresponding Fowler–Nordheim (FN) plots showed a higher intercept and lower slope for paste 4 than the others, indicating that the

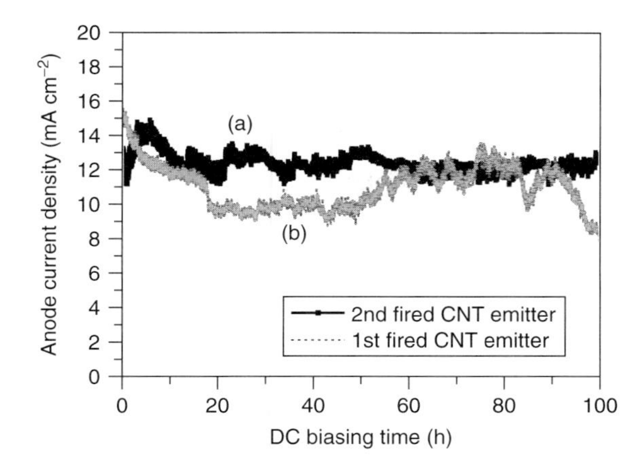

Figure 22.10 Anode current density as a function of time under a DC biasing for 100-h consecutive operation for the optimized CNT emitters: (a) second fired CNT emitter after first firing and surface treatment process and (b) first fired and then surface treatment. The first firing process was executed at 260 °C in an air ambient and the second firing was done at around 400 °C in a vacuum chamber of below 10^{-5} Torr after the surface treatment.

CNT emitters from paste 4 has more effective emission sites with lower work function than those from the pastes 1–3, as shown in Figures 22.7 and 22.8. Also, Figure 22.10 exhibits the lifetime test of CNT emitters from the optimized paste 4 under a constant DC operation at a very high current density of over 10 mA cm^{-2}. The field emission current was very stable over long periods, about 100 h, under the severe environment such as a high-level DC bias. We also could not observe any detachment of CNT emitters involving an arcing damage during the field emission measurements for the paste 4. The nano metal in the CNT paste may enhance the adhesion of CNT emitters to the cathode electrode. It is noted that field emission properties strongly depend on the CNT paste and emitter preparation process as well as the CNT itself. In order to ensure a high field emission current and good reliability, a clean surface should be maintained after the CNT emitter fabrication.

Most of applications of field emission devices need a triode configuration in which the gate induces electron emission from the field emitters and the anode receives the emitted electrons. For an ideal triode operation, the gate should shield the CNT emitters from the anode electric field perfectly, showing no anode-field-induced emission. Furthermore, the emitted electrons should be directed to only the anode, showing no gate leakage current. The Spindt tip has been an ideal triode structure because only a single emitter tip is located at the center of the gate opening and also the height of the tip apex can be adjusted to the outlet of the gate opening by controlling its aspect ratio to be nearly 1 [62]. In case of CNT emitters, however, it has been very difficult to make an ideal triode device because of the many irregular emitters in a gate hole. The triode configuration and the operation of CNT emitters largely depend on the fabrication process of gate opening along with an insulator. Desirably, the gate opening should be prepared on a thick insulator, resulting in

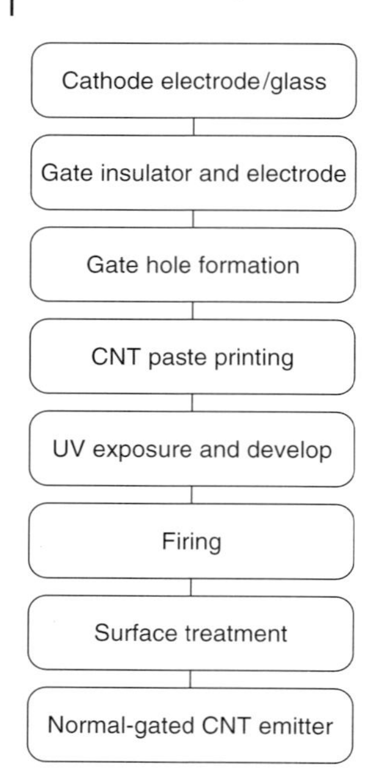

Figure 22.11 Process flow of normal-gated CNT emitters.

a higher gate position than the CNT tip heights. This leads to the suppression of uncontrollable field emission caused by a high anode bias.

Now, we introduce CNT emitter processes with the normal and the common gate structures. The normal-gated CNT emitters could be fabricated by a thick-film process used in PDPs. Figure 22.11 shows a process flow of the normal-gated CNT emitters [20]. Cathode electrodes with a stripe shape were formed on a substrate of soda lime glass. For the use of a photolithographic CNT paste, a transparent film such as of ITO can be used as the cathode electrode material. An insulator film with a thickness of above 10 μm was prepared on the cathode glass plate by screen printing of a glass paste mainly composed of $PbO-SiO_2-B_2O_3$, followed by a firing process at a high temperature of over 550 °C. Then, as the gate electrode a thin film of metal was deposited on the insulator by sputtering. The gate holes were formed on the gate metal and insulator by using photolithography and then etching process. Here, a UV-blocking layer might be adopted inside the gate holes using photoresist to define the size of CNT emitters at the desired positions [63]. After a screen printing of the CNT paste with a negative-type photopatternable resin, the backside of the cathode glass was exposed to UV. The development process that followed formed one paste dot per gate hole. The cathode was then fired at a proper temperature to burn out the organic binders in the CNT paste. Finally, a physical surface treatment was carried out using a weak adhesive roller or liquid elastomer to form CNT emitters with vertical alignment. The surface treatment using a liquid elastomer is a very effective method to form a fine CNT emitter in a small gate hole. The small gate hole means that the turn-on and operation voltage applied to the gate can be reduced, which, in turn, reduces the driving power and

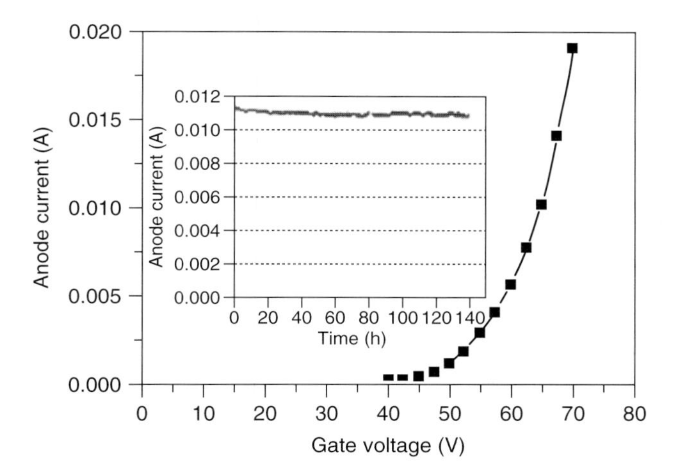

Figure 22.12 Anode current versus gate voltage curve of a sealed 32 in. diagonal CNT-FEL with a normal gate structure. The inset represents the emission stability, anode current versus operation time, of the panel. (Reproduced from [20] by Y. C. Choi, *et al.* (2008) *Nanotechnology*, **19**, 235306.)

the driver circuit cost. In general, the driving power is proportional to the square of the operation voltage for a capacitive load, and the cost of semiconductor integrated circuits strongly depends on the operation voltage.

Figure 22.12 shows the anode current (I_a) versus gate voltage (V_g) curve of a sealed 32 in. diagonal CNT-FEL with the normal gate structure fabricated by Choi, *et al.* [20]. The electron emission was turned on at $V_g = 40$ V and then increased steeply to 19.02 mA by increasing V_g to 70 V. The inset in Figure 22.12 represents the I_a versus operation time with the initial current of 11.3 mA. As shown in the figure, the CNT-FEL panel shows long-term emission stability, from which no emission decay for nearly 120 h was observed after an initial decay for about 20 h. They reported that $I_a = 11.3$ mA was high enough for the operation of the 32 in. diagonal CNT-BLU with sufficient luminance (6000 cd m^{-2}).

On the other hand, the gated CNT emitter with the common gate structure can be easily fabricated by using a metal mesh. Also, it has an additional advantage of high immunity to a high acceleration voltage compared with the normal-gated CNT emitters. In this device, the CNT emitters and the metal-plate gate were prepared separately, and then they were joined together during a vacuum packaging process. A spacer (gate spacer) was inserted between the CNT emitters and the metal-plate gate for spacing and insulation. The CNT emitters were formed on a plane surface of cathode electrodes like a diode configuration, while the mesh holes were formed on a metal plate such as a nickel–cobalt ferrous alloy called Kovar. It is to be noted that the matching of thermal expansion coefficient between the cathode glass and the metal is very crucial. Since the metal-plate gate cannot carry the image (dimming) signals in contrast to the normal gate structure, a special driving method of current limiting technique is needed as described later.

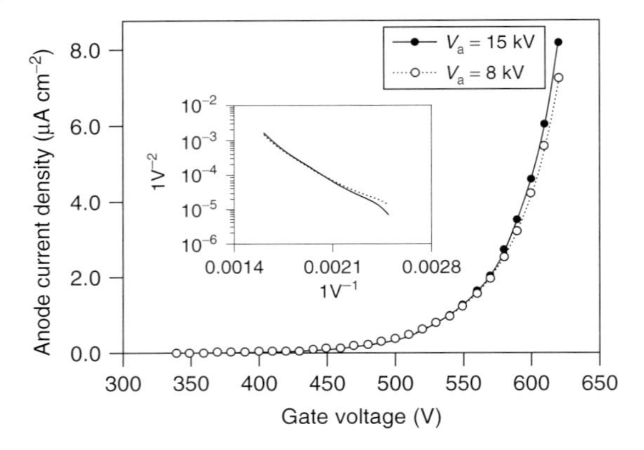

Figure 22.13 Anode current density versus gate voltage curve of a sealed 5 in. diagonal CNT-FEL with common gate (metal mesh) structure for two anode voltages (V_a) of 8 and 15 kV. The inset represents FN plot of the emission characteristics.

Figure 22.13 shows the anode current density (J_a) versus V_g curves of a sealed 5 in. diagonal CNT-FEL with the common gate (metal mesh) structure for two anode voltages (V_a) of 8 and 15 kV. The thickness of the metal mesh was about 0.2 mm and the gate hole is rectangular with an opening of 0.2 mm by 0.7 mm. The gate spacer maintains a fixed distance of about 0.2 mm between the metal mesh and the cathode substrate. The large gap maintained by the gate spacer reduces the gate to cathode capacitance and the driving power of FEL; meanwhile, the gate voltage for field emission increases. The FN plot of I_a versus V_g curves shown in the inset of the figure exhibited a linear behavior, suggesting a typical field emission characteristic. It was also found that the J_a versus V_g curve remains nearly constant upon the change of anode voltage up to 15 kV along with a negligible gate leakage current, which is attributed to the perfect shielding of CNT emitter from anode-induced electric field by the metal mesh. The developed common-gated CNT emitters have a good triode operation, showing the possibility of their application to FEL.

22.4 Anode

As previously stated in Section 22.2, low-voltage phosphors for FEL have still not developed in the efficiency and lifetime to the extent of commercialization in spite of several approaches to oxide-based CL phosphors [33, 34]. Therefore, most of FEL developments make use of high-voltage phosphors taken from CRT technology. The electron penetration depth into the phosphors is known to be proportional to V_a^{λ}, where V_a is the anode voltage and λ is a material constant: for example, λ for ZnS is 2.4 [64]. Therefore, higher V_a is preferred for higher luminous efficiency. However, it was found that a V_a of higher than 15 kV causes the device to generate

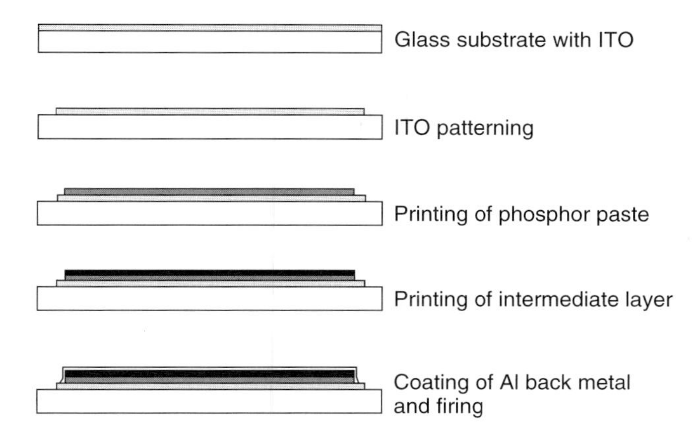

Figure 22.14 Anode process using a paste-printing technique.

large amounts of harmful X-rays [20, 24], which may limit the anode voltage in FEL applications. Furthermore, the surface coating for the protection of phosphors from physical and chemical attacks like abrasion, shock, and oxidation at the surface should be optimized considering the anode voltage for acceleration of electron beams. Since the electrical property of high-voltage CL phosphors is insulating, a metal layer should be provided at the back of the phosphor layer to discharge the charges on the phosphors, requiring more energetic electron beams to obtain sufficient luminance, efficiency, and lifetime.

A CL phosphor layer can be processed by a variety of methods such as photolithography, electrophoresis, and paste-printing [65]. Paste-printing is mainly used in FEL applications because of its large-area capability with a low cost, while the photolithographic method has been commonly used in the CRT technology. CL phosphors such as Y_2O_3:Eu for red, ZnS:Cu,Al for green, and ZnS:Ag,Al for blue were mixed in appropriate proportions to produce white color. Later, the mixture was blended with an organic binder with/without inorganic particles to make a phosphor paste. In the special application of color-tunable lighting or a field sequential LCD, the primary color paste of R, G, and B phosphors are separately prepared. Figure 22.14 shows an anode process for CNT-FEL using the paste-printing technique. A sodalime glass coated with ITO was used as the anode substrate and the ITO was patterned by a photolithography and etching process for the anode electrode. Then, the phosphor paste was screen-printed onto an active area of the anode substrate. After printing an intermediate layer composed of an organic binder with/without inorganic particles, a back Al metal layer was coated on the intermediate layer by thermal evaporation or by laminating an Al film. Here, the intermediate layer plays the role to make the surface of the anode substrate plane with the phosphor patterns. Finally, the anode substrate was fired at around 400 °C in an air ambient to burn out the organic binder in the phosphor and intermediate layers.

Figure 22.15 shows the cross-sectional scanning electron microscope (SEM) image of the anode structure by Choi *et al.* [20], showing the Al back-layer (100 nm

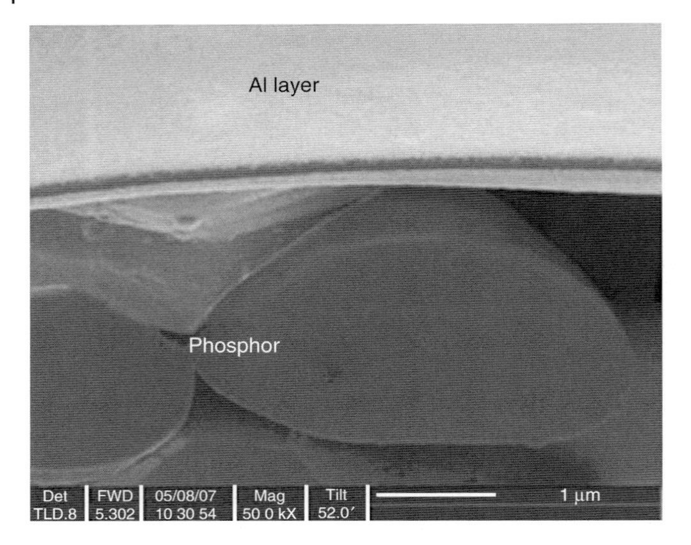

Figure 22.15 Cross-sectional SEM image of the anode structure showing phosphors and Al back metal layer. (Reproduced from [20] by Y. C. Choi, *et al.* (2008) Nanotechnology **19**, 235306.)

thick) and phosphor particles. The voids that were occupied with the organic binder before the firing can be clearly seen from the SEM micrograph. They reported that the optimized thickness of the phosphor layer was about 10 μm, which was estimated while taking the high luminance into consideration. The luminance and efficiency of the phosphor layer could be affected by its fabrication process as well as the phosphor material itself. Specifically, the inorganic particles mixed with the phosphor or intermediate layer may enhance the out-coupling of the trapped light in the phosphor layer, improving the luminescence efficiency.

22.5 Vacuum Packaging

A vacuum packaging technology should accompany the CNT-FEL device as in other field emission applications. Up to now, most of vacuum packaging processes are as follows: (i) hermetic sealing of the outer area of the FEL panel using frit glass (outer-sealing step); (ii) evacuation at an elevated temperature through an evacuation tube attached to the cathode plate; and (iii) tip-off of the evacuation tube. This method is a familiar and simple process used in CRT and PDP, but has the demerit of CNT degradation during the hermetic sealing. So, the outer sealing needs to be executed in a proper inert ambient of argon or hydrogen gas.

A novel vacuum packaging process of tubeless sealing in a vacuum chamber is introduced to avoid the degradation of CNT field emitters during the sealing

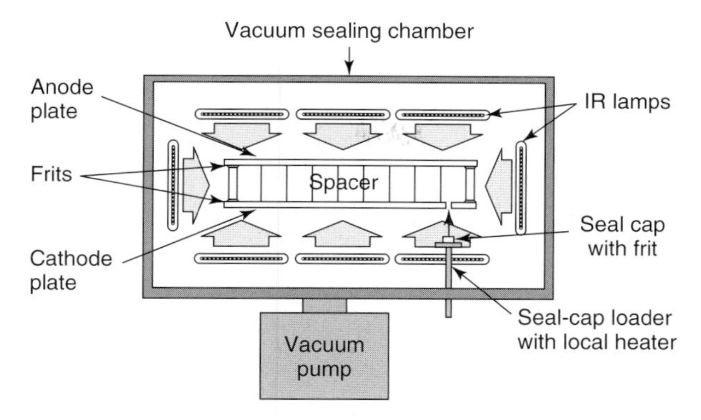

Figure 22.16 Apparatus for tubeless sealing of a CNT-FEL panel in a vacuum chamber.

process. In this technique, the outer sealing and tip-off processes are performed in an ultra high vacuum chamber system. An apparatus for tubeless sealing of the CNT-FEL panel in a vacuum chamber is shown in Figure 22.16. The vacuum sealing chamber has a heating element of IR lamps to heat the cathode and anode plates to a sealing temperature uniformly and a movable seal-cap loader having a local heater to cap the evacuating hole formed on the cathode glass. Figure 22.17 shows the process sequence of tubeless sealing for the CNT-FEL. At first, the spacers of glass or ceramic in the shape of pillars were loaded on the gate electrodes in the cathode plate. It may be possible to load the spacer automatically in the FEL panel because

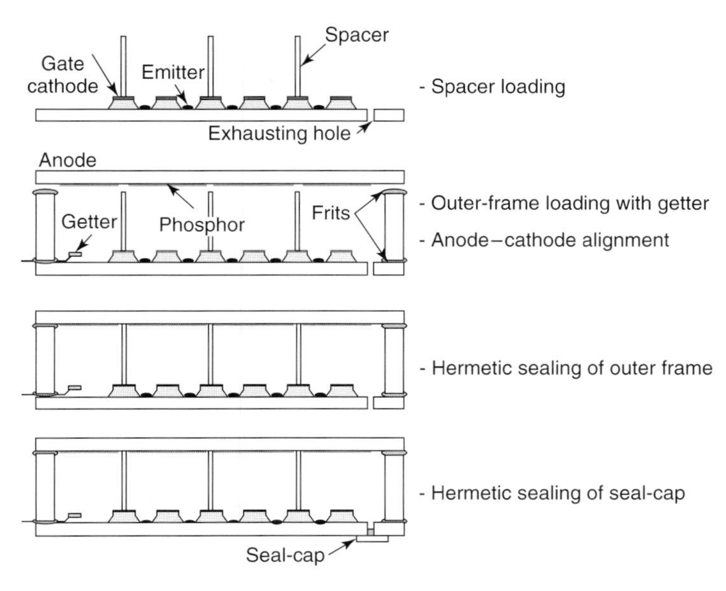

Figure 22.17 Process sequence of tubeless sealing. The hermetic sealing processes are performed in a vacuum chamber.

the size of the spacer can be large enough to be handled easily compared to the FED application. The prepared outer frame made of the same glass as the cathode was loaded on the cathode plate with frits on its surfaces to be sealed to the cathode and anode plates. During the loading of the outer frame a nonevaporable getter with a resistive electrode was also positioned on a proper side of the cathode plate. The evaporable getter of Ti–Zr–V–Fe alloy [66] can preserve the high vacuum state after vacuum packaging. After loading the outer frame with the getter, the anode plate with the phosphor and back metal was aligned with the cathode plate and they were clamped and then put into the vacuum sealing chamber. The vacuum sealing chamber was evacuated down to a pressure level below 10^{-6} Torr and then heated to the melting temperature (around 450 °C) of the frits for hermetic sealing of the outer frame with the cathode and anode. Finally, a hermetic capping of the evacuating hole on the cathode was performed in course of cooling of the sealing chamber by a movable seal-cap loader, leading to the complete fabrication of the fully sealed CNT-FEL. The nonevaporable getter inside the panel was activated by using a laser or by resistive heating after the sealing of the seal-cap.

The sealing process in the vacuum chamber can preserve the field emission properties of CNT emitters with no degradation during the vacuum packaging. It also prevents the getter from oxidation upon the exposure of CNT cathode to a high temperature ambient for sealing, which can enhance the adsorption ability of the getter for residual gas or out-diffused gas during the operation of the CNT-FEL. Another merit is in the removal of the evacuating tube, giving a real flat panel.

22.6 Driving and Characterization

A driving method of CNT-FEL for a local dimming BLU strongly depends on its cathode and gate architecture. We have introduced the two types of CNT FELs with normal and common gate structures in Section 22.2. Figure 22.18 shows a sequential-access addressing of local blocks in the CNT-FEL with normal gates. In this method, the row bus selects a row of local blocks consecutively and the column bus delivers a dimming signal to each block in the selected row and vice versa as in a conventional matrix addressing. This driving method controls directly the voltage biased to the cathode and gate electrodes in each block and the dimming is achieved by pulse width modulation (PWM), which is common and easy in CNT FELs. However, it has the disadvantages of a high addressing voltage depending on the operation voltage of the CNT emitters and a shortened duty with increase in the number of local blocks. If the image is refreshed by a frequency of f and the number of rows of the local blocks is N, then the duty time to each row is given by $1/fN$, which limits the whole luminance and the number of local blocks for a given panel size. Since the luminescent efficiency of the CL phosphor is still low compared to that of a photoluminescent one for CCFL, the shortened duty may give rise to a limitation in total luminance and hence local brightening and impulse-type

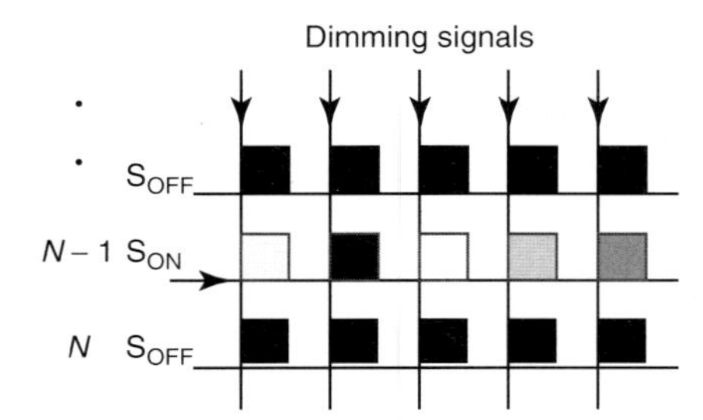

Figure 22.18 Sequential access addressing of local blocks in the CNT-FEL with normal gates. The S_{ON} and S_{OFF} represent on and off signals, respectively.

scanning. Another demerit of conventional voltage driving is that it hardly makes up for the degradation of CNT emitters, and so the lifetime of CNT-FEL may be improved only by the improvement in intrinsic property of CNT emitters.

The other driving method of random or direct access addressing for the CNT-FEL with a common gate is shown in Figure 22.19. It addresses the dimming signal to each local block randomly, not sequentially, and directly controls the field emission current through the addressing unit of a voltage-controlled current source like a transistor. Since the addressing transistor is connected to the cathode electrode directly as shown in Figure 22.19b, it should have a high breakdown voltage to

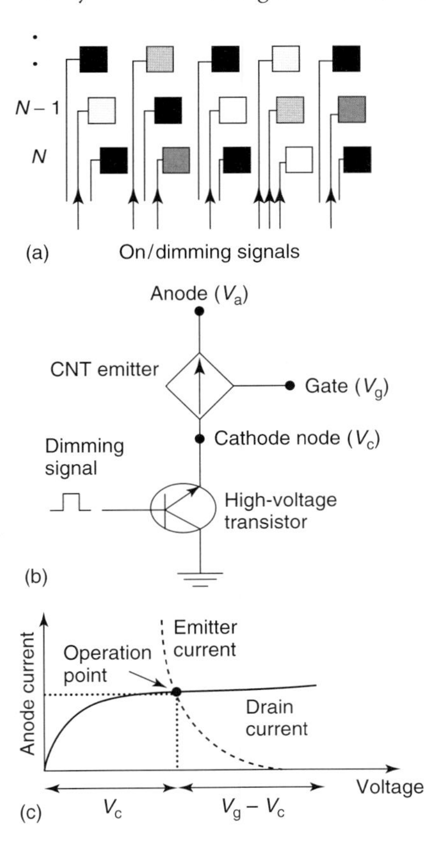

Figure 22.19 (a) Random access addressing of local blocks in the CNT-FEL with a common gate, (b) equivalent circuit of each block with a high-voltage transistor as an addressing unit, and (c) graphical determination of operation point of a single block CNT emitter for a constant voltage biased to the common gate (V_g).

endure the high voltage induced by the common gate of the CNT emitters. In general, the common gate of the CNT emitters is biased to a high voltage above 50 V depending on the distance between the CNT tip and the gate electrode. An equivalent circuit of the local block with the addressing unit is very similar to that of an active-matrix (AM) FED [67–71] in that the addressing transistor connected to the cathode electrode in series controls the field emission current from the CNT emitters actively. The addressing transistor is constructed in the peripheral region outside the active panel of CNT-FEL, while it is integrated into each pixel in the AM FED panel. The dimming or brightening signal is given to the gate of the addressing transistor with a low voltage of less than 5 V. When the device was operated, a constant DC voltage was applied to the common gate of the CNT emitters to induce sufficient field emission. The determination of the operation point of a single block CNT emitter with the addressing unit could be easily conjectured from a graphical method, as shown in Figure 22.19c. A voltage applied to the common gate (V_g) is divided into the addressing transistor and the CNT emitter by the cathode node voltage (V_c) and so the voltage V_c is biased to the drain of addressing transistor and ($V_g - V_c$) to the CNT emitter. As V_c increases, the biasing voltage to the CNT emitter decreases for a fixed V_g, which is presented by a reverse biasing to the CNT emitter and forward one to the addressing transistor in the figure. In the random-access addressing with a voltage-controlled current source, the dimming is easily obtained by pulse amplitude modulation (PAM) as well as PWM. Also, the signal voltage could be less than 5 V irrespective of the required voltage for field emission from CNT emitters because it was addressed to the gate of addressing transistor described above.

The advantages of the random-access addressing could be clearly seen from the driving principle shown in Figure 22.19. First, the duty time to each local block can be enlarged to nearly the refresh time of the image, $1/f$, enhancing or maintaining the total luminance of the CNT-FEL irrespective of the number of local blocks. Secondly, the active controlling of field emission current can improve the uniformity, stability, and reliability of the CNT-FEL. The addressing transistor with a wide saturation regime in the output curve (drain current vs. drain voltage) provides more uniform, stable, and reliable emission current than a linear resistive layer commonly adopted as a ballast in an FED application [72, 73]. Specifically, the degradation of CNT emitters can be compensated by the wide saturation behavior of emission current against the gate bias voltage. Furthermore, the common gate of metal mesh may protect the CNT emitters from electrical shocks caused by the high acceleration voltage applied to the anode. On the other hand, there are also some demerits in the CNT-FEL with a common gate, such as difficulty in alignment between the CNT cathode and the metal gate and limitation in the number of local blocks. The alignment problem may be more severe when the size of CNT-FEL panel increases. Since each local block in the CNT-FEL with a common gate requires an individual addressing bus and the cathode electrode is the only addressing bus, there may be limitation on local blocks in a single plane. However, this limitation can be solved by a double-metal technique with contact holes, as shown in Figure 22.20. The figure shows an array architecture

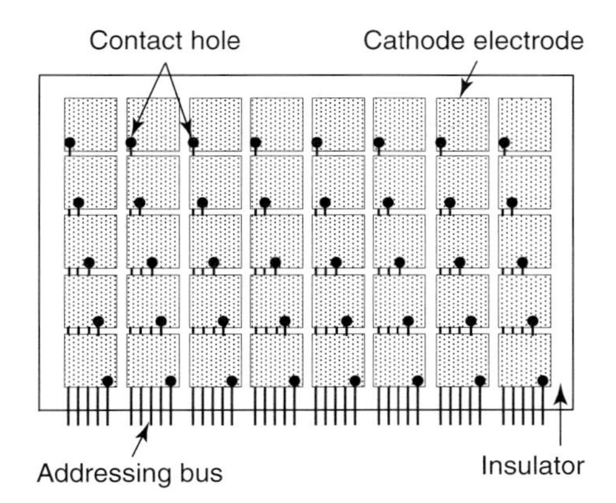

Figure 22.20 Array architecture of random-access-addressed local blocks with a common gate by the double-metal technique.

of random-access-addressed local blocks by the double-metal technique, in which an insulator is interposed between the addressing bus and cathode electrode and the through hole is formed on the insulator for electrical contact between the two layers. The double-metal structure for a dense local block is easily achieved by a screen-printing process, providing a large-area CNT-FEL with low cost.

From now on, the performances of CNT-FEL will be discussed from a vacuum-packaged panel. Figure 22.21 shows a vacuum-packaged, 15 in. diagonal CNT-FEL panel with a common gate of metal mesh. The CNT cathode had 42 local blocks and the size of each block was about $4 \times 4\,\text{cm}^2$. The anode had a white phosphor layer composed of red (Y_2O_3:Eu), green (ZnS:Cu,Al), and blue (ZnS:Ag,Al) CL particles and an Al back-layer on the phosphor. Pillar-type spacers with a height of 10 mm were positioned to maintain a uniform spacing between the metal mesh and the anode. The upper photograph in Figure 22.21 shows resistive heating of a nonevaporative getter inside the panel for activation. In general, the getter is activated just once during or after vacuum packaging, which limits its gettering capacity for gases. Outgassing was very severe in the aging step of the CNT-FEL and constantly occurred during the operation of CNT-FEL. It was found that the post-activation process for the getter is a prerequisite to

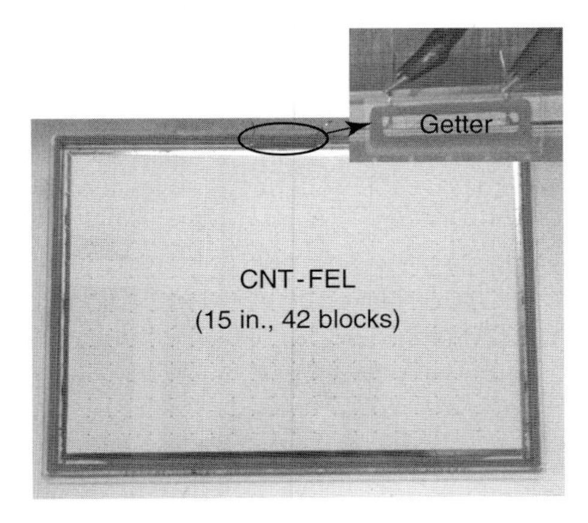

Figure 22.21 Vacuum-packaged 15 in. diagonal CNT-FEL panel with 42 local blocks and resistive heating of a non-evaporative getter inside the panel for activation.

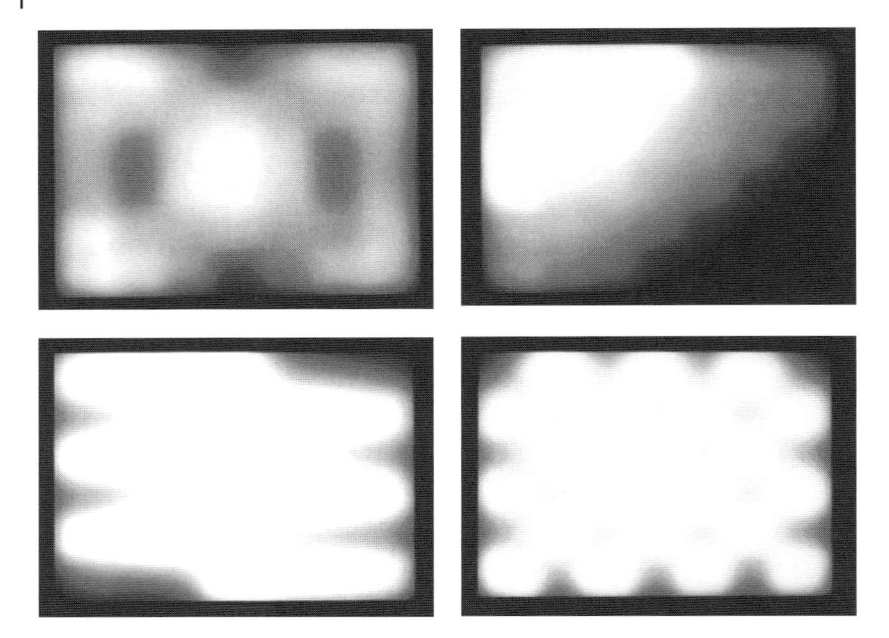

Figure 22.22 Local dimming images from the 15 in. diagonal CNT-FEL with 42 local blocks.

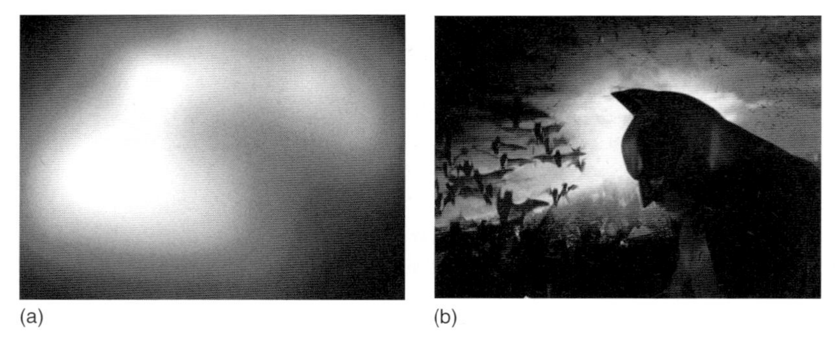

Figure 22.23 (a) Local dimming image from the 15 in. diagonal CNT-FEL with 42 local blocks for an LCD image (b).

absorb out-gases and to maintain the vacuum level of the CNT-FEL. Whenever the vacuum deteriorated by some out-gases, the getter might be activated through the resistive electrode connected to the getter inside the panel.

Figure 22.22 shows the local dimming images from the 15 in. diagonal CNT-FEL with 42 local blocks. Local dimming operation of the CNT-FEL for a specific LCD image is also shown in Figure 22.23. As seen in these figures, the brightness of each block was independently controlled with 64 (6 bits) gray scales, which was accomplished by varying the duty ratio of the PWM. In addition to the dimming, local brightening could be also obtained by simply increasing the duty ratio on the

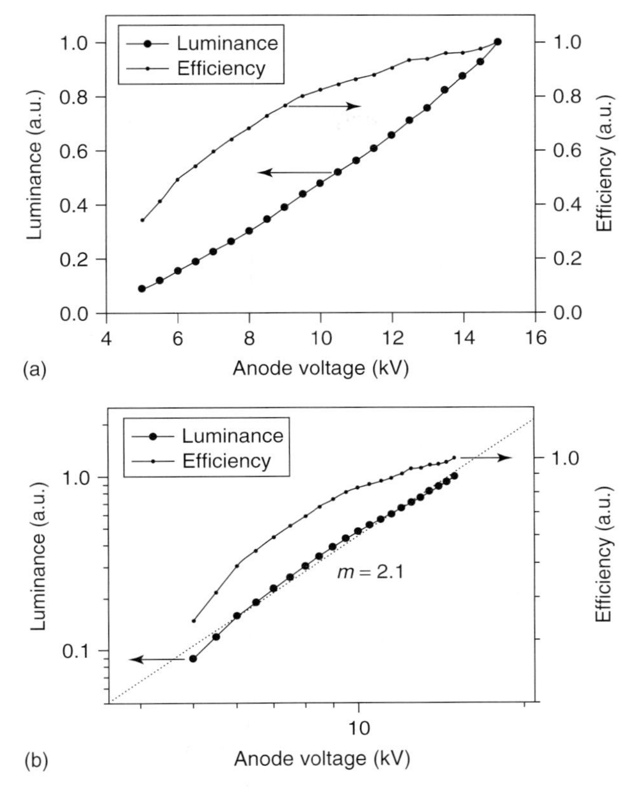

Figure 22.24 (a) Linear and (b) log–log plots of luminance (L) and efficiency (η) versus anode voltage (V_a) for a fixed anode current density of about 8 μA cm^{-2}. The dotted line shows the power relation of $L = C_v V_a^m$, where C_v is a constant and m is a power factor of about 2.1.

desired blocks [20]. The peak luminance of the CNT-FEL determines the highest brightness level considering the load ratio of the image size, in which the load ratio indicates the relative area representing the image over the whole display area. Both local dimming and brightening of BLU using a CNT-FEL greatly enhance the contrast ratio of the LCD. The optimum size of the local block is controversial in that a smaller local block gives a finer dimming image but requires a more expensive driver circuit and fabrication process. It may depend on the size of display panel and the objects displayed on the panel. A detailed study on the size of local block in CNT-BLU is required taking account of human factors on the display.

The luminance (L) and average luminescence efficiency (η) as a function of anode voltage for a fixed anode current density of about 8 μA cm^{-2} is shown in Figure 22.24. In the evaluation of η, we assume that the CNT-FEL has a Lambertian distribution as a surface emitting light source and L is measured at a vertical position to the panel plane. As a result, the average efficiency is given by $\eta = (\pi L/V_a J_a)$

[74], in which the driving power for local addressing is not considered because it is negligible compared to the anode power. The driving power, for example, was calculated to be about 17.3 mW for the 15 in. diagonal CNT-FEL at a driving frequency of 60 Hz, a cathode-to-gate capacitance of 2.3 nF for the gap of 100 μm between the cathode and the metal gate, and a cathode-to-gate modulation voltage of about 500 V induced by addressing of dimming signals.

The luminescence efficiency followed a sublinear dependence on the anode voltage and might be nearly saturated on a higher anode voltage than 14 kV. The luminance increased with anode voltage with a power law dependence, $L = C_v V_a^m$, where C_v is a constant and m is a power factor of about 2.1, as shown in Figure 22.24b. The power relation between the luminance and the anode voltage might vary with the anode current. The peak luminance was observed to be over

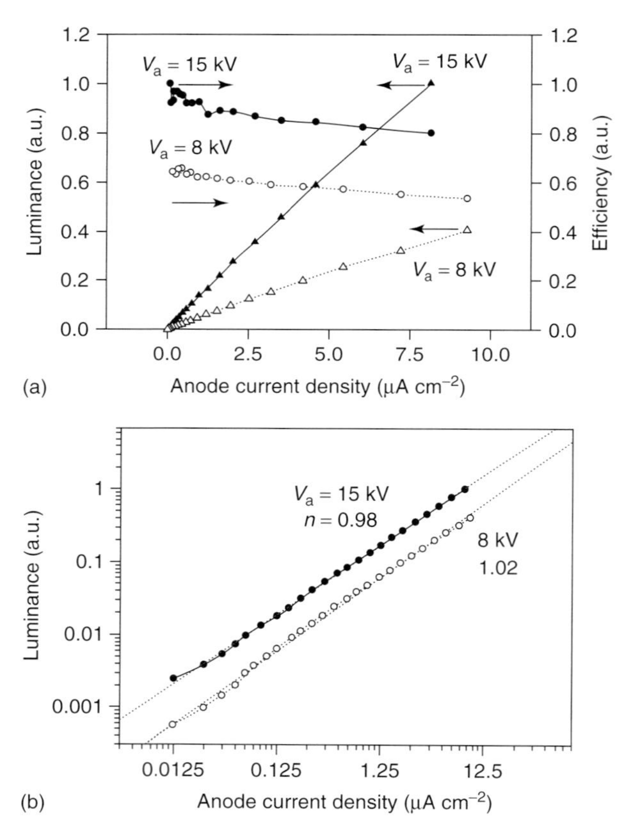

Figure 22.25 (a) Luminance and efficiency as a function of anode current density (j_a) for two anode voltages of 8 and 15 kV and (b) log–log plot of luminance versus anode current density. The two dotted lines follow the power relation of $L = C_j j_a^n$, where C_j is a constant and n is a power factor of 1.02 for $V_a = 8$ and 0.98 $V_a = 15$ kV.

12 000 cd m^{-2} and the average efficiency was estimated to be around 35 lm W^{-1}. The luminescence efficiency of CNT-FEL was lower compared to that of CCFL, resulting from the low efficiency of CL phosphors and a nonoptimized phosphor layer for FEL. We expect that the luminescence efficiency could be enhanced through the modification of phosphor layer by changing its thickness, particle size, and/or Al reflection layer. Figure 22.25 shows the luminance and average luminescence efficiency as a function of anode current density for two anode voltages of 8 and 15 kV. The efficiency decreased with the anode current density slowly. The luminance increased straightforwardly as the anode current density and followed a power law dependence, $L = C_j J_a^n$, where C_j is a constant and n is a power factor. It is noted that the power factor n is nearly 1.0 in the measured range of J_a for both V_a's. However, if we measure L in a higher J_a, the power factor might be decreased due to the well-known saturation effect of the CL phosphor [75]. From the results shown in Figures 22.24 and 22.25, a higher anode voltage and lower current density give a larger efficiency. The anode voltage and current, therefore, should be considered from the early design step of a CNT-FEL in order to achieve a high luminescence efficiency.

The response time of the CNT-FEL was known to be short enough to remove the motion blur phenomenon of LCD. Figure 22.26 shows response characteristics (white to black) of LCD TVs with a CNT-BLU and a CCFL by Choi *et al.* [20]. In the case of a CNT-BLU, the transient time for the variation of relative luminance

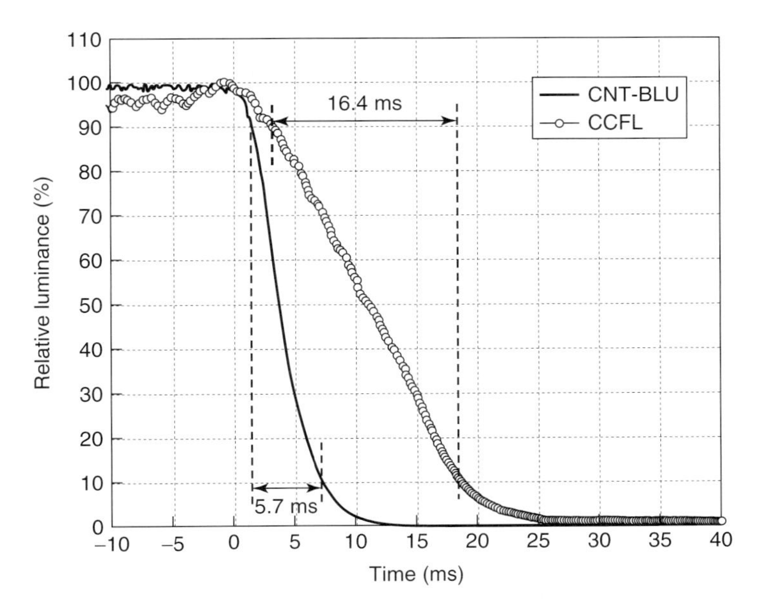

Figure 22.26 Response characteristics (white to black) of LCD TVs with a CNT-BLU and a CCFL. The response time is defined as the duration for the decrease of relative luminance from 90% to 10%. (Reproduced from [20] by Y. C. Choi *et al.* (2008) *Nanotechnology*, **19**, 235306.)

from 90% to 10% was as fast as 5.7 ms. This is almost three times faster than that observed from a CCFL (16.4 ms). They found that the motion blur did not occur even in fast-moving images. The CNT-BLU can solve the critical motion blur in LCD, giving an ideal image as with the CRT.

22.7 Future Works

The CNT-FEL technology for BLU, including the basic structure, fabrication process, driving scheme, and characterization, was reviewed. The local dimming/brightening CNT-BLU could improve the image quality of LCD such as contrast ratio, motion blur, and power consumption. Although a possibility of a CNT-FEL was shown recently, further work is required in the following areas: (i) enhancing the luminescence efficiency of CNT-FEL through the improvement in CL phosphor materials and optimization of the phosphor layer including intermediate and back metal layers; (ii) confirming and enhancing the lifetime of CNT-FEL by making progress in CNT emitters and phosphors, optimizing the structure and driving scheme, and maintaining the vacuum level of the panel; and (iii) developing a large-area and low-cost CNT-FEL process with screen-printing technology or other new methods.

Acknowledgments

The authors would like to thank their field emission research members, Dong-Il Kim, Jun-Tae Kang, Ji-Seon Kim, and Jae-Woo Kim for technical assistance; Dr Kyung-Ik Cho and Dr Kwang-Yong Kang for encouragement; and Senior Engineer In-Soo Choi and President Jae-Hong Park of Epion for details of the vacuum packaging process.

References

1. Rinzler, A.G., Hafner, J.H., Nikolaev, P., Nordlander, P., Colbert, D.T., Smalley, R.E., Lou, L., Kim, S.G., and Tománek, D. (1995) *Science*, **269**, 1550–1553.
2. Wang, Q.H., Setlur, A.A., Lauerhaas, J.M., Dai, J.Y., Seelig, E.E., and Chang, R.P.H. (1998) *Appl. Phys. Lett.*, **72**, 2912–2913.
3. Saito, Y. and Uemura, S. (2000) *Carbon*, **38**, 169–182.
4. Mirakami, H., Hirakawa, M., Tanaka, C., and Yamakaea, H. (2000) *Appl. Phys. Lett.*, **76**, 1776–1778.
5. Seelaboyina, R., Boddepalli, S., Noh, K., Jeon, M., and Choi, W. (2008) *Nanotechnology*, **19**, 065605–065608.
6. Choi, W.B., Chung, D.S., Kang, J.H., Kim, H.Y., Jin, Y.W., Han, I.T., Lee, Y.H., Jung, J.E., Lee, N.S., Park, G.S., and Kim, J.M. (1999) *Appl. Phys. Lett.*, **75**, 3129–3132.
7. Ho, J.-C., Chang, Y.-Y., Liao, J.-H., Cheng, H.-C., Sheu, J.-R., Hsiao, M.-C., Lee, C.-D., Huang, S.-M., Cho, C.-S., Huang, W.-K., Lin, W.-Y., and Lee, C.-C. (2002) *SID Symp. Digest*, **33**, 372–375.

8. Kang, S., Bae, C., Son, W., Kim, M.H., Yi, J., Chang, S.T.A., Kim, J.J., Lee, C.R., Moon, J.H., Lim, S.H., Kim, H.S., and Jang, J. (2003) *SID Symp. Digest*, **34**, 802–805.
9. Choi, Y.C., Jeong, K.S., Han, I.T., Kim, H.J., Jin, Y.W., Kim, J.M., Lee, B.G., Park, J.H., and Choe, D.H. (2006) *Appl. Phys. Lett.*, **88**, 263504–263506.
10. Dijon, J., Fournier, A., Levis, M., Meyer, R., Bridoux, C., Montmayeul, B., Muller, F., Nicolas, P., Sarrasin, D., Adamski, J.R., Bellanger, J.L., Bellissens, D., Lefort, M., and Ricaud, J.L. (2007) *SID Symp. Digest*, **38**, 1313–1316.
11. Lin, B.-N., Hsiao, M.-C., Chang, Y.-Y., Lin, W.-Y., Jiang, L.-Y., Lin, M.-H., Chan, L.-H., Jiang, Y.-C., Tsou, T.-H., and Lee, C.-C. (2006) *SID Symp. Digest*, **37**, 71–73.
12. Chou, L.-E., Lin, B.-N., Jiang, Y.-C., Tsou, T.-H., Fu, C.-H., Hsiao, M.-C., Chang, Y.-Y., Lin, W.-Y., Lin, M.-H., and Lee, C.-C. (2006) International Meeting on Information Display (IMID) Digest, 2006, pp. 150–155.
13. Yoo, H.-S., Sung, W.-Y., Yoon, S.-J., Kim, Y.-H., and Joo, S.-K. (2007) *Jpn. J. Appl. Phys.*, **46**, 4381–4385.
14. Yoo, H.-S., Sung, W.-Y., Son, S.-W., and Joo, S.-K. (2007) *Jpn. J. Appl. Phys.*, **46**, 7581–7585.
15. Wei, Y., Xiao, L., Zhu, F., Liu, L., Tang, J., Liu, P., and Fan, S.S. (2007) *Nanotechnology*, **18**, 325702–325706.
16. Cho, W.-S., Lee, H.-J., Lee, Y.-D., Park, J.-H., Kim, J.-K., Lee, Y.-H., and Ju, B.-K. (2007) *IEEE Elect. Dev. Lett.*, **28**, 386–388.
17. Kim, H.S., Lee, J.W., Lee, C.S., Lee, C.S., Jung, K.W., Lim, J.H., Moon, J.W., Hwang, M.I., Kim, I.H., Kim, Y.H., Lee, B.G., Choi, Y.C., Seon, H.R., Lee, S.J., and Park, J.H. (2007) International Meeting on Information Display (IMID) Digest, 2007, pp. 277–280.
18. Kim, Y.C., Jung, D.S., Song, B.K., Bae, M.J., Kang, H.S., Han, I.T., Kim, J.M., Choi, Y.C., Hwang, M.I., Kim, I.H., and Park, J.H. (2007) International Meeting on Information Display (IMID) Digest, 2007, pp. 1045–1048.
19. Jeong, J.-W., Kim, D.-J., Kang, J.-T., Kim, J.-S., and Song, Y.-H. (2007) Proceedings of the 14th International Display Workshops (IDW), 2007, pp. 2189–2190.
20. Choi, Y.C., Lee, J.W., Lee, S.K., Kang, M.S., Lee, C.S., Jung, K.W., Lim, J.H., Moon, J.W., Hwang, M.I., Kim, I.H., Kim, Y.H., Lee, B.G., Seon, H.R., Lee, S.J., Park, J.H., Kim, Y.C., and Kim, H.S. (2008) *Nanotechnology*, **19**, 235306–235310.
21. Huang, J.X., Chen, J., Deng, S.Z., She, J.C., and Xu, N.S. (2008) *J. Vac. Sci. Technol.*, **B26**, 1700–1704.
22. Chung, S.-Y., Liang, C.-C., Huang, C.-N., Huang, C.-F., and Pan, C.-T. (2008) *SID Symp. Digest*, **39**, 70–73.
23. Wang, H.-X., Harazono, H., Jiang, N., Hiraki, H., Harada, Y., Haba, M., and Nakamoto, M. (2008) *SID Symp. Digest*, **39**, 74–76.
24. Jeong, J.-W., Kim, D.-I., Kang, J.-T., Kim, J.-S., and Song, Y.-H. (2008) Proceedings of the 15th International Display Workshops (IDW), 2008, pp. 2019–2020.
25. Hiraki, H., Harazono, H., Onozawa, T., Nakamoto, M., and Hiraki, A. (2008) International Meeting on Information Display (IMID) Digest, 2008, pp. 1591–1593.
26. Chen, C.J. and Shen, Y.R. (2006) US Patent Application Publication, 20060,033,698 A1.
27. Hwang, I., Park, C.W., Kang, S.C., and Sakong, D.S. (2001) *SID Symp. Digest*, **32**, 492–493.
28. Shiga, T. and Mikoshiba, S. (2003) *SID Symp. Digest*, **34**, 1364–1367.
29. Kim, K.D., Baik, S.H., Sohn, M.H., Yoon, J.K., Oh, E.Y., and Chung, I.J. (2004) *SID Symp. Digest*, **35**, 1548–1549.
30. Shiga, T., Kuwahara, S., Takeo, N., and Mikoshiba, S. (2005) *SID Symp. Digest*, **36**, 992–995.
31. Kerofsky, L. and Daly, S. (2006) *SID Symp. Digest*, **37**, 1242–1245.
32. Peng, H.J., Zhang, W., Hung, C.-K., Tsai, C.-J., Ng, K.-W., Chen, S.-I., Huang, D., Chueng, Y.-L., and Liu, Y. (2007) *SID Symp. Digest*, **38**, 1336–1338.

33. Nagata, T., Nakayama, T., and Murakami, H. (2007) *SID Symp. Digest*, **38**, 1328–1331.
34. Yasuoka, Y., Kitada, M., Obara, Y., Mori, T., Naito, Y., Tamura, K., and Shinya, S. (2008) Proceedings of the 15th International Display Workshops (IDW), 2008, pp. 2025–2028.
35. Pan, L.S., Schropp, D.R. Jr., Chakarov, V.M., O'Reilly, J.K., Hopple, G.B., Spindt, C.J., Barton, R.W., Nystrom, M.J., Ramesh, R., Dunphy, J.C., Pei, S., and Narayanan, K.S. (2005) US Patent No. 6,861,798 B1.
36. Kim, S.-H., Kim, D.U., and Lee, S.K. (2006) *Curr. Appl. Phys.*, **6**, 766–771.
37. Hagiwara, K., Sakai, T., Ushirozawa, M., and Saito, N. (2008) *Jpn. J. Appl. Phys.*, **47**, 8534–8536.
38. Sim, H.S., Lau, S.P., Ang, L.K., You, G.F., Tanemura, M., Yamaguchi, K., Zamri, M., and Yusop, M. (2008) *Appl. Phys. Lett.*, **93**, 023131–023133.
39. Bagge-Hansen, M., Outlaw, a.R.A., Miraldo, P., Zhu, M.Y., Hou, K., Theodore, N.D., Zhao, X., and Manos, D.M. (2008) *J. Appl. Phys.*, **103**, 014311–014319.
40. Deng, J., Zhang, L., Zhang, B., and Yao, N. (2008) *Thin Solid Films*, **516**, 7685–7688.
41. Malesevic, A., Kemps, R., Vanhulsel, A., Chowdhury, M.P., Volodin, A., and Van Haesendonck, C. (2008) *J. Appl. Phys.*, **104**, 084301–084305.
42. Sohn, J.I., Lee, S., Song, Y.-H., Choi, S.-Y., Cho, K.-I., and Nam, K.-S. (2001) *Appl. Phys. Lett.*, **78**, 901–903.
43. Chen, Z., den Engelsen, D., Bachmann, P.K., van Elsbergen, V., Koehler, I., Merikhi, J., and Wiechert, D.U. (2005) *Appl. Phys. Lett.*, **87**, 243104–243106.
44. Uemura, S., Yotani, J., Nagasako, T., Kurachi, H., Yamada, H., Ezaki, T., Maesoba, T., Nakao, T., Saito, Y., and Yumura, M. (2003) *J. SID*, **11**, 145–153.
45. Yotani, J., Uemura, S., Nagasako, T., Kurachi, H., Nakao, T., Ito, M., Sakurai, A., Shimoda, H., Ezaki, T., Fukuda, K., and Saito, Y. (2007) *SID Symp. Digest*, **39**, 151–154.
46. Asano, M., Tsuruoka, Y., and Tanabe, H. (1999) US Patent No. 6,008,582.
47. Choi, H.-N., Kim, H.-S., Kim, Y.-S., and Joe, S.-W. (2002) *SID Symp. Digest*, **34**, 740–743.
48. Zhi, C.Y., Bai, X.D., and Wang, E.G. (2002) *Appl. Phys. Lett.*, **81**, 1690–1692.
49. Takai, M., Zhao, W.J., Sawada, A., Hosono, A., and Okuda, S. (2003) *SID Symp. Digest*, **37**, 794–797.
50. Vink, T.J., Gillies, M., Kriege, J.C., and van de Laar, H.W.J.J. (2003) *Appl. Phys. Lett.*, **83**, 3552–3554.
51. Ahn, K.S., Kim, J.S., Kim, C.O., and Hong, J.P. (2003) *Carbon*, **41**, 2481–2485.
52. Kim, K.-B., Song, Y.-H., Hwang, C.-S., Chung, C.-H., Lee, J.H., Choi, I.-S., and Park, J.-H. (2004) *J. Vac. Sci. Technol.*, **B22**, 1331–1334.
53. Liu, Y., Liu, L., Liu, P., Sheng, L., and Fan, S. (2004) *Diamond Relat. Mater.*, **13**, 1609–1613.
54. Kim, D.H., Kim, C.D., and Lee, H.R. (2004) *Carbon*, **42**, 1807–1812.
55. Kim, Y.C., Sohn, K.H., Cho, Y.M., and Yoo, E.H. (2004) *Appl. Phys. Lett.*, **84**, 350–352.
56. Gohel, A., Chin, K.C., Zhu, Y.W., Sow, C.H., and Wee, A.T.S. (2005) *Carbon*, **43**, 2530–2535.
57. Lee, C.-C., Lin, B.-N., Hsiao, M.-C., Chang, Y.-Y., Lin, W.-Y., and Jiang, L.-Y. (2005) *SID Symp. Digest*, **37**, 1716–1719.
58. Lee, H.J., Lee, Y.D., Cho, W.S., Ju, B.K., Lee, K.S., Lee, Y.-H., Kim, J.K., Pak, J.J.H., and Hwang, S.W. (2006) *SID Symp. Digest*, **38**, 638–640.
59. Kyung, S.J., Park, J.B., Voronko, M., Lee, J.H., and Yeom, G.Y. (2007) *Carbon*, **45**, 649–654.
60. Cho, H.J., Lee, N.S., Jang, I.G., Uh, H.S., and Hong, J.P. (2007) *J. Korean Phys. Soc.*, **50**, 1848–1853.
61. Lee, H., Jeon, J., Goak, J., Kim, K.B., Lee, N., Choi, Y.C., Kim, H.S., Sun, J., Park, Y., and Park, J. (2009) *J. Korean Phys. Soc.*, **54**, 185–189.
62. Spindt, C.A., Brodie, I., Humphrey, L., and Westerberg, E.R. (1976) *J. Appl. Phys.*, **47**, 5248–5263.
63. Chung, D.-S., Park, S.H., Lee, H.W., Choi, J.H., Cha, S.N., Kim, J.W., Jang, J.E., Min, K.W., Cho, S.H., Yoon, M.J., Lee, J.S., Lee, C.K., Yoo, J.H., Kim,

J.-M., Jung, J.E., Jin, Y.W., Park, Y.J., and You, J.B. (2002) *Appl. Phys. Lett.*, **80**, 4045–4047.

64. Feldman, C. (1960) *Phys. Rev.*, **117**, 455–459.

65. Tomita, Y. (1988) in *Phosphor Handbook* (eds S. Shionoya and W.M. Yen), CRC Press, pp. 337–341.

66. Lee, N.S., Chung, D.S., Han, I.T., Kang, J.H., Choi, Y.S., Kim, H.Y., Park, S.H., Jin, Y.W., Yi, W.K., Yun, M.J., Jung, J.E., Lee, C.J., You, J.H., Jo, S.H., Lee, C.G., and Kim, J.M. (2001) *Diamond Relat. Mater.*, **10**, 265–270.

67. Gamo, H., Kanemaru, S., and Itoh, J. (1998) *Appl. Phys. Lett.*, **73**, 1301–1303.

68. Song, Y.-H., Hwang, C.-S., Cho, Y.-R., Kim, B.-C., Ahn, S.-D., Chung, C.-H., Kim, D.-H., Uhm, H.-S., Lee, J.H., and Cho, K.-I. (2002) *ETRI J.*, **24**, 290–298.

69. Kim, D.-H., Song, Y.-H., Cho, Y.-R., Hwang, C.-S., Kim, B.-C., Ahn, S.-D., Chung, C.-H., Uhm, H.-S., Lee, J.H., Cho, K.-I., and Lee, S.-Y. (2002) *IEEE Trans. Elect. Dev.*, **49**, 1136–1142.

70. Song, Y.-H., Kim, K.-B., Hwang, C.-S., Park, D.-J., Lee, J.H., Kang, K.-Y., Hur, J.H., and Jang, J. (2005) *J. SID*, **13**, 241–243.

71. Jeong, J.-W., Kim, D.-J., Cho, K.-I., and Song, Y.-H. (2009) *J. Vac. Sci. Technol.*, **B27**, 1097–1100.

72. Levine, J.D. (1996) *J. Vac. Sci. Technol.*, **B14**, 2008–2010.

73. Baptist, R., Bachelet, F., and Constancias, C. (1997) *J. Vac. Sci. Technol.*, **B15**, 385–390.

74. Kim, Y.C. and Yoo, E.H. (2005) *Jpn. J. Appl. Phys.*, **44**, L454–L456.

75. Ohno, K. (1988) in *Phosphor Handbook* (eds S. Shionoya and W.M. Yen), CRC Press, pp. 489–498.

23
High-Current-Density Field Emission Electron Source

Shigeki Kato and Tsuneyuki Noguchi

23.1
Introduction

It is no exaggeration to state that carbon nanotubes (CNTs) have raised high expectations since they were considered to be a sort of ideal emitters on the basis of their unique properties. However, deficiencies in the field emission performance of CNTs have been identified. A high emission current of up to approximately 200 μA was reported for a single CNT emitter, corresponding to a current density of up to 10^9 A cm^{-2} [1]. This excellent result demonstrates the exceptional properties of CNTs. However, for many years, bulky emitters or film emitters comprising large numbers of randomly or regularly oriented CNTs, which were expected to be adopted in many fields of applications, have not yielded satisfactory emission currents. For instance, a film emitter comprising 10 000 or more CNTs with an apparent emission area of approximately 0.1 mm^2 cannot provide a total current of 1 A (the sum of the emission currents of the individual CNTs) even if considerable efforts are made to realize ideal conditions with regard to CNT characteristics, substrate junction, and CNT distribution, that is, distribution of electric field over the film emitter. Therefore, it is often mentioned that the field emission characteristics of CNTs are limited by unstable, deficient, and nonuniform brightness; short lifetime; and a high threshold voltage for electron emission.

What accounts for this difference between a single emitter and the film emitter? It is relatively easy to manipulate a single CNT and form a junction between the single emitter and a substrate, which in many cases is a metal wire, by using *in situ* scanning electron microscopy (SEM) observation techniques. Meanwhile, to this day, it remains incredibly difficult to produce a satisfactory film emitter using a large number of CNTs wherein each CNT works properly as an individual emitter [2]. The most significant reason is that, during emitter fabrication, *in situ* observation and feedback control techniques are incapable of producing adequate magnifications when forming junctions over a substrate area that is much larger than the area occupied by one CNT emitter. Neither sequential observation of individual CNT junctions nor simultaneous observation of all the junctions can be carried out within a reasonable production time. Consequently, it would be considerably difficult to

Carbon Nanotube and Related Field Emitters: Fundamentals and Applications. Edited by Yahachi Saito

ISBN: 978-3-527-32734-8

control all the individual CNTs on a substrate. Thus, realizing efficient film emitter operation is definitely a different challenge from realizing the efficient operation of a single emitter. Accordingly, it should be emphasized that the field emission of currently available film emitters is not restricted by the unique properties of individual CNTs; rather, it is restricted primarily by the junctions formed between a large number of CNTs and the underlying substrate. In this chapter, the challenges to be overcome for producing CNT film emitters with higher current densities and longer lifetimes will be described on the basis of guiding principles and practical methods.

23.2 Guiding Principles and Practical Methods for High-Performance Emitter

CNT film emitters with higher current densities and longer lifetimes can be realized by following the guiding principles and practical methods shown in Table 23.1.

Table 23.1 Guiding principles and practical methods to realize film emitters with higher current densities and longer lifetimes.

Guiding principles	Practical methods
Elicitation of inherent emission properties of individual CNTs	Selection of CNT: SW, MW, purity, diameter aspect ratio, chirality, open or close end, the number of defects
Increase in field enhancement factor at CNT surface	Formation of nanostructures projecting from the CNT surface: impregnation of atom clusters, vacuum deposition
Optimization of electric field distribution on film emitter surface	Control of CNT density: control of CNT growth conditions, pattering during CNT growth, control of CNT concentration in a dispersion, postremoval of CNTs such as burning, taping-off, and plasma or gas etching
Reduction in work function of the emitter	Selection of CNT, material deposition at a lower work function
Increase in thermal conduction from CNT to emitter substrate and mitigation of joule heating at CNT junction	Selection of junction and substrate material so as to keep high thermal and electric conduction: CNT rooting
Prevention of CNT disappearance	Reinforcement of CNT junction to substrate: selection of junction and substrate material, CNT rooting to substrate
Mitigation of ion sputtering and reactive etching of emitter surface	Lower base and operation pressures, control of residual gas quality, optimization in accelerating distance and voltage, material coating through ion sputtering with a lower yield

23.2.1
Elicitation of Inherent Emission Properties of Individual CNTs

It is clear that efforts to elicit inherently high field enhancement factors from individual CNTs produce good results. The inherent emission properties of individual CNTs depend on the selection of established high-quality CNTs such as single-walled nanotubes (SWNTs) or multiwalled nanotubes (MWNTs), the purity, diameter, aspect ratio, chirality, and selection of open- or close-end CNTs, as well as the number of defects on the CNTs. The quality of commercially available CNTs is variable. Even if CNTs are purified to high standards in the factory, defects on CNTs or contamination due to the presence of nonmetallic CNTs may not necessarily be considered in the purification process since the factory produces CNTs on the scale of kilograms.

Such oversights in CNT criteria can lead to problems in field emission. For instance, semiconductive CNTs, which have resistivities that are up to 200 times higher than those of metallic CNTs, cause undesired problems due to joule heating of the CNTs themselves during emission and due to broadening of the energy distribution of the emitted electrons.

CNT modification can provide better field emission performance. For example, highly pure and highly oriented SWNTs can exhibit excellent field emission only after they are crushed using a pestle and mortar before dispersion on a substrate. The highly oriented nature of the original SWNTs can be attributed to their strong bundling characteristics. However, the crushing process brings about a random orientation and eliminates the bundling characteristics of the SWNTs, thereby allowing the exploitation of their inherent properties.

23.2.2
Increase in Field Enhancement Factor at the CNT Surface

While CNTs evidently have small diameters and considerably large aspect ratios, the sharp tips are located only at the ends of the CNTs. If extra protruding points can be introduced on the CNT surface, mainly on the tubular portion of the surface, the field enhancement factors of individual CNTs can be increased. The extra emission points can be realized by the formation of nanostructures projecting from the CNT surface. Several methods have been proposed toward this end, for example, impregnation or vacuum deposition of atom clusters on the CNT surface and exposure of CNTs to particle beam irradiation (mainly ion beam irradiation), plasma, or reactive gases [3]. The techniques of particle beam irradiation, plasma exposure, or reactive gas exposure involve the introduction of defects on the CNT surface or the deformation and breakage of the surface; these techniques can be also used to control the CNT density on an emitter substrate (Section 23.2.3).

It would be encouraging if the emission current could be improved to some extent by using these techniques. However, at higher currents on the order of a few amperes per square centimeter, the properties of the attachment or defect would adversely affect the performance of these methods. Therefore, it is better to avoid the use of these methods in some applications since emitter degradation could occur at very high currents, such that in case of attachments, the atoms might migrate on the surface to produce a larger cluster, diffuse into the CNT, or initiate an evaporation process, and in case of defects, the constituent atoms might migrate to repair the damage or the damage might spread further.

Section 23.3 provides details of the challenges to be overcome when nanoparticles of RuO_2 and OsO_2 are impregnated on the CNT surface to enhance the emission properties of the CNTs [4].

23.2.3 Optimization of Electric Field Distribution on Film Emitter Surface

It is vital to control the CNT surface density in order to optimize the electric field distribution over the film emitter surface [5]. This consideration does not apply to the case of a single CNT emitter. Many methods have been proposed to control the CNT density; these methods can be divided into three categories.

The first category includes methods in which the CNT density is controlled during the process of direct growth of CNTs on an emitter substrate; in this case, the CNT density can be controlled simply by tuning the CNT growth conditions and/or patterning CNTs on the substrate in advance using techniques such as using catalyst pattering [6]. The second category includes methods in which the dispersion of synthesized CNTs from a bottle on the substrate is controlled during the emitter formation process. The methods in the third category are based on posttreatment techniques to partially remove CNTs from the prepared film emitters through burning of CNTs in air or pure oxygen, taping-off of CNTs, and reactive gas etching or plasma etching of CNTs [7].

A noteworthy point is that it might be difficult to judge the effects of these methods on the electric field distribution since most of the methods would influence the other improvement principles.

23.2.4 Reduction of Work Function of Emitter

Reducing the work function of CNTs is a conventional means to improve emitter performance, since the work function of CNTs is not as low as that of other field emitters. In most cases, an alkali metal coating was used; this resulted in metal evaporation in the high-current region. However, when the entire surface of a single MWNT was coated with a 2-nm-thick layer of LaB_6, which has a work function of 2.6 eV and a high melting point of 2900 K, the emission current was increased several fold up to 70 μA and the threshold electric field was halved to a value of 0.9 V μm^{-1} [8].

23.2.5
Improvement of Thermal Conduction and Mitigation of Joule Heating at CNT Junction

Both the electric resistance and the thermal resistance at the CNT junctions pose serious obstacles to efficient emitter performance at high emission currents. If the junction temperature is raised, a negative chain reaction of heating, electric resistance, and thermal resistance will damage the CNTs and lead to disappearance of working CNTs from the substrate [9].

Focusing on a single junction between a single CNT emitter and the substrate allows the effective preparation and management of the junction. However, when fabricating a film emitter with, say, 100 000 CNTs, group control of CNTs is the only option. From the viewpoint of eliciting the highly favorable inherent properties of CNTs as field emitters, it can be considered that the effect of (i) increasing thermal conduction from CNTs to the emitter substrate through the junction and (ii) mitigating joule heating at the junction would be most apparent in the case of a film emitter; this can be attributed to the fact that increasing the scale of the device would offer advantages in terms of the number of improved junctions. As mentioned previously, it would be impossible to control all the individual junctions between the CNTs and the substrate. However, it should be possible to form improved junctions amounting to 1% of the total CNT junctions, which would provide a very high emission current. For instance, a film emitter comprising 100 000 CNTs on a substrate area of 1 mm^2 with low junction quality can be improved to form an emitter with 1000 vivid CNTs that yield an emission current of 10 mA when the individual CNTs provide a current of 10 μA, which is a reasonable value for a single CNT emitter. Hence, to fabricate field emitters with high-current densities, it is crucial to select appropriate junction and substrate materials so as to maintain high thermal and electric conduction for a very large group of CNTs. The detailed method is mentioned in Section 23.4 [4, 10].

23.2.6
Restraint of CNT Disappearance

An important consideration in the group control of a very large number of CNTs is that not all the working CNTs are sufficiently robust to withstand Coulomb repulsion forces arising from the high electric field; thus, some CNTs disappear from the substrate when the field strength reaches a high value. However, it is essential to fix CNTs on a substrate as firmly as possible with tight junctions in order to achieve a high emission current density. The effect of this guiding principle, which involves the prevention of CNT disappearance, is closely related to the improvement in thermal conduction and mitigation of joule heating at the CNT junctions.

In general, CNTs grown on a substrate such as a Si wafer using a catalyst such as Fe or Ni do not have strong junctions; this is because the junctions thus formed are not designed to maintain high tensile strength, high thermal conductivity, and electric conductivity. Regardless of how conductive the CNT junctions are to heat and electricity, weak CNT junctions will damage the emitter. Even if good thermal and electric conduction and good tensile strength of CNT junctions are obtained at a low temperature, the junctions would weaken at high emission currents if they are made of some metals. This is because all materials have lesser tensile strength at higher temperatures, and metals ordinarily show positive temperature coefficients of electric resistivity. Therefore, if the emitter temperature were to increase, the field emitter would be damaged through the combined effects of increased heating, increased electric resistance, and decreased tensile strength. However, transition metal carbides may be suitable as junction materials since they do not generally show a rapid increase in electric resistivity or a rapid decrease in thermal conductivity with an increase in temperature. Section 23.4 describes the process of direct metal carbide formation, which utilizes CNTs and a transition metal substrate, to realize reasonable electric conduction, thermal conduction, and tensile strength without excessive CNT disappearance.

23.2.7 Mitigation of Ion Sputtering and Reactive Etching of Emitter Surface

It is obvious that high-current emitters require very low residual gas pressure even if CNT emitters are generally supposed to be more stable than metallic emitters at high residual gas pressures. This is because unavoidable damage of the tip surface becomes more serious either at higher emission current or at high pressure due to both heavier ion sputtering and reactive etching of the emitter tip surface. Ionized and/or dissociated residual gas including metastable states formed by collisions between the gas molecules and emitted electrons would attack CNT surfaces and destroy the emission points or change the composition of the CNTs, thus degrading the emission characteristics. Physical sputtering of the tip surface mainly depends on molecular weight of the gas species, electron accelerating voltage, and the angle of incidence of sputtering ions. In case of reactive gas species such as H_2, CO, H_2O, CO, and hydrocarbons, which are the usual gas species in reality, serious chemical sputtering occurs especially depending on the gas reactivity and the emitter temperature which are high in high-current operation.

Simultaneous adsorption and subsequent reaction in most cases would change the surface state, particularly work function of the surface, though the consequent small and stable work function can contribute the larger emission current.

Therefore, lower base and operation pressure, control of residual gas quality, optimization in the accelerating distance and voltage, material coating with lower ion sputtering yield, and so on, are recommended to mitigate those undesired phenomena. Attention should be paid to the fact that unexpected, high local pressure in the vicinity of an emitter due to the narrow space makes the problem worse even if a vacuum gauge shows a reasonable pressure. Hydrogen as dominant residual gas is ideal in every case as the quality control. Optimization in gap distance of electrodes and applied voltage to get required electric field allows mitigating the sputtering considering both electron energy dependence of ionization cross sections of residual gas species and ion energy dependence of the sputtering yields.

Preliminary investigation on the influence of residual gas species in a vacuum chamber on field emission current like in [11] is described in Section 23.7 [4].

23.3 Impregnation of RuO_2 and OsO_2

23.3.1 Properties of RuO_2 and OsO_2

In transmission electron microscopy (TEM) or SEM, OsO_4 is widely used to provide images of the specimen with high contrast, as well as to produce conductive films that can be used as biological staining agents for insulated specimens and to preferentially stain copolymers in polymer studies. RuO_4, which is a more powerful staining agent than OsO_4, is also used in polymer studies. OsO_4 and RuO_4 are used as staining agents because of their excellent properties, which are explained as follows. OsO_4 and RuO_4 react with the double bonds of unsaturated polymers to form thin films or sub-nanoclusters of osmium dioxide (OsO_2) and ruthenium dioxide (RuO_2), respectively; these films do not interfere with SEM observations of the polymer specimen. Moreover, these dioxides have low specific resistances closer to those of metals: the specific resistance of RuO_2 is 3.5×10^{-5} Ωcm at 300 K [2], and that of OsO_2 is 6.0×10^{-5} Ωcm at 300 K. In addition, these dioxides show good chemical and physical stability at ambient temperature on account of the high melting point of RuO_2 (1600 K) and the high pyrolysis temperature of OsO_2 (730 K). Therefore, it is expected that the electric characteristics of CNTs can be improved by impregnating OsO_4 or RuO_4 on the CNT surface.

23.3.2 The Method of RuO_2 Impregnation

RuO_2 impregnation on the CNT surface is a simple process. The CNT sample is first preserved in a RuO_4 solution for 8–20 h at ambient temperature. The sample is then dried in air until there is no weight loss. The quantity of RuO_2 impregnated on the CNT surface is estimated by measuring the residual RuO_2 concentration in the solution using ultraviolet adsorption spectroscopy. As shown in Figure 23.1, the quantity of RuO_2 impregnated on the CNTs was found to be about 80 mg g^{-1} of CNTs, as confirmed by inductively coupled plasma-optical emission spectroscopy (ICP-OES) measurements.

23.3.3 Observation of CNTs with Impregnation of RuO_2

Figure 23.2 shows the TEM images of the CNTs with and without RuO_2 sub-nanoclusters impregnated on the CNT surface. X-ray photoelectron spectroscopy (XPS) confirmed the stoichiometry of RuO_2 to be 4 at.%, on average.

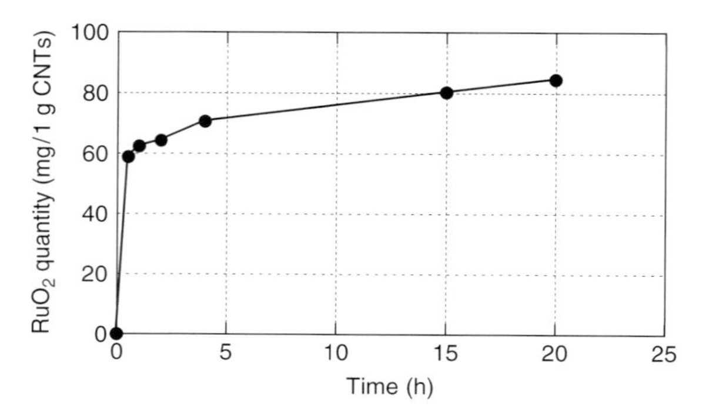

Figure 23.1 Quantity of RuO_2 impregnated on CNTs.

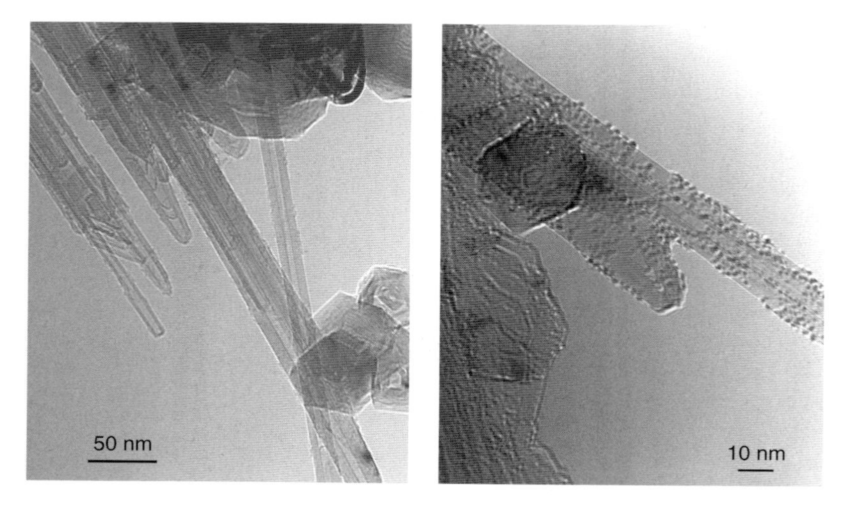

Figure 23.2 TEM images of CNTs impregnated with and without nanosized RuO_2.

Figure 23.2 shows that sub-nanosized RuO_2 clusters impregnated the CNT surface, probably along with a chemical reaction. The same method could be used to impregnate the CNT surface with OsO_2. The SEM images of CNTs impregnated with OsO_2 were similar to those of CNTs impregnated with RuO_2.

23.4
CNT Rooting

After CNTs impregnated with nanosized RuO_2 particles were subject to high emission currents in experiments, the disappearance of CNTs from the substrate was observed. It was assumed that the weak bond between the CNTs and the substrate caused the disappearance of the CNTs. Figure 23.3a and b shows CNTs on a Si substrate before and after measurements were taken. Carbide formation of

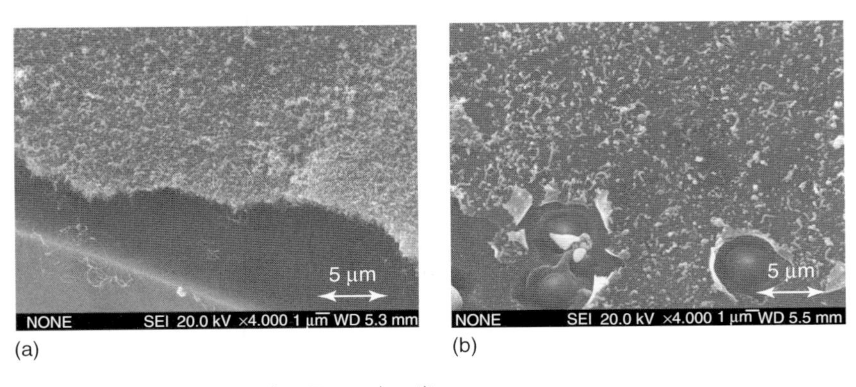

Figure 23.3 SEM images of CNTs on the silicon substrate (a) before and (b) after measurement.

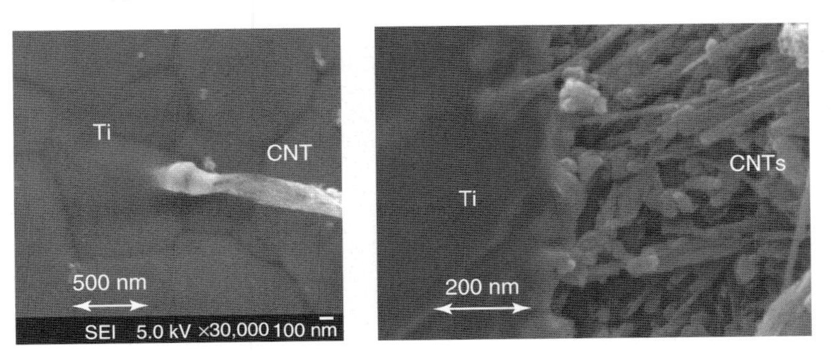

Figure 23.4 SEM images of CNTs rooted on the titanium film.

CNTs was performed using titanium metal to prevent the disappearance of CNTs. CNT powder was set on a 0.1-mm-thick titanium foil which was then placed on a 1-mm-thick graphite plate used as the base substrate; the arrangement was then heated to about 1400 K in a vacuum chamber at a pressure of 10^{-6} Pa or less for some time (this is the so-called "rooting" process) [4]. In another experiment, the graphite plate was replaced with a 0.2-mm-thick tantalum plate; in this case, the thickness of the titanium film was about 2 µm. The rooting process was carried out under the same conditions as those used for the graphite plate and titanium foil. Figure 23.4 shows an SEM image of CNTs rooted to the titanium film.

23.5 Effect of Impregnation on Field Emission Properties

Four kinds of samples were prepared for the measurement of field emission properties in order to demonstrate effects of the impregnation and the rooting [4].

1) The first sample comprised CNTs impregnated with RuO_2 by the rooting process using titanium foil on a graphite plate (RuO_2/rooted CNTs).

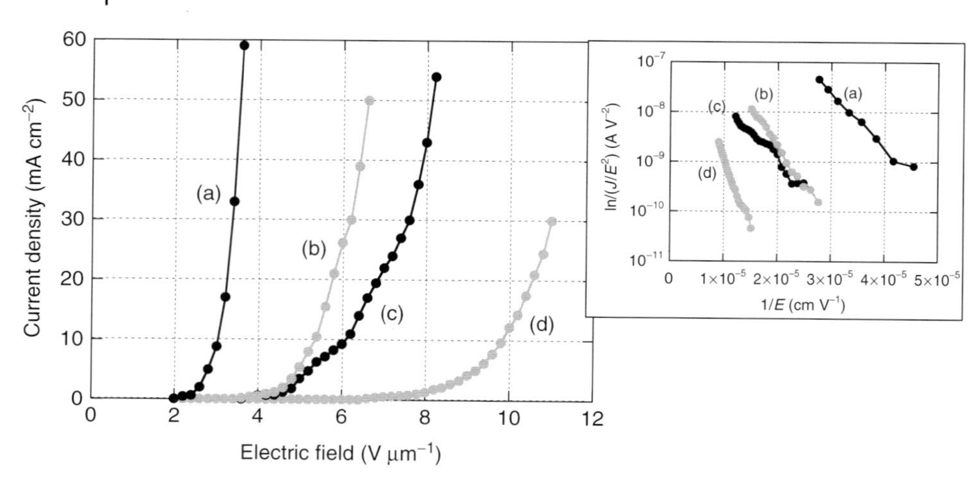

Figure 23.5 *E–I* characteristics: (a) RuO_2/rooted CNTs, (b) RuO_2/CNTs on Si, (c) rooted CNTs, and (d) CNTs on Si.

2) The second sample comprised CNTs impregnated with RuO_2 that were grown on a Si plate (RuO_2/CNTs on Si).
3) The third sample comprised CNTs rooted to the titanium foil on the graphite plate (rooted CNTs).
4) The fourth sample comprised CNTs grown on a silicon plate (CNTs on Si).

The second sample was prepared to show the effect of rooting, and the third and fourth samples were prepared for comparing normal CNTs to impregnated CNTs. The $E-I$ plots of these samples are shown in Figure 23.5. In this graph, it is found that for both types of substrates, the $E-I$ characteristics of the RuO_2-impregnated samples are far superior to those of the samples without RuO_2 impregnation. The impregnation of RuO_2 onto CNTs leads to considerable changes in the $E-I$ plots, decreasing the threshold field from 5.4 to 3.1 V μm^{-1} (Figure 23.5a and b). Furthermore, RuO_2 impregnation reduces the turn-on field to about half the corresponding value for CNTs without RuO_2 as shown in Figure 23.5c and d. Here, the turn-on and threshold fields denote the macroscopic fields needed to obtain current densities of 10 $\mu A\ cm^{-2}$ and 10 $mA\ cm^{-2}$, respectively.

Figure 23.6 shows the results of the endurance running test performed on RuO_2/rooted CNTs (the first sample). The running test was performed at a constant current density of 50 $mA\ cm^{-2}$, and the endurance of CNTs was evaluated for an increase in the electric field. In this graph, the electric field increased from 3.6 to 3.8 V μm^{-1} over 72 h; therefore, it was expected that the lifetime of the CNTs would be about 400 h if the *endurance time* were to be defined as the time required for the electric field to increase by 20%. A current density of more than 200 $mA\ cm^{-2}$ was obtained in this sample at an electric field of 5 V μm^{-1}. The characteristics of field electron emission were confirmed by the linear relationship between J/E^2 (A V^{-2}) and $1/E$ (cm V^{-1}) on the basis

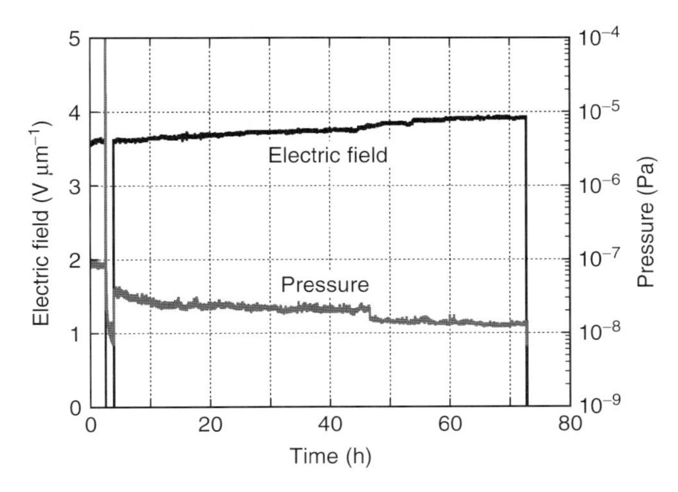

Figure 23.6 Endurance running test of RuO_2/rooted CNTs at current density of 50 mA cm^{-2}.

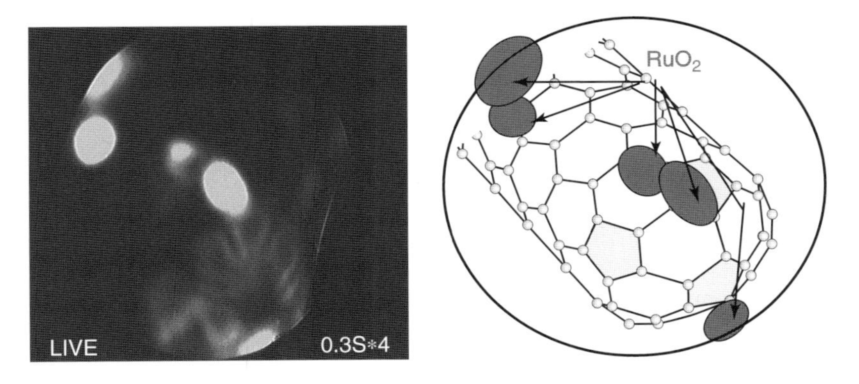

Figure 23.7 FEM image of CNTs impregnated with RuO_2.

of the Fowler–Nordheim (FN) equation, as shown in the FN plots (inset of Figure 23.5). Figure 23.7 shows a field emission microscopy (FEM) image of CNTs impregnated with RuO_2 (RuO_2/CNTs) [13]. Because the bright emission sites did not move during the FEM observation, it was assumed that the emission sites would contain RuO_2 particles bonded stably on the CNTs without any absorbed gas molecules. We could conclude that the stable and high electron emission characteristics of RuO_2/CNTs were due to the chemical bonding of RuO_2 to CNTs; the enhanced emission characteristics were obtained not by any reduction in the work function of the CNTs but by an increase in the field enhancement factor, because the work function of RuO_2 was 5.0 eV, a value similar to that of CNTs. In another experiment, RuO_2 was substituted with OsO_2 and the same measurements were carried out. As with CNTs impregnated with RuO_2, CNTs impregnated with OsO_2 showed superior $E-I$ characteristics to those without impregnation.

23.6
Effect of Rooting on Field Emission Properties

Figures 23.8 and 23.9 show the effect of rooting [4]. These graphs were obtained for samples prepared by the rooting process using a graphite substrate and a 0.1-mm-thick titanium film. Further improvements in the $E-I$ characteristics and endurance time were evident when tantalum was used as the substrate and a

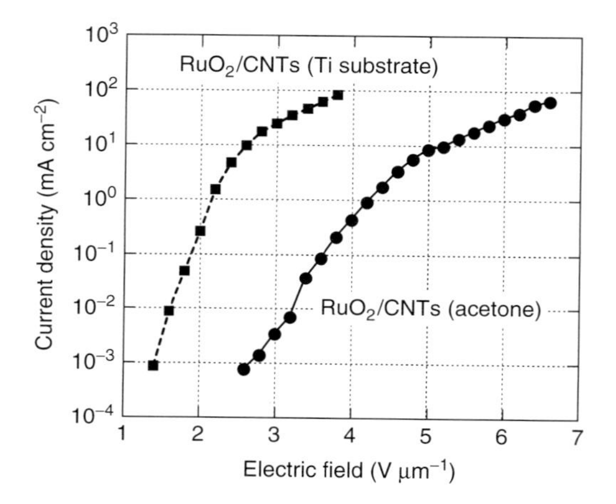

Figure 23.8 E-I characteristics of RuO_2 impregnated CNTs rooted in titanium and RuO_2 impregnated CNTs dispersed in acetone.

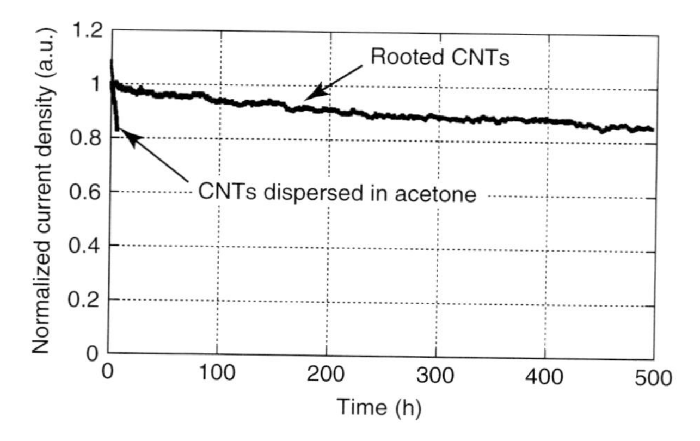

Figure 23.9 Comparison of endurance and stability between CNTs rooted in titanium and CNTs dispersed in acetone.

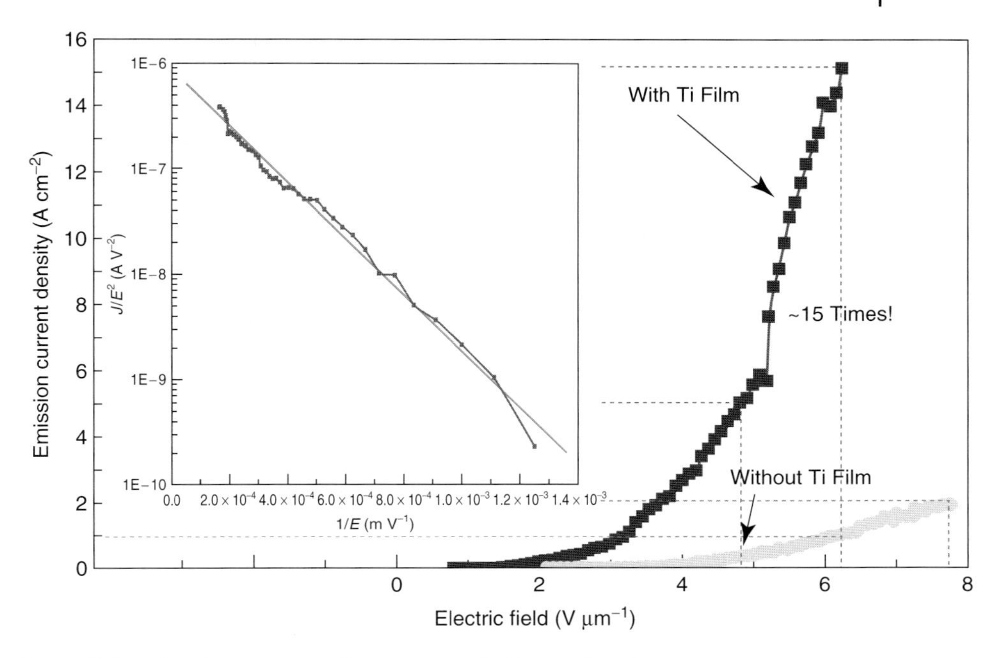

Figure 23.10 *E–I* characteristics of CNTs rooted in titanium film (2 μm).

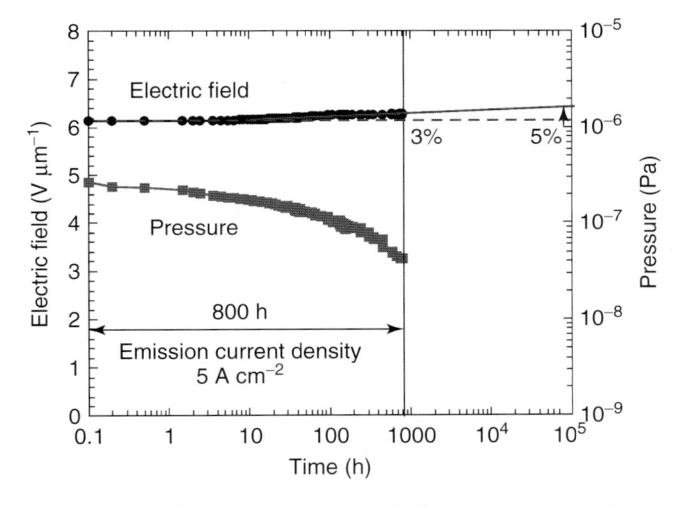

Figure 23.11 Endurance running test at high emission current density.

2-μm-thick titanium film was employed (Figures 23.10 and 23.11) [10]. In the $E-I$ plots of Figure 23.10, the emission current density increased to 5 A cm^{-2} at 4.8 V μm^{-1} and reached a maximum value of 15 A cm^{-2} at only 6.3 V μm^{-1}. The results of the endurance running test (Figure 23.11) in UHV where the major residual gas is H_2, revealed a very small increase of the electric field, implying an expected

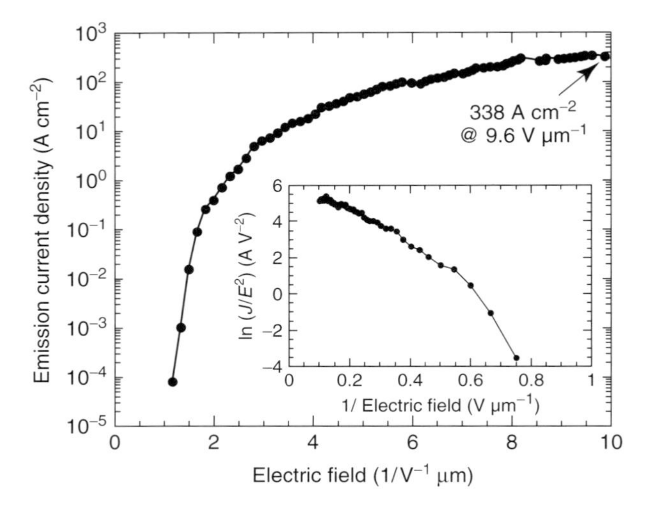

Figure 23.12 *E*–*I* characteristics and FN plot of MWNTs [10] rooted in titanium film (2 μm) and titanium micropowders.

lifetime of more than 10^5 h. Figure 23.12 shows a trial result of the achieved current density more than 300 A cm^{-2} by fabricating a film emitter of MWNTs rooted in a titanium film and titanium micropowders to enhance the reaction sites for better carbide formation.

23.7 Influence of Residual Gas

For examining the influence of gas introduction, a series of gas exposure tests were performed to observe the emission characteristics of the CNTs impregnated with RuO_2. In these experiments, the initial current density was set to be 40 mA cm^{-2} in the constant voltage mode with which electron accelerating voltage at 2 kV. After 1 or 2 h of idle time, gas exposure was started at a pressure of 1.0×10^{-5} Pa and was stopped when the emission current reduced by 20% with reference to the initial emission current. Gas introduction was also stopped if the desired 20% was not achieved within 24 h. The experiments using six different gas species were carried out, with results obtained to categorize three types of the emission property. Figure 23.13 shows the results of neon and water vapor for two of these types.

The first type comprised neon and hydrogen; this type property led to a very gradual decrease in the emission current. In the case of neon, emission showed a monotonic decrease down to +7% until the introduction of water vapor was stopped. The second type comprised methane and water vapor. In the case of water vapor, emission increased gradually up to +20% at the beginning of gas introduction and then showed a decreasing tendency down to −20% until the

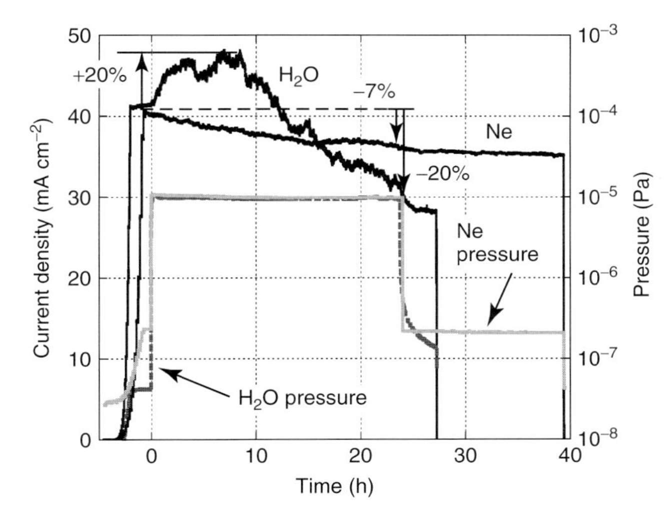

Figure 23.13 Results of Ne and water vapor introduction tests.

introduction of water vapor was stopped. When the vapor flow was stopped, the decreasing tendency became more pronounced. Though neon and water vapor have similar molecular weights, these two gas species showed a remarkably different behavior, where both sputtering and work function change or adsorbate tunneling [6] should be involved.

The reason for the decrease is physical sputtering of the CNTs or the RuO_2 nanoclusters. In the case of water vapor introduction, through chemical reaction of decomposed gas with the surface, chemical sputtering would play a dominant role in degradation of the emission characteristics and the emission decay was fast in comparison with the corresponding decay when neon was used. On the other hand, the gradual increase in emission might be explainable on the basis of decrease in the work function of the RuO_2/CNTs surface or the appearance of adsorbate tunneling state on account of gas adsorption. While the consequent net decrease due to these two effects would be attributed to be 20%, the effect of chemical sputtering on the decay could be decoupled to be 40%, which is a considerable large value compared with that of neon case.

The third type comprised oxygen and carbon monoxide. Figure 23.14 [4] shows the results obtained with oxygen. When the gas exposure was started, the emission current first decreased rapidly and then more gradually. When gas introduction was stopped on account of the 20% reduction in emission, the emission began to recover. This behavior is in stark contrast to those of the first and the second type properties that showed no recovery. This result can be interpreted as follows. The rapid decrease is due to an increase in the work function at the gas-adsorbed surface of the RuO_2/CNTs. This is followed by the gradual decrease in emission due to permanent damage caused by the chemical sputtering, while these two effects are concurrent during the gas exposure. Desorption of O_2 brings about gradual emission recovery when the gas introduction stops. Based on the results obtained

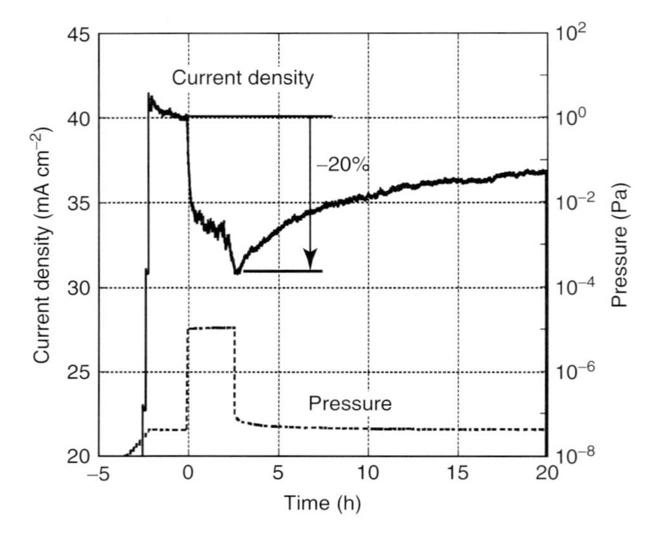

Figure 23.14 O_2 gas introduction test.

from the gas exposure experiments, it is clear that in operation of a high-current density, for example, as shown in Figure 23.11, one can neglect the influence of the operational pressure on the order of 10^{-7} Pa, with hydrogen the dominant residual gas which is considered to be of good quality vacuum and the extremely gradual degradation is due to the very gradual loss of good working CNTs at their junctions.

References

1. Loiseau, A., Launois, P., Petit, P., Roche, S., and Salvetat, J.P. (2006) *Understanding Carbon Nanotubes*, Springer, Berlin and Heidelberg.
2. Bonard, J.-M., Dean, K.A., Coll, B.F., and Klinke, C. (2004) *Phys. Rev. Lett.*, **89**, 197602.
3. Green, J.M., Dong, L., Gutu, T., Jiao, J., Conley, J.F., and Ono, Y. Jr. (2006) *J. Appl. Phys.*, **99**, 094308.
4. Noguchi, T. (2004) Doctoral Thesis; Study on enhancement of field emission characteristics for multiwalled carbon nanotubes, The Sokendai University.
5. Nilsson, L., Groening, O., Emmenegger, C., Kuettel, O., Schaller, E., Schlapbach, L., Kind, H., Bonard, J.-M., and Kern, K. (2000) *Appl. Phys. Lett.*, **76**, 2071.
6. Read, M.E. (2001) In Proceedings of the 2001 Particle Accelerator Conference, Chicago, 2001, p. 1026.
7. Chen, Z., Den Engelsen, D., Bachmann, P.K., Van Elsbergen, V., Koehler, I., Merikhi, J., and Viechert, D.U. (2005) *Appl. Phys. Lett.*, **87**, 243104.
8. Wei, W., Jiang, K., Wei, Y., Liu, P., Liu, K., Zhang, L., Li, Q., and Fan, S. (2006) *Appl. Phys. Lett.*, **89**, 203112.
9. Huang, N.Y. and She, J.C. (2004) *Phys. Rev. Lett.*, **93**, 075501.
10. Huarong, L. (2007) Doctoral Thesis, The Sokendai University.
11. Saito, Y., Nishianca, T., Kato, T., Kondo, S., Tanaka, T., Yotani, J., and Uemra, S. (2002) *Mol. Crys. Liq.*, **387**, 79.
12. Tong, K.Y. *et al.* (2001) *J. Mater. Sci. Lett.*, **20**, 699–700.
13. Saito, Y. (2004) Private communication.
14. Dean, K.A., von Allmen, P., Chalamala, B. (1999) *J. Vac. Sci. Technol. B*, **17**, 1959.

24
High-Resolution Microfocused X-ray Source with Functions of Scanning Electron Microscope

Koichi Hata and Ryosuke Yabushita

24.1
Introduction

W. C. Röntgen discovered a new radiation with a high transmission in 1895 and introduced a perspective photograph of a human hand taken using this radiation [1]. It was the beginning of X-ray radiography. In 1912, X-ray diffraction by crystals was demonstrated by M. von Laue to show that X-ray was an electromagnetic wave. It was expected that a microscope with X-ray illumination would give higher resolution than an optical microscope because of the shorter wavelength of X-rays, but much progress was not made because the fabrication of a lens by using the refraction effect like in the case of visible light or electron beam was difficult. For the purpose of overcoming this limitation, von Ardenne proposed making a higher resolution X-ray image by forming a very small X-ray spot by illuminating a metal target with a focused electron beam [2]. In the 1950s, W. C. Nixon developed a high-resolution X-ray microscope using a magnetic field-type lens with low aberration in order to form an electron probe with a much smaller diameter on a metal target [3]. Shinoda *et al.* [4] and Komoda *et al.* [5] also tried developing an X-ray microscope in Japan. In 1980, Yada *et al.* developed an X-ray microscope by modifying a scanning electron microscope (SEM) to get very small X-ray spot easily, and application to biology advanced [6]. However, these high-resolution X-ray microscopes showed low output intensity and were difficult in real-time observation because of limited probe current.

For increasing the performance of the X-ray microscope, the diameter of the electron probe has to be made small while keeping the probe current large. From this viewpoint, field emission cathode with a high brightness has been attempted for use in X-ray excitation. The field emission cathode has features beneficial to form a very small electron probe on the target because it has brightness 3–4 orders of magnitude higher than the conventional thermionic cathode as shown in Table 24.1 [7], and the effective source size is extremely small. Yada *et al.* obtained a resolution of 100 nm by combining a ZrO/W(100) thermal field emission cathode with brightness equal to that from a field emission cathode and a gun lens superposed with a magnetic field [8].

Carbon Nanotube and Related Field Emitters: Fundamentals and Applications. Edited by Yahachi Saito

ISBN: 978-3-527-32734-8

Table 24.1 Comparison of properties for various electron sources.

	Brightness (A (cm^2 sr)$^{-1}$)	Emission current (μA)	Effective source size (μm)	Working vacuum (Pa)	Lifetime (h)
Tungsten hairpin	10^4	100~300	20	10^{-3}	~100
LaB_6	10^5	100~300	10~15	10^{-5}	~500
ZrO/W Schottky	10^8	10~200	0.02~0.05	10^{-7}	>1000
Field emission	10^8	1~30	0.01~0.02	10^{-8}	>1000

On the other hand, the carbon nanotube (CNT) has recently attracted attention [9, 10] as a material with brightness characteristic a step higher than conventional tungsten, and its applications to the X-ray source have been reported by some groups, but their resolutions are mostly about 30 μm [11, 12] and the highest resolution is 5 μm or so [13]. This is due to the insufficient knowledge of CNT as cathode material, but also largely by the fact that electron optics to focus the electron beam has not been optimized. For the purpose of studying practical utilization of the CNT field emission cathode, we are carrying on with the development of a high-resolution X-ray microscope mounted with a Butler-type electrostatic lens and a CNT field emission cathode based on a commercial desktop SEM [14, 15]. In this chapter, the configuration and performance of this X-ray microscope with SEM function currently obtained are reviewed, and problems to be studied from now on are also presented.

24.2 Multiwalled CNT Field Emission Cathode

De Jonge *et al.* have reported, with regard to an electron beam emitted from a single multiwalled CNT, that brightness normalized with acceleration voltage is 10^5 A(cm^2 sr V)$^{-1}$, which is one order higher than the ordinary field emission cathode using tungsten, and that its effective source size (diameter of virtual spot) is about 2 nm [16]. However, a single multiwalled CNT cathode is rarely fabricated with high reliability from the aspects of contact resistance and adhesion strength with the substrate though various fabrication methods have been reported to date [17, 18]. We have thus employed a bundle of multiwalled CNT, which was made with the arc discharge method in helium ambience, superior in crystallinity, and with mean diameter about 20 nm [19]. This bundle was used as the cathode by adhering it with heat-resistant conductive paste at the tip of a tungsten hairpin filament of 100 μm diameter. The CNT and conductive paste can be degassed by passing a current to heat the hairpin. Figure 24.1 shows a photograph of the multiwalled CNT cathode assembly and an SEM image of the CNT cathode tip. As seen from Figure 24.1b, it is a multicathode in which many CNTs project in various directions, and when the voltage applied to the cathode is gradually raised,

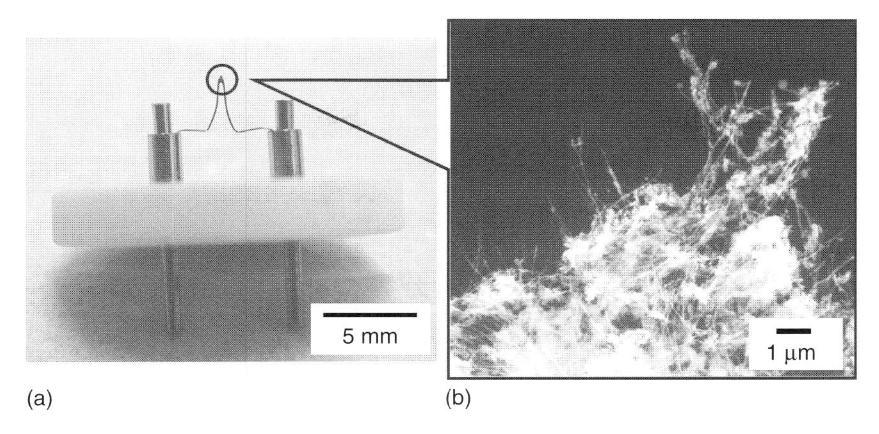

Figure 24.1 (a) Picture of multiwalled CNTs cathode assembly and (b) SEM image of a CNT bundle tip.

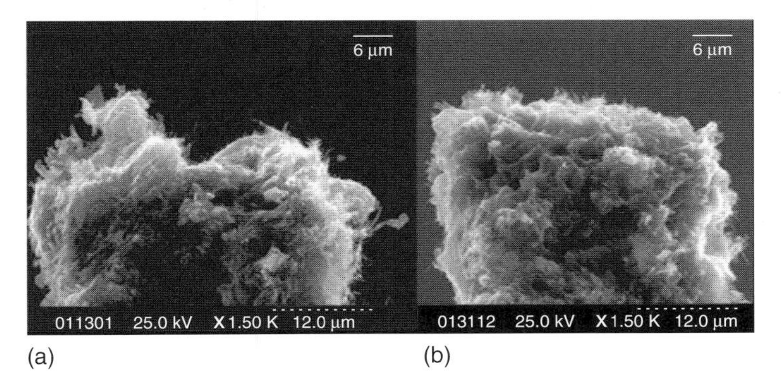

Figure 24.2 SEM images showing a typical change in shape of multiwalled CNT bundle by the aging process: (a) Before and (b) after the aging for 20 min at a fixed voltage of −1.5 kV.

electron emission starts from the most projecting CNT and their number increases as the applied voltage is raised. From among them, we selected only the electron beam emitted along the optic axis to be used as the probe for exciting X-rays.

The emission current from a CNT bundle shows a large reduction with time and spiky noise in the initial stage of extraction. These may be caused by the fact that field emission focused only from the projecting CNTs in the initial stage because CNTs are unevenly projecting at the bundle tip. When using the CNT cathode, therefore, aging for stabilizing the emission current is required. As an example of the shape change of the CNT bundle due to aging, Figure 24.2 shows the SEM images of a CNT bundle before and after aging for 20 min, with the initial emission current at 500 μA and cathode voltage at −1.5 kV fixed. It is found from

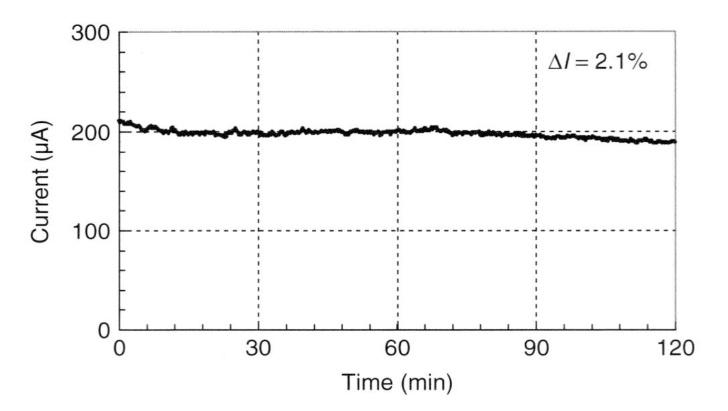

Figure 24.3 Time trace of emission current from a CNT field emission cathode obtained after aging for 5 h. The self-bias resister of 10 MΩ was connected to the first anode of the Butler lens for the measurement.

Figure 24.2b that the ups and downs at the bundle tip that existed before aging have disappeared and the CNT lengths have become nearly even. In observation of X-ray images, a 5-h aging was carried out under similar conditions. The emission current continued to decrease at a rate of about 60 μA h^{-1} during the aging, but became stable at about 200 μA after the aging finished. Shown in Figure 24.3 is change of emission current with passage of time measured after the aging finished. The current fluctuation coefficient ΔI, which is estimated with the equation:

$$\Delta I = \frac{I_{sd}}{I_{av}} \tag{24.1}$$

showed a good value at 2.1%. Here, I_{sd} and I_{av} are the standard deviation and the average value of the emission current, respectively.

The emission current is considered to become stable by eliminating the projecting CNTs to make the applied electric field on each CNT nearly even, resulting in reduced emission current per CNT, because of sublimation due to Joule heating [20] accompanied by microdischarge while aging and large current extraction.

24.3 Construction of High-Resolution Transmission X-ray Microscope Equipped with the Function of SEM

With the lens effect produced when an electron beam is extracted from the multiwalled CNT cathode and accelerated, aberration appears and this causes an increase in the virtual source diameter and a decrease in the brightness. In order to suppress this, we newly designed a Butler-type electrostatic lens [21] by means of ELFIN, software for electromagnetic field analysis, and BEAM, software for

Figure 24.4 Picture of Butler-type electrostatic lens newly designed by CAD for CNT field emission cathode.

charged particle orbital analysis, made by ELF Ltd. The appearance of the Butler lens that we experimentally fabricated is shown in Figure 24.4. This Butler lens has been constructed with three anode plates, and alignment of each plate is carried out by fitting a jig to the center holes. Mechanical setting accuracy by this is better than ±10 μm. After removing a dual-stage condenser lens, this Butler lens was installed at a distance of 10 mm from the cathode.

The metallic target material to let the X-ray emitted by illuminating with a focused electron probe has to be chosen by taking factors such as melting point, heat conductivity, chemical stability, and X-ray absorbance into consideration. We selected, first, tungsten with a large atomic number ($Z = 74$) and a high melting point as the target material from the viewpoint of X-ray intensity. Tungsten was deposited by sputtering to a thickness of 300 nm as the target on a beryllium window of 100 μm thickness, also working as vacuum seal. A hole of 2 mm diameter was drilled on the specimen stage for SEM observation, beneath which the target was installed.

Figure 24.5 shows the construction of a transmission-type X-ray microscope developed on the basis of the commercial desktop SEM (Tiny-SEM 1540 made by TECHNEX Craft Center). It is capable of secondary electron image observation, beam alignment, and astigmatic correction as the original functions of SEM, and the X-ray microscopic image can be detected while confirming the focusing state of the electron beam on the metallic target. The electron source chamber was kept at a pressure of 3.0×10^{-5} Pa through differential pumping with a turbomolecular pump with an evacuation speed of $60\ \mathrm{l\ s^{-1}}$, and emission current was stabilized also by connecting a 10-MΩ self-stabilization resistor [22] to the first anode of the Butler lens. The objective lens of the Tiny-SEM has a fixed focal length through the use of a permanent magnet, and focusing has to be done by adjusting the acceleration voltage (15 kV maximum); but on the other hand, this gives the advantage that the column is compact and the control system is simple. For the detection of the X-ray microscopic image, we have used an X-ray image intensifier for real-time observation and photographic film for electron microscopy for high-resolution recording.

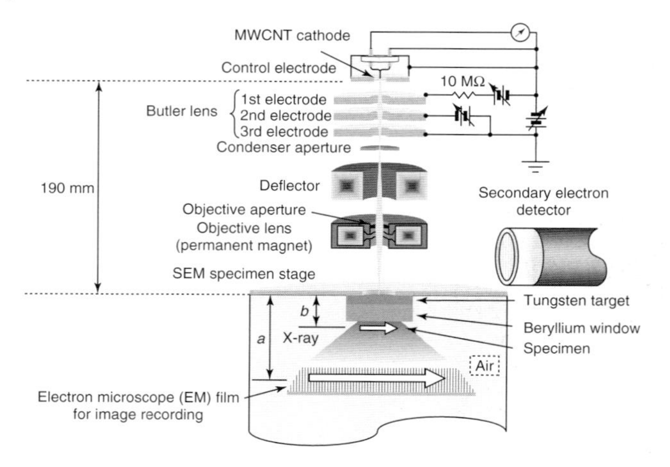

Figure 24.5 Schematic drawing of our transmission-type X-ray microscope equipped with a CNT field emission cathode. Conventional SEM functions, that is, SEM image, stigmator, and beam aligner are also shown.

24.4 Characteristic Evaluation of High-Resolution X-ray Microscope Provided with SEM Function

24.4.1 Resolution of SEM

Shown in Figure 24.6 are the SEM images of a 300-mesh copper surface taken at an acceleration voltage of 9.8 kV. In Figure 24.6b, showing a high-magnification image, a gap of about 20 nm between grains is found to have been resolved though shot noise and mechanical vibration are also seen. It has generally been shown in SEM that two points separated by a length equal to the electron probe diameter divided by the natural logarithmic base e can be resolved. Thus, we assume that the resolution of the SEM image δ complies approximately with

$$\delta \approx \delta_e / e \tag{24.2}$$

Here, δ_e is the electron probe diameter, which is given by the half-width of the intensity profile of the electron probe complying with normal distribution. From Eq. (24.2), the electron probe width δ_e is estimated at about 54 nm.

In the SEM observation of Figure 24.6b, specimen absorbing current was about 11 nA. The absorbing current I_{absorb} and probe current I_p have the following relation:

$$I_p = \frac{I_{absorb}}{[1 - (\eta + \sigma)]} \tag{24.3}$$

Here, η and σ are the coefficients of backscattering for incident electrons and of secondary electron emission [23]. Because the coefficients of backscattering and

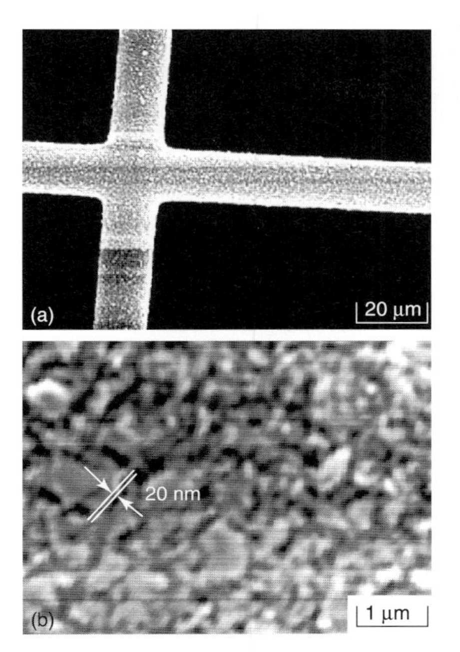

Figure 24.6 (a) SEM images of Cu 300 mesh with a line width of 7.5 μm and (b) the high-magnification image. A grain boundary with about 20 nm is successfully resolved.

secondary electron emission for copper ($Z = 29$) at an acceleration voltage of 10 kV are 0.3 [24] and 0.1 [25], respectively, the probe current can be estimated to be about 20 nA. Besides, the fluctuation coefficient of specimen absorbing current for every 25 s of image taking time for one frame is 10–20%. This fluctuation becomes a serious problem for image quality because fluctuation of the probe current during raster scanning results in variation of image luminance in the SEM. On the other hand, the X-ray image involves the integration of X-ray intensity over image taking time, so that influence of the current fluctuation is smaller in comparison with the SEM image.

24.4.2
Resolution of Transmission X-ray Microscope

Shown in Figure 24.7 is a transmission image obtained with this X-ray microscope by using an X-ray microchart (made by Japan Inspection Instruments Manufacture's Association) as a sample and a line profile of its contrast. The observation conditions are acceleration voltage of 10.6 kV and direct observing magnification of about 55 × because the distance between the target and the detector is $a = 12.325$ mm and that between the target and the specimen is $b = 0.225$ mm (Figure 24.5). In Figure 24.6a, the transmitted image, each slit of 1 μm width, is found clearly resolved. The resolution of about 400 nm has been obtained if we take the rising width from 10 to 90% of contrast at the slit edge as the resolution. The major factors that determine the resolution of the X-ray image are the diameter of the electron probe to excite X-rays, lateral distribution of X-ray generating

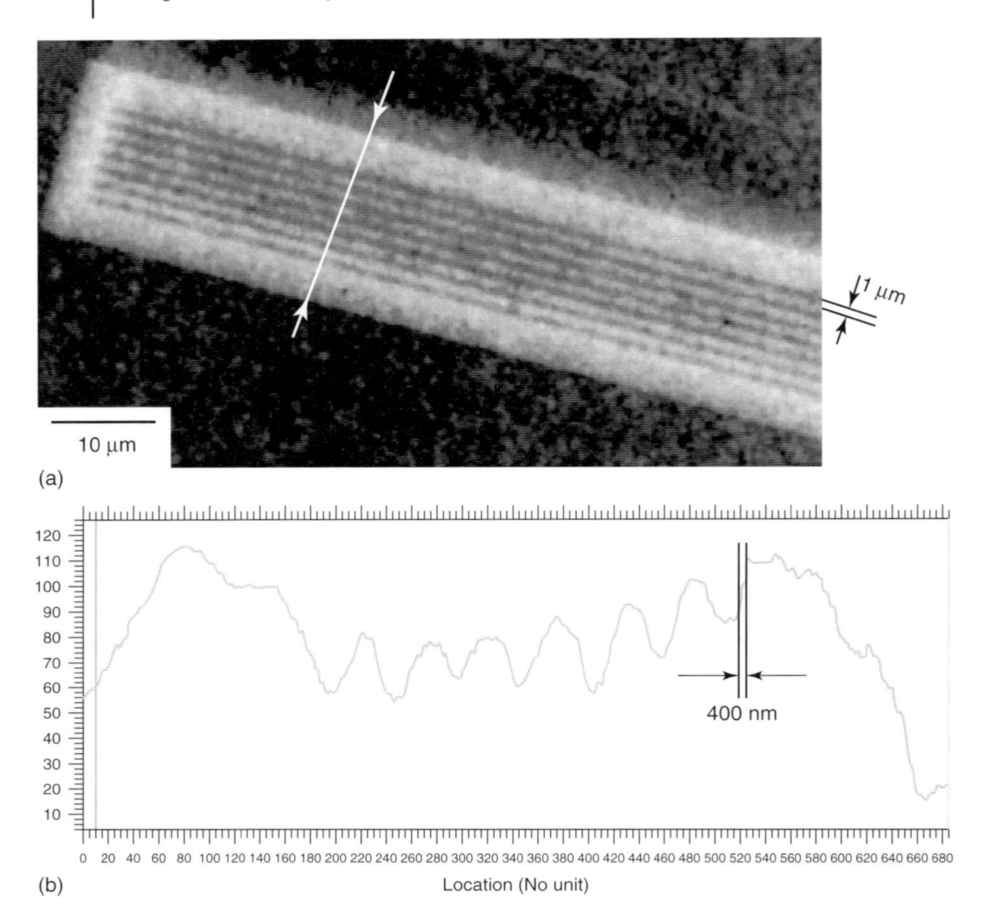

Figure 24.7 (a) Transmission X-ray micrograph of a microchart with a spacing of 1 μm and (b) line profile of image contrast at the position indicated by an arrow in (a).

region within the target, and blurring at edge of the specimen caused by Fresnel diffraction. Thus, we provide further insights into these factors hereunder.

24.5 Factors Limiting Resolution of X-ray Transmission Image

24.5.1 Lateral Distribution d_S of X-ray Generating Region

In estimating the lateral distribution of the X-ray generating region within the target, we first consider the energy of X-rays contributing to the image formation. The spectrum of the X-rays generated from the target is mainly dependent on the acceleration voltage of the electron beam and the target material. X-rays generated under the present experimental conditions are characteristic X-rays of tungsten Mα line (1.775 keV), Lα line (8.398 keV), and continuous X-rays accompanied by bremsstrahlung of electrons. The absorbed quantity of these X-rays until they are taken out can be estimated with the equation below [26]:

$$I = I_0 \exp\left[-\left(\mu/\rho\right)\rho t\right] \tag{24.4}$$

Here, ρ is density (g cm^{-3}) of absorber, (μ/ρ) is mass absorption coefficient (cm^2 g^{-1}) of the absorber with respect to the X-ray energy in question, and t is thickness (cm). It is found from Eq. (24.4) that the absorption of the Mα line X-ray is very small in a target of 300 nm thickness, but almost all the Mα line is absorbed in the beryllium window with 100 μm thickness. Absorption due to the air is several percentages at the most. Therefore, what mainly contributes to image formation is the Lα line and continuous X-rays. Further, taking the mass absorption coefficient of the beryllium window and the attenuation rate of X-rays in air with respect to the total energy into consideration, the X-ray image seems to have been formed mainly with the X-rays above 5 keV.

Electrons penetrating the target spread over the interaction volume through elastic and inelastic scattering processes. This interacting volume depends on the acceleration voltage of the electron beam and parameters of the material (atomic number Z, atomic weight A, and density ρ) [27]. Yamaguchi *et al.* conducted Monte Carlo simulation of lateral distribution of continuous X-rays generated in a gold (19.3 g cm^{-3}) target with a density nearly equal to that of tungsten (19.25 g cm^{-3}) at an acceleration voltage of 10 kV, and reported that the lateral distribution of continuous X-rays above 4 keV is about 50 nm [28]. Atomic weights A of gold ($Z = 79$) and tungsten ($Z = 74$) are 197 and 184 g mol^{-1}, respectively, so Z/A values of the both are nearly equal. Therefore, the lateral distribution δ_S of the X-ray generating region in our experiment is also considered as about 50 nm.

24.5.2
Blurring δ_F Caused by Fresnel Diffraction

As shown in Figure 24.8, Fresnel diffraction at the edge of the specimen that is not transparent optically is considered, where the distances from the X-ray spot to film for image recording and to the specimen are a and b, respectively. Distance p to the first maximum of the diffraction fringe intensity is given by

$$p = \left\{a\left(a + b\right)\lambda/b\right\}^{1/2} \tag{24.5}$$

from diffraction theory [28]. Here, λ is the wavelength of the X-rays. Because $a \gg b$ for a conventional transmission-type X-ray microscope, blurring caused by Fresnel diffraction converted on the specimen plane δ_F is given by

$$\delta_F = p/M \approx \left(\lambda b\right)^{1/2} \tag{24.6}$$

by dividing p by photographic magnification $M = a/b$. Under present experimental conditions, $\lambda = 0.3$ nm and $b = 0.225$ mm, δ_F becomes about 260 nm.

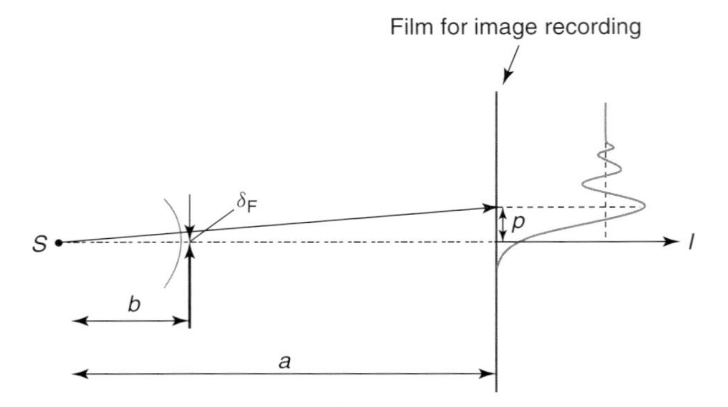

Figure 24.8 Formation of Fresnel fringes.

24.5.3
Evaluation of Theoretical Resolution δ_X

By using the estimated value of each factor mentioned above and the empirical evaluation formula

$$\delta_X = \left(\delta_e^2 + \delta_s^2 + \delta_F^2\right)^{1/2} \tag{24.7}$$

the value of resolution δ_X is obtained as about 270 nm, which is in good agreement with the experimental value. From this fact, it is found at this stage that resolution of this X-ray microscope is limited by the Fresnel diffraction δ_F. Reduction of δ_F requires making b smaller according to Eq. (24.6). This is possible by using a thinner beryllium window, or by using the target metallic thin film itself as the vacuum sealing window.

It is also important to reduce δ_e. Because this X-ray microscope has an objective lens with a permanent magnet that cannot change the excitation intensity, the aperture diameter must be reduced to make the electron probe diameter smaller. However, the normalized brightness obtained on the target plane is about 10^1 A $(\mathrm{cm}^2\,\mathrm{sr}\,\mathrm{V})^{-1}$, so it is currently difficult to reduce aperture diameter further while keeping sufficient X-ray intensity. The reason for this low brightness is largely the inaccurate alignment of the optic axis of the electron beam emitted from CNTs dispersed at random, so it seems effective to use a single CNT cathode [29] with a well-defined direction instead of the bundle.

24.6
Conclusion

In this paper, we introduced a compact, high-resolution X-ray microscope equipped with SEM function, which is under development, for the purpose of industrial application of CNT field emission cathode which is expected to show a high brightness. Although the resolution we have obtained at present is higher than that of X-ray source instruments mounting the CNT cathode recently reported, it is hard to say to have fully taken advantage of its high brightness. To obtain a higher resolution than 50 nm, which has been obtained with an X-ray microscope mounting a conventional ZrO/W Schottky cathode [30], however, we are pursuing further research with the hope of achieving technical innovation toward the next-generation devices.

References

1. Röntgen, W.C. (1896) *Nature (London)*, **53**, 274.
2. von Ardenne, M. (1939) *Naturwiss*, **27**, 485.
3. Nixon, W.C. (1955) *Proc. R. Soc. London, Ser. A*, **232**, 475.
4. Shinoda, G. and Yamanaka, S. (1954) *OYO BUTURI Jpn.*, **23**, 64.
5. Morito, N. and Komoda, T. (1958) *J. Electron Microsc.*, **6**, 12.
6. Yada, K. and Ishikawa, H. (1980) *Rep. Res. Inst. Sci. Meas. Tohoku Univ.*, **29**, 25.
7. Saito, Y., Ohashi, K., Fujino, S., Kai, H., and Yada, K. (2004) Extended Abstracts of the Autumn Meeting of Japanese Society for Non-Destructive Inspection, 2004, p. 73 (in Japanese).
8. Saito, Y., Kai, H., Shirota, K., and Yada, K.. (2006) Proceedings of the 8th International Conference on X-ray Microscopy IPAP Conference Series 7, 2006, p. 35 (in Japanese).
9. de Heer, W.A., Chatelain, A., and Ugarte, D. (1995) *Science*, **270**, 1179.
10. Hata, K., Takakura, A., Ohshita, A., and Saito, Y. (2004) *Surf. Interface Anal.*, **36**, 506.
11. Liu, Z., Yang, G., Lee, Y.Z., Bordelon, D., Lu, J., and Zhou, O. (2006) *Appl. Phys. Lett.*, **89**, 103111.
12. Kawakita, K., Hata, K., Sato, H., and Saito, Y. (2006) *J. Vac. Sci. Technol. B.*, **24**, 950.
13. Heo, S.H., Ihsan, A., and Cho, S.O. (2007) *Appl. Phys. Lett.*, **90**, 183109.
14. Yabushita, R., Hata, K., and Sato, H. (2007) *Vac. Sci. Technol. B*, **25**, 640.
15. Yabushita, R. and Hata, K. (2008) *Vac. Sci. Technol. B*, **26**, 702.
16. de Jonge, N., Lamy, Y., Schoots, K., and Oosterkamp, T. (2002) *Nature (London)*, **420**, 393.
17. Akita, S., Nakayama, Y., Mizooka, S., Takano, Y., Okawa, T., Miyatake, Y., Yamanaka, S., Tsuji, M., and Nosaka, T. (2001) *Appl. Phys. Lett.*, **79**, 1691.
18. Matsumoto, S., Pan, L., Tokumoto, H., and Nakayama, Y. (2001) *Nanotube*, **10**, 114.
19. Saito, Y., Mizushima, R., Kondo, S., and Maida, M. (2000) *Jpn. J. Appl. Phys.*, **39**, 4168.
20. Saito, Y., Seko, K., and Kinoshita, J. (2005) *Diam. Relat. Mater.*, **14**, 1843.
21. Monro, E. (1972) Proceedings of the 15th European Congress on Electron Microscopy, 1972, 1, p. 22.
22. Suga, H., Abe, H., Tanaka, M., Shimizu, T., Ohno, T., Nishioka, Y., and Tokumoto, H. (2006) *Surf. Interface Anal.*, **38**, 1763.
23. Goldstein, J., Newbury, D., Joy, D., Lyman, C., Echlin, P., Lifshin, E., Sawyer, L., and Michael, J. (2003) *Scanning Electron Microscopy and X-ray Microanalysis*, Chapter 2, Springer, New York, pp. 12–13.
24. Goldstein, J., Newbury, D., Joy, D., Lyman, C., Echlin, P., Lifshin, E., Sawyer, L., and Michael, J. (2003) *Scanning Electron Microscopy and X-ray Microanalysis*, Chapter 2, Springer, New York, p. 77.
25. Goldstein, J., Newbury, D., Joy, D., Lyman, C., Echlin, P., Lifshin, E., Sawyer, L., and Michael, J. (2003) *Scanning Electron Microscopy and X-ray Microanalysis*, Chapter 2, Springer, New York, p. 90.
26. Goldstein, J., Newbury, D., Joy, D., Lyman, C., Echlin, P., Lifshin, E., Sawyer, L., and Michael, J. (2003) *Scanning Electron Microscopy and X-ray Microanalysis*, Chapter 3, Springer, New York, p. 289.

27. Goldstein, J., Newbury, D., Joy, D., Lyman, C., Echlin, P., Lifshin, E., Sawyer, L., and Michael, J. (2003) *Scanning Electron Microscopy and X-ray Microanalysis*, Chapter 2, Springer, New York, pp. 61–72.
28. Yamaguchi, Y., Shimizu, R., Ikuta, T., Kikuchi, T., and Takahashi, S. (2006) *J. Surf. Anal.*, **13**, 223.
29. Zhang, J., Tang, J., Yang, G., Qiu, Q., Qin, L.-C., and Zhou, O. (2004) *Adv. Mater.*, **16** (14), 1219.
30. Saito, Y., Kai, H., Shirota, K., and Yada, K. (2007) Extended Abstracts of the 68th Autumn Meeting of Japan Society of Applied Physics, No. 2, 2007, p. 714 (in Japanese).

25
Miniature X-ray Tubes

Fumio Okuyama

25.1
Introduction

Since the invention by Coolidge of the so-called Coolidge tube [1], the generation of X-rays in conventional X-ray devices has been solely based on the principle of striking a metal target with accelerated thermionic electrons.

In thermionic emission (TE), the solid electron source, or the cathode, is incandescently heated to allow free electrons to escape from the cathode surface. This so-called hot cathode, commonly a heated tungsten (W) filament, works stably even in non-ultrahigh vacuum (non-UHV) ambiences, where a vast number of gaseous molecules impinge on the cathode during its operation.

The most popular among the existing X-ray technologies is X-ray radiography (XR) [2]. In recent years, the demand has been increasing in biological scientific communities for miniature X-ray tubes (MXTs), which can image biological tissues of micrometric dimensions. Unfortunately, it is near impossible to miniaturize TE X-ray tubes because the tube wall cannot be thermally shielded from the heated cathode for the miniature tube dimensions. This difficulty can be overcome by generating X-rays through the field emission (FE) process [3].

Using FE electrons to generate X-rays dates back to 1956, when Dyke and Dolan struck a copper target with FE electrons extracted from linearly arrayed W pins [4]. Regrettably, this pioneering work did not see any significant progress. The reason is as follows.

The FE process is exponentially dependent on the chemical and morphological states of the electron-emitting surface [3], which vary continuously in non-UHV. This means that FE operation in non-UHV destabilizes the electron emission, and sometimes ends in the fatal damage of the emitter through "cathode sputtering" (see the following section). This current instability is striking for metallic emitters, which are so reactive chemically that they undergo a strong chemical interaction with the residual gas molecules. (Dyke and Dolan carried out their experiments in pressures of 10^{-5} Pa.)

Carbon Nanotube and Related Field Emitters: Fundamentals and Applications. Edited by Yahachi Saito

ISBN: 978-3-527-32734-8

25.2
Our Technical Basis for Miniaturizing X-ray Tubes

Chemically, carbon is far more stable than metals; FE electron sources made of carbon or a related material might have a long lifetime even in non-UHV. Indeed, numerous articles published so far have demonstrated promising FE behavior of carbon nanotubes (CNTs) in non-UHV [5]. We were thus encouraged to equip X-ray tubes with CNT field emitters.

Commonly, X-ray tubes are operated in the pressure region of 10^{-5} Pa. In such a non-UHV ambience, residual gas molecules in the proximity of the emitter tip are polarized and clustered around the tip, which is biased negatively. These clustered gas molecules collide with FE electrons emitted from the emitter tip, thereby being ionized to strike the tip. From unknown reasons, this sputtering effect is especially enhanced on metal emitters. A single-point FE tip of W, for example, is morphologically damaged by a few minutes of operation. To develop FE X-ray tubes, therefore, nonmetallic emitter materials resistant to sputtering need to be employed. CNTs are sure to meet this demand (see above).

Our first FE X-ray tube was a simple metal chamber, inside which a Cu target was kept facing an FE emitter comprised of dense CNTs (Figure 25.1a) [6]. Since the Be window was connected to the chamber through a rubber flange, the chamber could not be entirely baked out; the ultimate chamber pressure was $\sim 4 \times 10^{-5}$ Pa.

CNTs were vertically grown on a cobalt-coated W wire. These CNTs amounted to $\sim 6 \times 10^{7}$ mm^{-2} in site density (Figure 25.1b) and were always topped with a metallic crystallite (Figure 25.1) (see also the transmission electron image inset in Figure 25.3). The crystallites were covered with a thin graphite layer, ensuring that the tips of the respective CNTs functioned as a graphitic electron source [6].

By repeating experiments, we confirmed that our CNTs were highly robust in non-UHV; they continue to field-emit electrons for 60–80 min in the above chamber pressure.

For soft samples like plants, the energy of X-rays must be lowered to the soft X-ray region. In terms of electron energy, the soft energy region is below 10 keV. Imaging at such a low electron energy inevitably prolongs the exposure time. For example, Figure 25.2a shows the contact X-ray image of a leaf recorded at an exposure time as long as 60 min and at the electron energy of 10 keV. The leaf was just plucked from the tree and hence still not dehydrated. Nevertheless, capillary veins for nutrition transportation are recognized in the image. (The capillary veins were around 0.1 mm across.) We emphasize that X-ray imaging at such low electron energy is generally impossible with metallic FE emitters, due to their limited lifetime in non-UHV.

Commonly, the current fluctuation is due to physical or chemical interactions of the residual gas molecules with the electron-emitting area on the tip surface. This process occurs independent of the emitter material, so our emitters, too, were not entirely stable but fluctuated at an amplitude of $\pm 10\%$ at the pressure employed (Figure 25.2b). Fortunately, the image on the photoplate is the result of integrated detection of X-rays, and therefore this amount of current fluctuation has

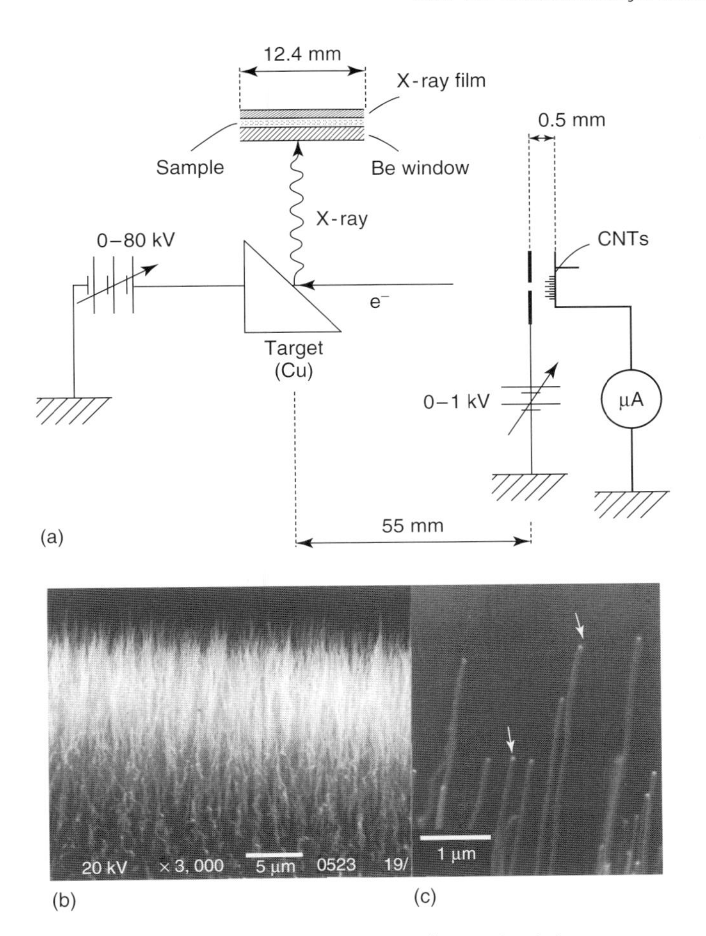

Figure 25.1 (a) Diagram of the X-ray producing circuit in our first FE X-ray device. No function of electron beam focusing was equipped on this device. (b) Typical CNTs used for a series of experiments with this device (SEM image), and (c) enlarged image of CNTs picked up from (b). The arrows in (c) indicate metal crystallites atop CNTs.

no negative effect on imaging the subject as it is. When using an image intensifier, the current stability has to be improved dramatically, because an unstable electron current would make a very short single shot less reliable. The best way to reduce the current fluctuation will be to control the electron current with a feedback technique.

The investigation stated above dealt with contact XR. A more developed version of XR is projection X-ray microscopy [2]. In this technique, the shadow image is formed on the screen with radially propagating X-rays. Theoretically, the more the X-ray emitting area is reduced, the more the image resolution is improved [2]. A common recognition in X-ray technology is that miniaturizing the tube

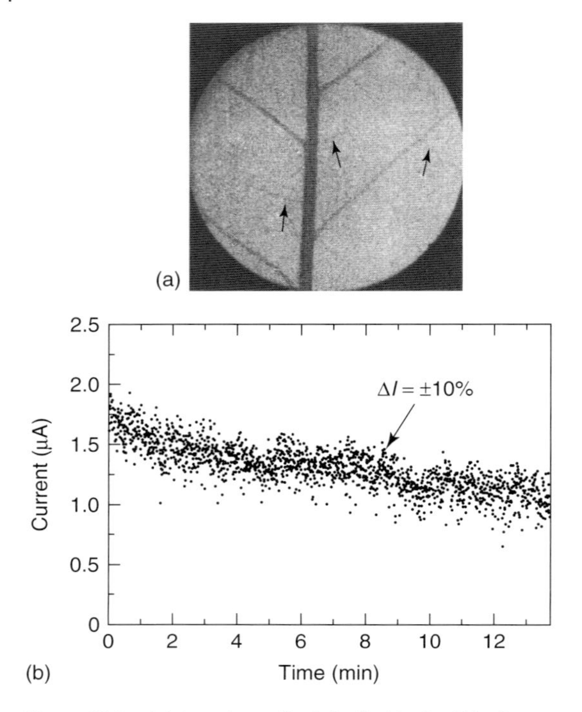

Figure 25.2 (a) Imaging a fresh leaf with the FE tube shown in Figure 25.1a. (b) electron-current variation during imaging. Electron energy: 10 keV.

dimension would ease the focusing of electron beams and accordingly improve the resolution. Our next concern was thus to devise an MXT equipped with a CNT emitter.

25.3
The Pd Emitter

The emitters comprising Co-induced CNTs were far more stable in non-UHV than metallic emitters, but the electron current that they provided decreased gradually with time while fluctuating at ±10%, thereby limiting their lifetime to around 80 min. In searching for more stable emitter materials, we noticed an intriguing fact that CNTs catalyzed by palladium (Pd) were enormously robust in non-UHV. Specifically, Pd-induced CNTs continued to emit electrons for 10 h or longer, with no discernible decline in intensity. Although it is still unclear why Pd-induced CNTs are less sensitive to impinging gas molecules, their practical significance is profound. Certainly, it was the emitters comprising these CNTs, henceforth referred to as the *Pd emitters* [7], that led to the success in the experiments on MXTs.

25.4
Devising X-ray Tubes with Miniature Dimensions

25.4.1
The 10-mm-Diameter Tube

Roughly speaking, the term *miniature* is adopted to tube diameters smaller than 10 mm.

Potential applications of MXTs are miscellaneous, ranging from microelectronics to medicine. Especially in biological communities, MXTs have long been awaited because they could image biological tissues in a micrometric resolution. To begin with, therefore, we tried to construct a 10-mm-diameter X-ray tube equipped with a Pd emitter.

The first hurdle that we had to cross was to prepare CNT emitters mountable within the narrow space of the MXT. To clear this crucial issue, we grew CNTs on one end of a Pd wire. Its procedure was as follows.

A polycrystalline Pd wire 1 mm across and ~2 mm long was vertically spot-welded on a horizontally stretched tantalum (Ta) wire around 1 mm in diameter (see the drawing inset in Figure 25.3). CNTs were then made to grow on the upper end of the Pd wire through the plasma-enhanced chemical vapor deposition (PECVD) process that we had established earlier [8]. Prior to CVD, the upper end, or the tip, of the Pd wire was mechanically polished into a conical shape (Figure 25.3). The CNTs thus grown were always topped with a strongly facetted single crystal of Pd (see the transmission electron image inset). After CVD, the Pd wire was carefully detached from the Ta wire to serve as the electron emitter in the MXT [9].

Figure 25.4a shows the cross section of our 10-mm-diameter X-ray tube (schematic) [9]. The tube was machined out from a Kovar rod, inside which the target and the emitter were kept facing each other at a distance of ~2 mm. The target was a W rod 2 mm in diameter and half embedded in a Cu rod. The target surface was machined hemispherical in order to electrostatically attract the electrons leaving the emitter (see the encircled enlargement in Figure 25.4a). The CNTs' substrate, or the Pd wire, was shielded in a stainless steel tube, with its tip slightly protruding (see the rectangle in Figure 25.4a). In this tube design, rough electron focusing was automatically achieved since the grounded tube wall deflected obliquely emitted electrons toward the target. The open end of the stainless

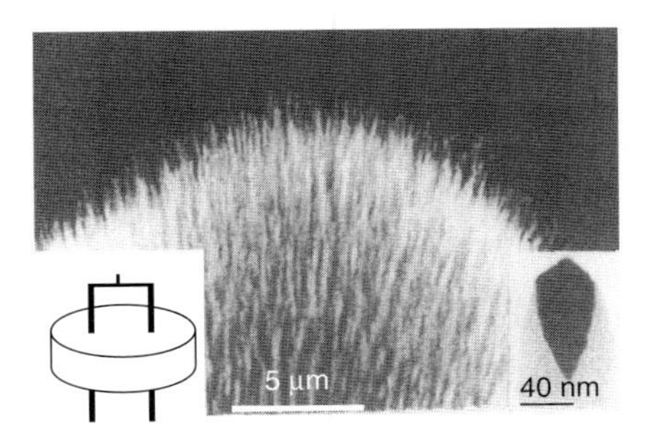

Figure 25.3 CNTs grown on the conically shaped tip of a Pd wire (SEM image). Inset are the emitter support for CVD and the transmission electron image of a Pd crystallite atop a CNT. The surface of the Pd crystallites was covered with a thin layer of graphite.

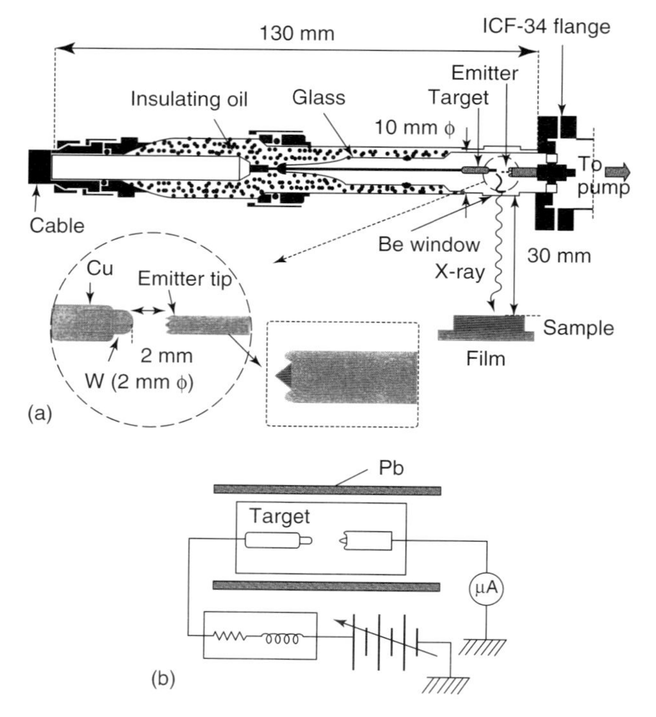

Figure 25.4 (a) Cross-sectional diagram of the 10-mm-diameter tube and (b) the X-ray producing circuit. Encircled in (a) is the schematic illustration of the emitter–target geometry, whereas enclosed by the rectangle in (a) is the emitter assembly depicted schematically.

tube also served to weakly converge the electron beam. The target potential and the electron current were +30 kV and 50 μA, respectively, at their highest level, corresponding to the maximum power of 1.5 W. Shown in Figure 25.4b is the X-ray producing circuit (diagram). The *RL* coupling between the target and the power supply was to prevent a current surge due to arcing. With the aid of a sputter-ion pump (ULVAC PST-030CU), the tube pressure was lowered to 2×10^{-6} Pa via a mild bakeout using a ribbon heater.

Although the electron current fluctuated at an amplitude of ±7–8%, the current–voltage (I–V) characteristics exhibited little change unless arcing took place between the emitter and the target. At pressures in the 10^{-6} Pa region, an appreciable current fluctuation is unavoidable irrespective of emitter material (see earlier). The fluctuations of 7–8% for the present emitters were comparable to those for other types of nanotube emitters operated under similar conditions [6]. Arcing was bound to occur when target potential was increased too quickly, so raising the potential must have been as slow as possible. After arcing, the electron current was well stabilized within the above range as long as the potential was kept constant. The first arcing never destroyed the entire cathode. Very likely, only a

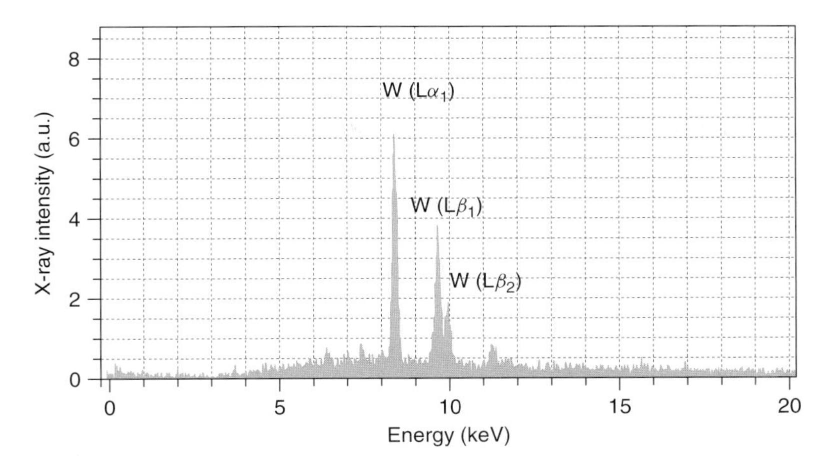

Figure 25.5 Spectrum of X-rays emitted from the target in Figure 25.4a, displaying no Cu signal. X-ray energies are calibrated in kiloelectronvolts on abscissa. Electron energy: 15 keV. Spectrometer: Rontec 2001.

limited number of CNTs were damaged upon the arcing, with the surviving CNTs making up for the relevant current loss.

For electron energies higher than 15 kV, the energy spectrum of X-rays radiated from the target was exclusively composed of $L\alpha$ and $L\beta$ signals, with no signal from Cu involved (Figure 25.5). This attests that most electrons emitted from the emitter arrived at the target, regardless of its small dimension and the closeness of the target and the emitter. When lowering the target potential to 10 kV, the spectrum was of a continuous line due to bremstrahlung.

An important role of XR is the nondestructive inspection of electronics devices, or large-scale integrated circuits (LSIs). Since the circuit patterns of LSIs are generally too fine to be resolved with X-rays, the current role of XR in this field is to inspect the quality of electrical connections. Shown in Figure 25.6a is an X-ray image of an LSI memory recorded with our MXT operated at 15 kV. The image is very clear, looking like a photographic image of an artistic pattern. As seen in Figure 25.6b, the respective lead wires (about 20 μm in diameter) and their welded spots have been fully resolved. Moreover, no distortion is recognized in the image, and the geometry of the imaged area and their contrasts are perfectly symmetrical with respect to the image's center.

Figure 25.7a shows a carpus excised from the carcass of a mouse. Note the joint structures sharply resolved at the end of the craw digits. The photographically enlarged image in Figure 25.7b reveals balloon-like soft tissues (see arrows). These might be the so-called foot-beat absorbers, or the pads, specific to quadrupeds.

The tip dimension, number density, and average length of CNTs, together with the tip–anode distance, critically affect the average electric field on the cathode. At present, we have no means to control these parameters. In the present MXT

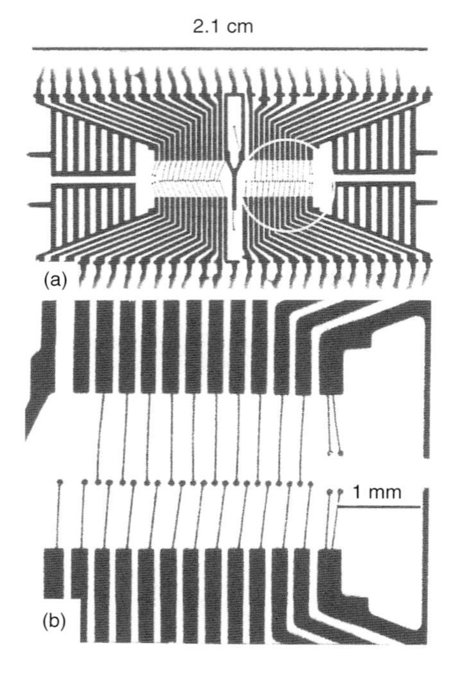

Figure 25.6 (a) XR image of an LSI and (b) enlarged view of the encircled area in (a). Electron energy: 15 keV; exposure time: 3 min.

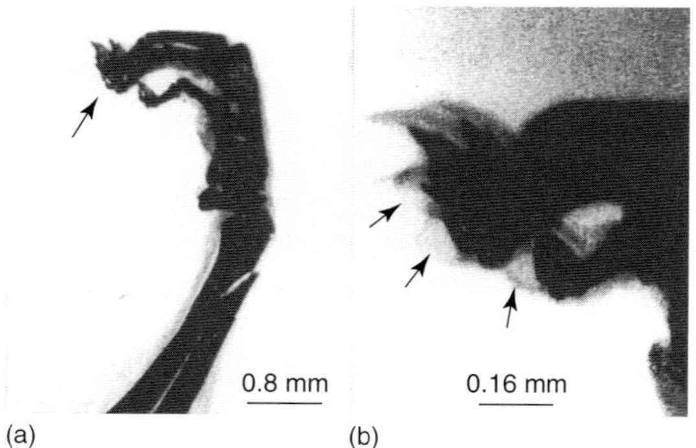

Figure 25.7 (a) Carpus of a mouse hind foot claw imaged at 15 keV. (b) Enlarged image of the arrow-indicated claw area in (a), revealing balloon-like tissues (arrows). Exposure time: 30 s.

prototype, therefore, the electron emission characteristics differed from emitter to emitter. For the diode structure, the electron emission is directly governed by the target potential, which has to be regulated when the control of electron currents is required. A means of controlling the electron current without changing the target potential would be to place a gate electrode just before the emitter; regulating the gate potential could keep the electron current at a desired level while fixing the electron energy, independently of the emitter's intrinsic quality. Developing a triode-type MXT with a built-in field emitter will thus be a future challenge. To this end, we will have to solve the technical issues of how to install a microgate electrode within the MXT.

25.4.2 The 5-mm-Diameter Tube

The 10-mm-diameter tube described above was equipped with a reflection-type target. For an MXT to serve as a medical tool such as an X-ray endoscope, the tube diameter has to be around 5 mm or smaller. Unfortunately, reflection-type MXTs cannot meet this demand. We were thus urged to develop a transmission-type MXT far smaller in overall tube dimension than the prototype. In this section, we demonstrate the design and performance of such a "super-miniature X-ray tube (SMXT)" assembled in our laboratory [10]. What is crucial in designing an SMXT is to devise an emitter assembly that can be accommodated within its very narrow inner space. As described already, Pd-induced CNTs continue to field-emit electrons for around 10 h at 2×10^{-6} Pa with little current loss [7]. In the succeeding study, FE from Pd-induced CNTs was found to maintain this stable nature even when the pressure was increased to the 10^{-5} Pa region [11]. We therefore employed the Pd-induced CNTs as the electron source for SMXT. For an emitter of this kind to be built within an SMXT, CNTs must be grown on the tip of a thin metal wire. Pd wires themselves cannot be sharpened into a fine point by simple electropolishing because Pd is very stable chemically. Our decision was to grow CNTs on a Pd overcoat sputter-deposited onto an electropolished molybdenum (Mo) tip. In our tube design, the Mo wire with a sharpened tip had to be loaded inside a stainless steel pipe around 0.5 mm in inner diameter (Figure 25.10). To do this, we chose a Mo wire 0.5 mm in cross-sectional diameter. Since the emitter is a core element in SMXT, its preparation procedure is described below very briefly.

The PECVD chamber used for this purpose was a simple vacuum diode [8], inside which the substrate of CNTs was set at a distance of around 10 mm from the grounded disc electrode of Pd. (As to the detailed structure of the CVD chamber, see [8].) The substrate, or a Mo wire ~2 mm long, was spot-welded onto a Ta wire, prior to which the upper end of the Mo wire was electropolished into a tapering point. The chamber's base pressure was kept in the 10^{-3} Pa region. Figure 25.8 illustrates the procedure of growing CNTs on the Mo tip. In short, positively biasing the substrate at 4 kV while letting argon (Ar) gases (~10 Pa) flow through the chamber induced a glow discharge to produce Ar^+ ions. These Ar ions bombarded the Pd plate to eject Pd atoms (Figure 25.8a), which deposited on the substrate and formed a Pd layer thereon (Figure 25.8b). (The Pd layer thickness was typically 100 nm for 10 min of sputtering.) Ar gas was then pumped out, followed by the process to align CNTs on a metal substrate (Figure 25.8c) [8].

After the above serial processes, the substrate was carefully detached from the Ta wire to be loaded in the SMXT. Figure 25.9 shows a sample before (inset) and

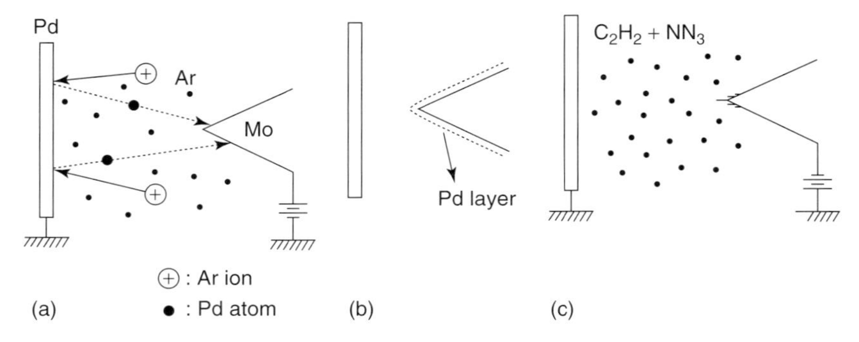

Figure 25.8 Schematic representation of the serial procedure of growing CNTs on a tip surface. (a) Sputter deposition of Pd atoms, (b) Pd film formation, and (c) CNT growth.

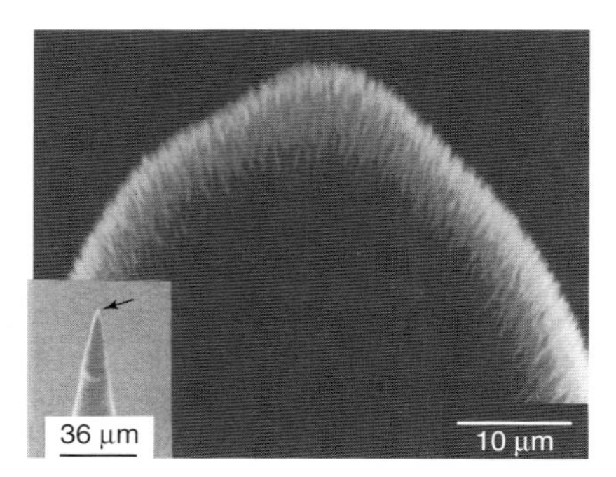

Figure 25.9 CNTs grown on a Pd-deposited Mo tip (SEM image). The inset is the as-polished substrate (SEM image), at the upper end (arrow) of which were grown CNTs through plasma CVD.

after PECVD. It is seen that CNTs are densely aligned on the Mo tip. This type of CNT emitters has been called the *CNTs-on-Tip* emitters [9].

Figure 25.10a shows a cross-sectional view of our SMXT system [10]. Basically, the X-ray tube was made up of two components: emitter assembly and grounded planar target located right before the electron source. The key component of the emitter assembly was a Kovar tube 2 mm in diameter and 5 mm in length, to the front of which had been welded a stainless steel tube around 0.5 mm in inner diameter. Inside this stainless steel tube, a Mo wire was equipped with CNTs at its electropolished tip (Figure 25.10b). The open end of the stainless steel tube accommodating the Mo wire was machined so as to roughly converge the electron beam emitted from the CNTs (Figure 25.10b). The emitter assembly was entirely shielded with a glass tube to be fixed to the vacuum flange.

The target was a Cu film ~3-μm-thick laser-deposited onto the inner surface of an aluminum (Al) foil 0.1 mm thick. (It was not possible to use Be as the window material because we had no technique to prepare a Be film of this thickness.) The cathode potential was variable between 10 and 15 kV, and the electron current was around 50 μA at the maximum level, corresponding to a maximum power of 0.75 W. The emitter operation beyond the maximum power heavily damaged

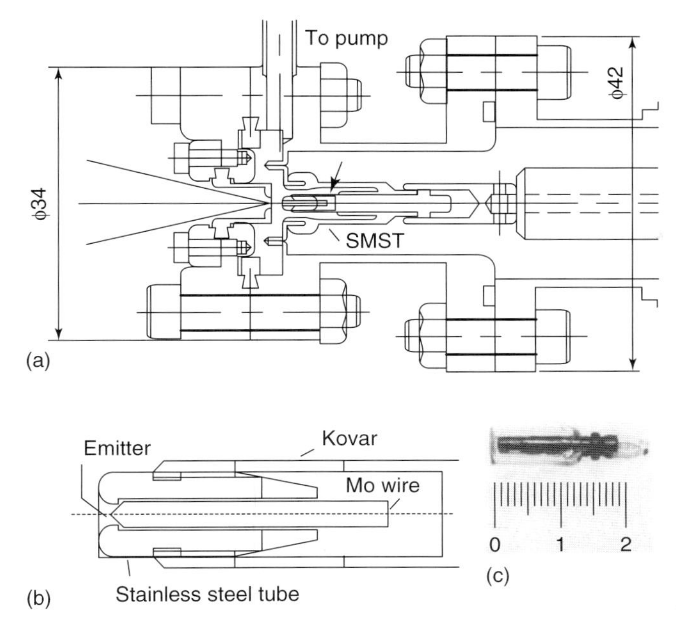

Figure 25.10 (a) Cross-sectional diagrams of the SMXT system and (b) the emitter assembly. (c) Photo image of a full-scale model of the SMXT, demonstrating its small dimension. The arrow in (a) indicates the emitter assembly. The distance between emitter and target was around 2 mm.

the target. As previously, the *RL* coupling was arranged between the target and the power supply to prevent arcing-induced current surge. The main pumping of the tube was done with a sputter-ion pump, reducing the pressure to $\sim 2 \times 10^{-5}$ Pa or lower via mild bakeout. Shown in Figure 25.10c is a full-scale model of our SMXT, which attests to its miniature dimension.

At the above pressure level, the electron current fluctuated at an amplitude of ±5% at a fixed emitter potential. Nevertheless, the $I-V$ characteristics showed little variation provided the emitter potential was kept constant. Also, the emitted electrons always satisfied the Fowler–Nordheim relationship.

As pointed out already, the resolution in X-ray imaging depends on the focal point size; the smaller the size, the higher the resolution. A conventional means to estimate the focal point size is to computer-simulate the electron beam trajectories, but this was not possible for our SXMT because of the complex geometry of the emitter assembly. We were therefore forced to estimate the focal point dimension in a simple experimental manner. Specifically, the operation of a CNT emitter was prolonged to around 2 h. The target was then removed from the vacuum chamber, and the electron-bombarded area thereon was visually inspected. We recognized a dark spot on the target. The formation of this dark spot, which was around 100 μm in diameter, might have been caused by the carbonization of the electron-bombarded area; perhaps, the electron beam decomposed residual hydrocarbon deposited on

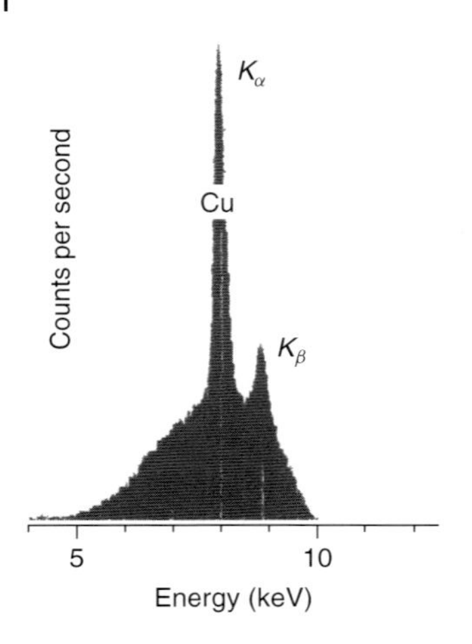

Figure 25.11 X-ray spectrum at 10 keV.

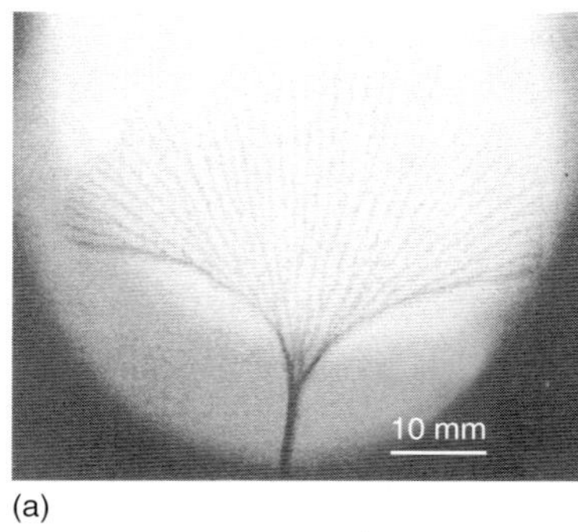

(a)

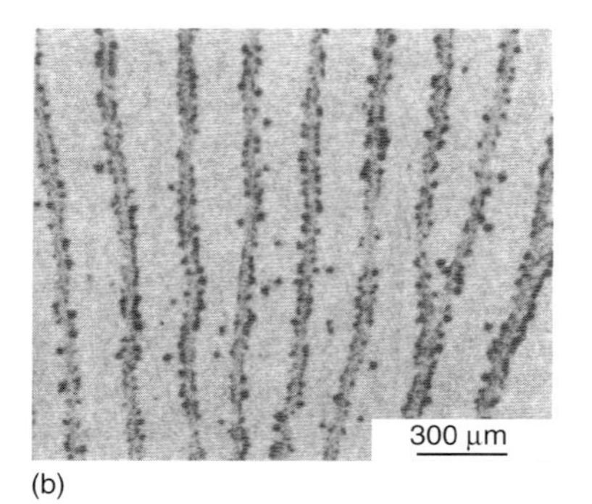

(b)

Figure 25.12 (a) Contact X-ray image of a fresh gingko leaf recorded with the SMXT, revealing its capillary veins and (b) a photographically enlarged image of the central area in (a). Electron energy: 12 keV; exposure time: 5 min. The image was recorded directly on a photoplate.

the target, thereby carbonizing the bombarded area. If so, the electron beam coming out from the emitter could have been converged into a microscopic spot.

The energy spectra of X-rays radiated from the target always involved intense Cu Kα and Cu Kβ signals superimposed on a continuous signal, at the emitter potentials employed (10–15 kV) (Figure 25.11). These characteristic signals are in the soft X-ray region in energy, suggesting that SMXT might even be a device for soft X-ray science and technology.

It was noted earlier that the prime role of XR is to observe the inner structure of biological tissues. Figure 25.12a shows the X-ray image of a gingko leaf just plucked from the tree. Note the respective veins resolved perfectly. In the photographically enlarged image in Figure 25.12b, dotted tissues along the veins are clearly seen. These dotted tissues were 30 μm in average dimension; our SMXT can resolve nondehydrated plant tissues of microscopic dimensions without extrinsic electron focusing. In our first study on field emission X-ray radiography (FE-XR) [6], the exposure time had to be prolonged to 60 min or longer to image a nondried leaf. For Figure 25.12a, by contrast, the imaging was completed within several minutes; SMXT may certainly evolve into a promising device for visualizing plant tissues in microscopic detail.

In the preceding work, the electron emission characteristics varied from emitter to emitter, because the growth parameters of CNTs could not be controlled [9]. Thanks to recent progress in emitter fabrication techniques [11, 12], CNTs can now be grown on a metal tip in rather uniform density, length, and diameter; CNTs-on-tip emitters will be fabricated in a well-controlled manner in the foreseeable future.

Here we emphasize once again that our SMXT could offer a focal point diameter as small as 100 μm with no external electron beam focusing. By placing electromagnetic lenses just outside SMXT, electrons discharged from a CNTs-on-tip emitter could be focused into a much finer spot. Such a high-resolution XR, which has long been awaited in life sciences, now seems just around the corner. Unfortunately, Cu films are prone to be damaged by bombarding electrons, something also true of tungsten films. For growing SMXTs into truly practical tools, therefore, synthesizing target materials highly resistant to electron beam impact is an urgent task.

25.5 Status Quo of Our MXT Technique

Around 90% of X-ray tubes marketed annually are for medical use. For MXTs, too, their major demand will be in medical areas. As a matter of fact, our project to develop MXTs was based on the prospect that MXTs would pioneer the direct X-ray irradiation of superficial lesions. Frankly speaking, our MXT technique is still at an initial stage and hence far from practical. In view of our recent study, however, MXTs may serve as new tools for radiation therapy of cancers in future [13]. Medical basis for this prospect is as follows.

The prerequisite to devising FE X-ray tools for radiation therapy is to confirm that FE X-rays are certain to induce biological processes. In specific terms, for their applications to cancer therapy, it must be proven that FE X-rays actually cause the DNA double-strand breaks (DSBs) in normal cells and apoptosis, or programmed death, of cancerous cells. In what follows, indisputable evidence is presented that FE X-rays, like TE X-rays, actually induce DSB and apoptosis for cultivated cells of mice [13].

25.5.1 DSB

The samples used in the experiments were the mouse thymic lymphoma 3SB cells. Prior to X-ray irradiation, the cells were subjected to biological treatments, which have been described in detail in the original article [14]. The FE X-ray system employed was a slightly modified version of our FE X-ray radiography system [7].

For comparison, some samples were also exposed to TE X-rays using a conventional TE system.

The operating conditions of the FE and TE systems are summarized in Table 25.1. The FE X-rays used here were mostly comprised of W L lines (Figure 25.5). TE experiments were done in Nagoya City University, where no X-ray spectrometer was available. The spectral structure of the TE X-rays, generated with an MCN™ 225 X-ray tube (Philips Industrial), was therefore unknown. Taking account of the common view in radiological physics that TE X-rays emitted from W are generally

Table 25.1 Operating conditions of FE and TE X-ray tubes.

Characteristic	FE type	TE type
Tube voltage (kV)	50	50
Current (mA)	1	1
Target	W	W
Window	Be	Be
Filter	None	None

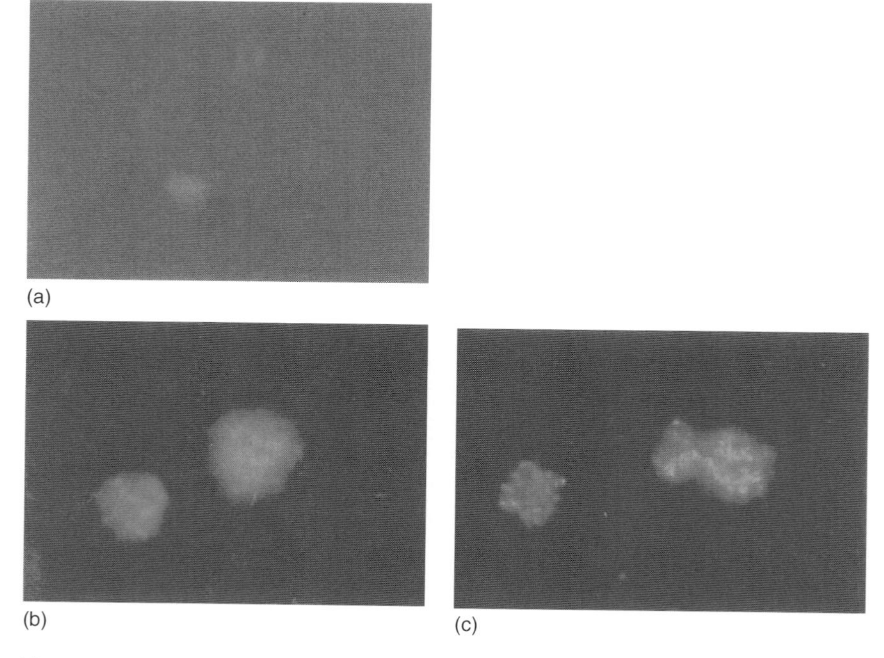

Figure 25.13 Foci formation of γ-H2AX in 3SB cells caused by TE and FE X-ray irradiations. (a) 0 Gy, (b) and (c) 3 Gy with TE and FE irradiations, respectively. Light green luminescence is a direct indication of DNA damage.

continuous, the present TE X-rays might have been close to continuous, with less intense W L lines involved.

H2AX (anti-phospho-ser139 histone) is a well-known biological marker of DSBs. To check whether or not FE X-ray irradiation leads to DSB, 3SB cells were irradiated by FE and TE X-rays, and the DSBs induced were analyzed by foci formation of phospho-H2AX (γ-H2AX) using antibody specific to γ-H2AX. As seen in Figure 25.13, foci formation of γ-H2AX was evident as early as 1 h after irradiation of FE X-rays (3 Gy), and the number of γ-H2AX foci in the cells treated with FE X-rays was nearly equal to that for TE X-ray irradiation. The same was true of the cells irradiated at 1 and 5 Gy (data not shown). FE X-rays were thus confirmed to damage DNAs.

25.5.2 Apoptosis

The TUNEL assay [14] is a useful tool for identifying programmed cell death and for quantifying the process in cell populations. With this assay, we examined whether FE X-rays cause apoptotic cell death or not. As shown in Figure 25.14, a time-dependent increase in the number of TUNEL-positive cells was detected in 3SB cells subjected to FE and TE X-ray irradiations. TUNEL-positive cells started to be detected 1 h after 3 Gy irradiation ($13.9 \pm 0.6\%$ with TE and $9.6 \pm 0.6\%$ with FE), and continued to increase in number for 5 h ($25.4 \pm 2.7\%$ with TE and $27.4 \pm 2.2\%$ in FE). The results at 1 and 5 Gy were quite similar to those at 3 Gy stated above. These provide unmistakable evidence that FE X-ray irradiation triggers and drives the apoptosis of cancer cells in a time- and dose-dependent manner.

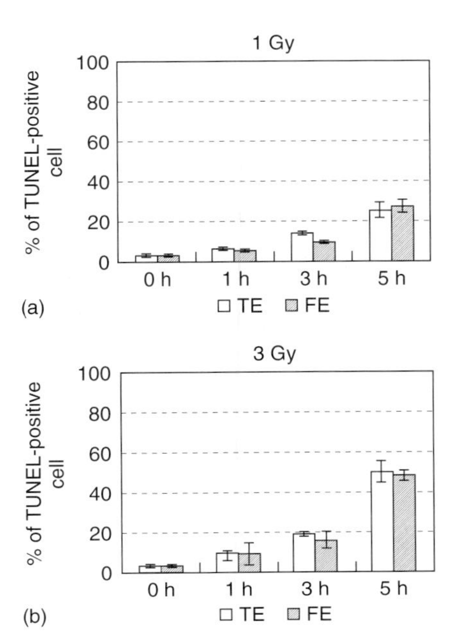

Figure 25.14 Induction of apoptosis by TE and FE X-ray irradiations at 1 and 3 Gy. Relative quantities of TUNEL-positive cells are displayed as a function of irradiation time.

25.6
Future Prospect of MXTs in Radiation Therapy

Based on the medical finding cited above, we may conclude that FE X-ray irradiation stimulates the apoptosis of cancerous cells, as well as DNA damage in normal cells. This suggests a new strategy for radiation therapy. Precisely, the entire dimension of FE X-ray tubes can now be reduced to a miniature level (Figure 25.10c), and by coupling with an endoscope and narrowing the radiation field, such an MXT may pave the way to targeted X-ray irradiation of an early stage tumor or a superficial lesion. Although the maximum tube voltage allowed for MXTs is 50 kV at the highest and hence the emitted X-rays will not be very intense, this would be practically insignificant because the intensity decline due to exhalation will be negligible owing to the close spacing between the X-ray emitting spot and the diseased area.

References

1. Coolidge, W.D. (1913) A powerful Röntgen ray tube with a pure electron discharge. *Phys. Rev.*, **2**, 409–420.
2. Michette, A.G. (1993) in *X-ray Science and Technology*, Chapter 1 (eds A.G. Michette and C.J. Buckley), Institute of Physics, Bristol.
3. Fowler, R.H. and Norddheim, L.W. (1928) Electron mission intense electric fields. *Proc. Roy. Soc. London, Ser.*, **A119**, 173–181.
4. Dyke, W.P. and Dolan, W.W. (1956) Field emission. *Adv. Electron. Electron Phys.*, **8**, 89–185.
5. de Jonge, N. and Bonard, J.-M. (2004) Carbon nanotube electron sources and application. *Phil. Trans. R. Soc. London A*, **362**, 2239–2266.
6. Sugie, H., Tanemura, M., Filip, V., Iwata, K., Takahashi, K., and Okuyama, F. (2001) Carbon nanotubes as electron sources in an x-ray tube. *Appl. Phys. Lett.*, **78**, 2578–2580.
7. Senda, S., Tanemura, M., Sakai, Y., Ichikawa, Y., Kita, S., Otuka, T., Haga, A., and Okuyama, F. (2004) New field emission x-ray radiography system. *Rev. Sci. Instrum.*, **75**, 1366–1368.
8. Tanemura, M., Iwata, K., Takahashi, K., Fujimoto, Y., Okuyama, F., Sugie, H., and Filip, V. (2001) Growth of aligned carbon nanotubes by plasma-enhanced chemical vapor deposition:optimization of growth parameters. *J. Appl. Phys.*, **90**, 1529–1533.
9. Haga, A., Senda, S., Sakai, Y., Mizuta, Y., Kita, S., and Okuyama, F. (2004) A miniature x-ray tube. *Appl. Phys. Lett.*, **84**, 2208–2210.
10. Senda, S., Sakai, Y., Mizuta, Y., Kita, S., and Okuyama, F. (2004) Super miniature x-ray tube. *Appl. Phys. Lett.*, **85**, 5679–5681.
11. Kita, S., Sakai, Y., Fukushima, T., Mizuta, Y., Ogawa, A., Senda, S., and Okuyama, F. (2004) Characterization of field-electron emission from carbon nanofibers grown on Pd wire. *Appl. Phys. Lett.*, **85**, 4478–4480.
12. Sakai, Y., Haga, A., Sugita, S., Kita, S., Tanaka, S.-I., Okuyama, F., and Kobayashi, N. (2007) Electron gun using carbon-nanofibers field emitter. *Rev. Sci. Instrum.*, **78**, 013305–013310.
13. Nakazato, T., Nakanishi, M., Kita, S., Okuyama, F., Shibamoto, Y., and Otuka, T. (2007) Biological effects of field emission-type x-rays generated by nanotechnology. *J. Radiat. Res.*, **48**, 153–161.
14. Loo, D.T. and Rillema, J.R. (1998) Measurement of cell death. *Methods Cell Biol.*, **57**, 251–261.

26
Carbon Nanotube-Based Field Emission X-ray Technology

Otto Zhou and Xiomara Calderon-Colon

26.1
Introduction

26.1.1
Current Thermionic X-ray Technology

X-rays radiations are used today in a wide range of applications. Some examples include medical imaging [1], radiotherapy [2], homeland security [3, 4], and industrial inspection. Simple projection imaging techniques are routinely used for screening purposes in hospitals and at the airport checkpoints. Computed tomography (CT) provides the internal structure of an object by reconstruction of projection images collected from different viewing angles [5]. CTs are commonly used for diagnostic imaging, for image guidance for radiotherapy, and for explosives detection.

A conventional X-ray tube comprises a metal filament or a film (cathode) which emits electrons when it is resistively heated to over 1000 °C and a metal target (anode) that emits X-ray when bombarded by the accelerated electrons [1]. The spatial resolution of an X-ray source is determined by the size of the focal spot – the area on the X-ray anode that receives the electron beam. The intensity of the X-rays is proportional to the electron beam current and the square of the acceleration voltage, and is limited primarily by heat dissipation of the anode.

The use of thermionic cathodes leads to several inherent limitations for the current X-ray source technology. Because of the high operating temperature and power consumption, essentially all current commercial X-ray tubes are single-pixel devices where X-rays are emitted from single focal spots on the anodes. This limits the performances of advanced X-ray imaging systems including CT. Current CT scanners collect the projection images by rotation of either the X-ray/detector pair or the object, which reduces the scanning speed and resolution, complicates the mechanical design, and restricts the scanner configuration [5, 6]. Thermionic emission in general has a slow response time and is difficult to program electronically. This makes gated imaging of moving objects difficult, especially non-periodic

Carbon Nanotube and Related Field Emitters: Fundamentals and Applications. Edited by Yahachi Saito

ISBN: 978-3-527-32734-8

motions such as respiration and heart beat. Motion blurring significantly degrades the image quality.

26.1.2
Previous Studies of Field Emission X-ray

The possibility of using a field emission electron source to generate X-ray has been investigated in the past [7, 8]. In these early studies, metal tips were used as the cathodes. Electrons were extracted by applying a pulsed high voltage between the target and cathode. X-rays were generated when the field-emitted electrons bombarded on the target. The advantages of the field emission X-ray tubes compared to the conventional thermionic X-ray tubes have been demonstrated in clinical studies in terms of resolution and exposure time [8]. However, the metal-tip emitters were inefficient. To obtain the high electrical field required to extract the electrons, the Max generator, which uses a series of discharging capacitors, was employed [9]. The X-ray tubes had a limited lifetime of 200–300 exposures and slow repetition rate. With the diode design, the acceleration voltage and the tube current could not be independently controlled. Field emission X-ray tubes using other types of emitters such as the Spindt tips and diamond emitter have also been investigated [10–12]. The highest electron current obtained in these X-ray tubes is only on the order of microamperes [10].

26.1.3
Carbon Nanotube-Based Field Emission X-ray

Carbon nanotubes (CNTs) are excellent electron field emitters [13]. Due to their atomically sharp tips and large aspect ratios ($\sim 10^3$), the CNTs have much larger field enhancement factors (β) and thus lower threshold fields required for emission than conventional emitters. They are stable at high emission currents. A stable emission current of $>1\,\mu A$ has been observed from an individual SWNT [14] and a CNT bundle [15]. We have achieved stable emission current density of over $1\,A\,cm^{-2}$ from *macroscopic* cathodes (Figure 26.1). This is due to their unique physical properties including high thermal and electrical conductivities, high temperature and chemical stability, and reasonable resistance to oxidation. These properties make them attractive "cold" cathode materials for X-ray tubes.

CNT-based field emission X-ray has recently been demonstrated [16–20]. A typical CNT X-ray source comprises a CNT cathode, a gate electrode, an electrostatic focusing unit, and a metal anode in a vacuum housing. Field-emitted electrons are extracted from the CNT cathode by applying a bias voltage on the gate electrode and are subsequently focused by the focusing lenses before reaching the anode. Compared to the conventional thermionic X-ray tubes it has several intrinsic advantages that make it attractive for X-ray imaging. Because of the nature of field emission mechanism, the source has fast response time and electronic programmability, which are highly desirable for gated imaging of moving objects [21]. The ability to switch on and off the X-rays by readily turning on or off the

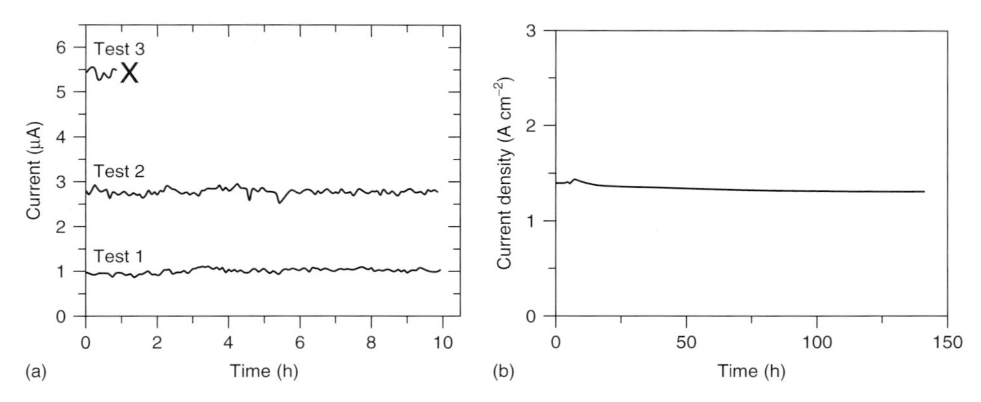

Figure 26.1 (a) Stability of field emission current from a single CNT bundle measured at three different current levels. (b) High current stability from a 1-cm-diameter CNT cathode. (The high current data is unpublished result from Xintek, Inc.)

electron beam enables efficient utilization of X-rays and better management of the anode heat load. The use of "cold" cathode rather than "hot" filament makes it possible to manufacture field emission cathode arrays with individually controllable electron emitting pixels which can generate X-rays from different focal points on the X-ray anode. By sequentially switching on and off the individual pixels, a scanning X-ray beam can be generated to image an object from different viewing angles without any mechanical motion [19, 22, 23]. The new source technology can potentially lead to a truly stationary and gantry-free CT with high scanning speed and high resolution. In addition, the distributed source array technology makes multiplexing – simultaneous collection of multi-images at the same time – a reality. This is accomplished by using either the binary multiplexing algorithm – activating a subset of the X-ray beams at a time according a pre-selected multiplexing algorithm – or the frequency division multiplexing algorithm [24, 25]. Multiplexing can significantly increase the data collection speed, although the trade-off between the imaging speed and the signal-to-noise ratio needs to be further investigated.

The CNT field emission X-ray technology is now being actively investigated by a number of academic groups and major companies around the world. Applications including dynamic micro-computed tomography (micro-CT) for preclinical imaging [26, 27], stationary digital tomosynthesis for detection of human breast tumors [28–30], stationary tomosynthesis for image guidance in radiation therapy [31], new volumetric CT design [32], and stationary CT for explosives detection are being considered. Here, in this chapter, we summarize our research on (i) high-performance CNT cathodes for X-ray generation, (ii) microfocus field emission X-ray source for dynamic micro-CT, and (iii) distributed multibeam field emission X-ray (MBFEX) source. We will also briefly introduce the two imaging systems under development at the University of North Carolina (UNC) based on this novel X-ray source technology.

Table 26.1 Some practical requirements for X-ray tubes.

High anode voltage	160–180 kV for airport security inspection 30–140 kV for medical imaging
High tube current	0.5–1.0 A at 1 × 1 mm focal spot for clinical CT ~1 mA at ~50 μm focal spot for microfocus tube
Nonideal vacuum	Ion bombardment, arcing
Lifetime	Over three years

26.2 Fabrication of CNT Cathodes for X-ray Generation

While it is relatively straightforward to demonstrate the feasibility of generating X-rays using the CNT field emitters in a laboratory environment, achieving the performance parameters required for practical devices is a challenging materials science problem. X-ray tubes require high current densities on the order of 10^2–10^3 mA cm^{-2} and high acceleration voltage ranging from 30 kV for mammography to over 180 kV for airport baggage inspection. Because of engineering constrains, X-ray tubes operate at nonideal vacuum environments where ion bombardment, arcing, and oxidation can seriously reduce the lifetime of the CNT emitters and cause catastrophic failures (Table 26.1).

Most of the published CNT field emission research so far has focused on field emission display. The behavior of the macroscopic CNT cathodes under high emission current and high voltage has not been adequately investigated except in a few published studies with less than satisfactory results [33–35]. Although an individual CNT can emit over 1 μA current [14, 15, 36], obtaining high currents from a *macroscopic* cathode is difficult [37]. Limited by the emitter non-uniformity and electrical screening effect, the emission site densities of macroscopic cathodes are typically low. This leads to a high current from each active emitter even under a moderate extraction current density, which causes failure of the nanotubes within a relatively short time. Adhesion of the CNTs to the substrate surface also plays a significant role in the emission stability, especially under high voltage. Removal of the CNTs under high potential is a common cause of arcing, which is detrimental to vacuum electronic devices. One of the most critical issues for the CNT X-ray technology is to develop field emission cathodes that can deliver high current and current density under high anode voltage and nonideal vacuum environment, along with long-term stability.

26.2.1 Fabrication Process

Several techniques are currently used for fabrication of CNT field emission cathodes including screen printing and direct surface chemical vapor deposition (CVD). We have developed a liquid-phase room temperature electrophoretic deposition (EPD) method for fabrication of patterned CNT films [38–40]. Recent results demonstrated that by optimizing the surface orientation and density of the CNTs and improving the adhesion, CNT cathodes with long-term stability under high current and high voltage conditions can be manufactured [41].

In our experiments, small-diameter multiwalled CNTs synthesized by a thermal CVD process developed at Jie Liu's lab at Duke were used as the staring material [42] (provided by XinNano Materials, Inc. of Taiwan). After purification, they were dispersed in alcohol together with $MgCl_2$, which served as the charger. To improve the adhesion of the CNTs to the substrate, glass frits were added to the EPD ink as a binder [39]. Glass substrates with printed Ag contact lines or metal plates were used as the substrates. The deposition procedure is illustrated in Figure 26.2. Under a DC electrical field, the electrically charged CNTs and the binders were driven to and deposited onto the substrate surface to form a composite film composed of

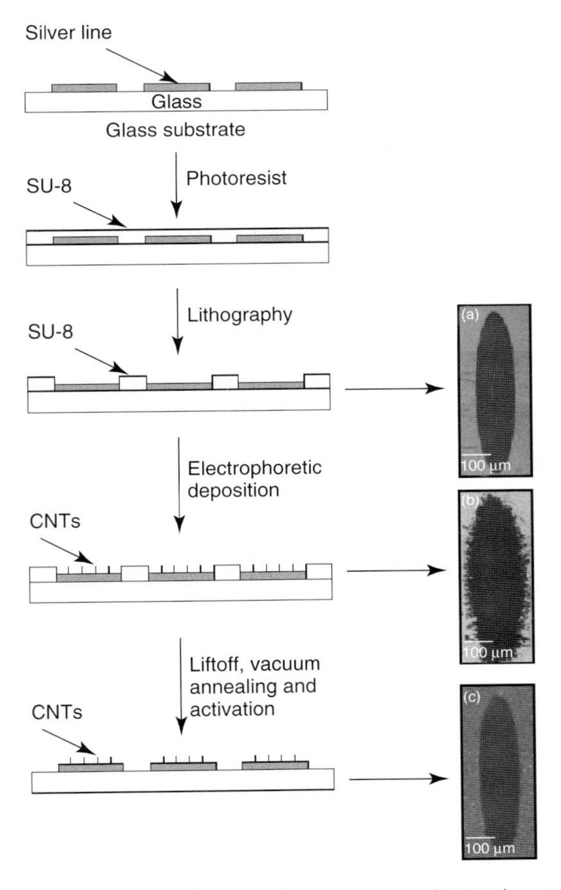

Figure 26.2 A schematic showing patterned CNT deposition by EPD. SEM images show the morphologies of the cathodes at different stages of the deposition process.

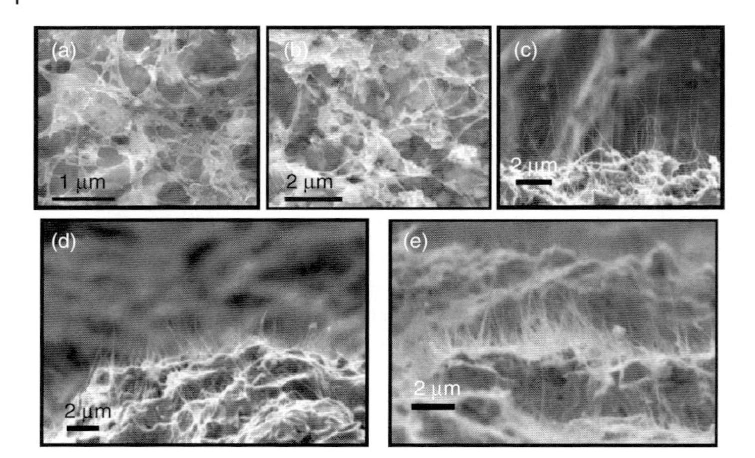

Figure 26.3 SEM images showing the top surface of the composite CNT film (a) before and (b) after vacuum annealing. The CNTs are randomly oriented on the surface. (c) After the activation process, the surface CNTs are aligned in a direction perpendicular to the substrate surface. Cathode shown in (e) was made using an ink with four times the CNT concentration of the cathode shown in (d).

the CNTs dispersed in the binder matrix (Figure 26.3). The film thickness was controlled by the deposition time and the voltage applied, and the CNT density by the composition of the EPD ink.

Most of the materials were deposited into the exposed area on the conducting substrate, although some were also found around the edges of the SU-8 photoresist, as shown by the optical images in Figure 26.2. The bonding between the CNT composite film and the substrate was strong enough so that most of the CNTs remained on the substrate surface after the photoresist liftoff. SEM micrographs show that the CNTs were randomly oriented after EPD and after vacuum annealing at 10^{-6} Torr at 500 °C. The surface CNTs became vertically aligned with one end embedded inside the matrix and the other end protruding from the surface after mechanically removing a top surface layer of the composite film, as shown by the cross section SEM images in Figure 26.3. This is the desired morphology for field enhancement and adhesion. The protruding length of the CNTs is remarkably uniform considering the large length variation of the raw CNTs. After annealing, the composite film got bonded strongly to the substrate surface and could not be removed by taping. This is drastically different from the typical CNT films directly grown by CVD which have very weak interface bonding. A strong adhesion is achieved in these films because they are composites rather than bare CNTs and the protruding CNTs are partially embedded inside the matrix.

The density of these protruding CNTs can be controlled by the starting concentration of the CNTs in the EPD ink. A series of cathodes were made using inks with increasing CNT concentrations while keeping the other parameters the same. Although the exact value is difficult to determine, cross-sectional SEM images

(Figure 26.3) clearly show that the density of vertically aligned CNTs increases with increasing CNT concentration of the EPD ink.

The film thickness depended on the deposition conditions used, such as the current, deposition time, and voltage applied. For cathodes made under the same conditions, the average film thickness is 15 μm, with less than 10% variation. Measurement by a profilometer shows that the film was thicker around the edge than in the center. The variation is about 5–10 μm over ~2000 μm distance. This is attributed to the electrical field concentration around the edge of the exposed metal contact line during the EPD process which causes a high rate of CNT deposition. Edge effect is known to cause preferential emission from the edges [43], which reduces the overall performance. This problem was partially mitigated by increasing the thickness of the photoresist. The thickness variation was reduced by 40% when the thickness of the photoresist was increased from 10 to 100 μm.

Using this combined EPD and photolithography methods, we have fabricated CNT cathodes with variable geometries and dimensions, and also cathode arrays. An example of a CNT cathode array patterned on a glass substrate with Ag contact lines is shown in Figure 26.4. The variation from cathode to cathode and between the individual pixels within the cathode array are in general small [41].

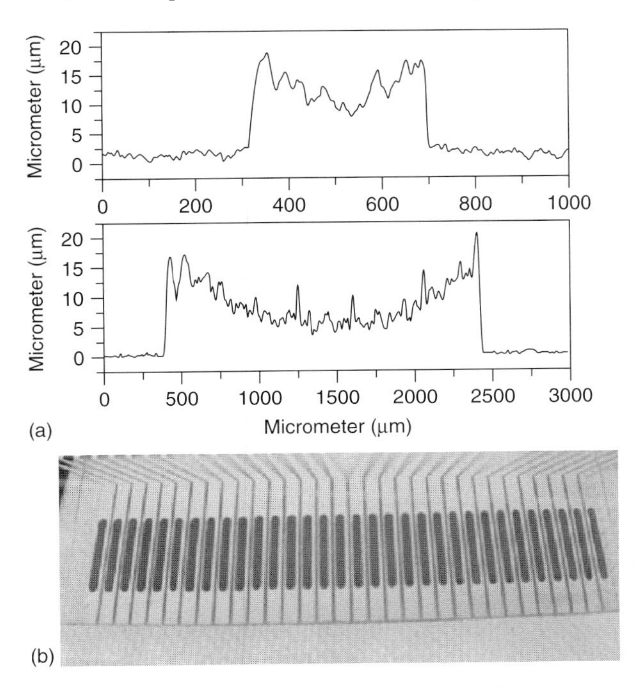

Figure 26.4 (a) Height profiles along the short and long axes of an elliptical CNT cathode (2.35 × 0.50 mm) measured by a profilometer. (b) CNT cathodes patterned on a glass substrate with printed Ag contact lines.

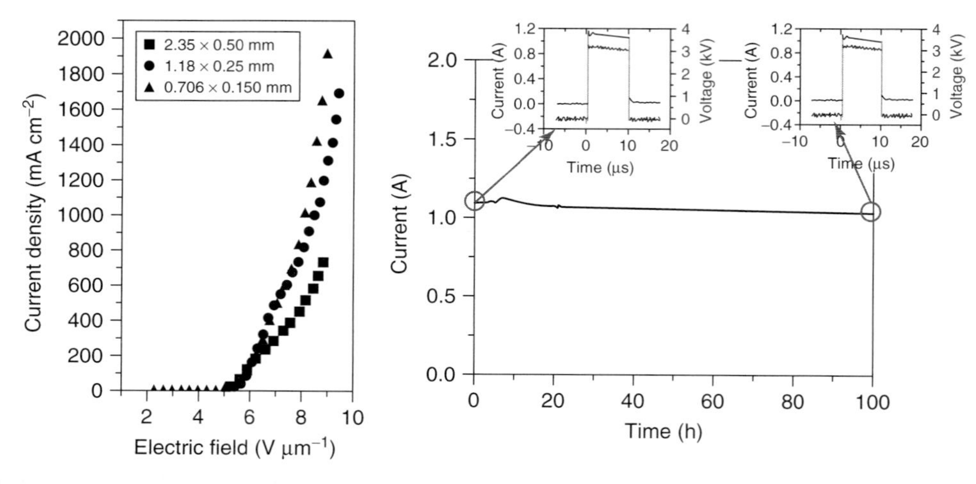

Figure 26.5 The field emission current density versus applied electrical field from different size cathodes measured in the parallel plate geometry. The voltage applied to the anode has a square waveform with 10 ms pulse width and 1 Hz repetition rate [41]. The plot on the left shows a pulsed emission current of 1 A (10 ms pulse width) at 0.1% duty cycle from a 2 × 15mm CNT cathode measured in the diode mode with a fixed extraction voltage. (Unpublished results from Xintek, Inc.)

26.2.2 Field Emission Properties

Stable and reasonably high emission current and current density have been achieved from the CNT cathodes fabricated by the EPD process. Figure 26.5 plots the experimentally measured emission current densities versus the applied electrical fields for cathodes of three different sizes measured in the parallel plate geometry. The voltage applied to the anode has a square waveform with 10 ms pulse width and 1 Hz repetition rate. The cathode to anode distance is 150 µm. The cathode areas are 0.08, 0.23, and 0.92 mm^2, respectively, for the three cathodes. As shown, over 1500 mA cm^{-2} density was readily achieved. At the same applied electrical field, the smaller the cathode area, the higher the current density achieved. For example, the smallest cathode (0.08 mm^2) reached a current density of over 1400 mA cm^{-2}, while the large one (0.92 mm^2) generated ~400 mA cm^{-2} density at the same applied field of 8.5 V µm^{-1}. This is attributed to the edge effect observed in the profilometer measurement. The results also indicate that there is still room for improvement in terms of the sample non-uniformity.

In a separate study, a total emission current of 1 A was obtained from a 2 × 15 mm CNT cathode. As shown in Figure 26.5, there is a direct correspondence between the onset of the extract voltage and the emission current, with no visible delay at the 10 µs pulse width. The emission current was stable during the course of the measurement. The 1 A value is important because it is the output power of the X-ray tube used in high-end clinical CT scanners.

26.3
Field Emission Microfocus X-ray Tube

Micro-computed tomography (micro-CT) is an important tool for non-invasive imaging of small animals such as mice and rats for preclinical biomedical studies [44, 45]. Current commercially available micro-CT scanners offer the capability of imaging objects *ex vivo* with high spatial resolution. But performing *in vivo* micro-CT on small animals is still challenging largely because their physiological motions are much faster than those of humans. Motion-induced artifacts blur the micro-CT images and significantly deteriorate the spatial resolution. The objective of this research was to develop a dynamic micro-CT scanner with enhanced temporal and spatial resolutions that were sufficient to resolve the cardiopulmonary organs of free-breathing mice. The key to our approach was the CNT field emission microfocus X-ray source technology [26, 46, 47]. In this section we introduce the design and performance of the X-ray tube. The dynamic micro-CT scanner will be introduced later in this chapter.

26.3.1
Tube Design

The microfocus X-ray source developed in our lab for dynamic micro-CT imaging is illustrated in Figure 26.6 [46]. It comprises a CNT cathode, a gate electrode, a modified Einzel-type electrostatic lens, and a metal anode housed in a metal vacuum housing. Field-emitted electrons are extracted from the CNT cathode by applying a bias voltage on the gate electrode and are subsequently focused by the focusing lenses before reaching the anode. The modified Einzel electrostatic lens comprises three electrostatic focusing electrodes. The focusing electrodes 1 and 3 are made of planar metal diaphragms. The central focusing electrode 2 has the shape of a truncated cone. The electrode 1 is at equal potential as the gate electrode;

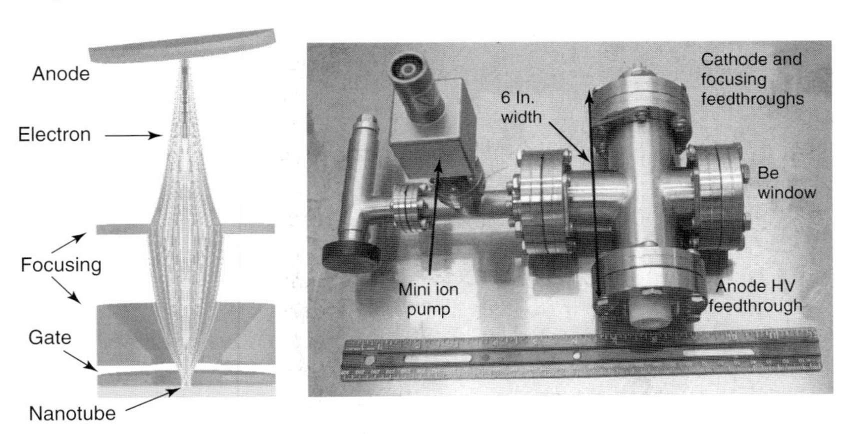

Figure 26.6 A compact CNT-based microfocus X-ray source developed at UNC.

electrodes 2 and 3 have independently controllable potentials. The dimensions of the focusing lenses were optimized by simulating the trajectory of the field-emitted electrons. To obtain an effective isotropic focal spot size, the CNT emission surface has an elliptical geometry. The ratio between the elliptical long axis and the short axis is determined by the anode tilting angle.

The electron gun assembly and anode are placed inside a stainless steel vacuum housing with a Be X-ray window, a mini ion pump, a valve, and a high-voltage feed-through for the anode and a multipin feed-through for the focusing and gate electrodes, as shown in Figure 26.6. During operation, the vacuum is maintained in the 10^{-8} Torr range by the mini ion pump. A constant high DC voltage is applied to the anode. Electrons from the CNT cathode are extracted by applying a variable electrical field to the gate electrode, usually on the order of 0–10 $\mathrm{V\,\mu m^{-1}}$. A feedback circuit is built into the design so that a constant emission current and therefore X-ray flux can be maintained by automatically adjusting the extraction electrical field. Two DC power supplies are used for the two focusing electrodes. In this particular design, up to three CNT cathodes can be installed. Electron emission and the emission current from these cathodes can be independently programmed by varying the gate electrical field.

26.3.2 Tube Current and Lifetime

The maximum X-ray tube current that can be achieved, and therefore the maximum X-ray flux, is primarily determined by two factors: the maximum emission current from the cathode and the anode heat load. The maximum power $P_{\max}$ (in watts) of a fixed-anode microfocus X-ray tube can be estimated as $P_{\max} \approx 1.4(X_{\mathrm{f,FWHM}})^{0.88}$ [21], where $X_{\mathrm{f,FWHM}}$ is the focal spot size in micrometers. At the targeted effective focal spot size of $\sim 100 \times 100\,\mu\mathrm{m}$, $P_{\max}$ is ~80 W or ~3 mA cathode current (with ~67% transmission rate) at 40 kV anode voltage. For the CNT-based X-ray tube, the question becomes whether the field emission current can reach this upper limit at a given focal spot size.

The long-term emission stability was evaluated by measuring the variation of the applied gate voltage needed to maintain a constant cathode current in the X-ray tube at 40 kV anode voltage. A stable cathode current of 3 mA (325 $\mathrm{mA\,cm^{-2}}$) was readily obtained at a gate voltage (V_g) of ~1650 V from a 2.35×0.50 mm elliptical CNT cathode (for $100 \times 100\,\mu\mathrm{m}$ focal spot size). Seventy-four percent of the cathode current reached the X-ray anode, with the rest leaking through the gate and focusing electrodes. As shown in Figure 26.7, after three days, V_g essentially stayed the same, indicating no measurable cathode degradation. The cathode current of 3 mA is the maximum current for a fixed-anode X-ray tube under the DC mode. This result proves that at least for this microfocus X-ray tube, the X-ray flux from the CNT cathode can be at least as high as that afforded by a conventional thermionic X-ray tube operating at the same focal spot size.

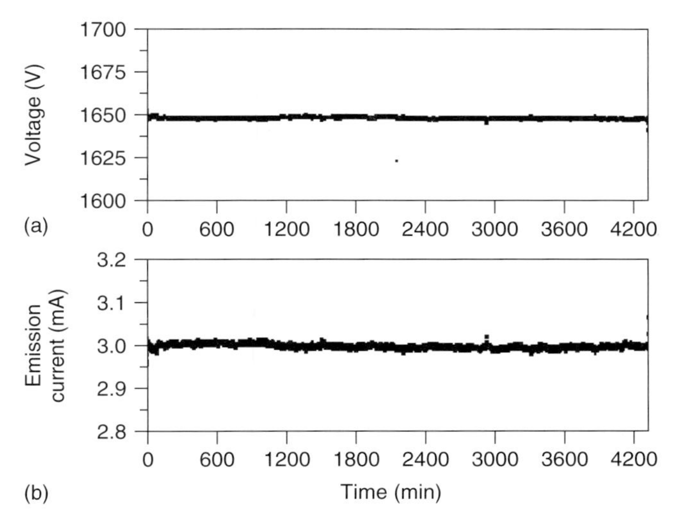

Figure 26.7 Emission lifetime measurement of a 2.35 × 0.50 mm CNT cathode at the constant current mode in triode geometry at 40 kV anode voltage. The emission current was fixed at 3 mA. The distance between the gate and the CNT cathode was 190 μm. The gate transmission rate (percentage of the current that passed through the gate electrode) was 74%. In both measurements, the waveform of the applied voltage was 10 ms pulse width and 1 Hz.

26.3.3 Focal Spot Size

The focal spot size was experimentally measured following the European Standard (EN 12543-5), by calculating the geometric unsharpness of a crosswire phantom placed between the X-ray source and the detector. The resolutions in two orthogonal directions were obtained on the basis of the line profiles of the transmitted X-ray intensity of the crosswire phantom. The voltages applied to the two independent focusing electrodes were adjusted to optimize the focusing power. Electrostatic simulation results indicate that the present design can give an optimum focusing factor of about 5 in linear dimension. The effective focal spot size scales linearly with the linear dimension of the cathode. This means that for an isotropic, effective focal spot size of 100 μm, an elliptical CNT cathode of 2.35 mm × 0.5 mm is needed at 12° anode tilting angle.

Figure 26.8 plots the effective focal area measured at 40 kV anode voltage for CNT cathodes of seven different sizes using the microfocus X-ray tube. The targeted effective spot size of 100 μm is readily achieved using the 2.35 mm × 0.5 mm elliptical cathode, as predicted by the simulation. For smaller cathodes, the focusing power is, however, smaller than 5. The lens becomes ineffective for a focal spot size of less than 50 μm with the present design. The focusing power also depends on the anode voltage applied. A slightly smaller focal spot size can be achieved when the anode voltage increases from 40 to 50 kV. The electron beam is stable during

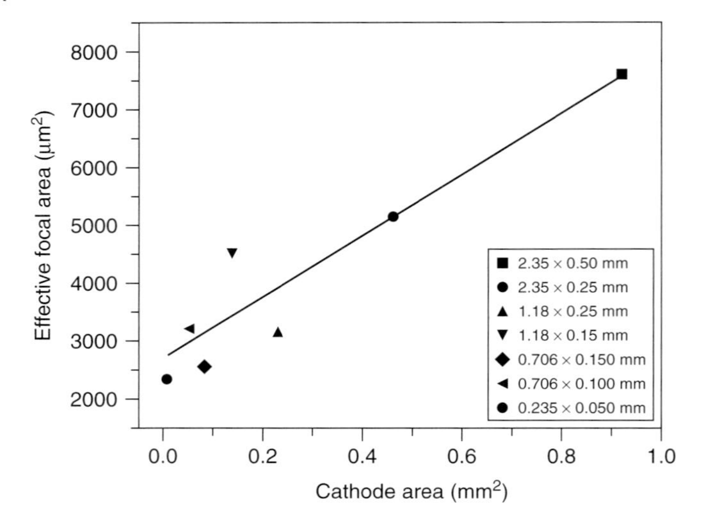

Figure 26.8 Effective focal area as a function of the CNT cathode area at 40 kV anode voltage.

the operation. No observable drifting of the beam was found. The change of the focal spot size is also insignificant when the gate voltage changes by 20–30% to maintain the output current over the lifetime of the X-ray tube.

26.4 Distributed Multibeam Field Emission X-ray

The concept of spatially distributed X-ray sources has been proposed and demonstrated in the past using the conventional thermionic technology. The motivation is primarily for improving the performances of imaging systems where information from multiple viewing angles is required such as CT. Several approaches have been considered. One is to use an electromagnetic field to scan a single electron beam to different locations on the X-ray anode. This is very similar to the working mechanism of the cathode-ray tube. Examples of devices built using this approach include the electron beam computed tomography (EBCT) [48] and the scanning beam digital X-ray (SBDX) [49, 50]. Such X-ray tubes are, in general, large and have a limited viewing range because of the difficulties in steering the high-energy electron beam. EBCT has been discontinued as a commercial product. The second is to use multiple conventional single-beam (single focal spot) X-ray tubes to form a scanning array. Examples of systems based on this approach include the dynamic spatial reconstructor (DSR) developed at the Mayo Clinic and the recently introduced dual-source CT scanner. In the DSR, a large number of X-ray tubes (~30) are placed around the CT gantry, which significantly increases the scanning speed and the temporal resolution of the system. The scanner is, however, impractical for clinical use because of its size, cost, and maintenance issues. Recently, a dual-source clinical CT scanner was introduced that essentially doubles the scanning speed compared to the regular single-source scanner and enhances the dual energy imaging capability. The third approach is to construct an X-ray source array with multiple and individually controllable thermionic cathodes to produce a scanning X-ray beam [51, 52]. The high operating temperature and power consumption, as well as the relatively large sizes of the control units of the thermionic cathodes, make this approach technically challenging.

The CNT field emission X-ray technology enables the fabrication of spatially distributed X-ray sources with multiple, independently controlled X-ray emitting pixels [23]. Compared to the conventional X-ray tube technology, it affords great flexibility in the source design in terms the dimension, number of pixels, and pixel configuration. Figure 26.9 shows the schematic of a CNT-based MBFEX source with a linear array of field emission cathodes. This particular source, the first prototype built by our group, comprises a field emission cathode with a linear array of five gated CNT emitting pixels, focusing electrodes, and a molybdenum target in the reflection mode [19, 22, 23]. Activation of the individual pixel is achieved by an n-channel metal–oxide–semiconductor field-effect transistor (MOSFET) circuit. The X-ray flux from each focal point is controlled by the electrical field applied between the gate the corresponding CNT pixel. By programming the gate voltage,

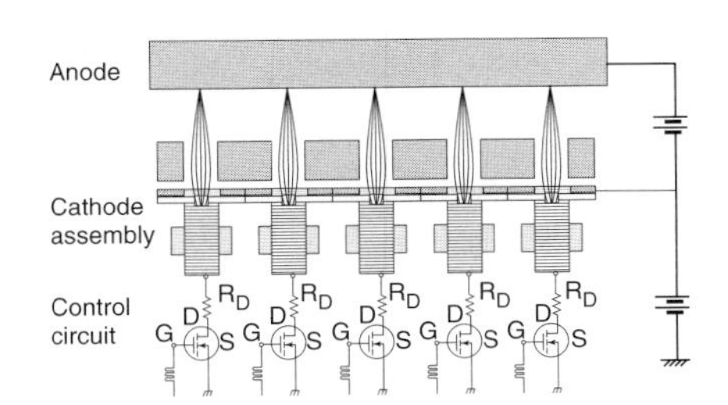

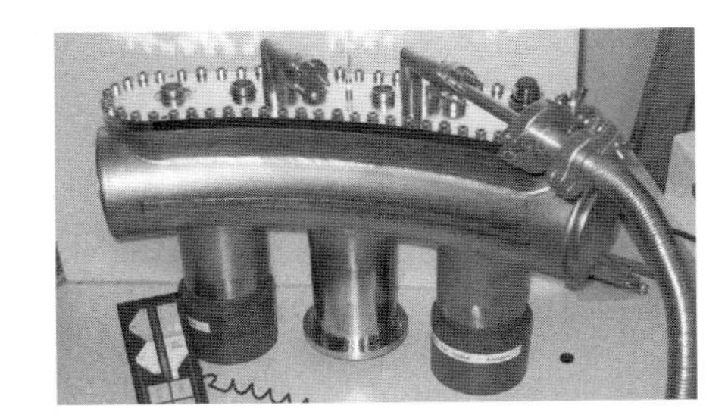

Figure 26.9 A schematic illustration of a distributed multibeam field emission X-ray (MBFEX) source with five independently controllable X-ray "pixels" (top right). (Photos of MBFEX tubes with different configurations manufactured by XinRay Systems LLC.)

a uniform emission current and thus X-ray flux can be obtained from each focal point.

Since the first publication, we have further improved the design and have now demonstrated a prototype system with a linear array of 25 individual controlled X-ray beams for digital breast tomosynthesis (DBT) [30], which we will introduce briefly in this section. There is also ongoing effort to commercially develop and manufacture the CNT MBFEX sources based on this technology for medical and security applications [31]. Three prototype MBFEX sources manufactured by XinRay Systems – a Joint Venture of Siemens and Xintek – are shown in Figure 26.9. The curved MBFEX source contains over 100 individually controllable X-ray beams covering 40° angular range. The linear MBFEX source contains 75 beams operating at the anode voltage up to 140 kV. Dedicated control electronics have also been developed to switch, scan, and regulate the intensity of the X-ray beams.

26.5 Imaging Systems

26.5.1 Dynamic Micro-Computed Tomography

We have developed a high-resolution, dynamic micro-CT scanner, called *Charybdis*, at UNC using the CNT field emission microfocus X-ray tube [26, 27]. The Charybdis system provides high spatial ($\leqslant$100 μm) and temporal (10–20 ms) resolutions, stationary and horizontal mouse bed configuration, and *prospective free-breath* gating capability. The cone-beam micro-CT scanner consists of a rotating source and detector pair and a stationary sample stage, as shown in Figure 26.10. In a typical CT scan, 400 projections are acquired over a circular orbit of 199.5° with a stepping angle of 0.5°. Reconstruction is done using the Feldkamp algorithm.

To demonstrate the capabilities of this scanner for *in vivo* imaging of small animals with high spatial and high temporal resolutions, the prospective cardiac and respiratory gated micro-CT imaging of *free-breathing* mouse is shown here. The experiments were performed following protocols approved by the UNC at Chapel Hill. Animals were anesthetized with 1–2% isoflurane at a flow rate of 1.5–2 l min^{-1} from a vaporizer. The animal breathing rates after anesthetization were typically in the range of 80–100 breaths per minute. The respiration signal was obtained from a respiration sensor pad placed on the surface of the animal bed approximately in the position where the abdomen of the animal would rest to gain maximum respiratory movement. The ECG signal was obtained by affixing three ECG electrodes to the paws of the animal. The mice were not intubated and were allowed to breathe freely throughout the experiment. Prior to imaging, an iodinated contrast agent was infused via tail vein injection (0.02 ml g^{-1} of mouse).

Projection images were acquired using prospective gating. By running the detector at 1 frame per second (camera integration time = 500 ms), the scan time

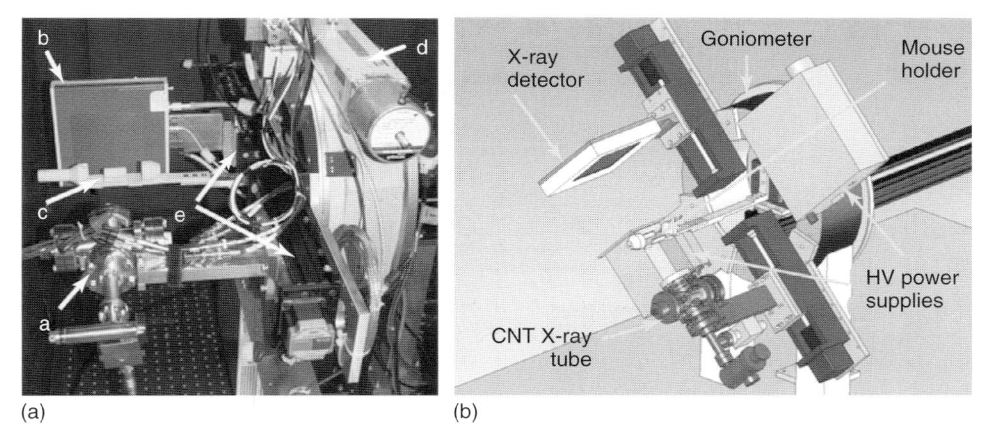

Figure 26.10 (A) Photo of the Charybdis scanner using a compact CNT microfocus X-ray source. It consists of (a) a compact CNT X-ray tube, (b) a flat-panel X-ray detector, (c) a mouse positioning stage, and (d) a goniometer. (B) CAD drawing of the same scanner.

for a cardiac and respiratory gated mouse micro-CT was typically 15–30 min, depending on the mouse's respiration and heart rates. Images were collected at 100 μm spatial resolution and 20 ms temporal resolution. A representative result from respiratory and cardiac gated micro-CT of mouse heart is shown in Figure 26.11. As we can see, structures inside the heart are clearly delineated. The motion-induced artifacts are not present, indicating a proper gating.

26.5.2 Stationary Digital Breast Tomosynthesis

Breast cancer is the most common cancer type among women. Approximately over 10% of women in the United States will develop breast cancer during their lifetime, and 30–40% of these patients die from it. Early detection is considered the best hope to decrease breast cancer mortality. Mammography is currently the most effective screening and diagnostic tool for early detection of breast cancer. It has been attributed as a major factor in reduction of breast cancer mortality rate in recent years [53, 54]. However, the current two-view mammography method lacks sensitivity and has a very high false alarm rate.

DBT is a limited-angle CT technique that can distinguish tumors from its overlying breast tissues and has potentials for detection of cancers at a smaller size and earlier stage [55–57]. Currently, several DBT scanners from commercial vendors are under clinical trial. The designs of all current DBT scanners are based on a full-field digital mammography (FFDM) system [57–62] and require partial isocentric motion of an X-ray tube over certain angular range to record the projection views. This prolongs the scanning time and in turn degrades the imaging quality because of motion blur [62].

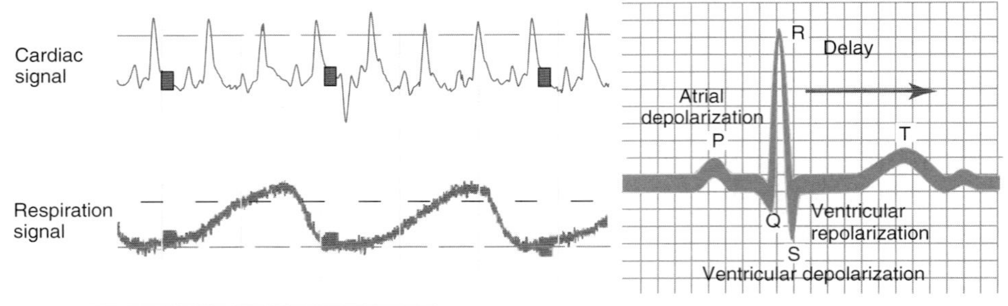

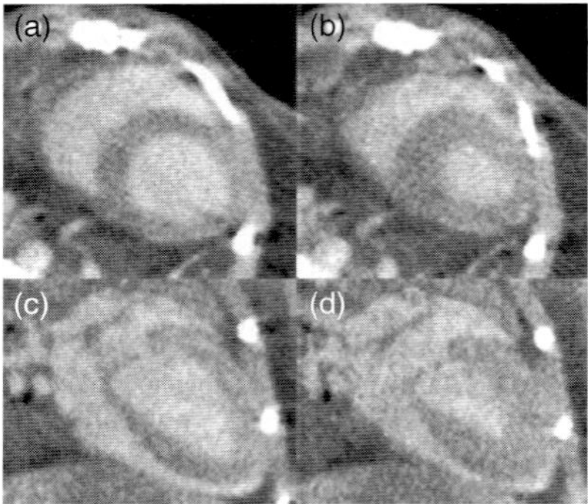

Figure 26.11 Representative cardiac and respiration signals during mouse cardiac imaging. The black solid boxes indicate the physiological triggers that were generated on the basis of the temporal coincidence between the selected phases of the cardiac signal. The physiological triggers were selected at 0 (systole) and 50 ms (diastole) delay after the first R wave of the ECG which occurred within the acquisition window. The reconstructed slice images show a clear difference between systole (a) and (c), and diastole (b) and (d) in the axial and coronal views, respectively.

To mitigate the above limitations, we have developed a proof-of-concept stationary digital breast tomosynthesis (s-DBT) scanner based on the newly developed spatially distributed multibeam field [28–30]. Figure 26.12 shows the configuration of the proof-of-concept system, called *Argus*. It is composed of a CNT MBFEX source array, a flat-panel X-ray detector, and a control unit. The system geometry follows the typical values for the Siemens DBT scanner. The X-ray pixels are arranged linearly to reduce the system complexity, with even angular distribution and a 2° increment. The source comprises 25 individually controllable X-ray beams that operate at 30 keV and cover 48° angular range. The system generates all the projection images by electronically activating the multiple X-ray beams from different viewing angles without any mechanical motion. Switching and scanning of the X-ray beam are accomplished through the MOSFET-based control unit we have described previously [28].

Tomosynthesis imaging was performed using a 3D tissue equivalent compressible breast phantom (Model 013, CIRS, Inc.). Twenty-five projection images were obtained from the system at a total exposure of 100 mA (4 mA per beam). The experimental condition was as follows: 28 kVp, molybdenum filter, and molybdenum

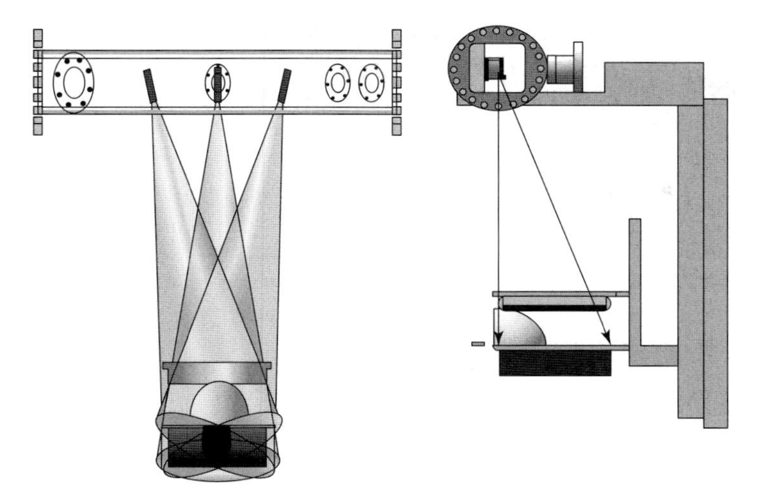

Figure 26.12 (a) The orientation of the individual X-ray source assembly with respect to the object. (b) Side view CAD drawing of the s-DBT scanner showing the positions of the source, the detector, and compression pad. X-ray anodes are positioned above the edge of the detector.

target. These projection images were then reconstructed using the modified ordered subsets convex (MOSC) method and design geometry parameters to yield 60 slices through the phantom. The slice distance is 1 mm. Figure 26.13 shows four slices at the depths of 6, 11, 16, and 21 mm from the top of the phantom. These slices clearly demonstrate the different masses getting focused at different depths.

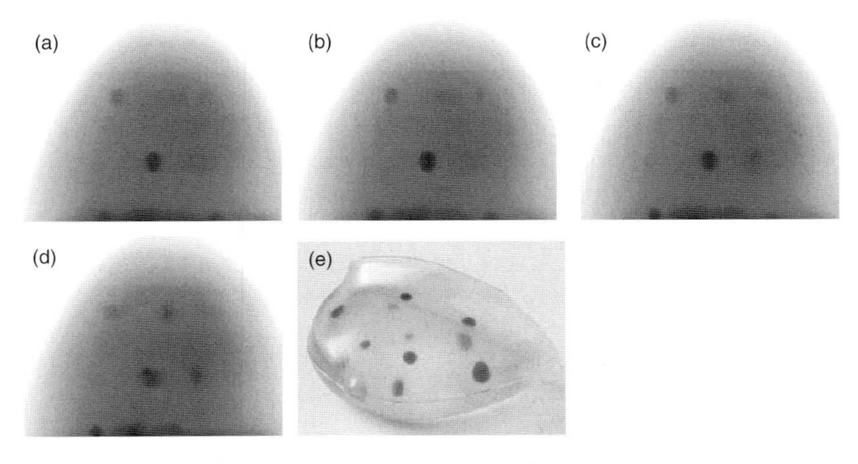

Figure 26.13 Four slices of the reconstructed breast phantom image. These slices are at the depths of (a) 6, (b) 11, (c) 16, and (d) 21 mm. The CIRS stereotactic needle biopsy breast phantom used in this study is shown in (e).

The s-DBT design has the potential to provide a significant increase in the scanning speed and to improve the system spatial resolution by eliminating the motion blur of the X-ray tube in the current systems and reduce the patient motion and discomfort from prolonged compression. This is an ongoing project at UNC. Key issues including the X-ray flux, spatial resolution, scanning time, beam-to-beam consistency, and reliability have recently been evaluated [30]. We hope to be able to validate this technology in clinical tests in the next few years.

26.6 Summary and Outlook

After several years of intense R&D efforts, the CNT X-ray technology has moved from an academic curiosity to commercial development. The technology offers considerable advantages over the current thermionic X-ray sources and has features that are uniquely suited for advanced X-ray imaging and potentially radiotherapy applications. Prototype imaging systems have been developed at academic labs and by leading commercial system vendors. In the coming years, we expect to see some of these novel systems moving into hospitals for clinical tests and to the airports for security inspections. There are still, however, plenty of challenges and opportunities remaining. Despite the increasing complexity of these imaging and treatment systems, the heart of the technology is the carbon nanotube field emitters. They have not reached their full potential. Further improvement of their performances is critical for this embryonic research field and hopefully industry.

Acknowledgments

The research results described in this chapter are the works of an interdisciplinary team of engineers, physicists, radiologists, and medical physicists at the University of North Carolina at Chapel Hill including Prof. Jianping Lu, Prof. Sha Chang, Prof. Guohua Cao, Prof. Jian Zhang, Prof. Yueh Lee, and Prof. David LaLush. A large number of current and former students, postdoctoral fellows, and staff in our group have worked on this interdisciplinary project over the last few years. In particular, we acknowledge contributions from Lei An, David Bordelon, Laurel Burk, Rui Peng, Tuyen Phan, Xin Qian, Ramya Rajaram, Christy Redmon, Shabana Sultana, and Guang Yang. Without them, this work would not have been possible. We have also been fortunate to be able to collaborate with a diverse group of colleagues from the biomedical research community, including Prof. E. Pisano, Prof. E. Hoffman, Prof. Y. Xiong, Prof. B. Grubb, Prof. C. Patterson, and Prof. I. Rusyn, whose participation and feedback have made this research more meaningful and interesting. We acknowledge supports and collaborations from colleagues at Xintek and XinRay Systems.

The research on CNT field emission X-ray at UNC has been generously supported by grants from the National Cancer Institute, the National Institute of Biomedical Imaging and BioEngineering, the University Cancer Research Fund at UNC, the Department of Homeland Security, the Department of Defense, and Xintek.

References

1. Bushong, S.C. (1997) *Radiologic Science for Technologist*, Mosby.
2. Hendee, W.R. and Hendee, G.S.I.E.G. (2004) *Radiation Therapy Physics*, Wiley-Liss.
3. National Research Council (1996) *Airline Passenger Security Screening: New Technologies and Implementation Issues*, National Academy Press.
4. Harding, G. (2004) X-ray scanner tomography for explosive detection. *Radiat. Phys. Chem.*, **71**, 869–881.
5. Kalender, W.A. (2005) *Computed Tomography*, Publicis Corporate Publishing.
6. Hsieh, J. (2003) *Computed Tomography: Principles, Design, Artifacts, and Recent Advances*, SPIE Press.
7. Charbonnier, F.M., Barbour, J.P., and Dyke, W.P. (1974) Resolution of field emission x-ray sources. *Radiology*, **117**, 165.
8. Hallenbeck, G.S. (1974) Clinical evaluation of the 350-KV chest radiography system. *Radiology*, **117**, 1–4.
9. Crooks, H.E., Sangster, J., and Ardran, G.M. (1972) The fexitron 848 portable x-ray apparatus. *Radiology*, **38** (456), 311.
10. Rangstein, P. *et al.* (2000) Field-emitting structures intended for a miniature x-ray source. *Sens. Actuators*, **82**, 24–29.
11. Baptist, R. (2001) X-ray tube comprising an electron source with microtips and magnetic guiding means. US Patent 6, 259,765.
12. Whitlock, R.R. *et al.* (2001) Transmission cathodes for x-ray production. US Patent 6, 333,968.
13. Zhu, W. (2001) *Vacuum Micro-Electronics*, John Wiley & Sons, Inc.
14. Wei, Y. *et al.* (2001) Stability of carbon nanotubes under electric field studied by scanning electron microscopy. *Appl. Phys. Lett.*, **79** (27), 4527–4529.
15. Zhang, J. *et al.* (2004) Efficient fabrication of carbon nanotube point electron sources by dielectrophoresis. *Adv. Mat.*, **16** (14), 1219–1222.
16. Zhou, O. and Lu, J.P. (2003) New x-ray generating mechanism using electron field emission cathode. US Patent 6, 553,096.
17. Sugie, H. *et al.* (2001) Carbon nanotubes as electron source in an x-ray tube. *Appl. Phys. Lett.*, **78**, 2578.
18. Yue, G.Z. *et al.* (2002) Generation of continuous and pulsed diagnostic imaging x-ray radiation using a carbon-nanotube-based field-emission cathode. *Appl. Phys. Lett.*, **81** (2), 355.
19. Zhang, J. *et al.* (2005) A stationary scanning x-ray source based on carbon nanotube field emitters. *Appl. Phys. Lett.*, **86**, 184104.
20. Zhou, O. and Lu, J.P. (2006) X-ray generating mechanism using electron field emission cathode. US Patent 6, 850,595.
21. Cheng, Y. *et al.* (2004) Dynamic x-ray radiography using a carbon nanotube field emission x-ray source. *Rev. Sci. Inst.*, **75** (10), 3264.
22. Zhang J. *et al.* (2006) in *A Multi-Beam X-ray Imaging System Based on Carbon Nanotube Field Emitters*, Proceedings of SPIE, Vol. 6142 (eds J.M.Flynn, and H., Jiang), Medical Imaging, p. 614204.
23. Zhou, O., Lu, J.P., and Qiu, Q. (2005) Large-area individually addressable multi-beam x-ray system and method of forming same. US Patent 6, 876,724.
24. Zhang, J. *et al.* (2006) Multiplexing radiography using a carbon nanotube based x-ray source. *Appl. Phys. Lett.*, **89**, 064106.
25. Lu, J.P., Zhou, O., and Zhang, J. (2007) X-ray imaging systems and methods using temporal digital signal processing for reducing noise and for obtaining

multiple images simultaneously. US Patent 7, 245,692.

26. Cao, G. *et al.* (2009) A dynamic micro-CT scanner based on a carbon nanotube field emission x-ray source. *Phys. Med. Biol.*, **54**, 2323–2340.
27. Cao, G. *et al.* (2009) A dynamic micro-CT scanner with a stationary mouse bed using a compact carbon nanotube field emission x-ray tube. *SPIE Med. Imaging: Phys. Med. Imaging.*
28. Yang, G. *et al.* (2008) Stationary digital breast tomosynthesis system with a multi-beam field emission x-ray source array. *SPIE Proc. Med. Imaging.*
29. Otto, Z. *et al.* (2009) Stationary x-ray digital breast tomosynthesis systems and related methods. US Patent 2, 009,022,264.
30. Qian, X. *et al.* (2009) Design and characterization of a spatially distributed multi-beam field emission x-ray source for stationary digital breast tomosynthesis. *Med. Phy.*, **36** (10), 4389–4399.
31. Maltz, J.S. *et al.* (2009) Fixed gantry tomosynthesis system for radiation therapy image guidance based on a multiple source x-ray tube with carbon nanotube cathodes. *Med. Phys.*, **36** (5), 1624.
32. Zhang, T. *et al.* (2009) Tetrahedron beam computed tomography(TBCT): a new design of volumetric CT system. *Phys. Med. Biol.*, **54**, 3365.
33. Zhu, W. *et al.* (1999) Very high current density from carbon nanotube field emitters. *Appl. Phys. Lett.*, **75** (6), 873–875.
34. Milne, W.I. *et al.* (2006) Aligned carbon nanotubes/fibers for applications in vacuum microwave amplifiers. *J. Vac. Sci. Technol. B*, **24**, 345.
35. Shiffler, D. *et al.* (2004) A high current, large area, carbon nanotube cathode. *IEEE. Trans. Plasma. Sci.*, **32** (5), 2152.
36. Wang, Z.L. *et al.* (2002) In-situ imaging of field emission from individual carbon nanotubes and their structural damage. *Appl. Phys. Lett.*, **80** (5), 856–858.
37. Zhou, O. *et al.* (2002) Materials science of carbon nanotubes: fabrication, integration, and properties of macroscopic structures of carbon nanotubes. *Acc. Chem. Res.*, **35**, 1045–1053.
38. Oh, S.J. *et al.* (2004) Liquid-phase fabrication of patterned carbon nanotube field emission cathodes. *Appl. Phys. Lett.*, **87** (19), 3738.
39. Oh, S. and Zhou, O. (2008) Deposition method for nanostructured materials. US Patent 7, 455,757.
40. Zhou O. *et al.* (2001) Deposition methods for nanostructure materials. US Patent 7, 252,749.
41. Calderon-Colon, X. *et al.* (2009) Carbon nanotube field emission cathode with high current density and long-term stability. *Nanotechnology*, **20**, 325707.
42. Qian, C. *et al.* (2006) Fabrication of small diameter few-walled carbon nanotubes with enhanced field emission property. *J. Nanosci. Nanotechnol.*, **6**, 1346.
43. Choi, W.B. *et al.* (2001) Electrophoresis deposition of carbon nanotubes for triode-type field emission display. *Appl. Phys. Lett.*, **78** (11), 1547.
44. Paulus, M.J. *et al.* (2000) High resolution x-ray computed tomography: an emerging tool for small animal cancer research. *Neoplasia*, **2** (1-2), 62–70.
45. Ritman, E.L. (2004) Micro-computed tomography:current status and development. *Annu. Rev. Biomed. Eng.*, **6**, 185–208.
46. Liu, Z. *et al.* (2006) Carbon nanotube based microfocus field emission x-ray source for microcomputed tomography. *Appl. Phys. Lett.*, **89**, 103111.
47. Guohua, C. *et al.* (2008) Respiratory-gated micro-CT using a carbon nanotube based micro-focus field emission x-ray source. *SPIE. Proc. Med. Imaging*
48. Lipton, M.J. *et al.* (1984) Cardiac imaging with a high-speed cine-CT scanner: preliminary results. *Radiology*, **152** (3), 579–582.
49. Speidel, M.A. *et al.* (2006) Scanning-beam digital x-ray (SBDX) technology for interventional and diagnostic cardiac angiography. *Med. Phys.*, **33** (8), 2714.
50. Schmidt, T.G. *et al.* (2006) A prototype table-top inverse-geometry volumetric CT system. *Med. Phys.*, **33** (6), 1867.
51. Luggar, R.D. *et al.* (1999) An electronically gated multi-emitter x-ray source

for high speed tomography. *SPIE. Conf. Radiat. Sources. Radiat. Interact.*, **3771**, 44.

52. Frutschya, K *et al.* (2009) X-ray multi-source for medical imaging. *SPIE. Med. Imaging*.
53. Pisano, E.D. *et al.* (2005) Diagnostic performance of digital versus film mammography for breast-cancer screening. *New Engl. J. Med.*, **353**, 1773–1783.
54. Bassett, L.W. *et al.* (2005) *Diagnosis of Diseases of the Breast*, 2nd edn, Pennsylvania Elsevier Saunders, Philadelphia.
55. Niklason, L. (1997) Digital tomosynthesis in breast imaging. *Radiology*, **205**, 399–406.
56. Dobbins, J.T. and Godfrey, D.J.III (2003) Digital x-ray tomosynthesis: current state of the art and clinical potential. *Phys. Med. Biol.*, **48**, R65–R106.
57. Wu, T. *et al.* (2003) Tomographic mammography using a limited number of low-dose cone-beam projection images. *Med. Phys.*, **30**, 365–380.
58. Chen, Y., Lo, J.Y., and Dobbins, J.T. (2005) Impulse response analysis for several digital tomosynthesis mammography reconstruction algorithms. *Proc. SPIE*, **5745**, 541–549.
59. Bissonnette, M. *et al.* (2005) Digital breast tomosynthesis using an amorphous selenium flat panel detector. *Proc. SPIE*, **5745**, 529–540.
60. Zhang, Y. *et al.* (2006) A comparative study of limited-angle cone-beam reconstruction methods for breast tomosynthesis. *Med. Phys.*, **33** (10), 3781–3795.
61. Maidment, A.D. *et al.* (2006) Evaluation of a photon-counting breast tomosynthesis imaging system. *Proc. SPIE*, **6142**, 89–99.
62. Ren, B. *et al.* (2005) Design and performance of the prototype full field breast tomosynthesis system with selenium based flat panel detector. *Proc. SPIE, Phys. Med. Imaging*, **5745**.

27
Microwave Amplifiers

Pierre Legagneux, Pierrick Guiset, Nicolas Le Sech, Jean-Philippe Schnell, Laurent Gangloff, William I. Milne, Costel S. Cojocaru, and Didier Pribat

27.1
Introduction

A vacuum amplifier includes three subassemblies, which are an electron gun, an interaction structure to transfer the kinetic energy of the electron beam to the wave being amplified, and an electron collector. Because of these different elements, they are bulky compared to solid-state amplifiers (SSAs) but deliver higher powers and exhibit a higher power yield as well as a higher reliability. They are preferred to SSAs for several applications in the range of power (up to a few tens of watts) and frequencies (up to a few tens of gigahertz) where they are highly competitive. At higher frequency, their performances are unequaled. A particularly appropriate application is satellite telecommunications, where amplifiers exhibiting very high power yields (close to 70%) and long lifetime (15 years) are required. The tubes of interest are traveling-wave tubes (TWTs). This chapter aims to describe the state of research on carbon nanotube (CNT) cold cathodes to replace thermionic cathodes in electron tubes and in particular in TWTs.

TWTs use thermionic cathodes as electron sources. In continuous mode, such cathodes deliver current densities of 1–2 $\mathrm{A\,cm^{-2}}$ and exhibit a 15 year lifetime. Using a dedicated electron gun design invented by Pierce [1], electron beams with current densities between 50 and 150 $\mathrm{A\,cm^{-2}}$ are typically obtained. However, due to the emission mechanism, the emitted beam can not be modulated at frequencies above a few gigahertz. The beam is modulated afterwards during its interaction with the wave that is injected at the entrance of the interaction structure. This "post" modulation that takes place along the interaction structure increases significantly the tube length and consequently its weight. It has long been recognized that the use of cold cathodes that directly deliver a modulated beam would allow the fabrication of compact, light, and highly efficient TWTs. Thus, a large research effort

Carbon Nanotube and Related Field Emitters: Fundamentals and Applications. Edited by Yahachi Saito

ISBN: 978-3-527-32734-8

has been devoted to cold cathodes which can be modulated at very high frequency.

Most cold cathodes studied since the beginning of the 1970s are based on arrays of microtips called field emission arrays (FEAs). Microelectronic-type technologies have been extensively applied to fabricate them. Microtips are made of refractory metals or semiconductors. Among them, the most popular, with the highest performances in terms of current density, is the so-called Spindt cathode. It consists of an array of molybdenum cones, each emitter being deposited in a gated structure. As a result, the cathode–grid capacitance is high and the microwave excitation of such cathodes at high frequency has been difficult to implement.

Since 1995, CNTs have been studied as electron sources. Their cylindrical shape is almost ideal to efficiently enhance the electric field at their apices. Moreover, they are arguably the most stable and robust cold field emitters known. The grid can be spaced away from the cathode and coupling an electromagnetic wave appears more efficient.

It is interesting to compare these new CNT cathodes to the thermionic cathodes they aim to replace. For this purpose, Section 27.2 describes briefly the thermionic cathodes used in TWTs and proposes a methodology to evaluate the competing CNT cathodes.

Section 27.3 presents the state of the art of CNT-based electron guns as high current electron sources. Their capability to deliver electron beams that can be focused to generate high current density beams will also be reviewed.

Section 27.4 describes the various methods used to modulate the emitted electron beam at high frequency. When the applied field is directly modulated, this section will show that resonant cavities are an efficient way to provide microwave modulation: modulated beams at frequencies as high as 32 GHz have been demonstrated. However, electron guns based on this modulation method are limited to narrow bandwidth applications.

For large bandwidth operation, two different types of optically controlled CNT cathodes have been proposed. The first type is based on the optical modulation of the current supplied to the CNTs. For this purpose, p-i-n photodiodes which act as optically controlled current sources are located under the CNTs. The second type, more exploratory, depends upon the direct action of the light at the nanotube apex, to modulate the electron emission. This method is compatible, in principle, with terahertz modulation.

Taking advantage of the large bandwidth and parallelism of the optical drive, optically controlled cathodes should satisfy the requirements for a new generation of TWTs.

As a conclusion, Section 27.5 will indicate future directions of research, and perspectives.

27.2
State of the Art of Thermionic Cathodes and Methodology to Review CNT Cathodes

27.2.1
State of the Art of Thermionic Cathodes Used in Traveling-Wave Tubes

The most used thermionic cathode in microwave industry is the M-type osmium-coated cathode, invented by Zalm and van Stratum [2] in the 1960s. It is an "impregnated" cathode in which an alkaline-earth mixture of BaO, CaO, and Al_2O_3 is distributed throughout the pores realized in a (porous) tungsten matrix (Figure 27.1). After cathode heating, the tungsten surface is covered by a layer of Ba and O. As Ba and O are, respectively, electropositive and electronegative species, a dipole is formed. This dipole lowers the work function of the cathode from 4.5 eV for a pure tungsten material to around 2.1 eV. As the cathode ages, the barium evaporation is compensated by barium that diffuses from the bulk to the surface using the impregnated pores. When an osmium layer is deposited on the impregnated cathode, the resultant work function is further reduced from 2.1 to 1.9 eV. This allows reduction of the operating temperature and thus barium evaporation. As a consequence, the lifetime is increased. The typical lifetime for M-type cathodes delivering 1–2 A cm^{-2} is around 15 years. This lifetime is reduced to a few years when the current density is increased to 10 A cm^{-2}.

For current densities of 1–2 A cm^{-2}, the emission of M-type cathode is space charged limited [2]. This regime is particularly interesting as the electron emission is highly uniform. The emitted electrons also exhibit a low initial velocity (below 1 eV) and thus a low transverse velocity. These two features (high emission

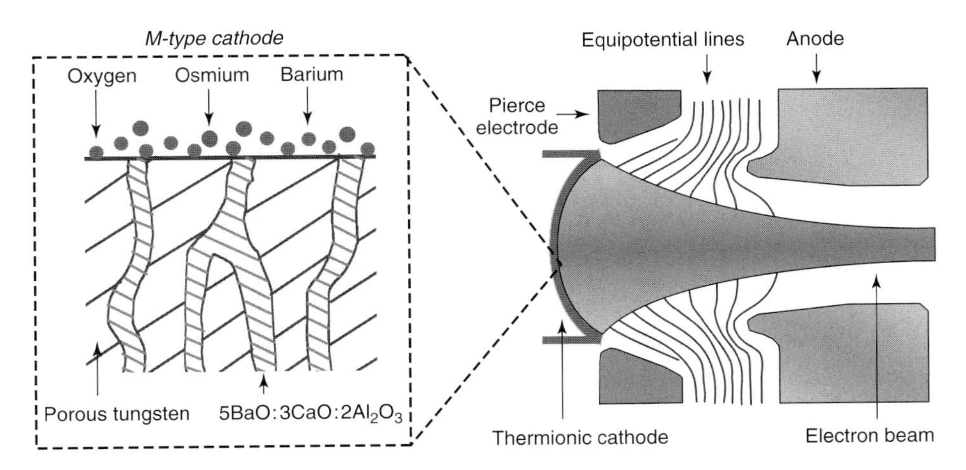

Figure 27.1 Schematic description of an M-type cathode [2] integrated in a Pierce-type electron gun [1].

uniformity and low transverse electron velocity) have allowed the design and the fabrication of electron guns exhibiting a high convergence factor (ratio between the beam surface and the cathode surface). Figure 27.1 shows also a schematic description of the electron gun which was invented by Pierce in 1940 [1]. It is based on a curved thermionic cathode, a focus electrode, and an anode. Pierce showed that, for an appropriate angle of the focus electrode with respect to the axis of the electron beam and for an adequate design of the anode, a focused beam can be obtained. The convergence factor values are typically 50–75. That means that the current density in the beams used in TWTs varies from 50 to 150 A cm^{-2}. The electron beam then interacts with the slow wave structure which is generally a helix in TWTs. To maintain the beam diameter during its interaction with the helix (to avoid beam spreading due to the charge in the electron beam), the use of an axial magnetic field was also proposed by Pierce in 1949 [3].

27.2.2 Interest in Cathodes Delivering a High-Frequency Modulated Electron Beam

In a TWT, the amplification process is based on the interaction between a modulated electron beam and an electromagnetic wave. However, up to now, only thermionic cathodes emitting a continuous electron beam have been used. Therefore, the beam has to be postmodulated and this increases the size/weight of the TWTs. Thus, TWTs are bulky and heavy, and take up a valuable budget in a telecommunications satellite (particularly for the satellite launch).

The idea of direct current modulation of the electron beam has been around for many decades [4, 5]. The direct modulation of thermionic cathodes has been performed by the use of a grid that modulates the electric field applied to these cathodes. However, due to the cathode temperature (~1000 °C), the minimum cathode–grid distance is around 100 μm. This results in a high cathode–grid transit time which limits the modulation frequency to a few gigahertz.

The FEAs studied since the beginning of the 1970s solve the transit time issue. They are based on arrays of tips with integrated gate electrodes. The cathode–grid distance is below 1 μm and the transit time becomes negligible. Among these FEAs, the most popular with the highest performances in terms of current and current density is the Spindt cathode [6, 7]. Due to their high current density capabilities and instant turn-on feature, field emission cathodes have been recognized to be a promising alternative to thermionic cathodes [8]. Calame [9] and Jensen [10] studied the design of microwave amplifiers based on Spindt cathodes delivering a modulated electron beam. Jensen showed that 10 GHz TWTs using a very short (1.5 cm long) helix and exhibiting a 15 dB gain could be fabricated. However, with gated cathodes, the cathode–grid capacitance is high and the parallel microwave excitation of all the emitters of the cathode is difficult to implement at high frequency. To solve the gate signal propagation issue, ring cathodes with outer diameters of 600 μm and "widths" as low as 20–60 μm were proposed [11]. As the peak current needed is high (~200 mA for a 50-W X-band amplifier [11]) and the actual grid surface low (to minimize propagation losses), this imposes the use of cathodes exhibiting very high performance (i.e., cathodes delivering a peak current density above 100 A cm^{-2}). However, different types of ring cathodes have been mounted in a 10 GHz klystrode and the maximum output power was only 4 mW [11]. No further developments seem to have been performed, perhaps because the fabrication and microwave excitation of ring cathodes is difficult to implement.

In 2002, Whaley *et al.* studied a cathode design that minimizes parasitic capacitances, the cathode capacitance, and also in-line resistance [12]. The cathode area is a 300 μm diameter disk having 10 000 emitter tips. The cathode–grid distance was about 2 μm, that is, twice the normal distance of a "standard" cathode. This complicates the fabrication process but reduces by a factor of 2 the cathode capacitance. The substrate was also thinned under the cathode area to reduce in-line resistance. The emission-gated 6.8 GHz TWT prototype was operated up to a current of 5 mA and RF output power of 280 mW when tested in a single-pulse mode using 100 μs pulses.

As a conclusion, the use of FEAs delivering a modulated electron beam allows one to significantly reduce the length of the interaction structure of microwave amplifiers such as TWTs. They appear to be the solution for the fabrication of compact and light microwave amplifiers. Among the different FEAs, Spindt-type cathodes exhibit the highest performances in terms of current density. But, due to the high cathode–grid capacitance, the microwave operation of such cathodes is difficult to implement. Recently, Whaley *et al.* [13] have focused their efforts on developing a 100 W, 5 GHz TWT based on a "nonmodulated" cathode. In such a configuration, the TWT size cannot be reduced, but the use of a cold cathode allows instant turn-on and easy pulsed operation.

27.2.3 Methodology of Reviewing CNT Cathodes

To evaluate the performances of CNT cathodes as electron sources for microwave amplifiers, we will review the capabilities of CNT cathodes to deliver

- a high current density electron beam;
- an electron beam that can be modulated at high frequency.

To deliver high current density electron beams, CNT cathodes have to be integrated in an electron gun whose role is to focus the beam emitted by the cathode. Thus, the current density in the electron beam is equal to the current density at the cathode level times the beam convergence factor.

In Sections 27.3.1 and 27.3.2

- the current density at the cathode level and
- the current density in the electron beam after focusing will both be analyzed.

The major advantage of cold cathodes compared to thermionic cathodes is their ability to deliver a beam modulated at high frequency.

In Section 27.4, three different approaches to modulate the emitted beam will be presented:

- the modulation of the applied electric field;
- the optical modulation of the current supplied to the CNTs;
- the optical modulation of the electric field at the nanotube apex.

27.3
CNT-Based Electron Guns as High Current Electron Sources

To be used in microwave amplifiers, CNT-based electron guns have to deliver beams with current densities above 50 A cm^{-2}. This current density is the current density at cathode level times the convergence factor (ratio between beam cross section and cathode surface area).

27.3.1
Current Density at Cathode Level

27.3.1.1 Currents Emitted by Individual CNTs

Since the first field emission experiments with CNTs performed in 1995 [14, 15], uniform arrays of vertically aligned multiwalled carbon nanotubes (MWCNTs) have been extensively studied for field emission applications [16–21]. MWCNTs are excellent field emitters. They are whisker-like in shape with very high aspect ratio, leading to high field enhancement at their apices. They are arguably the most stable and robust cold field emitters known to date [22, 23]. The emission from MWCNTs is stable even if the temperature of the apex attains 2000 K [22]. On the other hand, for metal tips, the combination of high temperature and field permits the well-known mechanism of field-driven sharpening. The tip sharpening increases the field at metallic tip apex and thus the emitted current, which in turn increases the tip temperature. This leads to the destruction of metal tips. MWCNTs are not subjected to this mechanism [22], first because the resistance of nanotubes decreases with temperature, and second because surface diffusion is much slower for covalent carbon, which inhibits the field-driven sharpening. Due to these intrinsic characteristics, a maximum emission current of 100–200 μA per individual MWCNT has been demonstrated [24, 25].

27.3.1.2 Current Density Emitted by CNT Cathodes

To fabricate a CNT cathode, two different approaches are currently studied. The first one consists in growing the nanotubes by any known method (arc discharge, laser vaporization, high-pressure disproportionation of CO, or CVD) to harvest them and subsequently deposit them on the cathode substrate. This deposition step is generally performed at room temperature from a suspension [26] or a paste [27]. For the second approach, the cathode substrate is loaded in a CVD system and the CNTs are directly grown on this substrate. In this case, high-temperature (500–700°C) heating of the cathode substrate is required.

Among the different teams working on the first approach, the most noteworthy are the results obtained by Calderon-Colon *et al.* at the University of North Carolina [28]. They fabricated cathodes delivering peak current densities above 1.5 A cm^{-2}. The voltage applied to the anode had a square waveform with 10 ms pulse width and 1 Hz repetition rate (duty cycle = 1%).

Teams working on the second approach have studied two different methods to grow vertically aligned MWCNTs. For the first method, CNTs are grown in a plasma-enhanced chemical vapor deposition (PECVD) reactor where the nanotubes are directly exposed to the plasma (Figure 27.2). Although Bower *et al.* [29] showed unambiguously the effect of the electric field on nanotube alignment in a microwave plasma, the mechanism is not fully understood [30].

The second method consists in growing CNTs in a CVD reactor or in a PECVD system where the CNTs are not exposed to the plasma. To obtain vertical CNTs, microarrays are grown (Figure 27.3). In this case, the vertical alignment is due to a crowding effect within the microarrays (nanotubes supporting each other by van der Waals attraction).

The first method allows the growth of ideal arrays (Section 27.3) that minimize the screening effect between neighboring CNTs [31, 32]. However, the CNTs grown

Figure 27.2 Uniform array of individual 5-μm-high and 50-nm-diameter multiwalled CNTs grown by PECVD [33].

Figure 27.3 Uniform CNT microarrays made of thin (~10 nm) multiwalled CNTs [34]. (Figure reproduced with permission from IOP Publishing, UK.)

by the corresponding PECVD method tend to contain some crystalline defects as a result of ion bombardment. The second method leads to the growth of higher crystalline quality CNTs but the cathode design is not ideal.

Among all the reported results with CNT cathodes, the largest current was obtained by the team at the University of Electronic Science and Technology of China [34]. They fabricated a $5\,\text{cm}^2$ cathode and tested it with a 5-mm-diameter anode. Using 10 μs width bias pulses with a 0.1% duty cycle, the peak current and the peak current density were, respectively, 0.71 A and $3.5\,\text{A cm}^{-2}$.

Using gated devices, Hsu and Shaw [35] from the Naval Research Lab. have also obtained current densities up to $3.5\,\text{A cm}^{-2}$.

With arrays of individual CNTs [33], the University of Cambridge associated with Thales have fabricated individual nanotubes emitting 80–120 μA [25] and CNT arrays delivering a current density of $1\,\text{A cm}^{-2}$ [21]. Both measurements have been performed in continuous mode. One can note that the use of low duty cycles and short pulses allows measurement of higher current densities. For example, Cambridge and Thales obtained from the same type of cathode a peak current density of $12\,\text{A cm}^{-2}$ [36].

As a conclusion, current densities of a few amperes per square centimeter are obtained from CNT cathodes, that is, values that are equivalent to those obtained with thermionic cathodes.

27.3.2
Convergence Factors Obtained with CNT-Based Electron Gun

To deliver very high current density beams, it is necessary to focus the beam emitted by the cathode and thus to integrate this cathode in a dedicated electron gun. Very few papers have been published on CNT-based electron guns for electron tubes. Among them, one can note the work performed by Kim *et al.* [39], who studied the integration of a CNT cathode in a Pierce-type electron gun, that is, the gun used for thermionic cathodes. For this purpose, they used a curved cathode and integrated an extraction grid (Figure 27.4). The role of the extraction grid is to generate the high electric field required for field emission.

27.3.2.1 Simulation of a CNT-Based Electron Gun

The gun design was optimized for TWT application. The cathode radius was 2.18 mm and the cathode curvature radius was 9.73 mm. The gun was based on a double grid configuration with the spacing between the cathode and the first grid electrode of 200 μm. The grid bias was 1 kV in order to apply a $5\,\text{V}\,\mu\text{m}^{-1}$ electric field on the cathode. As the anode voltage increased from 4 to 8 kV, the focused beam radius was reduced from 0.93 to 0.68 mm. With the periodic permanent magnet (PPM), a nonlaminar beam with a maximum beam radius of 0.92 mm was obtained. This radius is larger than the laminar beam radius (0.39 mm) obtained with thermionic cathodes. This was attributed to the fact that, with field-emission-based cathodes, a large voltage bias is needed on the first grid to extract electrons from the

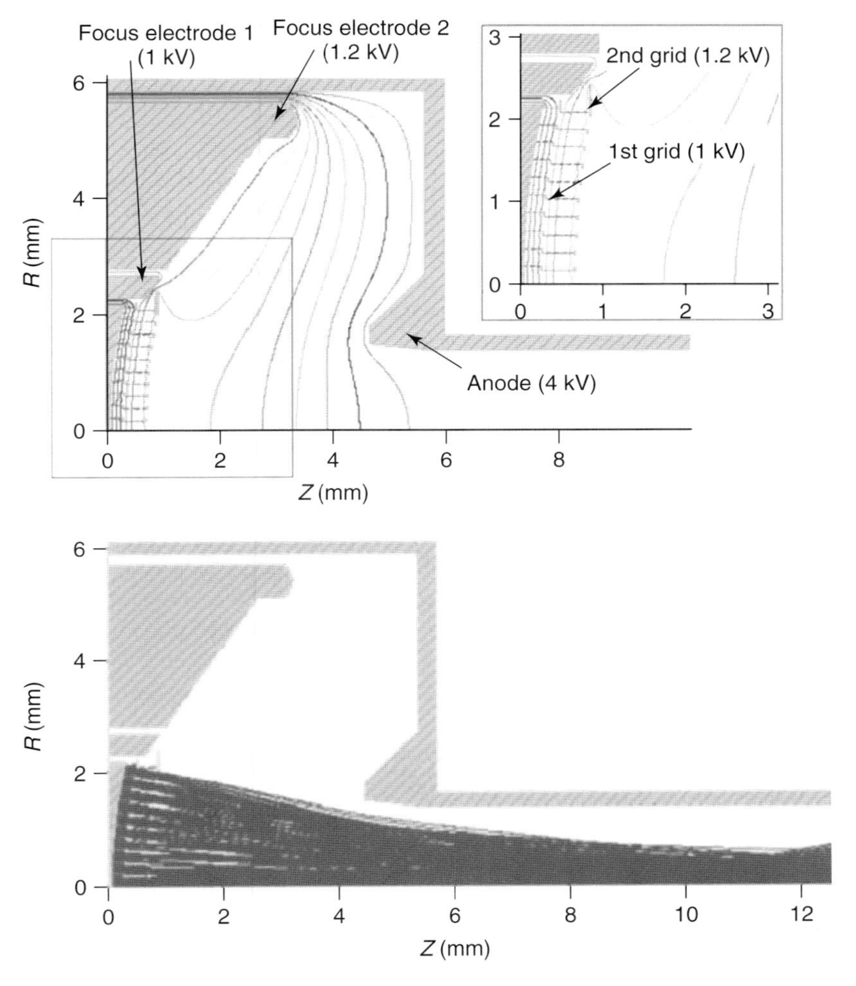

Figure 27.4 Pierce-type electron gun based on a curved CNT cathode [39]. (Figure reproduced with permission from the Institute of Electrical and Electronics Engineers (IEEE). © 2006 IEEE.)

cathode (for thermionic cathodes, only low applied fields are required). The first grid then acts as a diverging lens and electrons acquire large transverse velocities. This effect is reduced with the double grid configuration but the obtained beam convergence factor (ratio between beam cross section and cathode surface area) is lower than the one obtained with thermionic cathodes (6 instead of 30).

This electron gun was experimentally tested. A well-dispersed CNT paste was printed onto a curved cathode. In diode mode, the cathode emitted an 11 mA current corresponding to a current density of 73 mA cm^{-2}. The Pierce-type gun was then fabricated. In triode mode, the beam current from the "gridded" cathode was 4.6 mA. The current reduction was attributed to a handling process that could have deteriorated the CNT cathode. The size of the emitted beam was not experimentally analyzed.

As a first conclusion, on the basis of published work, the beam convergence factor obtained with CNT-based electron guns is smaller than the one obtained with thermionic cathodes. Thus, to obtain the same current density in the focused beam, CNT cathodes will have to deliver higher current densities than those emitted by thermionic cathodes ($1-2\,\mathrm{A\,cm^{-2}}$).

A second conclusion is that a new design of electron gun adapted to CNT cathodes has to be invented. This electron gun should be designed to minimize beam interception (e.g., by a grid) and to provide high beam convergence.

Electron beams emitted by CNT cathodes present large transverse electron velocities and thus are difficult to focus. To decrease these transverse velocities, it is necessary to accurately design the CNT cathodes.

27.3.2.2 Design of Cathodes Delivering Low Transverse Electron Velocities

Field emission from tips generates electrons exhibiting large transverse velocities. To illustrate this phenomenon, we display on Figure 27.5a a set of electron trajectories emitted by a nanotube presenting a hemispherical apex. As the electric field is perpendicular to the nanotube surface, electrons acquire transverse velocities (velocities perpendicular to the nanotube axis). The position of each emitting point of the nanotube apex can be defined by the angle θ as shown in Figure 27.5a. Figure 27.5b shows the transverse velocities of electrons emitted from nanotubes of 1, 2, and 5 μm height. For each nanotube, the aspect ratio (h/r) is equal to 200 and a value of 20 V μm^{-1} has been used for the applied field. Thus the electric

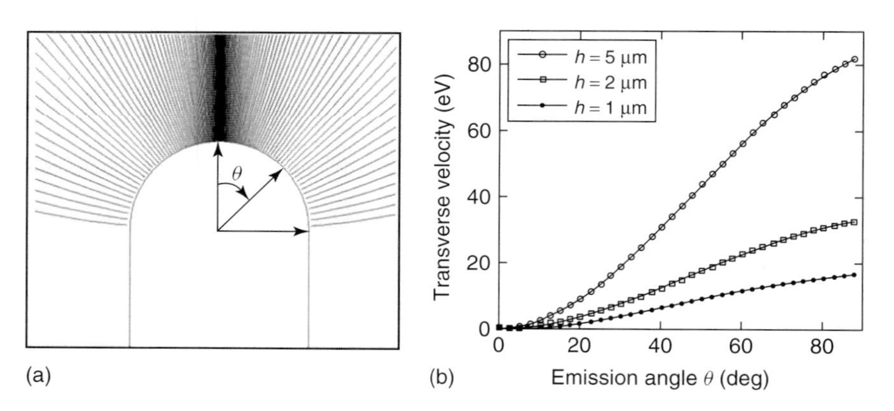

Figure 27.5 (a) Simulation of electron trajectories emitted by a CNT exhibiting an aspect ratio of 200. (b) Electron transverse velocities as a function of the position of the emitting point defined by the angle θ, for 1-, 2-, and 5-μm-high emitters.

field at the nanotube apex is constant. For the 5-μm-high nanotube, the transverse velocities attain very large values reaching up to 82 eV.

Due to the Fowler–Nordheim law, the emitted current varies exponentially with the local electric field, and thus the current emitted from points situated on the nanotube apex decreases when θ increases. To take into account this mechanism, we display on Figure 27.6a the normalized current density as a function of the transverse electron velocity. The obtained distributions are extracted from simulations using the ray-tracing software CPO [40]. One can see that the distribution is narrowed as the nanotube height is reduced. Moreover, as shown on Figure 27.6b, the mean transverse velocity decreases linearly with the nanotube height. For a 5-μm-high nanotube, the average transverse velocity is equal to 21 eV, whereas for a 1-μm-high nanotube the value becomes 4.2 eV.

The linear dependence of the electron velocities with the nanotube height can be qualitatively understood. Due to the shape of the equipotentials above the nanotube, the transverse electron velocities are acquired in an area that is very close to the nanotube apex. In this area, situated at a distance h from the substrate, the electrons are accelerated from the nanotube potential to a potential equal to roughly hE_{appl}, where E_{appl} is the applied field. Thus, in this region, the electron velocities are proportional to the nanotube height. The transverse electron velocities exhibit the same behavior.

In conclusion, 1-μm-high nanotubes exhibit transverse electron velocities five times lower than 5-μm-high nanotubes. Thus, in principle, higher convergence factors can be obtained with short nanotubes. For the same applied field, decreasing the height of the nanotubes will necessitate a proportional decrease in the nanotube radius to maintain the electric field at the nanotube apex.

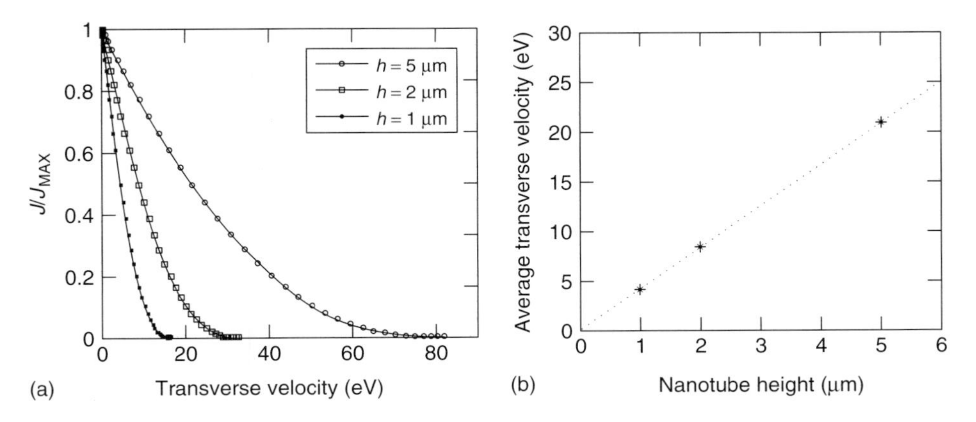

Figure 27.6 (a) Normalized current distribution emitted by nanotubes exhibiting the same aspect ratio as a function of the transversal velocity (three CNT heights are considered). (b) Average transverse velocity as a function of the nanotube height.

27.3.3
Potential of CNT Electron Guns as High Current Electron Sources

Today, the current density at the cathode level is similar to the one obtained by thermionic cathodes. However, as quoted above, CNT cathodes are characterized by large transverse electron velocities and thus only low convergence factors can be obtained.

To be used in microwave amplifiers, CNT-based electron guns will have to deliver beams with current densities above 50 A cm^{-2}. To attain such values, it will be necessary to

- improve current density at the cathode level;
- use short and thin nanotubes to reduce transverse electron velocities; and
- develop a dedicated electron gun adapted to the CNT cathode particularities.

Concerning lifetime, the few experiments performed thus far by the scientific community have confirmed the potential of CNT cathodes. However, this is still an issue, as the lifetime of microwave amplifiers is between tens of thousand hours and more than 100 000 h.

27.4
CNT Cathodes Delivering a Modulated Electron Beam

Emission from CNT cathodes can be modulated using

- the modulation of the applied electric field;
- the optical modulation of the current supplied to the CNTs; and
- the optical modulation of the electric field at nanotube apex.

27.4.1
Modulation of the Applied Electric Field

Three different approaches have been proposed. The first one consists in integrating a grid on the cathode substrate where the CNTs are deposited. The second one is to use an external grid electrode. The third one is to integrate the cathode in a resonant cavity.

27.4.1.1 Modulation of the Applied Electric Field with an Integrated Grid

For this approach, different solutions have been studied. The first one is very close to the well-known Spindt cathode [41], where one emitter per gate aperture is disposed [42, 43] (see "Spindt" cathode and CNT cathode in Figures 27.7 and 27.8). Current densities obtained with such CNT devices are typically low. In fact, CNTs are flexible. As the alignment between the nanotube and the gate aperture is rarely perfect, the CNTs are submitted to a lateral electric field and tend to touch the grid.

To fabricate cathodes delivering high current densities, research teams have focused their efforts on cathode designs that avoid any contact between the CNTs

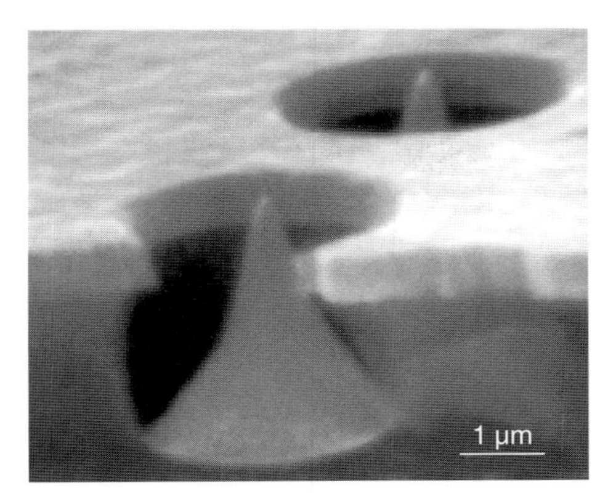

Figure 27.7 The well-known Spindt cathode [41]. (Figure reproduced with permission from the American Institute of Physics.)

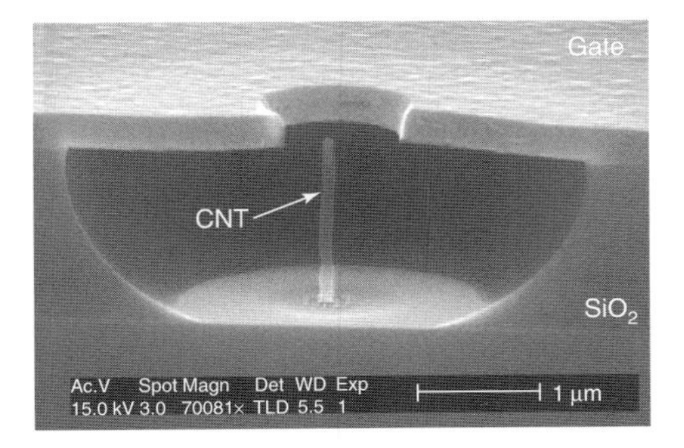

Figure 27.8 A CNT cathode with a self-aligned gate electrode [43].

and the gate. The maximum current and current density have been obtained by Hsu and Shaw [35] with an array of microgated cathodes based on multiple CNTs per open aperture in the gate (Figure 27.9).

The maximum current and current density delivered by the cathode were, respectively, 5.6 mA and 3.5 A cm^{-2}. However, these devices suffer from a high cathode–grid capacitance (C_g), typically a few nanofarads per square centimeter. As the specific transconductance (g_m) is around 0.5 S cm^{-2}, the transition frequency of such devices, which is given by $f_t = g_m/(2\pi C_g)$, is relatively low.

27.4.1.2 Modulation of the Applied Electric Field with an External Grid Electrode

This approach consists in utilizing an external grid electrode. An original configuration has been proposed by Bower *et al.* [44]. Using microelectromechanical system (MEMS) technology, they incorporated on chip the cold cathode (aligned and patterned CNTs), the grid, and the anode to create a miniature vacuum tube

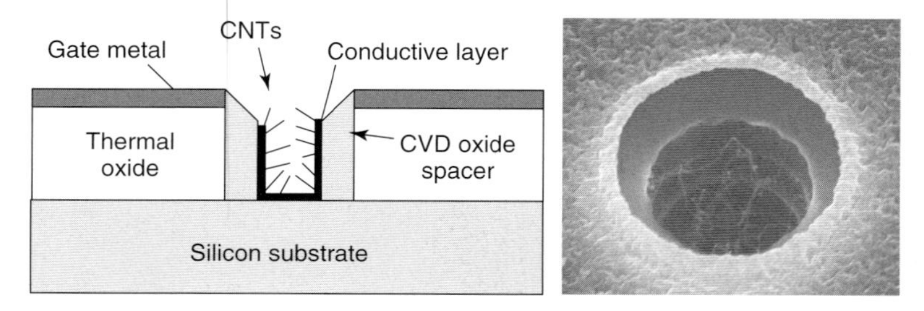

Figure 27.9 Microgated CNT cathode delivering 5.6 mA current corresponding to 3.5 A cm^{-2} current density [35]. (Figure reproduced with permission from the Institute of Electrical and Electronics Engineers (IEEE). © 2007 IEEE.)

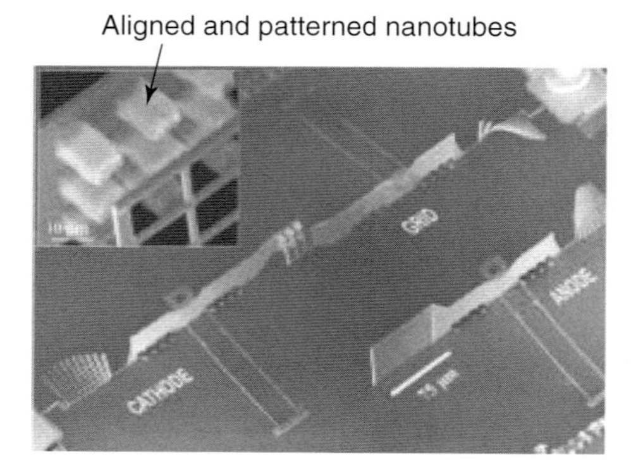

Figure 27.10 Miniature vacuum microtriode on chip fabricated using MEMS technology [44]. (Figure reproduced with permission from the Institute of Electrical and Electronics Engineers (IEEE). © 2003 IEEE.)

(Figure 27.10). They obtained a transconductance of 1.3 μS. Taking into account the cathode–grid capacitance, they estimated the transition frequency to be above 200 MHz. According to the authors, the device transconductance could be improved to attain 100 μS.

27.4.1.3 **Modulation of the Applied Electric Field with a Resonant Cavity**

The third approach consists in integrating the CNT cathode in a resonant cavity. This method has allowed the University of Cambridge and Thales to modulate a CNT cathode at 1.5 and 32 GHz.

The 1.5 GHz Diode The first modulation of a CNT cathode in the microwave regime has been performed by Teo *et al.* [36]. The cathode was integrated in a resonant cavity (Figure 27.11). The cavity height and diameter were, respectively, 40 and 100 mm. The reentrant part of the cavity was designed to ensure a resonance frequency near 1.5 GHz (1.47 GHz) and to generate the high microwave electric

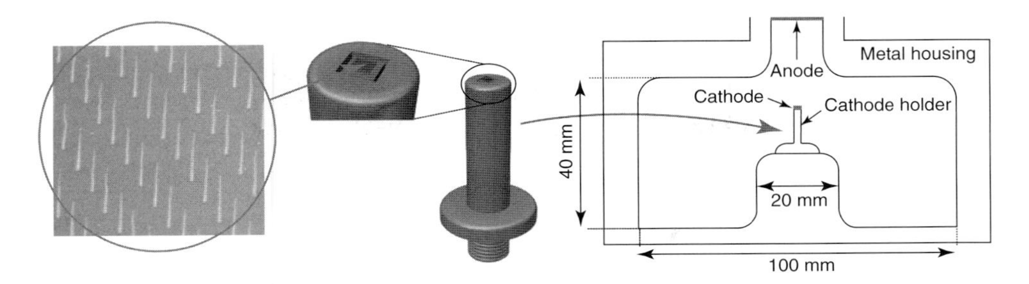

Figure 27.11 CNT cathode in a 1.5-GHz reentrant cavity.

fields required for field emission. It was composed of a 20-mm-diameter cylinder on which the 4-mm-diameter cathode holder is mounted. The total height of this reentrant part was close to the cavity height. The cathode substrate size was $1.7 \times 1.7\,\text{mm}^2$ and the emitting area was a $0.5 \times 0.5\,\text{mm}^2$ array of 5-μm-high and 50-nm-diameter CNTs with a pitch of 10 μm.

The experimental setup to measure electron emission is described in Figure 27.12. The RF electric field was produced by injecting the RF power from a 5 kW klystron. Simulations were performed to determine the electric field applied on the cathode as a function of the power dissipated in the cavity. A circulator and a water load protected the klystron against the reflected power. The delivered power was pulsed (pulse width = 0.2–1 ms and period = 100 ms) and the emitted electrons were collected by a current probe. Finally, a current amplifier measured the average current emitted during the RF excitation of the cavity. From these measurements, the peak current could be calculated.

Using this experimental setup, peak current densities up to $15\,\text{A}\,\text{cm}^{-2}$ and average current densities of $1.5\,\text{A}\,\text{cm}^{-2}$ have been demonstrated [37, 38]. Figure 27.13 shows the average current measured on the current probe as a function of the applied microwave peak electric field.

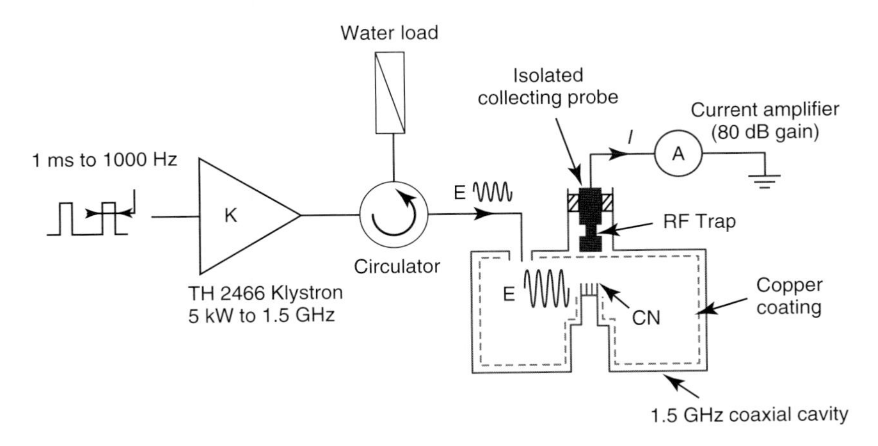

Figure 27.12 Experimental 1.5 GHz setup.

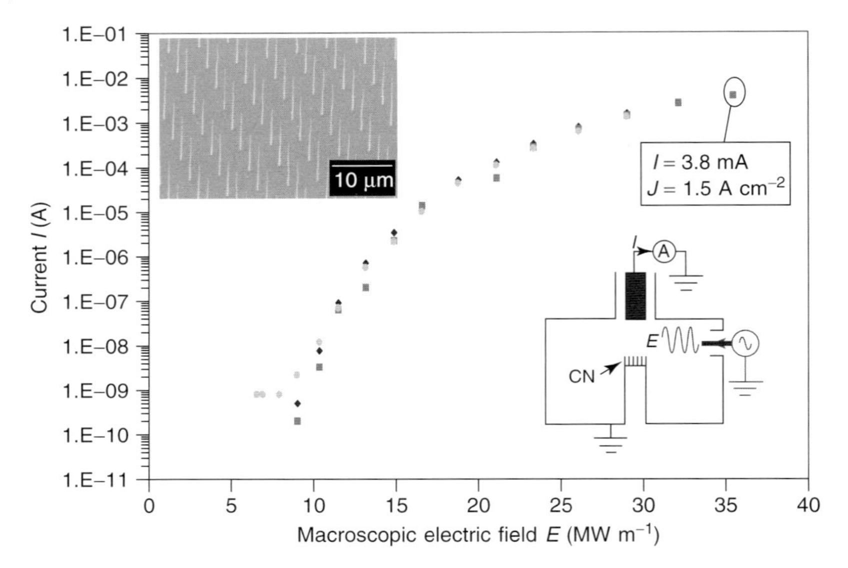

Figure 27.13 Emission current versus microwave electric field [37, 38].

The same type of cathode, which delivered a 1 A cm^{-2} current density in DC mode [21], could exhibit peak current densities of 15 A cm^{-2} in this 1.5 GHz diode. These densities could be obtained without damage, because, due to the exponential dependence of the current as a function of the applied field, the duration of the current pulses emitted by the cathode was very short (around 100 ps) [36].

The important results of this work are as follows:

- First demonstration of modulation of electron emission from CNTs in the microwave domain.
- Substantial peak current density up to 15 A cm^{-2}, with a 100% modulation ratio.

The 32 GHz Triode Using a cathode made of CNTs fabricated by the University of Cambridge, Hudanski *et al.* [38] from Thales have demonstrated a modulation of the emitted beam at 32 GHz.

The adopted structure is a triode with two reentrant cavities facing each other (Figure 27.14). It is chosen to produce a triode characterized by a resonant frequency of 16 GHz and to operate it at its second-order resonance (32 GHz).

This choice allows to enlarge the triode dimensions and to ease the device fabrication. The input and output cavity lengths are, respectively, 4.44 and 5.60 mm.

The reentrant part of the input cavity holds the cathode and that of the output cavity holds the anode. The cavities are separated by a grid which is grounded. Cathode–grid and anode–grid distances are 100 and 500 µm, respectively. Both the anode and the cathode are continuously biased with respect to the grid. For a cathode bias of 2 kV, the transit time is around 10 ps and compatible with high-frequency operation. The reentrant parts are mobile and allow one to tune

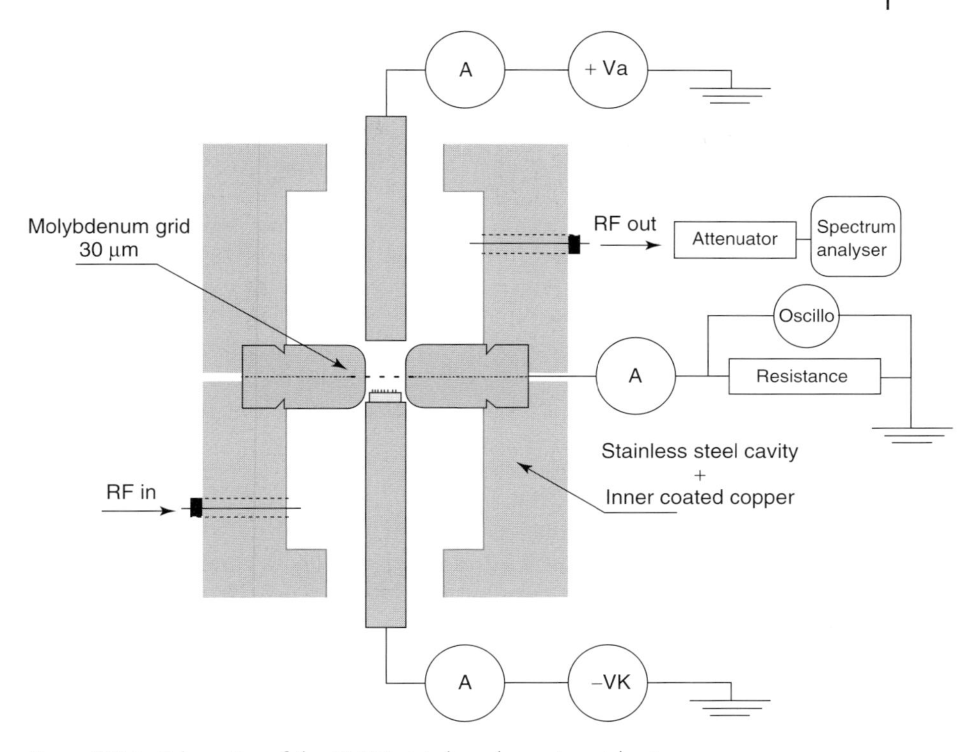

Figure 27.14 Schematics of the 32 GHz triode and experimental setup.

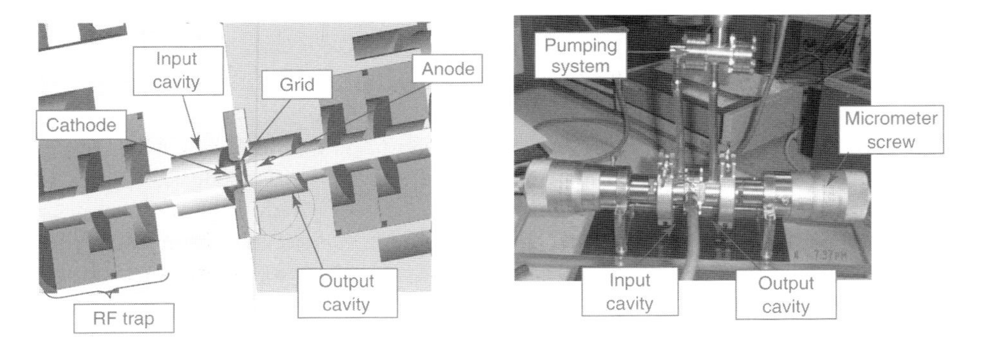

Figure 27.15 Actual 32 GHz triode.

the resonant frequencies of each cavity. Because these reentrant parts are not in contact with the cylindrical cavities, traps are used to prevent microwave leaks. Figure 27.15 shows the actual device which was fabricated, together with a cross section showing the internal parts. Vacuum is obtained by pumping through both microwave traps.

The experimental setup is shown schematically in Figure 27.14. The input microwave power is coupled to the input cavity. The resulting microwave field, enhanced in the cathode region induces a modulated emission of electrons. The beam crosses the grid and, accelerated by the anode DC bias, excites an amplified microwave field in the output cavity. The microwave power is extracted through the output port.

The cathode substrate size was $1.25 \times 1.25\,\text{mm}^2$ and the emitting area was an array of 5-μm-high and 50-nm-diameter CNTs.

In a first experiment, the grid was replaced by a disk acting as a current probe. The modulation has been demonstrated by the increase of the DC current induced by the microwave field. This current increase is due to the nonlinear characteristic of the Fowler–Nordheim emission law. With a cathode DC bias of 1920 V, a 2 mA current, corresponding to a $0.8\,\text{A}\,\text{cm}^{-2}$ current density was measured. With a superimposed 32 GHz, 8 W input power, pulsed at a duty cycle of 500 μs/60 ms, a 3.43 mA DC current was measured. From these values, a $1.4\,\text{A}\,\text{cm}^{-2}$ peak current density was calculated, with a 70% modulation ratio.

In a second experiment, a grid was used. The 32 GHz modulation of the emitted current was observed through the increase in the DC anode current induced by the microwave pulses. It was also directly observed with the output microwave power measured with a spectrum analyzer.

Two important features can be drawn from these experiments:

- The first is the demonstration of modulated emission from CNTs in the Ka band.
- The second is that the peak current density was $1.4\,\text{A}\,\text{cm}^{-2}$.

Improvement of the emitted current for a given input microwave power is expected with cavities operating at 32 GHz on their first-order resonance. Due to the reduction of the triode size, MEMS-type technology instead of traditional tube technology will have to be used.

27.4.1.4 Conclusion about the Approach Consisting in Modulating the Applied Field

In this chapter, we have compared three different approaches to modulate the electric field applied to the cathodes. The first approach consists in integrating a grid electrode on the cathode substrate, and the second one is to place an external grid in front of the cathode. However, high-frequency modulation has not been demonstrated probably because cathodes with integrated grids suffer from a high cathode–grid capacitance and those with an external grid exhibit low transconductances.

The third method consisted in integrating a CNT cathode in a resonant cavity. This approach was highly successful as 32 GHz modulation of a CNT cathode has been demonstrated. However, the use of resonant devices is only compatible with narrow bandwidth operation.

In conclusion, when high-frequency and large bandwidth amplifiers are targeted, modulating the applied field does not seem to be an adequate solution to control the emitted current. To solve this bandwidth issue, optical modulation of CNT cathodes has been proposed.

27.4.2
Optical Modulation of the Current Supplied to the CNTs

For this new approach, the applied electric field is constant and the emission is optically modulated. The CNT photocathode is an array of vertically aligned MWCNTs, each CNT being associated with one semiconducting p-i-n photodiode (Figure 27.16). Unlike conventional photocathodes, the functions of photon–electron conversion and subsequent electron emission are physically separated. Photon–electron conversion is achieved by the photodiodes and the electron emission occurs from the CNTs.

Schematically, the photodiode acts as a current source and delivers a current which is proportional to the illumination power. Thus, the frequency and bandwidth performances of these CNT photocathodes depend on the characteristics of the optical source and the photodiode structure. Due to their operating principles, laser sources as well as p-i-n photodiodes can operate from DC mode up to their cutoff frequencies [45]. Thus, in principle, the bandwidth issue is solved.

27.4.2.1 Design of a CNT Photocathode

Principle of Operation When an electric field is applied to the photocathode, the p-i-n diodes are reverse biased. Upon illumination by photons with energy exceeding the band gap of the semiconductor, electron–hole pairs are generated in or near the depletion region. Due to the electric field present in this region, the carriers are separated and electrons reaching the n-type electrode are subsequently emitted. Thus the electron current is proportional to the illumination power absorbed in the active area of the photodiode.

Figure 27.17 describes schematically the working principle of an element (p-i-n/CNT) of the photocathode. It shows the current emitted by the CNT (I_{CNT}) and the photodiode current (I_{diode}) as a function of the voltage drop that appears across the diode (ΔV). Under dark conditions (no illumination), the photodiode current corresponds to the diode leakage current which increases as a function of the diode bias, particularly at high biases where the avalanche regime takes place. Under illumination, the photodiode current is equal to $I_{OFF\ diode}$ plus a photogenerated current ($I_{ph\ diode}$).

1) **Steady state: ON or OFF state:** In ON or OFF state, the emitted current is equal to the photodiode current. Thus for each state, this graphical method allows us

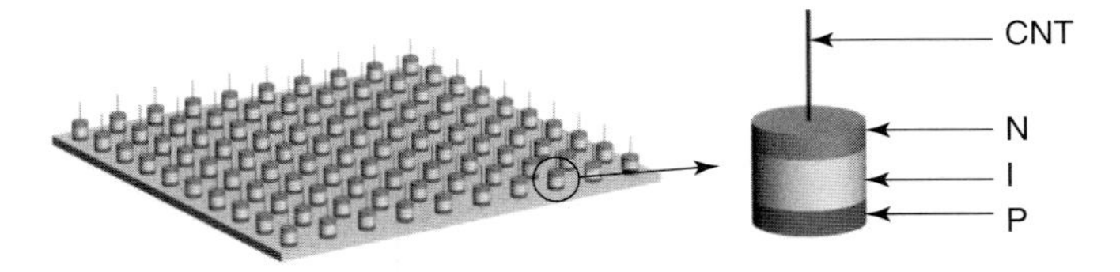

Figure 27.16 Schematic view of a CNT photocathode.

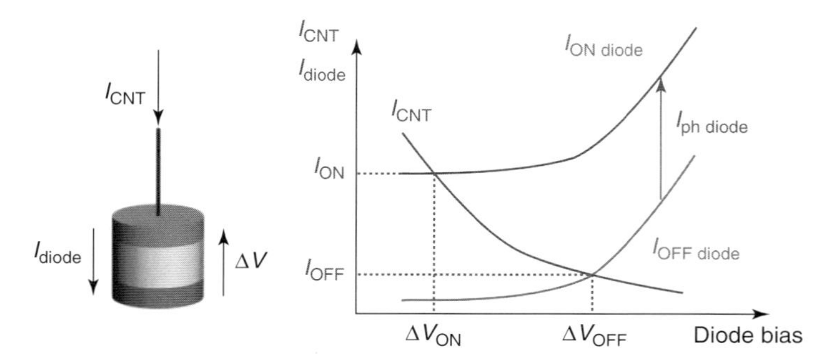

Figure 27.17 Schematic description of the working principle of an element (p-i-n/CNT) of the photocathode.

to determine both the emitted current and the diode bias. Then, the ON/OFF ratio for the current emitted by the photocathode can be graphically calculated.

2) **Transition from ON to OFF state:** When the light is turned off, the photodiode current becomes very small and equal to its leakage current ($I_{\text{OFF diode}}$). The electrons previously accumulated at the CNT apex are still field-emitted. The emission current being higher than the diode current, the electron density at the CNT apex decreases and the local electric field decreases also. Thus the emitted current decreases. During this transition, the emission current has been higher than the photodiode current. Consequently the bias of the n-type electrode of the p-i-n structure has increased positively compared to the substrate. Due to the increase of the diode bias, the diode leakage current also increases. Thus, during the ON–OFF transition, the emitted current decreases and the diode current increases. These two currents adjust themselves until they become equal.
3) **Photocathode design:** From this schematic description, one can deduce that a high ON/OFF ratio implies that high voltage biases can be applied to the diodes. It is thus particularly important to determine the impact of the ΔV value on the local field at the CNT apex and thus on the emission current.

In conclusion, it is of prime importance to design accurately the photocathode in order that

- in the OFF state, the voltage drop (ΔV) stays below the diode voltage breakdown;
- in the ON state, the p-i-n photodiodes delivers the required currents;
- the ON/OFF transition time is compatible with the targeted frequency.

Design of the Photocathode: OFF State To evaluate the impact of the voltage drop in the photodiode (ΔV) on the emission current, a design leading to an I_{ON}/I_{OFF} ratio of 100 is presented below. Such a high ratio is not required for the targeted application but this allows one to estimate the impact of the different parameters.

We consider an array of 5-μm-high MWCNTs with a pitch of 10 μm. According to Minoux *et al.* [25], such CNTs emit a current of 20 μA when the local field at the nanotube apex is around 7000 V μm^{-1}. When an array of CNTs field-emits, only the emitters presenting the best field enhancement factors contribute effectively to the total current. Figure 27.18 shows the emitted current as a function of the applied field for a $0.5 \times 0.5\ mm^2$ CNT array made of 5-μm-high and 50-nm-diameter CNTs [46]. The enhancement factor deduced from the emission curve is 350 and the maximum current density is equal to $1\ A\ cm^{-2}$. One can also see that the current reduces 100 times (from 2 to 0.02 mA) when the applied field is reduced by 30%.

To summarize, in the array, a 5-μm-high CNT with an enhancement factor of 350 emits 20 μA current when it is submitted to an applied field of 20 V μm^{-1}. A 30% decrease of this field causes the current to drop to 0.2 μA. We will use these values as a starting point.

Figure 27.19 shows a configuration where a full sheet p-i-n diode is used to modulate the emission. In this example, the anode is placed 100 μm from the cathode and is biased at 2 kV to obtain an electric field of 20 V μm^{-1}. For this geometry, a very high diode bias of 600 V is necessary to reduce the applied field by 30%. Such a high voltage is much larger than the breakdown voltage of high-frequency p-i-n diodes.

To overcome this problem, it is proposed to localize the p-i-n diodes under the CNTs. According to Minoux [25], when a voltage drop appears under a CNT, the

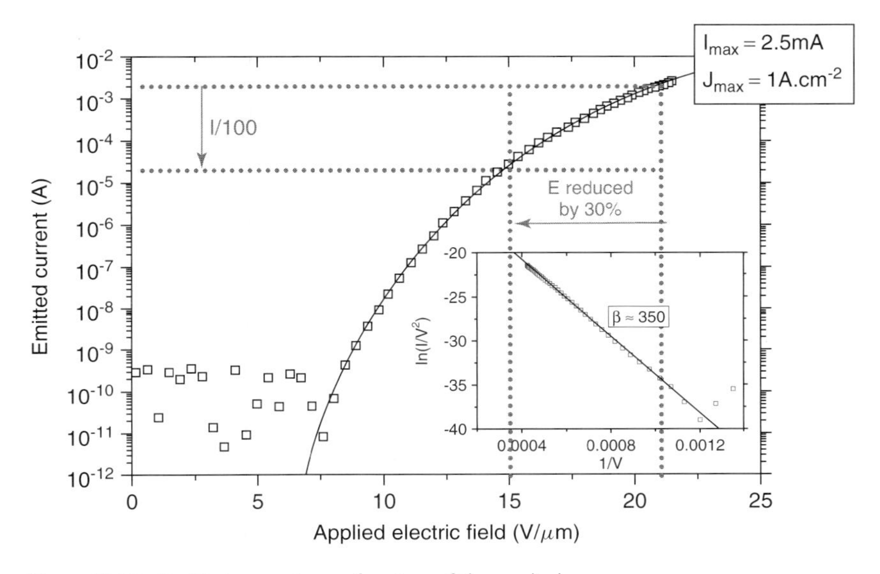

Figure 27.18 Emitted current as a function of the applied field for a $0.5 \times 0.5\ mm^2$ CNT array made of 5-μm-high and 50 nm CNTs [46].

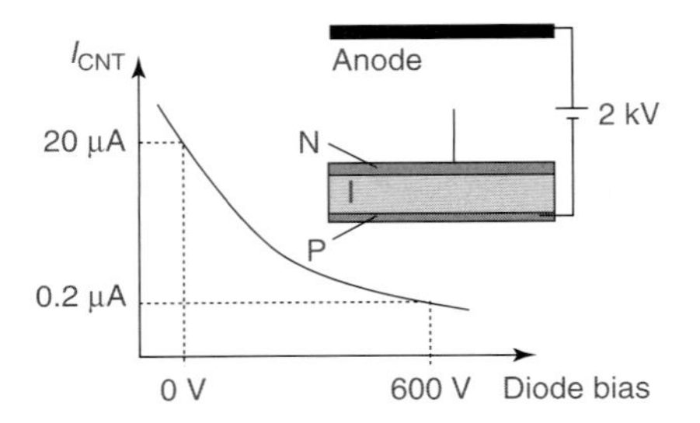

Figure 27.19 Current emitted by a 5-μm-high CNT as a function of the diode bias for a full sheet p-i-n diode.

effective enhancement factor obeys the following law:

$$\beta = \beta_0 \left(1 - \alpha \frac{\Delta V}{h E_{\text{appl}}}\right) \tag{27.1}$$

where β_0 is the enhancement factor without voltage drop (for $\Delta V = 0\,\text{V}$), E_{appl} is the applied field (here $20\,\text{V}\,\mu\text{m}^{-1}$), h is the nanotube height (here $5\,\mu\text{m}$), and α is a constant ($\alpha = 0.9$ for 5-μm-high CNTs). Using this equation and the Fowler–Nordheim equation, the emitted current can be schematically plotted as a function of the diode bias (Figure 27.20).

One can see that, for a localized diode, a small voltage drop induces a large reduction of the emitted current. For a 5-μm-high CNT, a diode bias of 33 V induces a reduction of the emitted current of two orders of magnitude. The explanation is shown on Figure 27.20: this voltage drop induces a local reconfiguration of the equipotentials, which reduces significantly the local field at nanotube apex.

From these figures, one can deduce that it is of prime importance to localize the p-i-n photodiodes under the CNTs. This localization allows the reconfiguration of the equipotential surfaces as shown on Figure 27.20. Typically, the photodiode diameter has to be equal to or smaller than the nanotube height. The current modulation is highly efficient as it uses an optically controlled reconfiguration of the electric field at the CNT locations.

In conclusion, for 5-μm-high CNTs, an ON/OFF ratio of two orders of magnitude implies that the onset of avalanche breakdown appears above a 33 V reverse bias.

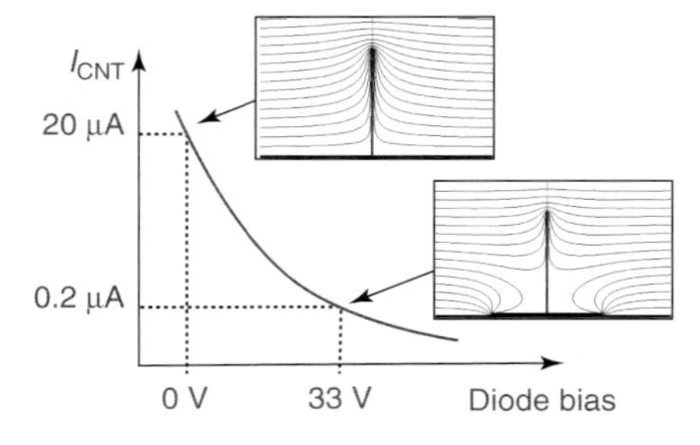

Figure 27.20 Current emitted by a 5-μm-high CNT as a function of the diode bias for localized p-i-n diodes.

For most semiconductors, this is compatible with a thickness of the intrinsic layer of a few micrometers.

One can note that the use of short and thin nanotubes is advantageous. In fact, according to Eq. (27.1), the voltage drop is proportional to the nanotube height. For a total p-i-n diode plus nanotube height of 2 μm, the voltage drop can be estimated to be only 13 V.

Design of the Photocathode: ON State TWTs operate with cathodes delivering between 10 and 200 mA. These currents are attainable with CNT photocathodes. For example, assuming a p-i-n responsivity of $0.8\,\mathrm{A\,W^{-1}}$ and an effective surface ratio (p-i-n surface divided by cathode surface) of 25%, one can estimate a 80 mA ON current with a 400 mW laser source (Figure 27.21).

This evaluation does not take into account the ratio of effective CNT/p-i-n elements. Generally, only a fraction of CNTs in an array (those with highest βs) contribute to the total current. In the case of a photocathode, the photodiode acts as a current saturator and thus a larger fraction of nanotubes should emit.

Design of the Photocathode: ON/OFF Transition Time Photodiodes used in telecommunications in optical fibers operate at frequencies up to 40 GHz. For CNT photocathodes, the main limitation will be the transit time in the intrinsic region and the charging time of the photodiode capacitance.

For a thickness of the intrinsic region around 0.8 μm, the transit time is compatible with cutoff frequencies above 10 GHz. In these conditions, the overall cutoff frequency is determined mainly by the charge and discharge of the PIN diode capacitance through the transconductance characterizing the emission process. This transconductance can be expressed by the ratio:

$$g_{\mathrm{m}} = \mathrm{d}I_{\mathrm{CNT}}/\mathrm{d}V_{\mathrm{PIN}} \tag{27.2}$$

From Figure 27.18, one can deduce that, for high delivered currents, a variation of 10% in the electric field at the nanotube apex induces a variation by a factor of 3 of the cathode emitted current. From this measure, one can deduce that a 10% variation of the applied field can modulate the beam emitted by individual nanotubes between 10 and 30 μA. Under these conditions, the average current is 20 μA and the modulation depth is 50%.

Assuming a p-i-n/CNT height of 2 μm and an applied field of $20\,\mathrm{V\,\mu m^{-1}}$, a 10% variation of the applied field implies (Eq. (27.1)) that the corresponding voltage variation in the photodiode is of the order of 4 V. Thus the transconductance is equal to 5 μS. For a 1-μm-diameter p-i-n photodiode, the capacitance is roughly

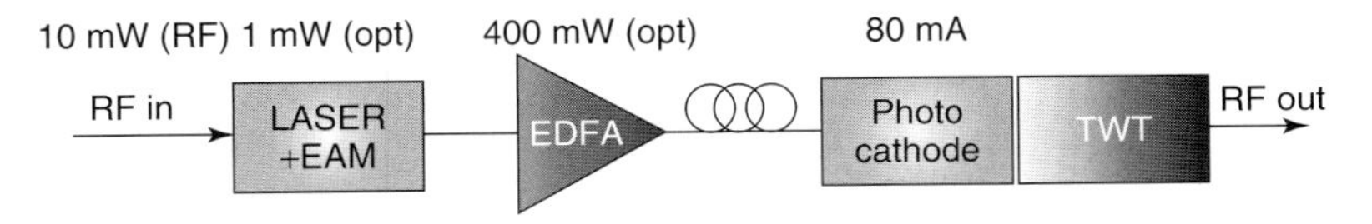

Figure 27.21 Optical control of a CNT photocathode delivering 80 mA.

0.8×10^{-16} F. Under these conditions, the transition frequency, which is equal to $g_m/2\pi C$, is close to 10 GHz.

Optimized designs allow us to increase this transition frequency, for example, by using uni-traveling carrier (UTC) photodiodes [47–49]. UTC photodiodes with a 1-μm-thick absorbing layer allows bandwidths above 20 GHz. As their saturation current is high (of the order of $400\,\mu A\,\mu m^{-2}$), the diameter of such photodiodes can be reduced. The result is a decrease of the p-i-n capacitance and an increase of the transition frequency.

27.4.2.2 **Demonstration of a 300 MHz CNT Photocathode**

To demonstrate this new photocathode concept, Hudanski *et al.* [50] have fabricated and tested a CNT photocathode based on silicon p-i-n diodes (Figure 27.22). The photocathode is an array of MWCNTs in which each MWCNT is electrically connected to a localized n+ doped area defined in a 3-μm-thick intrinsic layer grown on a p+ doped semiconducting substrate. Due to the localization of the n+ areas, the p-i-n diodes are independent of each other. Compared to the schematic description shown in Figure 27.16, several CNTs appear on each diode and the diodes are not physically localized. Both modifications were made to ease the fabrication process. For the first one, the impact is limited, as the highest CNT will emit most of the current. The second modification implies that the diameter of the depleted region will increase as a function of the voltage bias that appears across the diode.

The CNT photocathode was tested in a triode-type configuration. The cathode–grid distance was 100 μm and the anode was located a few centimeters away from the cathode. A 532 nm wavelength optical source was used to illuminate the photocathode. At this wavelength, the 3-μm-thick intrinsic layer absorbs 98% of the laser power. Figure 27.23 shows the emitted current as a function of the applied voltage and of the absorbed optical power. For emitted currents below 1 μA, the emitted current follows the classical Fowler–Nordheim behavior. At higher voltages, the curves deviate from this law and exhibit different saturations,

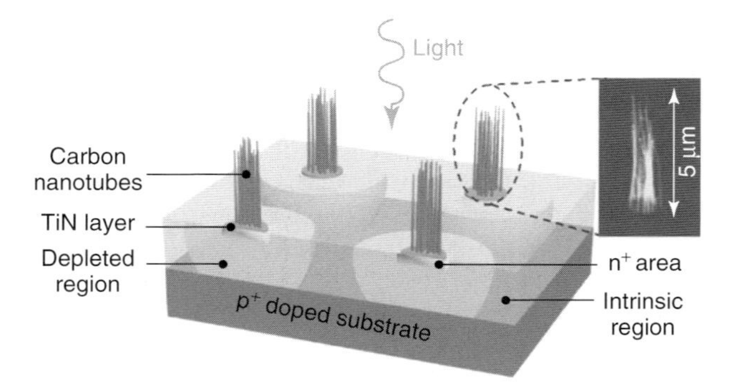

Figure 27.22 Schematic view of the CNT photocathode fabricated on a silicon substrate. (Figure reproduced with permission from IOP Publishing, UK.)

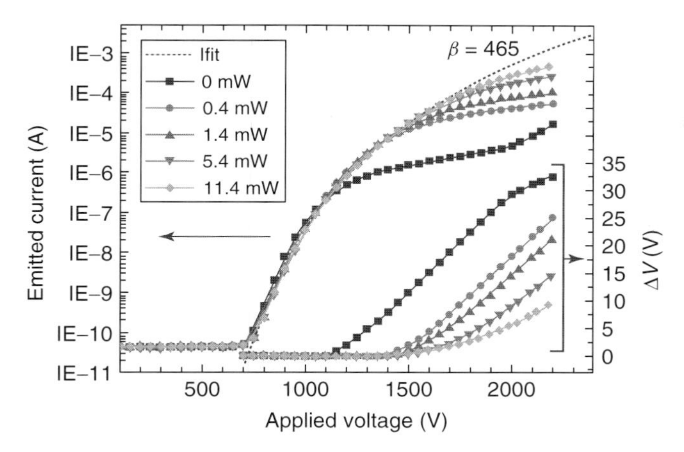

Figure 27.23 Emitted current and diode biases as a function of the applied cathode–grid voltage and absorbed optical powers [50]. (Figure reproduced with permission from IOP Publishing, UK.)

corresponding to the different absorbed optical powers. Using Eq. (27.1), the voltage drop ΔV developed across the p-i-n diode was calculated. It is plotted as a function of the applied voltage in Figure 27.23 (right-hand axis). This photocathode delivered a current ranging from 16 μA (0.006 A cm^{-2}) to 500 μA (0.2 A cm^{-2}) as the absorbed optical power is varied from 0 to 11.8 mW.

Under dark conditions and for applied voltages above 2000 V, the voltage drop exceeds 30 V. Due to the field in the intrinsic region of the photodiodes (above 10 V μm^{-1}), the onset of the avalanche regime appears. This explains the increase of the off current. Due to this phenomenon, for an applied voltage of 2200 V, the ON/OFF ratio was around 30 instead of 100.

Using a 658 nm wavelength laser source, the photocathode was experimentally operated at frequencies up to 300 MHz. Figure 27.24 shows the frequency response of the anode current (continuous curve) and of the used laser source (dotted curve). One can note that the bandwidth of the photocathode is the same as that of the laser bandwidth. Thus the high-frequency electron beam modulation of this silicon photocathode was limited by the laser source used.

27.4.2.3 **Development of High-Frequency CNT Photocathodes for Microwave Amplifiers**

The development of high-frequency CNT photocathodes requires the use of high-frequency modulated lasers. Due to the development of telecommunications in optical fibers, very high bandwidth 1.55-μm laser sources are available.

For this wavelength, InGaAs-based photodiodes have to be used. As they are fabricated on InP substrates which are transparent at 1.55 μm, they can be illuminated from the back. This is a major advantage because it eases their integration in electron tubes.

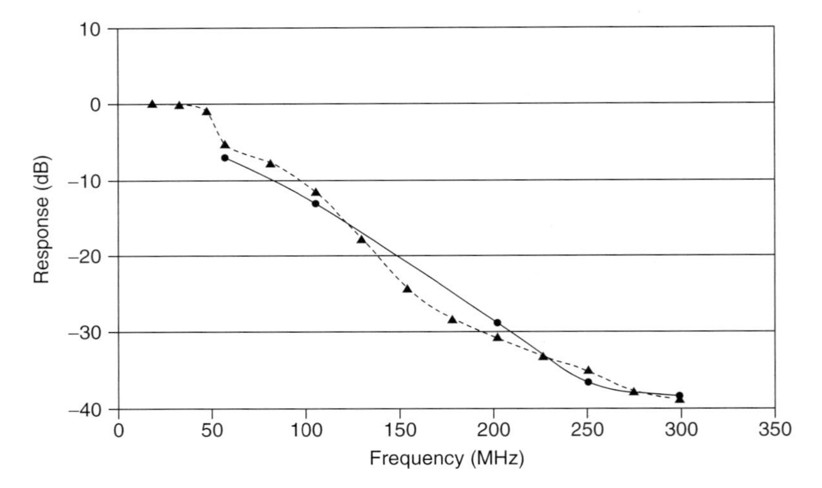

Figure 27.24 Comparison between the bandwidth of the silicon photocathode (continuous curve) and the laser source (dotted curve).

A schematic fabrication process (Figure 27.25) begins with the epitaxial growth of an intrinsic InGaAs layer and an n+-doped InP film on a p+-doped InP substrate. Then an array of p-i-n mesas is etched. A passivation layer is deposited to prevent leakage current along the mesa edges. This passivation layer is opened on top of the photodiodes to deposit the catalyst for CNT growth. A schematic cross section of the device is shown in Figure 27.25.

The growth of vertically aligned CNTs on III–V materials remains a big challenge. As the typical growth temperature for vertically aligned CNTs is around 650–700 °C, some interdiffusion of In, Ga, or P or dopant may occur. To prevent these mechanisms, we have developed a new water-based growth process where H_2O is employed [51]. The growth temperature is then reduced to 550 °C and CNTs are grown without any appreciable degradation of the photodiodes (Figure 27.26).

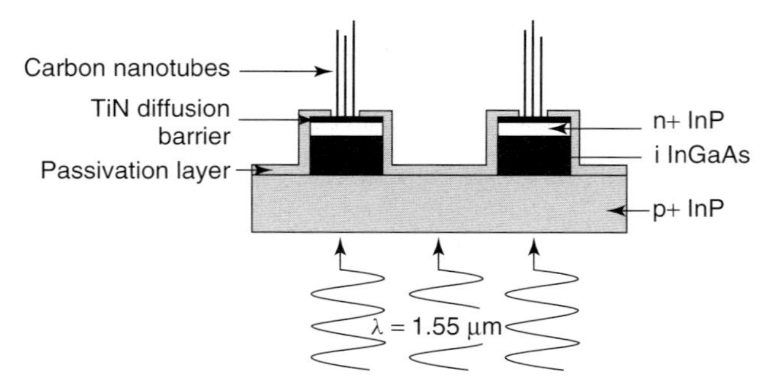

Figure 27.25 Schematic cross section of carbon nanotubes on InP–InGaAs–InP photodiodes. This structure is adapted to illumination from the back.

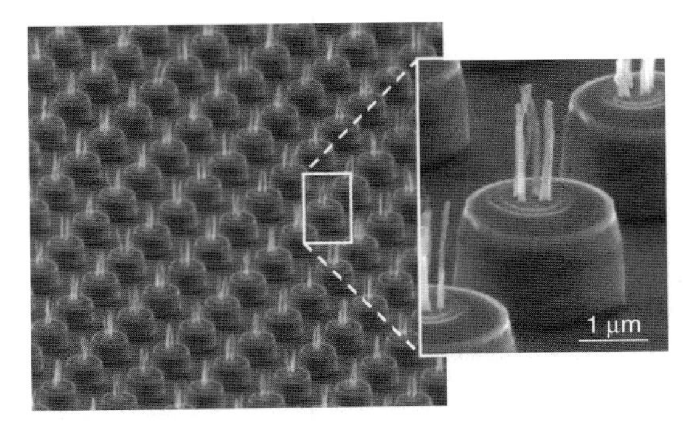

Figure 27.26 SEM picture of a CNT photocathode based on InGaAs photodiodes.

27.4.3 Optical Modulation of the Electric Field at Nanotube Apex: THz Cathodes

The terahertz part of the electromagnetic spectrum (0.3–10 THz) is of great interest in many fields such as space, security, medicine, and biology. For instance, it provides new ways to detect weapons in crowded areas, to image cancerous tissues, or to investigate the presence of drugs and explosives [52, 53]. Up to now, different terahertz sources have been developed (quantum cascade lasers, vacuum electronic devices, photomixing), but their output powers are still limited to a few milliwatts.

Our approach proposes to increase powers delivered in this spectral region by increasing the gain of traveling-wave tube amplifiers (TWTAs). This improvement is based on a premodulation of the electronic beam directly at the cathode, well known to enhance the TWTA efficiency [54]. To obtain such a modulation of the beam in the terahertz range, the CNT array is coupled to a terahertz dual-frequency laser operating at telecom wavelengths [55]. A mixing component is then generated in the emitted current thanks to the beating signal of the laser and the nonlinear I–V behavior of the field emission mechanism [56, 57]. The efficient driving of the cathode is then achieved with the integration of the CNTs on specific metallic 1D gratings as presented on the cathode pattern in Figure 27.27a.

This structure couples the incident laser light to surface plasmons polaritons (SPPs) by diffraction. As these surface waves present a polarization aligned with the CNTs, antenna coupling is then expected to occur [56, 58].

Acting like a photonic band gap medium for SPPs [58], this grating couples the incident laser light with a surface mode which localizes the incident energy around each field emitter. A numerical optimization of this concentration has been performed using a 2D dispersive finite difference in time domain code. The computation corresponding to the spatial distribution of the localized mode is given in Figure 27.27. Interestingly, the electric field is enhanced by factors of 10–20 at the center of the grating where the emitters are located.

A first performance assessment shows that, due to the evanescent nature of SPPs, this cathode should advantageously integrate relatively small nanotubes (~1 μm),

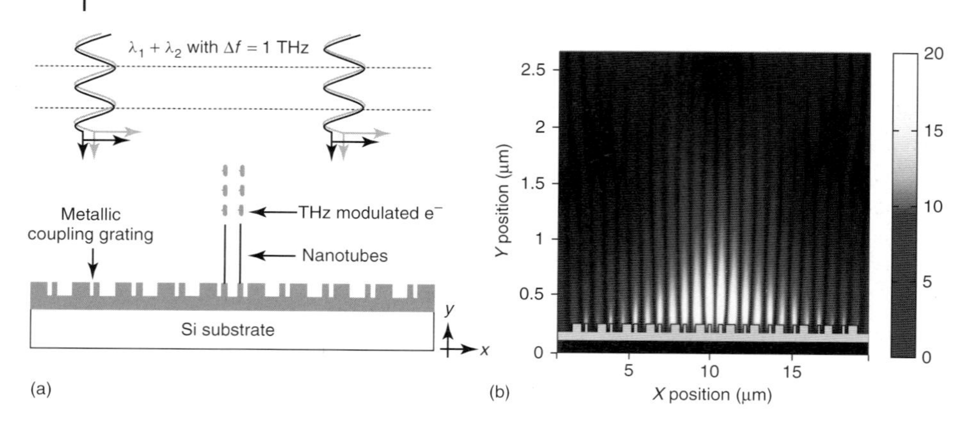

Figure 27.27 (a) Scheme of the terahertz cathode pattern (b) numerical map of the normal electric field (E_y) in the vicinity of the grating surface.

which also reduces beam divergence (Section 27.3.2.2). Integration of CNTs on the metallic gratings is currently being studied by Thales [59].

27.5 Conclusion

Compared to SSAs, vacuum microwave amplifiers offer higher powers, higher frequency operation, and higher power yields but are bulky and heavy. It has been known for a long time that the integration of cathodes delivering a modulated electron beam is a solution to reduce the size and weight of these tubes. For this purpose, FEAs have been extensively studied.

The most well-known field emission cathode is the Spindt cathode, which exhibits excellent performance in terms of current densities. Such a cathode delivering a current density of 15.4 A cm^{-2} was recently operated in a 5 GHz TWT delivering 100 W output power [13]. Compared to TWTs based on thermionic cathodes, this tube offers instant turn-on and easy pulsed operation. However, as the cathode is operated in a "nonmodulated" mode, the TWT remains heavy and bulky. TWTs based on Spindt cathodes and operating with a modulated beam have not delivered such high powers. In fact, Spindt cathodes present a high cathode–grid capacitance, and the microwave operation of such cathodes has been difficult to implement.

Since the first field emission experiments with CNTs performed in 1995 [14], uniform arrays of vertically aligned MWCNTs have been extensively studied for field emission applications [16–21]. MWCNTs are arguably the most stable and robust cold field emitters known [22, 23]. Due to these intrinsic characteristics, a stable and continuous emission current of 100–200 μA per individual MWCNT has been demonstrated [24, 25].

To review the performances of CNT cathodes as electron sources for microwave amplifiers, we have evaluated the capabilities of CNT cathodes to deliver high current densities and to emit an electron beam that can be modulated at high frequency.

CNT cathodes deliver current densities around a few amperes per square centimeter, that is, the same current densities as delivered by M-type thermionic cathodes used in TWTs for space applications. However, the beam emitted by cathodes having 5-μm-high CNTs exhibit high transverse electron velocities and thus is difficult to focus. To reduce significantly these transverse electron velocities, arrays of short (~1 μm) and thin (~10 nm diameter) nanotubes have been proposed. With such cathodes and with a dedicated electron gun (whose design is still to be found), one can anticipate that electron beams presenting current densities above 50 A cm^{-2} will be obtained in the next few years.

The major advantage of cold cathodes compared to thermionic cathodes is their ability to deliver a beam modulated at high frequency. For this purpose, research groups have studied three different approaches to modulate the emitted beam: the modulation of the applied electric field, the optical modulation of the current supplied to the CNT, and the optical modulation of the electric field at nanotube apex.

The modulation of the applied field has been performed using an integrated gate electrode (see e.g., [35]) or an external grid electrode using MEMS technology [44]. However, in the first case the cathode–grid capacitance is high, and in the second case the transconductance is low. Thus the cutoff frequency of fabricated devices was below 1 GHz.

To allow high-frequency modulation of the emitted beam, the use of resonant cavities has been proposed and the modulation of a CNT cathode at 32 GHz has been demonstrated (Section 27.4.1.3, The 32 GHz Triode). This is the first demonstration of modulated emission from CNTs in the Ka band.

Although the use of resonant cavities for narrow bandwidth amplifiers appears as an adequate solution, it is not adapted for TWTs that are used for large bandwidth applications such as telecommunications. To solve this bandwidth issue, two new approaches have been proposed: the optical modulation of the current supplied to the CNTs, and the optical modulation of the electric field at nanotube apex.

For the first approach, the applied electric field is constant and the emission is optically modulated. The CNT photocathode is an array of vertically aligned MWCNTs, each CNT being associated with one semiconducting p-i-n photodiode. Schematically, the photodiode acts as a current source and delivers a current which is proportional to the illumination power. Thus, the frequency and bandwidth performances of these CNT photocathodes depend on the characteristics of the optical source and of the photodiode structure. Due to their operating principles, laser sources as well as p-i-n photodiodes can operate from DC mode, up to their cutoff frequencies [45]. Thus, the bandwidth issue is solved. The optical control is particularly interesting, as all the p-i-n photodiodes are operated in parallel. In this mode, no RF input signal attenuation or dephasing occurs between the different CNT emitters. One can note that it is a major problem when electrical control of cold cathodes, in particular gated cathodes, is performed.

The first CNT photocathode has been demonstrated [50] and the development of high-frequency photocathodes is in progress (Section 27.4.2). Thus, using these photocathodes as modulated electron sources, the development of compact, light, and highly efficient TWTs can be envisaged in the future. The maximum operating frequency will depend on the performances of laser sources and photodiodes that are developed for telecommunications on optical fibers. This frequency is today around a few tens of gigahertz.

For very high-frequency operation and particularly in the terahertz regime, the direct coupling of an incident laser light with a CNT array has been proposed (Section 27.4.3). To obtain a modulation of the beam in the terahertz range, a terahertz dual-frequency laser can be used. A mixing component is then generated in the emitted current thanks to the beating signal of the laser and to the nonlinear $I-V$ behavior of the field emission mechanism [56, 57]. The efficient driving of the cathode is achieved with the integration of the CNTs on specific metallic 1D gratings. These gratings couple the incident laser light with a surface mode which localizes the incident energy around each field emitter. The computations corresponding to the spatial distribution of the localized mode have shown that the electric field can be enhanced by factors of 10–20. Although this approach is highly ambitious, it could allow the development of terahertz amplifiers delivering large powers, that is, $\geq$100 mW instead of a few milliwatts.

References

1. Pierce, J.R. (1940) *J. Appl. Phys.*, **11**, 548–554.
2. Zalm, P. and van Stratum, A.J.A. (1966) *Philips Tech. Rev.*, **27**, 69.
3. Pierce, J.R. (1949) *Theory and Design of Electron Beams*, D. Van Nostrand Company, Inc., New York.
4. Haeff, A.V. (1939) *Electron*, **12**, 30–32.
5. Lichtenberg, A.J. (1962) *IRE Trans. Electron Devices*, **ED-9**, 345–351.
6. Spindt, C.A. *et al.* (1976) *J. Appl. Phys.*, **47** (12), 5248–5263.
7. Spindt, C.A. (1998) *J. Vac. Sci. Technol. B*, **16**, 758–761.
8. Zaidman, E.G. and Kodis, M.A. (1991) *IEEE Trans. Electron Devices*, **38**, 2221–2228.
9. Calame, J.P., Gray, H.F., and Shaw, J.L. (1993) *J. Appl. Phys.*, **73**, 1485–1504.
10. Jensen, K.L. (1998) *J. Appl. Phys.*, **83**, 7982–7992.
11. Jensen, K.L., Abrams, R.H., and Parker, R.K. (1998) *J. Vac. Sci. Technol. B*, **16**, 749–753.
12. Whaley, D.R., Gannon, B.M., Heinen, V.O., Kreischer, K.E., Holland, C.E., and Spindt, C.A. (2002) *IEEE Trans. Plasma Sci.*, **30** (3), 998–1008.
13. Whaley, D.R., Duggal, R., Armstrong, C.M., Bellew, C.L., Holland, and Spindt, C.A. (2009) *IEEE Trans. Electron. Devices*, **56**, 896–905.
14. Rinzler, A.G., Hahner, J.H., Nikolaev, P., Lou, L., Kim, S.G., Tomanek, D., Nordlander, P., Colbert, D.T., and Smalley, R.E. (1995) *Science*, **269**, 1550–1553.
15. De Heer, W., Châtelain, A., and Ugarte, D. (1995) *Science*, **270**, 1179–1180.
16. Ren, Z.F., Huang, Z.P., Xu, J.W., Wang, J.H., Bush, P., Siegal, M.P., and Provencio, P.N. (1998) *Science*, **282**, 1105.
17. Fan, S., Chapline, M.G., Franklin, N.R., Tomber, T.W., Cassell, A.M., and Dai, H. (1999) *Science*, **283**, 512.
18. Teo, K.B.K., Chhowalla, M., Amaratunga, G.A.J., Milne, W.I., Hasko, G., Pirio, G., Legagneux, P., Wyczisk, F., and Pribat, D. (2001) *Appl. Phys. Lett.*, **79**, 1534.

19. Teo, K.B.K., Chhowalla, M., Amaratunga, G.A.J., Milne, W.I., Pirio, G., Legagneux, P., Wyczisk, F., Pribat, D., and Hasko, G. (2002) *Appl. Phys. Lett.*, **80**, 2011.
20. Teo, K.B.K. *et al.* (2003) *Nanotechnology*, **14**, 204.
21. Milne, W.I., Teo, K.B.K., Minoux, E., Groening, O., Gangloff, L., Hudanski, L., Schnell, J.-P., Dieumegard, D., Peauger, F., Bu, I.Y.Y., Bell, M.S., Legagneux, P., Hasko, G., and Amaratunga, G.A.J. (2006) *J. Vac. Sci. Technol., B*, **24**, 345–348.
22. Purcell, S.T., Vincent, P., Journet, C., and Binh, V.T. (2002) *Phys. Rev. Lett.*, **88** (10), 105502.
23. De Jonge, N., Allioux, M., Doytcheva, M., Oostveen, J.T., Teo, K.B.K., and Milne, W.I. (2005) *Appl. Phys. Lett.*, **87**, 133118.
24. Bonard, J.-M., Maier, F., Stockli, T., Chatelain, A., De Heer, W.A., Salvetat, J.-P., and Forro, L. (1998) *Ultramicroscopy*, **73**, 7.
25. Minoux, E. *et al.* (2005) *Nano Lett.*, **5** (11), 2135–2138.
26. Oh, S.J., Zhang, J., Cheng, Y., Shimoda, H., and Zhou, O. (2004) *Appl. Phys. Lett.*, **84** (19), 3738–3740.
27. Choi, W.B., Chung, D.S., Kang, J.H., Kim, H.Y., Jin, Y.W., Han, I.T., Lee, Y.H., Jung, J.E., Lee, N.S., Park, G.S., and Kim, J.M. (1999) *Appl. Phys. Lett.*, **75**, 3129.
28. Calderon-Colon, X., Geng, H., Gao, B., An, L., Cao, G., and Zhou, O. (2009) *Nanotechnology*, **20**, 325707.
29. Bower, C., Zhu, W., Jin, S., and Zhou, O. (2000) *Appl. Phys. Lett.*, **77**, 830.
30. Meyyappan, M., Delzeit, L., Cassell1, A., and Hash, D. (2003) *Plasma Sources Sci. Technol.*, **12**, 205–216.
31. Groening, O., Kuettel, O.M., Emmenegger, Ch., Groening, P., and Schlapbach, L. (2000) *J. Vac. Sci. Technol., B*, **18**, 665.
32. Nilsson, L., Groening, O., Emmenegger, C., Kuettel, O., Schaller, E., Schlapbach, L., Kind, H., Bonard, J.-M., and Kern, K. (2000) *Appl. Phys. Lett.*, **76**, 2071.
33. Teo, K.B.K., Lee, S.B., Chhowalla, M., Semet, V., Binh, V.T., Groening, O., Castignolles, M., Loiseau, A., Pirio, G., Legagneux, P., Pribat, D., Hasko, D.G., Ahmed, H., Amaratunga, G.A.J., and Milne, W.I. (2003) *Nanotechnology*, **14**, 204.
34. Chen, Z., Zhang, Q., Lan, P., Zhu, B., Yu, T., Cao, G., and den Engelsen, D. (2007) *Nanotechnology*, **18**, 265702.
35. Hsu, D.S.Y. and Shaw, J.L. (2007) Open aperture microgated carbon nanotube FEAs, IEEE 20th International Vacuum Nanoelectronics Conference, IVNC, 8–12 July 2007, pp. 80–81.
36. Teo, K.B.K., Minoux, E., Hudanski, L., Peauger, F., Schnell, J.-P., Gangloff, L., Legagneux, P., Dieumegard, D., Amaratunga, G., and Milne, W.I. (2005) *Nature*, **437**, 968.
37. Legagneux, P., Minoux, E., Hudanski, L., Teo, K., Griening, O., Peauger, F., Dieumegard, D., Schnell, J.P., Gangloff, L., Amaratunga, G.A.J., and Milne, W.I. (2005) GHz modulation of carbon nanotube cathodes for microwave amplifiers. Proceedings of 2005 5th IEEE Conference on Nanotechnology, 11–15 July, 2005, Nagoya, Japan, pp. 865–867.
38. Hudanski, L. (2005–2008) Modulation of the electron beam emitted by carbon nanotube based cathodes, application to microwave tubes. PhD thesis at Thales Research & Technology, Palaiseau, France, 2005-2008.
39. Kim, H.J. *et al.* (2006) *IEEE Trans. Electron Devices*, **53** (11), 2674–2680.
40. Charged Particle Optics Programs CPO Ltd, Registered Company Number 4012016, Manchester, UK. Available at *http://www.electronoptics.com/*
41. Spindt, C.A., Holland, C.E., Rosengreen, A., and Brodie, I. (1993) *J. Vac. Sci. Technol., B*, **11**, 468.
42. Guillorn, M.A., Melechko, A.V., Merkulov, V.I., Hensley, D.K., Simpson, Lowndes, D.H. (2002) *Appl. Phys. Lett.* **81**, 3660–3662.
43. Gangloff, L., private communication, see also Gangloff, L. *et al.* (2004) *Nano Lett.*, **4**, 1575.
44. Bower, C. *et al.* (2002) *IEEE Trans. Electron Devices*, **49** (8) See also Chen, L.H. and Jin, S. (2003) *J. Electron. Mater.*, **32** (12), 1360.

45. Bowers, J.E. and Burrus, C.A. (1987) *J. Lightwave Technol.*, **LT-5** (10), 1339–1350.
46. Minoux, E., Hudanski, L., Teo, K.B.K. *et al.* (2007) Carbon nanotube cathodes as electron sources for microwave amplifiers. Conference Information: 7th IEEE Conference on Nanotechnology, August 2007, Hong Kong, vol. 1–3, pp. 1259–1262.
47. Shi, J.W. *et al.*, (2010) *IEEE, J. Quatum Electron.*, **46**, 80–86.
48. Ishibashi, T., Furuta, T., Fushimi, H., Kodama, S., Hirochi, H., Nagatsuma, T., Shimizu, N., and Miyamoto, Y. (2000) *IEICE Trans. Electron.*, **E83-C**, 938–949.
49. Chtioui, M., Enard, A., Carpentier, D., Bernard, S., Rousseau, B., Lelarge, F., Pommereau, F., and Achouche, M. (2008) *Photonics Technol. Lett., IEEE* **20** (13) 1163–1165.
50. Hudanski, L., Minoux, E., Gangloff, L., Teo, K.B.K., Schnell, J.-P., Xavier, S., Milne, W.I., Pribat, D., Robertson, J., and Legagneux, P. (2008) *Nanotechnology*, **19**, 105201.
51. Cojocaru, C.S. *et al.*, to be published in the proceedings of the 15th International Display Workshop (IDW'08), Dec. 2008.
52. Federici, J.F., Schulkin, B., Huang, F., Gary, G., Barat, R., Oliveira, F., and Zimdars, D. (2005) *Semicond. Sci. Technol.*, **20**, S266–S280.
53. Woodward, R.M., Cole, B.E., Wallace, V.P., Pye, R.J., Arnone, D.D., Linfield, E.H., and Pepper, M. (2002) *Phys. Med. Biol.*, **47**, 3853–3863.
54. Whaley, R.W., Gannon, B., Smith, C.R., and Spinth, C.A. (2000) *IEEE Trans. Plasma Sci.*, **28** (3), 272–747.
55. Czarny, R., Alouini, M., Larat, C., Krakowski, M., and Dolfi, D. (2004) *IEEE Electron. Lett.*, **40** (15), 942–943.
56. Hagmann, M.J. (2004) *IEEE Trans. Microw. Theory Tech.*, **52** (10), 2361–2365.
57. Wang, Y., Kempa, K., Kimball, B., Carlson, J.B., Benham, G., Li, W.Z., Kempa, T., Rybczynski, J., Herczynski, A., and Ren, Z.F. (2004) *Appl. Phys. Lett.*, **85** (13), 2607–2609.
58. Carras, M. and De Rossi, A. (2006) *Opt. Lett.*, **31** (1), 47–49.
59. Guiset, P. (2007–2009) Optical control of CNT cathodes for THz sources. PhD thesis at Thales Research & Technology, Palaiseau, France, 2007–2009.

Index

a
alkali-development process, for CNT paste 293
alternating current (AC) arc 7
amorphous carbon buildup 24
anisotropic thermal conductivities, in aligned SWNT films 13
anodization process 18
arc discharge technique 7–8
array of CNTs, simulated electric field in
– for any shape of the CNT apex 157–158
– 9 × 9 CNTs 145, 149, 153, 146*f*, 147*t*
– computational model for electric field analysis 145–146
– effect of length 158–160
– effects of geometrical parameters 152–153, 155–156, 151*f*
– field analysis of VA-CNTs 148–157
– of network-structured CNTs 160–162
– simulation method 143–145
atomic force microscope (AFM) 12
– probes 18
Au-coated substrates, of nanowires 238

b
backlight units (BLUs), for liquid crystal displays (LCDs) 287
Ba getter 338
Bardeen transfer Hamiltonian method 49–50
Bessel functions 48
BN nanotubes 47
breast cancer imaging technique. *see* stationary digital breast tomosynthesis
brightness, of electron source 74–78, 185
Brillouin zone, of the nanotube 50

c
calcination method 19
carbon–carbon chemical bonds 13
carbon nanofibers (CNFs)
– Ar^+-induced 210
– CNF-tipped cones 213
– current–voltage (I–V) characteristics 208–209, 213, 209*f*
– flexibility 211–215
– formed on a Kapton polyimide foil 211, 212*f*
– Fowler–Nordheim (FN) plot 208, 209*f*
– growth in ion-irradiation method 207
– growth mechanism of ion-induced 208
– operation time of FE current 209–211
– Pd-catalyzed 209–210
– polyimide-based 213
– RT growth of ion-induced 206
– TEM images 207*f*
– in UHV condition 210
carbon nanotube–based field emission X-ray 418–419
– fabrication of cathode for 420–424
– field emission properties 424
carbon nanotube field emission displays, for low-power character displays
– advantages 272
– anode–emitter spacing 277*f*
– automatic CVD equipment 281*f*
– characteristics 284*t*
– CNT-deposited metal lead frame 277–279
– CNT electrodes 275–277
– current–voltage (I–V) characteristics 276–278*f*
– emission characteristics 275–276
– fabrication process of panel 279, 280*f*
– luminance characteristics 272*f*
– panel structure and rib design 273, 274*f*

Carbon Nanotube and Related Field Emitters: Fundamentals and Applications. Edited by Yahachi Saito

ISBN: 978-3-527-32734-8

carbon nanotube field emission displays, for low-power character displays (*contd.*)
– performance 280–281, 282*f*
– phosphor screen 273
– pixel design 274
– power consumption 273*f*
– SEM images of 279*f*
carbon nanotubes (CNTs)
– alignment along electric and magnetic field lines 17–18
– armchair type 4
– caps of 5–6
– chiral type 4
– electric properties 11–12
– electronic structure 10–11
– field emission microscopy (FEM) of 96
– mechanical properties 12
– metallic 10–11
– nucleated on Ni dot arrays 30*f*
– production methods 7–10
– scattering of conduction electrons 11–12
– semiconducting 10, 12
– single emitter *vs* film emitter 373–374
– on solid substrates 9
– structure 3–7, 5*f*
– thermal conductivity 13
– zigzag type 4
carbon nanowalls
– characterization 195–196
– current–voltage ($I-V$) characteristics 199, 198*f*
– effect of coating 200–202
– field emission properties of 199
– Fowler–Nordheim plots 201*f*
– future prospects 203
– G and D band peak of spectrum 196
– growth by rf capacitively coupled PECVD 197–198, 198*f*
– metal nanoparticles deposition and emission properties 200–202
– morphology of films 197–199
– N_2 plasma treatment effect 202
– Pt-deposited 201–202
– Raman spectra of 196, 197*f*
– SEM images of 197–198*f*
– synthesis techniques 194
– TEM images of 195*f*
– unique structure of 193, 194*f*
catalytic chemical vapor deposition technique
– advantages 9
– plasma-enhanced (PECVD) 9–10
– procedure 8
– thermal 9
catastrophic runaway phenomenon 81
cathode fall 228
cathode–grid capacitance 440
cathode ray tube (CRT) 287
cDreams 324, 315*f*, 320*f*
C_{60} fullerenes 98
Charybdis system 430
chemical vapor deposition (CVD) 6
chiral indices 3
chiral vector 3
clean surface MWNT, FEM studies 97–98, 97*f*
CNT-based electron guns
– convergence factors 446–449
– current density at cathodes 444–446
– potential 450
– simulation 446–448
CNT cathodes, capabilities of 443–444
– 1-μm-high nanotubes 449
– 5-μm-high nanotubes 449
– modulation of 450–456
CNT film emitters, high-current-density
– CNT disappearance from substrate 377–378
– CNT rooting 380–381, 381*f*
– effects of impregnation and rooting on emission 381–386
– $E-I$ characteristics of RuO_2-impregnated samples 382*f*
– electric resistance and thermal resistance at the CNT junctions 377
– endurance running test performance 382, 383*f*
– field enhancement factor of 375–376
– guiding principles and practical methods 374*t*
– impregnation of RuO_2 and OsO_2 379–380, 380*f*
– influence of gas introduction 386–388
– inherent emission properties 375
– ion sputtering and reactive etching 378–379
– optimization of electric field distribution on film emitter surface 376
– reduction of work function using coating 376
CNT photocathode, design
– high-frequency 463–464
– 300 MHz design concept 462–463
– OFF state 457–461
– ON/OFF transition time 461–462
– principle of operation 457–458
– ON state 461

CNT solution 293
CNTs-on-tip emitters 410
cold field emission gun (CFEG) 77
compact FE-SEM system 184–186
computed tomography (CT) 417
cone-beam micro-CT scanner 430
contamination 16
current–voltage *(I–V)* characteristics
– carbon nanofibers (CNFs) 208–209, 213, 209*f*
– carbon nanotube field emission displays 276–278*f*
– of carbon nanowall 199, 198*f*
– of clean SWNTs 124–126
– of CNT-FEDs 335, 336*f*
– of CNT lamp devices 264–265*f*
– field emission lamp (FEL), CNT-based 352–353, 353*f*
– graphite nanoneedle (GRANN) field emitter 182*f*
– high-resolution X-ray microscope 391–392
– 5-mm-diameter X-ray tube 411
– 10-mm-diameter X-ray tube 406
– of nanowires 249, 251, 250*f*
– of surface-coated CNT emitters 164, 173, 171*f*
CVD-grown MWNTs 12
CVD-grown SWNT tips 18

d
depth-profiled elemental analysis 34
diameter control, of nanowires 236–237
diamond emitters
– for cold-cathode fluorescent lamps (CCFLs) 228
– dependence of EDC of field emission electrons 223, 228
– Fowler–Nordheim equation 221
– intrinsic or p-type, field emission from 219–220
– low-temperature thermionic emitters based on a nitrogen-incorporated diamond film 229
– nitrogen-doped or n-type, field emission from 220–221
– nitrogen-incorporated ultra nanocrystalline diamond (UNCD) layer 229
– phosphorus-doped or n-type, field emission from 221–225
– pn-Junction, field emission from 225–228
– surface energy band diagrams (SEBDs) 221, 220*f*
– ultraviolet photoelectron spectroscopy (UPS) and field emission spectroscopy (FES) studies 220–221
diamond with H impurities, FE properties of 62–63, 64*f*
diode-type CNT-FED 333–335
direct current (DC) arc 7
direct-growth technique 18
discretization numbers, of CNTs 148–150, 149*f*, 148*t*
dispersion of emission, of CNT-FEDs 338
double-wall carbon nanotubes (DWNTs or DWCNTs) 7
– degradation processes of 111
dynamic micro-CT 430–431

e
ECG 430
elastic scattering within CNT 11–12
electron beam computed tomography (EBCT) 428
electron beam lithography 20
electron emitters, of CNT
– film 19–21
– point 15–19
electrophoresis 16, 20
emission currents from carbon nanotubes, calculation
– current–voltage characteristics 51
– Dirac-electron behavior 51
– extended states 46–47
– FN plots of 51, 52*f*
– localized state 45–46
– properties of open and closed nanotubes 49
– quantum mechanical methods 49
– semiclassical approaches 49–51
– time-dependent Schrödinger equation 46–47
– transfer matrix method 47–49
energy gap, of semiconducting CNTs 10
exchange-correlation effects 44

f
field amplification factor 164
field electron emission spectroscopy (FEES) 81
field emission arrays (FEAs) 440
field emission lamp (FEL), CNT-based
– anode 356–358
– application to backlight unit (BLU) in LCD 343–344
– cathode 348–356

field emission lamp (FEL), CNT-based (*contd.*)
- comparison of CCFL- and CNT-based BLUs 345*f*, 345*t*
- dimming signals 361–362, 361*f*, 364*f*
- driving method 360–368
- field emission from the CNT paste 349, 351–352, 350*f*
- Fowler–Nordheim (FN) plots 352–353, 353*f*
- future prospects 368
- indium tin oxide (ITO) electrode 346
- luminance 365–367, 365–366*f*
- phosphor layer 347
- process flow for CNT emitters 348–349, 354, 349*f*
- response time 367–368, 367*f*
- structure 346–348
- surface morphology of the CNT emitter 351
- vacuum packaging 358–360, 359*f*

field emission microfocus X-ray source technology
- effective focal spot size 427–428
- spatial distribution of multibeam 428–430
- tube current and lifetime 426
- tube design 425–426

field emission microscopy (FEM) studies
- of adsorption and desorption of molecules on the surface 98
- carbon dioxide-exposed MWNT 101–102, 101*f*
- carbon monoxide-exposed MWNT 101
- of CNTs 96, 287
- of deposited Al on MWNTs 103–105, 104*f*
- of FE patterns from adsorbates on MWNT 98–103
- hydrogen-exposed MWNT 99
- methane-exposed MWNT 102–103, 102*f*
- of MWNT emitters with clean surfaces 97–98, 97*f*
- nitrogen-exposed MWNT 99–100
- oxygen-exposed MWNT 101
- patterns depending on tip radius 98
- pentagon patterns 98
- resolution of 105–106
- schematic diagram of 96*f*

field emission theory. *see also* graphitic nanostructures, FE mechanism of
- from CNTs 44–52
- Fowler–Nordheim theory 43–44
- heat generation and losses in CNTs. *see* heating effects, in CNTs
- optical performance analysis. *see* optical performance measurement, of CNT field emitters

field emission X-ray 418

field-emitting geometry, of SWNT
- from adsorbates 131–133
- anomalous high-temperature behavior 130–131
- clean surface 121–126
- current degradation 137–140
- current fluctuations 137
- current–voltage $(I-V)$ measurements of clean surface 124–126
- emission current saturation at high fields 134–136
- emission stability 136–137
- energy distribution of emitted electrons 133–134
- field emission properties of 120–121
- field evaporation 128–130
- Fowler–Nordheim description 119
- Fowler–Nordheim tunneling behavior 136
- graphitic sheet walls 120
- lobe-type field emission images 131, 132*f*
- self-cleaning behavior 136
- thermally assisted field emission behavior 126–128
- under UHV conditions 136

field enhancement ratio (β) 43

film emitters, of CNT
- by CVD method 20–21
- by electrophoresis 20
- screen printing method 19
- by spray coating 19

Fowler–Nordheim (FN) equation
- carbon nanofibers (CNFs) 208, 209*f*
- carbon nanowalls 201*f*
- diamond emitters 221
- field emission lamp (FEL), CNT-based 352–353, 353*f*
- field-emitting geometry, of SWNT 119
- for a metallic field emitter 291
- 5-mm-diameter X-ray tube 411
- model of CNT field emitters 69–74
- tunneling behavior of SWNT 136

Fowler–Nordheim theory 43–44, 55, 61, 69, 86, 449

full-field digital mammography (FFDM) system 431

g

graphene sheets, properties of 61–62
graphite nanoneedle (GRANN) field emitter

– brightness B, 185
– as a cold cathode 184–186
– current intensity and fluctuations 186–187, 190, 189*f*
– fabrication and structure characterization 179–180, 181*f*
– field emission characteristics 181
– Fowler–Nordheim plot 182*f*
– graphene sheet structure 181*f*
– in a microwave plasma chemical vapor deposition equipment 180*f*
– physical adsorption energy 190
– physisorption energy 190
– pulse X-ray generation 182–183
– stabilization of emission current 186–188
– stochastic birth-and-death model 188–191
– time-resolved X-ray radiography 182–183
– transmission electron microscope (TEM) image of nanoneedle 181*f*
– typical etching condition 179–180
graphite platelet nanofibers (GPNs) 55
graphitic nanostructures, FE mechanism of
– diamond surfaces with hydrogen impurities 62–64, 64*f*
– graphene arrays 61
– graphene sheets 61–62
– H termination and field direction 57–60
– method and model 56–57
– π states and orbitals 60
– zigzag graphitic ribbons 57

h
heating effects, in CNTs
– diffusion equation 83–85
– experimental measurements of TEDs 88–91
– FEES experiments 88–89
– gradual degradation phenomenon 87
– Joule heating 81
– Nottingham effect 81–82, 84
– Nottingham/Joule ratio 84
– self-consistent solution of the thermal and FE equations 86–87
– simulations 85–87
– time-independent heat diffusion equation 84
hexandiol diacrylate (HDDA) 293
high-resolution electron beam lithography 27
high-resolution X-ray microscope 390
– characteristics of, with SEM functions 394–395
– commercial desktop SEM (Tiny-SEM 1540), 393
– Fresnel diffraction effects 397, 398*f*
– functioning as SEM 392–393, 394*f*
– interaction volume and elastic and inelastic scattering processes 397
– lateral distribution of the X-ray generating region 396–397
– limiting factors of image resolutions 396–397
– properties 390
– resolution of the X-ray image 395–396
– SEM images of a 300-mesh copper surface 395*f*
hot cathode 401
hot filament chemical vapor deposition (HFCVD) method 193
H-terminated boron-doped diamond 227
H-terminated diamond surface 63
H-terminated graphitic ribbons 60
H-terminated p-type diamond 221–222
H-terminated zigzag ribbon 57, 61, 58*f*
hydrogen impurities in subsurface, doping effect 63
hydrothermal synthesis, of nanowires 232, 240–241

i
impregnated cathode 441
inelastic scattering within CNT 11–12
in situ TEM observations, of CNT emitters
– degradation and failure at large current conditions 110–112
– electron holography experiments of individual field-emitting MWNTs 114–115
– Fowler–Nordheim plots 112
– relationship between field emission and gap width 113–114
– surface conditions at the tip 112–113
– at the tips of individual MWNTs with diameters of 14–55 nm 115–116
– welding process of an MWNT 113*f*

k
Kohn–Sham Hamiltonian 56
Kovar 355

l
lamp devices, using CNT emitters
– CNT cathode preparation 261–262
– current–voltage $(I–V)$ characteristics 264–265*f*

lamp devices, using CNT emitters (*contd.*)
– experimental devices 262*f*
– fabrication processes of FED lighting element 263, 265*f*
– life characteristics of the field emission 263*f*
– merits 271
– performance of lighting elements 265
– structure of lightning element 261–263, 262*f*
– use of CNT emitter 263–264
Landauer formula for 1D conductors 11
large-scale integrated circuits (LSIs) 407
lateral-gate cathode 297
Laudauer–Bütticker formalism 47, 49
Laue, M. von 389
light emitting diode (LED)-BLU 344
Lippmann–Schwinger equation 49
lithographic techniques 20, 25
luminance characteristics, of the CNT-FEDs 335–337*f*

m
magnetophoretic method 18
mammography 431
metal-organic chemical fluid deposition (MOCFD) 200
metal-organic compound (trimethyl(methylcyclopentadienyl) platinum: $MeCpPtMe_3$) 200
Meyer, Robert 288
Micro-CT 425
microelectromechanical system (MEMS) technology 451
microgated CNT cathode 452*f*
microtips 440
microwave amplifier 439
– 1.5 GHz diode 452–454, 453*f*
– 32 GHz triode 454–456, 455*f*
miniature X-ray tubes (MXTs)
– application of 413–415
– catalyzed by palladium (Pd) 404
– FE process in 401
– future prospects 416
– 5-mm-diameter tube 409–413
– 10-mm-diameter tube 405–409
– technical basis 402–404, 403*f*
modified ordered subsets convex (MOSC) method 433
molten flux synthesis 247
Motorola, Inc. 314, 319, 324, 326–328, 315*f*, 320*f*
M-type osmium-coated cathode 441
multibeam field emission X-ray (MBFEX) source 419
multi-wall carbon nanotubes (MWNTs) 6, 82, 87, 89, 266, 334, 444, 459
– alignment and purification of 20
– bamboo-structured 6
– CVD grown 18
– diameters of 6
– FEM studies 97–106
– gluing a bundle procedure 15–16
– grown by PECVD 10
– mounting inside SEM chamber 16, 17*f*
– production of 8, 10
– *in situ* TEM observations 110–116
– splitting process of 110
– TEM picture of 6*f*
– using radio frequency (RF) plasma heating technique 8*f*
– Young's moduli of 12
multiwalled CNT cathode 390–392

n
nanobelts 231
nanocluster, of catalyst
– average number and standard deviation 29, 31*f*
– calculated diameters 30–31
– distributions in 31
– field enhancement factor of 31
– nucleation of 28–29, 29*f*
– tip diameter and height distribution of the arrays 29–30, 32*f*
nanografibers (NGF)s 266–267, 266*f*
nanotube field emission displays (FEDs)
– approaches to integrate nanotubes 311
– challenges in developing 312
– color coordination 325–326, 327*f*
– design and drive voltage 312–316, 313*f*, 315*f*
– display performance 323–325, 326*f*
– fabrication cost 312
– fabrication of display 316–320
– "full-on" field emission conditions 313
– gate-to-nanotube spacing 314
– growth parameters on glass 316–320
– low-cost screen-printing fabrication techniques 314
– luminance uniformity and distributions 321–323
– mechanical properties 314–315
– Ni catalyst films 317–319, 318*f*
– "off" field emission conditions 313
– operating lifetime of 328–329
– pixel luminance 324

- role of ballast resistor network 321–322
- sealing of cathode plates 327–328
- structures of low-voltage and high-voltage anodes 316
- transition metal catalysts 317–319, 318*f*
- uniformity as a function of device design 322–323, 323*f*
- using chemical vapor techniques 311–312

n-channel metal–oxide–semiconductor field-effect transistor (MOSFET) circuit 429
NCPS97 56
n-doped polycrystalline diamond film 220
negative electron affinity (NEA), of diamond 219
nonevaporable getters 338–340, 339*f*
Nottingham effect 81–82
n-type diamond emitter 220–221

o

ODT-modified Au-coated substrates 238–239, 239*f*
optically modulated CNT 457
optical performance measurement, of CNT field emitters
- brightness measurement 74–76
- current density 69–70
- current–voltage characteristics 70–71, 71*f*, 72*t*
- emission models 69–74
- energy spectrum 72–73
- energy spread of the emitted electron beam 71–72
- field at the apex 70
- geometrical factors 76
- maximum theoretical brightness 77
- reduced brightness 74–75, 78*f*
- tunneling parameter 72, 74
- using nanomanipulator system in an SEM 68–69

orientation control, of nanowires 237–238

p

patterned emitters of CNT
- from catalyst film edges 25–27
- on diffusion barriers 27–28

Pd emitters 404
pentaerythritol triacrylate (PETIA) 293
periodic permanent magnet (PPM) 446
phonons 13
photoimageable CNT pastes 292–295
photolithography 20, 251, 314
Pierce-type electron gun 446–448
p-i-n capacitance 462
p-i-n photodiodes 440
Planck's law 81
plasma-enhanced chemical vapor deposition (PECVD) method 9–10, 193
plasma-enhanced chemical vapor deposition (PECVD) reactor 445–446
plasma-enhanced hot filament CVD 317
pn-junction diamond emitters 225–228
point emitters, fabrication methods of CNT 15–19
Poisson equation 44
polarized CNTs 17
pore growth 18
positional control, of nanowires 238–240
printed CNTs
- anode of FED 305–306
- cathode of FED 300–305
- field emission display based on 300–306
- formulation of CNT paste 292–295, 295*f*
- posttreatment of the surface 295–300
- SEM images of 294*f*, 296*f*
- viscoelastic properties (rheology) of the mixtures of paste 295

probe tips, of CNT 18
Pt-deposited carbon nanowall film 201
p-type diamond emitter 219–220

r

radical chain polymerization, in CNT paste formulation 294*f*
Rayleigh scattering 83
reduced brightness, of electron source 74–75, 78*f*
respiratory gatedmicro-CT imaging technique 430
Richardson–Dushman equation 229
Röntgen, W. C. 389
rooting process, on CNT film emitters 380–386
RuO_2 impregnation, on CNT surface 379–380, 382, 380*f*

s

scanning beam digital X-ray (SBDX) 428
SCF-MOCFD using supercritical carbon dioxide ($scCO_2$) 200
Schottky emitter 67, 77
seed materials 205
sharpening effect 25
Si-based integrated circuits 9
SiC nanowires 231
silicon (Si) tip 18
Si nanowires 231, 248–253
- aligned 250

Si nanowires (*contd.*)
– coarsening of the catalyst droplets during growth 235–236
– diameter control of 236–237
– effect of Au/Si droplets on the surface 235
– and gold diffusion 234
– hydrothermal reaction 232, 240–241
– orientation control of 237–238
– positional control of 238–240
– and postgrowth annealing 236
– SEM images of 235*f*
– sidewalls of the 234
– *in situ* growth of 234
– synthesis 231–241
single-wall carbon nanotubes (SWCNTs or SWNTs), 3–6, 82, 87, 375. *see also* field-emitting geometry, of SWNT
– alignment and purification of 20
– anisotropic thermal conductivities 13
– clean 121–126
– diameter of 4–5
– electronic density of states (DOS) of 10–11, 11*f*
– production of 7
– TEM picture of 5*f*
– translational symmetry of 4
soft lithography 20
solid-state amplifiers (SSAs) 439
Spindt, Capp 287
Spindt cathode 288, 440, 442, 450, 451*f*
– process flow 289*f*
Spindt FED panels 288–289, 290*f*
Spindt-tip arrays 289, 290*f*
Spindt-type field emitter arrays 219
stationary digital breast tomosynthesis 431–434
Stefan–Boltzmann law 83–84
substrate surface, CNT growth 32–37
– a-C : N layer 36–37
– Auger measurements 33–35, 37
– $C_2H_2 : NH_3$ gas composition 32–37
– composition depth profiles 35–36, 36*f*
– deposition conditions 35*t*
– interfacial layers 37
– N–C bonding 36
– N–Si bonding 36
– performance of insulating layers on the substrate 37–38
– Si substrate surface 33–34, 34*f*
super-high-luminance light source applications, of CNTs 268*f*
– device structure 267–268
– luminance 269–270*f*
– MWNT emitters 266–267
– nanografibers (NGF)s 266–264, 266*f*
– performance of device 268–270
– phosphor screen 267, 269–270
– total emission current from the NGF cathode 268, 269*f*
super-miniature X-ray tube (SMXT) 409–411, 413, 411*f*
surface charge method 143
surface coating, of CNT emitters
– BN layer 165–169*f*
– cesium (Cs) 165
– decreasing of potential barrier 165–167
– field emission properties 164–165
– Fowler–Nordheim plot 164, 167, 168*f*, 170*f*, 173*f*
– functions of coating layer 163
– with low work function 165
– MgO layers 169–172
– stabilization of emission current 167
– TiC layers 172–174
– tunneling probability 166
– with wide-band-gap layer (WBGL) 165–167, 166*f*
surface emission display (SED) 319
surface growth, of SWNT tips 18
surface plasmons polaritons (SPPs) 465
Suzuki–Trotter split operator method 56

t

Taylor expansion method 56
TEM tension test 12
tensile strength, of arc-grown MWNTs 12
thermal conductivities, of an isolated MWNT 13
thermal conductivity, of SWNT films 13
thermionic X-ray technology 417–418
thin-walled carbon nanotubes (MWNTs) 7
three-dimensional (3D) BCM 143–144
THz cathodes 465–466
Ti diffusion barrier 24
time-dependent Schrödinger equation 46–47
tip, field emissions
– adsorption and desorption of molecules 167
– Spindt-tip arrays 289, 290*f*, 291
– of surface-coated CNT emitters 164–165
tip, of MWNTs, *in situ* TEM observations 112–113, 115–116
tomosynthesis imaging 431–434
total energy distributions (TEDs) 81
– experimental measurements of 88–91
transparent-like CNT-FED panel 335*f*, 340*f*
traveling-wave tube amplifiers (TWTAs) 465

traveling-wave tubes (TWTs) 439, 442
– amplification process 442–443
– cathode–grid distance 442–443
– thermionic cathodes in 441–442
– use of FEAs 442–443
trimethylolpropane triacrylate (TMPTA) 293
triode field emission display system 288*f*
triode-type CNT-FED 333
Troullier–Martins scheme 56

u
under-gate cathode 297, 302, 303*f*
uni-traveling carrier (UTC) photodiodes 462
UV-curing system 293

v
vacuum amplifier 439
van der Waals interactions 16–17, 23, 445
vertically aligned carbon nanotubes (VA-CNTs) 143, 161
– computational model for electric field analysis 145–146
– discretization number for accuracy in electric field calculation 148–150
– field analysis 148–157
– nonuniform length, field analysis 154–157
– uniform length, field analysis 150–154
Vienna *ab initio* simulation package (VASP) 57
VLS processes 232–236

w
Wentzel–Kramers–Brillouin (WKB) approximation 43, 45, 49
WO nanowires 231
wrapping vector. *see* chiral vector

x
X-ray tubes, practical requirements for 420*t*

y
Young's modulus, of isolated MWNTs 12

z
ZnO nanobelts 247
ZnO nanopencils 248
ZnO nanorods 245
ZnO nanostructures with sharp tips 247
ZnO nanowires (NWs) 231
– with carbon powder 233
– coarsening of the catalyst droplets during growth 235–236
– crystalline 233
– diameter control of 236–237
– field emission 241–244, 245*t*–246*t*
– field enhancement factor β 245–246
– Fowler–Nordheim (FN) plots 245
– hydrothermal reaction 232, 240–241
– metal–organic chemical vapor deposition (MOCVD) synthesis 232
– morphological characteristic 245*t*
– orientation control of 237–238
– positional control of 238–240
– scanning electron microscope (SEM) image of 234*f*
– sidewalls of the 234
– synthesis 231–241
– template-directed synthesis 232
– transmission electron microscope (TEM) image 234*f*
– vapor–liquid–solid (VLS) chemical vapor deposition (CVD) synthesis 231–232
– vapor–solid (VS) synthesis 232–236
– zinc nitrate salt and growth of 240–241